Milestones in Drug Therapy

For further volumes:
http://www.springer.com/series/4991

Graham Molineux • Tara Arvedson •
MaryAnn Foote
Editors

Twenty Years of G-CSF

Clinical and Nonclinical Discoveries

Springer

Volume Editors
Dr. Graham Molineux
Amgen Inc.
Thousand Oaks, CA 91320
USA
grahamm@amgen.com

Dr. MaryAnn Foote
MA Foote Associates
Westlake Village, CA 91362
USA
fmawriter@aol.com

Dr. Tara Arvedson
Amgen Inc.
Thousand Oaks, CA 91320
USA
taraa@amgen.com

Series Editors
Prof. Michael J. Parnham, Ph.D.
Visiting Scientist
Research & Clinical Immunology Unit
University Hospital for Infectious Diseases "Dr. Fran Mihaljević"
Mirogojska 8
HR-10000 Zagreb
Croatia

Prof. Dr. Jacques Bruinvels
Sweelincklaan 75
NL-3723 JC Bilthoven
The Netherlands

ISBN 978-3-0348-0217-8 e-ISBN 978-3-0348-0218-5
DOI 10.1007/978-3-0348-0218-5

Printed on acid-free paper

Springer Basel AG is part of Springer Science + Business Media (www.springer.com)

Preface

On April 25, 1953, 58 years ago, JD Watson and FHC Crick published their article entitled "A Structure for Deoxyribose Nucleic Acid" in the journal *Nature*. This article has been cited for its brevity, only 1 page and 1 diagram. The impact of this article cannot be fully measured, but it is safe to suggest that recombinant DNA biopharmaceuticals, such as recombinant human granulocyte colony-stimulating factor (rmet-HuG-CSF), would not be available today without the basic knowledge of DNA structure.

A quick search of PubMed suggests that no articles had been published on the topic of rmet-HuG-CSF or even G-CSF as of 1953. Forward to April 2011 and a quick search of PubMed cites 31,965 articles tagged to "G-CSF," 1,753 tagged to "filgrastim," 350 tagged to "pegfilgrastim," 295 tagged to "lenograstim," and 13 tagged to "biosimilar filgrastim."

We have come a long way in 58 years since the publication of the proposed structure of DNA and further since the first approval of filgrastim by the US Food and Drug Administration in 1991 for the treatment of patients with chemotherapy-induced neutropenia. In the intervening 20 years since this first marketing approval, countless patients worldwide have been treated with a recombinant form of G-CSF for the treatment of chemotherapy-induced neutropenia; severe chronic neutropenia; neutropenia due to disease; to mobilize peripheral blood stem cells for transplantation, either autologous or allogenic; and for bone marrow recovery after bone marrow or stem cell transplantation, to name a few. rmet-HuG-CSF has been tried in the treatment of infections, diabetic foot ulcers, neonatal sepsis, and community-acquired pneumonia.

In almost all settings, it can be said that rmet-HuG-CSF ameliorated neutropenia, increased neutrophil counts, reduced the need for intravenous antibiotics, and/or reduced the need or duration for hospitalization. Thus, it is appropriate to celebrate 20 years of research and therapy with rmet-HuG-CSF.

The authors of several chapters are some of the early clinical investigators of rmet-HuG-CSF and staff of Amgen, which manufactures filgrastim and pegfilgrastim. The editors have allowed information in chapters to provide various perspectives on topics. We are hopeful that readers will find the presentations varied but balanced.

The editors have tried to obtain the necessary permissions and authorizations before publication, and great care has been exercised in the preparation of this volume. Nevertheless, errors cannot always be avoided. The editors, their employers or companies, and the publisher cannot accept responsibility for any errors or omissions that inadvertently occurred. The views and opinions expressed in the book are those of the participating individuals and do not reflect the views of the editors, the publisher, Amgen Inc., or any other manufacturer of pharmaceutical products named herein. The current package insert should be consulted before any pharmaceutical product is administered.

California, USA

Graham Molineux
Tara Avredson
MaryAnn Foote

Contents

Contributors

Mojtaba Akhtari Division of Hematology and Oncology, Department of Internal Medicine, University of Nebraska Medical Center, 987680 Nebraska Medical Center Omaha, NE 681980-7680, USA, mojtaba.akhtari@unmc.edu

Tara L. Arvedson Amgen Inc., Hematology Research, Thousand Oaks, CA USA, taraa@amgen.com

Isaac Ben-Bassat Institute of Hematology, Davidoff Cancer Centre, Tel-Aviv University, Tel-Aviv, Israel

Ashanka Beligaswatte Royal Adelaide Hospital and Institute of Medical and Veterinary Science, Adelaide, Australia

Roberto Bolli Institute of Molecular Cardiology, University of Louisville, Louisville, KY 40292, USA

Audrey Anna Bolyard Division of General Internal Medicine, University of Washington, Box 356422, , Seattle, WA USA

Melanie V. Cohen Department of Radiation Oncology, School of Medicine, University of Maryland, 10 South Pine Street, MSTF 6-34D, Baltimore, MD 21201, USA, melveirs@yahoo.com

Mario Cruciani Center of Community Medicine, Infectious Diseases Treatment and Surveillance Unit, Verona, and Infectious Diseases Service, G. Fracastoro Hospital, San Bonifacio, Verona, Italy, crucianimario@virgilio.it

David C. Dale Division of General Internal Medicine, University of Washington, 1959 NE Pacific St, AA 522 Health Sciences Building, Box 356422, Seattle, WA 98195, USA dcdale@u.washington.edu

Buddhadeb Dawn Division of Cardiovascular Diseases, Cardiovascular Research Institute, University of Kansas Medical Center, 3901 Rainbow Blvd, Room 1001 Eaton Hall, MS 3006, Kansas City, KS 66160, USA, bdawn@kumc.edu

Ann M. Farese Department of Radiation Oncology, School of Medicine, University of Maryland, 10 South Pine Street, MSTF 6-34D, Baltimore, MD 21201, USA, afarese@som.umaryland.edu

Napoleone Ferrara Genentech, South San Francisco, CA USA, nf@gene.com

MaryAnn Foote MA Foote Associates, Westlake Village, CA USA, fmawriter@aol.com

Anat Gafter-Gvili Institute of Hematology, Davidoff Cancer Centre, Tel-Aviv University, Tel-Aviv, Israel

Mike J. Giffin Amgen Inc., Hematology Research, Thousand Oaks, CA USA

Ronit Gurion Institute of Hematology, Davidoff Cancer Centre, Tel-Aviv University, Tel-Aviv, Israel

Vagish Hemmige Department of Medicine, Section of Infectious Diseases and Global Health, University of Chicago, 5841 S. Maryland Avenue, Chicago, IL 60637

Bradford R. Hirsch Department of Medicine, Duke University and the Duke Cancer Institute, 2424 Erwin Road, Suite 205, 27705, Durham, NC USA, bradford.hirsch@duke.edu

Fumihiko Ishikawa Research Unit for Human Disease Models, RIKEN Research Center for Allergy & Immunology, Yokohama, Japan, f_ishika@rcai.riken.jp

Letizia Leone Section of Dermatology and Venereology, Department of Biomedical and Surgical Science, University of Verona, Verona, Italy, laetitialeo@libero.it

Ian Lewis Royal Adelaide Hospital, Institute of Medical and Veterinary Sciences, and University of Adelaide, Frome Rd, Adelaide, SA 5000, Australia

W. Conrad Liles Divsion of Infectious of Diseases, University of Toronto, Toronto General Hospital, Eaton North Wing, 13th floor Room 13 EB 220, 200 Elizabeth St, Toronto, Ontario, Canada

Daniel C. Link Division of Oncology, Washington University, 660 S. Euclid Avenue, Campus Box 8007, St. Louis, MO 63110, USA dlink@dom.wustl.edu

Gary H. Lyman Duke Cancer Institute, Duke University, 2424 Erwin Road, Suite 205, Durham, NC 27705, USA, gary.lyman@duke.edu

Thomas J. MacVitte Department of Radiation Oncology, School of Medicine, University of Maryland, 10 South Pine Street, MSTF 6-34E, Baltimore, MD 21201, USA, tmacvitt@umaryland.edu

Lori Maness Division of Hematology and Oncology, Department of Internal Medicine, University of Nebraska Medical Center Omaha, NE 681980-7680, USA

Graham Molineux Amgen Inc., Hematology Research, Thousand Oaks, CA USA

George Morstyn Victorian Comprehensive Cancer Centre, Level 3, Royal Melbourne Hospital 3050, 766 Elizabeth Street, P.O. Box 2148, Melbourne, VIC 3000 Australia, rmorstyn@bigpond.net.au; georgemor@gmail.com

Pellegrino Musto Department of Onco-Hematology, IRCCS, Centro di Riferimento Oncologico della Basilicata, Rionero in Vulture (Pz), Italy, p.musto@crob.it; p.musto@tin.it

Theresa A. Neumann Menlo Park, CA USA

Hoainam T. Nguyen-Jackson Department of Immunology and Center for Inflammation and Cancer, The University of Texas M D Anderson Cancer Center, Houston, TX USA, The Graduate School of Biomedical Sciences, The University of Texas, Houston, TX USA

Howard Ozer Department of Medicine, Section of Hematology-Oncology, University of Illinois at Chicago, Chicago, IL USA, ozer@uic.edu

David L. Pitrak Department of Medicine, Section of Infectious Diseases and Global Health, University of Chicago, 5841 S. Maryland Avenue, Chicago, IL 60637 dpitrak@medicine.bsd.uchicago.edu

Pia Raanani Institute of Hematology, Davidoff Cancer Centre, Tel-Aviv University, Tel-Aviv, Israel, mojtaba.akhtari@unmc.edu

William Renwick Department of Haematology and Medical Oncology, Western Hospital, Footscray, Melbourne, VIC Australia, William.Renwick@wh.org.au

Lorin K. Roskos MedImmune, Gaithersburg, MD USA, RoskosL@MedImmune.com

Santosh K. Sanganalmath Division of Cardiovascular Diseases, Cardiovascular Research Institute, University of Kansas Medical Center, 3901 Rainbow Blvd, Room 1001 Eaton Hall, MS 3006, Kansas City, KS 66160, USA

Santosh Saraf Department of Medicine, Section of Hematology-Oncology, University of Illinois at Chicago, Chicago, IL USA

Ofer Shpilberg Institute of Hematology, Davidoff Cancer Centre, Tel-Aviv University, Tel-Aviv, Israel

Stephen J. Szilvassy Amgen Inc., Hematology/Oncology Research Therapeutic Area, Thousand Oaks, CA USA, sszilvas@amgen.com

Luen Bik To Royal Adelaide Hospital, Institute of Medical and Veterinary Sciences, and University of Adelaide, Frome Rd, Adelaide, SA 5000, Australia, Bik.To@health.sa.gov.au

Edmund K. Waller Department of Hematology and Molecular Oncology, Winship Cancer Institute, Emory University School of Medicine, Atlanta, GA USA, ewaller@emory.edu

Stephanie S. Watowich Department of Immunology and Center for Inflammation and Cancer, The University of Texas M D Anderson Cancer Center, Houston, TX USA, The Graduate School of Biomedical Sciences, The University of Texas, Houston, TX USA, swatowic@mdanderson.org

Karl Welte Department of Molecular Hematopoiesis, Children Hospital, Medical School, Hannover, Germany, Welte.Karl.H@mhannover.de

Bing-Bing Yang Amgen Inc., Thousand Oaks, CA USA, byang@amgen.com

Huiyuan Zhang Department of Immunology and Center for Inflammation and Cancer, The University of Texas M D Anderson Cancer Center, Houston, TX USA

Glossary and Abbreviations

A

ACE	Angiotensin-converting enzyme
ADCC	Antibody-dependent cell-mediated cytotoxicity
ADSCN	Autosomal dominant severe congenital neutropenia
aGVHD	Acute graft-versus-host disease
AIDS	Acquired immunodeficiency syndrome
ALDH	Aldehyde dehydrogenase
ALL	Acute lymphocytic leukemia/acute lymphoblastic leukemia
alloHCT	Allogeneic hematopoietic cell transplantation
AML	Acute myeloid leukemia
ANC	Absolute neutrophil count
Ang	Angiopoietin
APP	Amyloid precursor protein
Ara-C	Chemotherapy with idarubicin/daunorubicin and cytosine arabinocide
ARDS	Acute respiratory distress syndrome
ASCO	American Society of Clinical Oncology
ASH	American Society of Hematology
AUC	Area under the curve
AuSCT	Autologous stem cell transplantation

B

B-CLL	B-cell chronic lymphocytic leukemia
BFU-E	Erthyroid blast-forming units

C

CAE	Chemotherapy regimen of cyclophosphamide, doxorubicin, and etoposide
CAFC	Cobblestone area forming cells
CALGB	Cancer and Leukemia Group B
CDC	Centers for Disease Control and Prevention complement-dependent cytotoxicity
CFC	Colony-forming cell
CFU-C	Cell colony-forming unit
CFU-G	Granulocyte progenitor cell
CFU-GEMM	Granulocyte–erythrocyte–monocyte–macrophage progenitor cell
CFU-GM	Granulocyte–macrophage progenitor cell
cGVHD	Chronic GVHD
CHOEP	Chemotherapy regimen of cyclophosphamide, doxorubicin, vincristine, etoposide, and prednisone
CHOP	Chemotherapy regimen of cyclophosphamide, doxorubicin, vincristine, and prednisone
CHR	Cytokine-binding homology region
CIN	Chronic idiopathic neutropenia
CLL	Chronic lymphocytic leukemia
CML	Chronic myeloid leukemia
CMML	Chronic myelomonocytic leukemia
CMV	Cytomegalovirus
CNOP	Chemotherapy with cyclophosphamide, mitoxantrone, vincristine, and prednisone
CNTF	Ciliary neurotrophic factor
COR	Circulating opsonin receptor
CRH	Cytokine receptor homologous
CSF	Colony-stimulating factor
CSF-1	Another name for M-CSF
CSF-2	Another name for GM-CSF
CSF-3	Another name for G-CSF
CT	Computed tomography

E

ECOG	Eastern Cooperative Oncology Group
EBMT	European Group for Blood and Marrow Transplantation
ECOG	Eastern Cooperative Oncology Group
EGF	Epidermal growth factor
EGFR	Epidermal growth factor receptor
EORTC	European Organization for Research and Treatment of Cancer

eNOS	Endothelial nitric oxide synthase
EPO	Erythropoietin
EPOR	Erythropoietin receptor
ESA	Erythropoiesis-stimulating agent

F

5-FU	5-flurouracil
FAB	French-American-British
FACS	Fluorescence-activated cell sorting
FDA	Food and Drug Administration
FEC	Chemotherapy with flurouracil, epirubicin, and cyclophosphamide
FGF	Fibroblast growth factor
FIV	Feline leukemia virus
FL	Flt3 ligand
fMLP	*N*-formyl-methionyl-leucyl-phenylalanine
FNIII	Fibronectin type III-like

G

GALT	Gut-associated lymphoid tissue
G-CSF	Granulocyte colony-stimulating factor
G-CSFR	Granulocyte colony-stimulating factor receptor
GFP	Green fluorescent protein
Gfi-1	Growth factor independence-1
GH	Growth hormonc
GHR	Growth hormone receptor
GIST	Gastrointestinal stromal tumor
GM-CSF	Granulocyte–macrophage colony-stimulating factor
GMP	Granulocyte–monocyte committed progenitors
GSD 1b	Glycogen storage disease 1b
GVHD	Graft-versus-host disease

H

HAART	Highly active antiretroviral therapy
H-ARS	Hematopoietic subsyndrome of acute radiation syndrome
HCP	Hematopoietic cell phosphatase
HGF	Hepatocyte growth factor
HIES	Hyperimmunoglobulin E syndrome

HIF-1α	Hypoxia inducible factor-1α
HIV	Human immunodeficiency virus
HPC	Hematopoietic progenitor cell
HPLC	High-pressure liquid chromatography
HR	Hazard ratio
HSC	Hematopoietic stem cell
HSPC	Hematopoietic stem/progenitor cells

I

ICAM	Intercellular adhesion molecule
ICER	Incremental cost effectiveness ratio
IFN	Interferon
Ig	Immunoglobulin
IL	Interleukin
IL-1ra	IL-1 receptor antagonist
iNOS	Inducible nitric oxide synthase
IPSS	International Prognostic Scoring System
IST	Immunosuppressive therapy

J

Jak	Janus protein tyrosine kinases/Janus kinase

K

kDA	Kilodalton

L

LD	Lethal dose
LIF	Leukemia inhibitory factor
LIFR	Leukemia inhibitory factor receptor
LPS	Lipopolysaccharide
LSC	Leukemia stem cell
LTC-IC	Long-term culture-initiating cells
LVEDV	Left ventricular end-diastolic volume
LVESV	Left ventricular end-systolic volume

LVEF	Left ventricular ejection fraction
LYG	Life year gained

M

MAI	*Mycobacteria avium* infection
MAPK	Mitogen-activated protein kinase
MCM	Medical countermeasure
M-CSF	Macrophage colony-stimulating factor
MDS	myeLodysplastic syndromes
MDSC	Myeloid-derived suppressor cell
MGDF	Megakaryocyte growth and development factor
MIP	Macrophage inflammatory protein
MMP	Matrix metalloproteinase
MOR	Maximum opsonin receptor
MRD	Minimal residual disease
MT-1 MMP	Membrane type-1 MMP
mTOR	Mammalian target of rapamyicin
M-VAC	Chemotherapy with methotrexate, vinblastine, doxorubicin, and cisplatin

N

NCCN	National Comprehensive Cancer Network
NHL	Non-Hodgkin's lymphoma
NIH	National Institutes of Health
NK	Natural killer
NMR	Nuclear magnetic resonance
NSAA	Nonsevere aplastic anemia
NSCLC	Nonsmall-cell lung cancer

O

OSM	Oncostatin M

P

PBCT	Peripheral blood cell transplantation
PBMC	Peripheral blood mononuclear cell

PBPC	Peripheral blood progenitor cell
PBSC	Peripheral blood stem cell
PCR	Polymerase chain reaction
PCI	Percutaneous coronary intervention
PD	Pharmacodynamics
PDGF	Platelet-derived growth factor
PDGFR	Platelet-derived growth factor receptor
PECAM-1	Platelet/endothelial cell adhesion molecule-1
PEG	Polyethylene glycol
PET	Positron emission tomography
PlGF	Placenta growth factor
PK	Pharmacokinetics
PMNL	Polymorphonuclear leukocyte
PPO	Pluripotent colony-stimulating factor
PRL	Prolactin
PRLR	Prolactin receptor
PS1	Presinilin 1

Q

QALD	Quality-adjusted life day
QALY	Quality-adjusted life year

R

RA	Refractory anemia
RAEB	Refractory anemia with excess blasts
RAEB-T	Refractory anemia with excess blasts in transformation
RARS	Refractory anemia with ringed sideroblasts
RES	Reticuloendothelial system
rHuEPO	Recombinant human erythropoietin
rHuCSF	Recombinant human colony-stimulating factor
rHuG-CSF	Recombinant human granulocyte colony-stimulating factor
rHuGM-CSF	Recombinant human granulocyte–macrophage colony-stimulating factor
rHuIL-3	Recombinant human interleukin-3
RIT	Radioimmunotherapy
RR	Risk ratio; relative risk
RTKI	Receptor tyrosine kinase inhibitors
RT-PCR	Reverse transcription polymerase chain reaction

S

S1P	Sphingosine-1-phosphate
SAA	Severe aplastic anemia
SCF	Stem cell factor
SCLC	Small-cell lung cancer
SCN	Severe chronic neutropenia
SCNIR	Severe Chronic Neutropenia International Registry
SCT	Stem cell transplantation
SDF	Stromal cell-derived factor
SDS	Shwachman Diamond syndrome
SIV	Simian immunodeficiency virus
SLE	Systemic lupus erythematosus
SNS	Strategic National Stockpile
SOCS	Suppressor of cytokine signaling
SoS	Son of sevenless
STAT	Signal transducers and activators of transcription
SWOG	SWOG

T

TAF	Tumor-associated fibroblast
TA-GVHD	Transfusion-associated graft-versus-host disease
TAM	Tumor-associated macrophage
TAN	Tumor-associated neutrophil
TBI	Total body irradiation
TEM	Tie2 expressing monocyte
TGF	Transforming growth factor
TKI	Tyrosine kinase inhibitor
TNF	Tumor necrosis factor
tPA	Tissue plasminogen activator
TPO	Thromobopoietin
TRAIL	TNF-related apoptosis-inducing ligand

V

VAPEC-B	Chemotherapy regimen of vincristine, doxorubicin, prednisolone, etoposide, cyclophosphamide, and bleomycin
VCAM-1	Vascular cell adhesion molecule-1
VEGF	Vascular endothelial growth factor
VEGFR	Vascular endothelial growth factor receptor

VICE	Chemotherapy regimen of vincristine, ifosfamide, carboplatin, and etoposide
vIL	Viral interleukin
VLA-4	Very late antigen-4
VSAA	Very SAA

W

WBC	White blood cell
WHO	World Health Organization
WT	Wildtype

Z

ZDV	Zudovidine

Part I
Basic Science

Hematopoiesis in 2010

George Morstyn

1 Brief History of Hematopoietic Growth Factors

In 1987, the first clinical results of the use of hematopoietic growth factors were presented at a small meeting in Garmish-Partenkirchen [1]. It is timely, 23 years later, to review what we have learned since that first report.

Donald Metcalf reviewed for the 50th Anniversary of the American Society of Hematology (ASH) our knowledge of the regulation of hematopoiesis by specific growth factors [2], and we have previously reviewed the important features of hematopoiesis: the cell hierarchy, the movement of cells from multipotential progenitors to mature, committed cells with specific functions, and the many cytokines that regulate the process [3]. It was possible to purify the regulators and obtain protein-sequence data for cloning of the hematopoietic growth factors because of the development of various biologic assays in the preceding 50 years and the development of recombinant DNA technology in the 1980s [2].

The regulator we knew most about was erythropoietin (EPO), initially as an activity detectable in the urine of patients with aplastic anemia. Until the cloning and expression of EPO and the development of an immunoassay, monitoring of red cell-stimulating activity was cumbersome, and radioactive iron incorporation into red blood cells was used. The assays that were used to measure granulocyte–macrophage progenitor cells were carried out on semisolid cultures that allowed the counting of colonies of mature cells produced from myeloid precursors [4]. The assays were later adapted to identify red cells, megakaryocytes, and even earlier precursors.

Early work with fluorescent-activated cell sorting (FACS) allowed the identification, morphologically and functionally, of these precursors, and it became

G. Morstyn (✉)
Victorian Comprehensive Cancer Centre, Level 3, Royal Melbourne Hospital 3050, 766 Elizabeth Street, P.O. Box 2148, Melbourne, VIC 3000, Australia
e-mail: rmorstyn@bigpond.net.au; georgemor@gmail.com

G. Molineux et al. (eds.), *Twenty Years of G-CSF*, Milestones in Drug Therapy,
DOI 10.1007/978-3-0348-0218-5_1, © Springer Basel AG 2012

apparent that the production of mature cells in the blood, the red cells and granulocytes, was dependent on the presence of specific regulators such as EPO, granulocyte colony-stimulating factor (G-CSF), and granulocyte–macrophage colony-stimulating factor (GM-CSF). It was not until the 1990s that the megakaryocyte regulator was identified. The role of these factors was slightly different in murine models compared with their role in humans, but many of the biologic findings were directly comparable between the species.

GM-CSF (also known as CSF2), macrophage colony-stimulating factor (M-CSF) (also known as CSF1), and G-CSF (also known as CSF3) were identified as growth factors for myeloid progenitor cells (reviewed in [5]). The cytokines stimulate the proliferation, differentiation, maturation, and survival of granulocytes and macrophages. The CSF acts through specific receptors. The G-CSF receptor (G-CSFR) is a member of the type-1 cytokine receptor family; the GM-CSF receptor consists of a unique α chain and a common β chain through which signaling occurs.

The control of platelet production is different to that of granulocytes and macrophages. Platelets form by the fragmentation of mature megakaryocytes. The production of megakaryocytes is under the control of the *c*-Mpl receptor, and its ligand was identified as thrombopoietin (TPO). TPO is the primary regulator of platelet production and elimination of either TPO or the *c*-Mpl receptor results in severe thrombocytopenia. Importantly, TPO does not seem to accelerate platelet shedding and so its actions are slower than that of G-CSF that acts on increasing not only the production of myeloid precursors but also their maturation. Both G-CSF and TPO blood concentrations appear to be reduced by the mass of mature cells; granulocytes, megakaryocytes, and platelets increase, respectively, and this provides a feedback loop for control. G-CSF, TPO, and EPO are critical to the maintenance of hematopoiesis, and knock-outs of the genes for the ligand or receptor lead to profound neutropenia, thrombocytopenia, or anemia [6].

EPO is a 34.4-KD glycoprotein hormone and was cloned in 1985. EPO is regulated by hypoxia. It acts on erythroid precursors to enhance red blood cell production and thus the oxygen-carrying capacity of the blood. EPO, which is produced predominantly in the kidney, is required for the production and terminal differentiation of red blood cells. Like G-CSF and GM-CSF, EPO controls proliferation, maturation, and survival of red blood cells. The receptor exists as a dimer and when the ligand binds, a conformational change and a cascade of activation occur through transphosphorylation of JAK2.

Controversy exists about where the EPO receptor (EpoR) is expressed and on what cell types it is functional. This controversy has become important in evaluating reported nonclinical and clinical effects on the central nervous system and the cardiovascular system, and explaining adverse outcomes in the cancer setting.

The actual regulation of hematopoiesis, the feedback loops, the role of a plethora of cytokines in maintaining homeostasis in the hematopoietic system, and then creating an appropriate response to perturbations, such as sepsis, requires a broad approach. The complexity that could be investigated was reviewed [7] in the context of a systems biology approach.

In the clinic, beginning in the late 1980s, we generally did not exploit the complexity of multiple overlapping activities of some of the factors, other regulators such as stem cell factor (*c*-kit ligand), M-CSF, interleukin (IL)-11, multicolony-stimulating factor (IL-3), and IL-6. These cytokines also entered clinical development but have not found broad utility.

In this chapter, I focus on the trials and tribulations of the development of 3 families of regulation: the erythropoiesis-stimulating agents (ESA), the G-CSF, and the thrombopoietic agents.

Don Metcalf pointed out the value of 50 years of laboratory research before the initiation of the clinical development of each of these factors. It is apparent, however, that despite an extensive knowledge of murine biology and in vitro human studies, there were many surprises in the clinic and, in some cases, issues not strictly scientific, such as economic and legal issues, also impacted on the development and use of these agents.

The theoretical challenges encountered during the development of the ESA, G-CSF (filgrastim and lenograstim), and thrombopoietic agents had both common and unique features. Each was a critical regulator of an important cell lineage. Therefore, questions were raised whether accelerated depletion of the bone marrow would occur with prolonged use. This situation did not occur. There was concern that the receptors for each factor would be present on malignant cells either of the hematopoietic systems, such as the myeloid leukemias or on other cancers, and that this situation could have had an adverse outcome due to undesirable tumor cell stimulation. There was also concern that neutralizing antibodies to the recombinantly produced proteins would cross-react with the normal endogenous regulators and result in single lineage or multi-lineage aplasia. There were also concerns that the rate of rise in mature cells such as neutrophils, red blood cells, or platelets would cause harm or that the absolute high numbers of these cells could be harmful. During the development of these agents, some of these potential adverse events did become apparent, however, sometimes only after the agents entered clinical practice, and their doses and target populations were greatly expanded.

In general, millions of patients have received the hematopoietic agents with significant reductions in morbidity and mortality, and improvements in quality of life. The first study that identified the theoretical concerns that could occur was a randomized study of recombinant human EPO (rHuEPO) in patients who were receiving dialysis and who had heart disease in whom the concept of achieving high hemoglobin concentration to improve cardiac function resulted in significant adverse events [8]. It was reported that targeting a normal hematocrit significantly increased the incidence of thromboses and that there were more deaths in patients treated to obtain a normal hematocrit target than in patients treated to obtain a lower hematocrit target.

A second concern was realized during the development of a TPO (megakaryocyte growth and development factor, MGDF) when normal volunteers developed neutralizing antibodies after two or more doses that cross-reacted with endogenous TPO to produce prolonged thrombocytopenia. Another example of the potential harm of neutralizing antibodies was the development in a small number of patients receiving rHuEPO of pure red cell aplasia due to the development of cross-reactive neutralizing antibodies to endogenous EPO [9].

The concern about off-target stimulation of malignancies took longer to emerge. Large randomized studies in patients with cancer did appear to show in some studies poorer cancer outcomes – but the studies were not always well designed and were not stratified.

At the same time as the therapeutic window was narrowed, positive developments occurred including more convenient forms of rHuG-CSF (pegfilgrastim) and an ESA (darbepoetin alfa), and a new agent was developed that stimulated the TPO receptor but did not induce cross-reacting antibodies.

A new treatment paradigm, the use of peripheral blood progenitor cells (PBPC), was established and the risk of leukemia development did not appear to be significantly increased, although studies in severe chronic neutropenia and the myelodysplastic syndromes are still investigating the issue [10, 11]. In parallel to these developments, some of the clinical indications were expanded.

Not only did we learn the limits of the therapeutic agents, but the clinical settings also evolved. In oncology, the paradigm of using chemotherapeutic drugs to maximum tolerability thus causing the neutropenic complications reduced by rHuG-CSF was challenged. Guidelines appeared, although initially on the appropriate use of growth factors rather than the chemotherapy regimes (reviewed in chapter "Practice Guidelines for the Use of rHuG-CSF in an Oncology Setting" by Saraf and Ozer). The issue of cost benefits, cost offsets, and reimbursement dominated the development of the granulocyte-stimulating factors. Reimbursement also became important in determining the use of ESA and iron-replacement therapy, and this issue again led to guidelines that were modified as data emerged.

More recently, the cytokine area has attracted the development of biosimilars and discussion about whether given the challenges that have been identified during the development of cytokines, can other agents be approved without substantive clinical experience. I briefly discuss what we have learned about each of these agents.

2 Erythropoiesis-Stimulating Agents

Administration of rHuEPO is effective in increasing red blood cell counts. Anemic patients develop high concentrations of measurable endogenous EPO if they do not have renal disease, but in patients with renal failure or with malignancies, there can be inappropriately low amounts of endogenous EPO.

The first clinical use of rHuEPO was in patients with anemia who were relatively deficient in endogenous EPO due to renal disease. In early clinical trials of rHuEPO in patients with renal disease, there was a rapid reversal of the anemia, and although formal quality-of-life measurements were often not incorporated into the earliest studies, it seemed clear that patients developed improved states of well-being when their red blood cell counts recovered.

The increase in hemoglobin was observed in the first patients treated, and the agent was rapidly incorporated into therapy. Issues that arose included adverse effects such as thrombosis and hypertension in early studies, but were not perceived to be at a higher frequency than in control patients. It was also noted that patients needed to be replete with iron before the full effects of ESA were manifest.

After incorporation into therapy for renal disease, the anemia of cancer became a target for therapy. Initially, there was focus on patients who were receiving nephrotoxic chemotherapy such as cisplatin, but subsequently it was thought that patients with cancer who were receiving chemotherapy might have inappropriately low amounts of endogenous EPO for the degree of their anemia, and therapy with ESA was initiated to obviate the need for blood transfusions and also to improve quality of life.

The use of ESA became more complicated. There was much effort in trying to define optimal hemoglobin targets in both anemia of renal failure and anemia associated with cancer and cancer chemotherapy. It was suggested that higher hemoglobin concentrations could lead to a reduction in complications in the cardiovascular system of patients with chronic renal failure and in pre-dialysis patients. In addition, in oncology, the aim moved from preventing the need for red cell transfusions to improving the well-being of patients.

These studies led to an increase in the expenditure on ESA, particularly in the USA. An unexpected finding of the larger randomized studies, however, was that targeting a higher hemoglobin concentration seemed to lead to excess deaths. The phenomenon did not seem to depend on the level of hemoglobin reached but the increased dosing of ESA to reach the target. The basis for this remains unclear. There may also be a relationship between toxicity and the rate of rise in hemoglobin. Treatment guidelines and label warnings were adjusted for these findings [12]. In parallel, a new form of ESA which had additional glycosylation (darbepoetin alfa) was developed to reduce the frequency of dosing needed with rHuEPO and to improve convenience.

What began as a relatively clear benefit to anemic patients became much more complicated, and our assumptions about risk benefit had to be reviewed [12, 13]. It now seems that we have found the edges of the therapeutic window with attempts to normalize hemoglobin concentrations in pre-dialysis and dialysis patients, leading to increased adverse events and even mortality [8]. In the oncology setting, sometimes non-stratified randomized clinical trials have led to data suggesting reduced survival and loss of local cancer control. These findings were unexpected and have led to controversy about whether EPO receptors are present and functional on cancer cells and endothelium, and whether EPO acts directly on these cells to stimulate cancer growth [14, 15]. Others have suggested that while mRNA for the EpoR can be identified, the receptors are not functional [16].

Another unexpected aspect of the EPO story was its use in blood doping by cyclists to increase their red cell concentrations and endurance. In an episode in Europe, certain vials appeared to lead to immunogenicity due to the development of neutralizing antibodies and pure red cell aplasia in patients who received rHuEPO from this batch [9, 17]. This episode is often thought of in the context of quality control for biosimilar drug [18] development, particularly for agents that are glycosylated. A new agent has been developed that can stimulate the receptor but does not cross-react with neutralizing antibodies [19, 20].

The development of rHuEPO and ESA has taught us a great deal about how an agent that has been studied extensively non-clinically and for which there is a direct pharmacodynamic marker can lead to surprises when adopted broadly in clinical practice, and the need for appropriately designed phase 4 trials [21–23].

3 Granulocyte Colony-Stimulating Factors

The story of G-CSF has some similarities. The human molecule was first purified and cloned by a group in the USA (reviewed in chapter "Discovery of G-CSF and Early Clinical Studies" by Welte). It was not clear whether rHuG-CSF or rHuGM-CSF would prove more useful. In the mouse, rHuGM-CSF appeared to produce higher peripheral blood counts than rHuG-CSF; however, from the earliest clinical studies of rHuG-CSF [24–27], it was clear that rHuG-CSF produced significant increase in neutrophil counts and was well tolerated. Nonclinical studies suggested that rHuG-CSF could be used in patients including those with severe congenital neutropenia and those who had chemotherapy-induced neutropenia. Another application that was considered was in patients with normal neutrophil values who had sepsis and who might benefit from improved neutrophil function or higher neutrophil counts. A special setting that was also investigated was HIV-related infection and therapy that often led to neutropenia. In parallel to rHuG-CSF development, rHuGM-CSF was cloned and tested in the clinic, but will not be further discussed. Both agents were approved and incorporated into practice.

The early studies with rHuG-CSF produced some surprises [28]. The findings included that rHuG-CSF produced a transient decrease in circulating neutrophils in the first few minutes after injection, presumably due to tissue entry, and that the neutrophils were available to the tissue [26, 29]. The neutrophils were "left shifted" and rHuG-CSF not only stimulated production but also accelerated maturation. Studies also showed no change in frequency of progenitor cells in the bone marrow, but very rapid mobilization into the periphery [30]. The latter observation led to the practical widespread application of PBPC transplantation [31, 32; reviewed in chapter "Use of rHuG-CSF in Peripheral Blood Progenitor Cell Transplantation" by Beligaswatte et al.]. The basis for the mobilization is now better understood as disruption of the interactions between adhesion molecules and their ligands [33, 34].

The next set of agents to enter the clinic in 1986 was rHuG-CSF, rHuGM-CSF, and more recently, a pegylated form of rHuG-CSF (pegfilgrastim). The first indications that were approved were in the reduction of the infection complication of chemotherapy and as a consequence, the use of rHuG-CSF to intensify the doses of chemotherapy. These studies are reviewed extensively elsewhere. It was clear that in every setting, rHuG-CSF reduced the duration of neutropenia and the risk of febrile neutropenia by 40–50% [28].

The development of rHuG-CSF initially focused on the reduction of the consequences of infection, febrile neutropenia, antibiotic use, and hospitalization [35]. It however soon became apparent that cost–benefit studies were required. The optimal use of rHuG-CSF was as primary prophylaxis starting the day after chemotherapy, and its effect is diminished if its administration is delayed. Despite clear evidence, even today it is not used optimally in most patients [36, 37; reviewed in chapter "The Economics of the Recombinant Granulocyte Colony-Stimulating Factors" by Hirsch and Lyman].

Filgrastim and pegfilgrastim are used in patients who have a significant risk of febrile neutropenia. Well-developed guidelines in the USA and Europe distilled evidence and recommend usage, although some criticisms have been raised of the guideline processes [38].

One obvious indication for early development was in children with congenital neutrophil disorders who suffered from recurrent infections, developmental delays, and premature death [39]. The area of severe congenital neutropenia proved fascinating and remains challenging [40]. When the first studies of rHuG-CSF were initiated, the basis for the neutropenia was not clear. When the studies were initiated by David Dale and his colleagues, a registry was established to study the long-term consequences of daily administration of rHuG-CSF for potentially many years [41; reviewed in chapter "rHuG-CSF for the Treatment of Severe Chronic Neutropenia" by Dale and Bouylard].

It is clear that the clinical outlook of the patients was improved by rHuG-CSF but as patients lived longer, it also became apparent that the patients particularly requiring the higher pharmacologic doses of rHuG-CSF were at risk of developing myelodysplastic syndromes or acute myeloid leukemia. Whether this risk is in part due to rHuG-CSF administration or the manifestation of the underlying disorder remains unclear, and observation and a cautious approach continue.

Several randomized studies of rHuG-CSF were undertaken in infectious settings in which neutropenia was not an issue, including nosocomial and community-acquired pneumonia and liver transplantation. These studies [42–44] showed not only no increased risk from increased neutrophil counts up to 70×10^9/L in some patients, but also no benefit to the antibiotics.

4 Thrombopoietins

Platelet stimulation has two approved agents. Romiplostim is a peptibody that contains the human immunoglobulin IgG Fc domain covalently linked to two 14 amino acid peptides that bind to and stimulate the TPO receptor. It was developed when a pegylated recombinant human megakaryocyte factor (PEG-rHuMGDF) was withdrawn from development because of the appearance of neutralizing antibodies. The antibodies were still detected 2–3 years after administration and were more common in healthy volunteers than in patients with malignancy. The other challenge of development was that in general, the effect on platelets was a large overshoot and high concentration but not an abrogation of the duration that platelets values are $<10 \times 10^9$/L and the patient was at high risk of bleeding. The success of romiplostim is the chronic setting of immune thrombocytopenia in patients who failed to be helped by steroids [45–48].

5 Lessons Learned

What have we learned after 20 years of developing these agents? The erythropoietic-, granulocyte-, and thrombopoietic-stimulating agents have helped many patients. When they were purified and cloned, it was important to be able to modify each either to improve convenience with glycosylation or pegylation, or to alter completely the protein sequence to avoid neutralizing antibodies. Neutralizing antibodies can be serious and lead to consequences such as pure red cell aplasia or persistent thrombocytopenia and so must be avoided.

In the clinic, we can develop new biologic insights that can lead to new fields, such as peripheral progenitor cell transplantation. We also learned that as we treat more patients after marketing approval and commercialization, thousands and millions of patients instead of hundreds treated during clinical trials, and as we push the envelope to achieve greater benefit (or cynics might say to increase adoption), we can identify doses, schedules, and settings where the risk:benefit ratio can be diminished.

We found the immunogenicity of thrombopoietic agents early and re-engineered the molecule. In the severe chronic neutropenia setting with rHuG-CSF and in the cancer setting and pre-dialysis setting for ESA, we will need to continue research to understand the clinical observations. A challenge that was also faced was the duration of development, and the number of patients needed for safety studies continued to increase from several hundred in the regulatory submissions for rHuEPO and rHuG-CSF to over a thousand for darbepoetin alfa and pegfilgrastim. During the 5 years taken for development and the time taken for label extensions of the hematopoietic growth factors, the clinical settings and the hurdles for successful development changed.

In the period 1987–2000, AIDS became a treatable disease and neutropenia became a rare event, and the development of filgrastim in this setting became less feasible. In the setting of thrombocytopenia, the realization that lower platelet concentrations ($<10 \times 10^9$/L) were safe changed the endpoints for trials.

The inability to store platelets long term and the large demand for platelets led to a study of increasing platelets in normal volunteers to enhance collection. The immunogenicity of rHuMGDF led to cessation in this setting and it is unclear whether romiplostim could be used. In oncology, changes in the use of chemotherapy, the introduction of targeted agents, and the incidence of febrile neutropenia and data about dose intensive therapy led to new approaches and the need for more studies.

In the setting of anemia associated with renal failure, once improvements in quality of life were achieved, the importance of cardiac disease became apparent and led to dose intensification approaches in ESA use. In parallel, the safety of the red cell transfusion services improved and the impetus to avoid blood transfusions diminished.

The past 20 years of clinical development have been rich in achievements and challenges, and many opportunities remain.

References

1. Cosman D (1988) Colony-stimulating factors in vivo and in vitro. Immunol Today 9:97–98
2. Metcalf D (2008) Hematopoietic cytokines. Blood 111:485–491
3. Morstyn G, Foote M, Lieschke G (2004) Introduction to hematopoietic growth factors. In: Morstyn G, Foote M, Lieschke GJ (eds) Hematopoietic growth factors in oncology, vol 1. Humana Press, Totowa, NJ, pp 3–10
4. Bradley TR, Metcalf D (1966) The growth of mouse bone marrow cells in vitro. Aust J Exp Biol Med Sci 44:287–299
5. Hamilton JA (2008) Colony-stimulating factors in inflammation and autoimmunity. Nat Rev Immunol 8:533–544
6. Lieschke GJ, Grail D, Hodgson G et al (1994) Mice lacking granulocyte colony-stimulating factor have chronic neutropenia, granulocyte and macrophage progenitor cell deficiency, and impaired neutrophil mobilization. Blood 84:1737–1746
7. Whichard ZL, Sarkar CA, Kimmel M, Corey SJ (2010) Hematopoiesis and its disorders: a systems biology approach. Blood 115:2339–2347
8. Besarab A, Bolton WK, Browne JK et al (1998) The effects of normal as compared with low hematocrit values in patients with cardiac disease who are receiving hemodialysis and epoetin. N Engl J Med 339:584–590
9. Gershon SK, Ludsenburg H, Coté TR, Braun MM (2002) Pure red-cell aplasia and recombinant erythropoietin. N Engl J Med 346:1584–1586
10. Greenberg PL, Sun Z, Miller KB et al (2009) Treatment of myelodysplasic syndrome patients with erythropoietin with or without granulocyte colony-stimulating factor: results of a prospective randomized phase 3 trial by the Eastern Cooperative Oncology Group (E1996). Blood 114:2393–2400
11. Beekman R, Touw IP (2010) G-CSF and its receptor in myeloid malignancy. Blood 115:5131–5136
12. Kimmel PL, Malozowski S (2010) Reevaluating erythropoiesis-stimulating agents. N Engl J Med 362:1742–1743
13. Solomon SD, Uno H, Lewis EF et al (2010) Erythropoietic response and outcomes in kidney disease and type 2 diabetes. N Engl J Med 363:1146–1155
14. Spivak JL (1994) Recombinant human erythropoietin and the anemia of cancer. Blood 84:997–1004
15. Rossert J, Eckardt KU (2010) Erythropoietin receptors: their role beyond erythropoiesis. Nephrol Dial Transplant 20:1025–1028
16. Sinclair AM, Coxon A, McCafferty I et al (2010) Functional erythropoietin receptor is undetectable in endothelial, cardiac, neuronal, and renal cells. Blood 115:4264–4272
17. Bergman H, Danielson BG et al (1993) A case of anti-erythropoietin antibodies following recombinant human erythropoietin treatment. In: Bauer C, Koch KM, Scigalia P (eds) Erythropoietin: molecular physiology and clinical application. Marcel Dekker, New York, pp 266–275

18. Engert A, Griskevicius L, Zyuzgin Y, Ludenau H, del Giglio A (2009) XM02, the first granulocyte-stimulating factor biosimilar, is safe and effective in reducing the duration of severe neutropenia and incidence of febrile neutropenia in patients with non-Hodgkin lymphoma receiving chemotherapy. Leuk Lymphoma 50:374–379
19. Bussel JB, Kuter DJ, George JN et al (2006) AMG 531, a thrombopoiesis-stimulating protein, for chronic ITP. N Engl J Med 355:1672–1681
20. Macdougall IC, Rossert J, Cassadevall N et al (2009) A peptide-based erythropoietin-receptor agonist for pure red-cell aplasia. N Engl J Med 361:1848–1855
21. Erslev AJ (1991) Erythropoietin. N Engl J Med 324:1339–1344
22. Merchionne F, Dammacco F (2009) Biological functions and therapeutic use of erythropoiesis-stimulating agents: perplexities and perspectives. Br J Haematol 146:127–141
23. Rizzo JD, Brouwers M, Hurley P et al (2010) American Society of Hematology/American Society of Clinical Oncology clinical practice guideline update on the use of epoetin and darbepoetin in adult patients with cancer. Blood 116:4045–4059
24. Bronchud MH, Scarffe JH, Thatcher N et al (1987) Phase I/II study of recombinant human granulocyte colony-stimulating factor in patients receiving intensive chemotherapy for small cell lung cancer. Br J Cancer 56:809–813
25. Gabrilove JL, Jakubowski A, Scher H et al (1988) Effect of granulocyte colony-stimulating factor on neutropenia and associated morbidity due to transitional cell carcinoma of the urothelium. N Engl J Med 318:1414–1422
26. Morstyn G, Campbell L, Souza LM et al (1988) Effect of granulocyte colony-stimulating factor on neutropenia induced by cytotoxic chemotherapy. Lancet 1:667–672
27. Morstyn G, Campbell L, Lieschke G et al (1989) Treatment of chemotherapy induced neutropenia by subcutaneously administered granulocyte colony-stimulating factor with optimization of dose and duration of therapy. J Clin Oncol 7:1554–1562
28. Welte K, Gabrilove J, Bronchud MH, Platzer E, Morstyn G (1996) Filgrastim (r-metHuG-CSF): the first 10 years. Blood 88:1907–1929
29. Lieschke GJ, Ramenghi U, O'Connor MP, Sheridan W, Morstyn G (1992) Studies of oral neutrophil levels in patients receiving G-CSF after autologous marrow transplantation. Br J Haematol 82:589–595
30. Duhrsen U, Villeval JL, Boyd J, Kannourakis G, Morstyn G, Metcalf D (1988) Effects of recombinant human granulocyte colony-stimulating factor on hematopoietic progenitor cells in cancer patients. Blood 72:2074–2081
31. Sheridan WP, Begley CG, Juttner CA et al (1992) Effect of peripheral-blood progenitor cells mobilised by filgrastim (G-CSF) on platelet recovery after high-dose chemotherapy. Lancet 339:640–644
32. Copelan EA (2006) Hematopoietic stem-cell transplantation. N Engl J Med 354:1813–1826
33. Grigg A, Begley CG, Juttner CA et al (1993) Effect of peripheral blood progenitor cells mobilised by filgrastim (G-CSF) on platelet recovery after high-dose chemotherapy. Bone Marrow Transplant 11:23–29
34. Pusic I, DiPersio JF (2008) The use of growth factors in hematopoietic stem cell transplantation. Curr Pharm Des 14:1950–1961
35. Kelly S, Wheatley D (2009) Prevention of febrile neutropenia: use of granulocyte colony-stimulating factors. Br J Cancer 101:S6–S10
36. Renwick W, Pettengell R, Green M (2009) Use of filgrastim and pegfilgrastim to support delivery of chemotherapy: twenty years of clinical experience. Biodrugs 23:175–186
37. Ramsey SD, McCune JS, Blough DK et al (2010) Colony-stimulating factor prescribing patterns in patients receiving hemotherapy for cancer. Am J Manag Care 16:678–686
38. Haines IE, Olver I (2009) Are guidelines on use of colony-stimulating factors in solid cancers flawed? Intern Med J 39:259–262
39. Guba SC, Sartor CA, Hutchinson R, Boxer LA, Emerson SG (1994) Granulocyte colony-stimulating factor (G-CSF) production and G-CSF receptor structure in patients with congenital neutropenia. Blood 83:1486–1492

40. Dale DC, Boxer L, Liles WC (2008) The phagocytes: neutrophils and monocytes. Blood 112:935–945
41. Rosenberg PS, Alter BP, Bolyard AA et al (2006) The incidence of leukemia and mortality from sepsis in patients with severe congenital neutropenia receiving long-term G-CSF therapy. Blood 107:4628–4635
42. Nelson S, Belknap SM, Carlson RW et al (1998) A randomized controlled trial of filgrastim as an adjunct to antibiotics for treatment of hospitalized patients with community-acquired pneumonia. J Infect Dis 178:1075–1080
43. Hotchkiss RS, Karl IE (2003) The pathophysiology and treatment of sepsis. N Engl J Med 348:138–150
44. Root RK, Lodato RF, Patrick W et al (2003) A multicenter, double-blind, placebo-controlled study of the use of filgrastim in patients hospitalized with pneumonia and severe sepsis. Crit Care Med 31:367–373
45. Kaushansky K (1995) Thrombopoietin: the primary regulator of platelet production. Blood 86:419–431
46. Kuter DJ, Begley CG (2002) Recombinant human thrombopoietin: basic biology and evaluation of clinical studies. Blood 100:3457–3469
47. Kaushansky K (2008) Historical review: megakaryopoiesis and thrombopoiesis. Blood 111:981–986
48. Kuter DJ, Rummel M, Boccia R et al (2010) Romiplostim or standard of care in patients with immune thrombocytopenia. N Engl J Med 363:1889–1899

Discovery of G-CSF and Early Clinical Studies

Karl Welte

1 Introduction

In the 1960s, two groups simultaneously developed methods for growing colonies of granulocytes and monocytes from mouse bone marrow or spleen cells in semi-solid agar (for review, see [1]). The colony growth was dependent on the presence of unknown factors, which were given the operational name colony-stimulating factors (CSF). Efforts to biologically identify and biochemically purify these CSF kept many laboratories busy until the middle of the 1980s and revealed that there is no single CSF, but rather four quite biochemically different CSF with different colony-stimulating activities. The four CSF were given names dependent on the type of colonies: GM-CSF stimulated granulocyte and macrophage colonies; M-CSF, macrophage colonies; G-CSF, granulocyte colonies; and multi-CSF (interleukin [IL-3]), a broad range of hematopoietic cell colonies [1].

2 Purification and Biochemical Characteristics of G-CSF

Murine G-CSF was purified by Nick Nicola and colleagues in Melbourne, Australia, in 1983 [2], and human G-CSF was purified independently by a group, which included Erich Platzer and myself, in New York in 1983/1984 [3] (Fig. 1). The murine G-CSF was purified from mouse lung-conditioned medium, and the human G-CSF from the human bladder carcinoma cell line 5637. Intriguingly, not knowing the results of the work of Nicola et al. when we started to purify our CSF in 1983, our goal was initially to purify human IL-3, as we had been successful in

K. Welte (✉)
Department of Molecular Hematopoiesis, Hannover Medical School, Carl-Neuberg-Str. 1, 30625 Hannover, Germany
e-mail: Welte.Karl.H@mh-hannover.de

G. Molineux et al. (eds.), *Twenty Years of G-CSF*, Milestones in Drug Therapy,
DOI 10.1007/978-3-0348-0218-5_2,

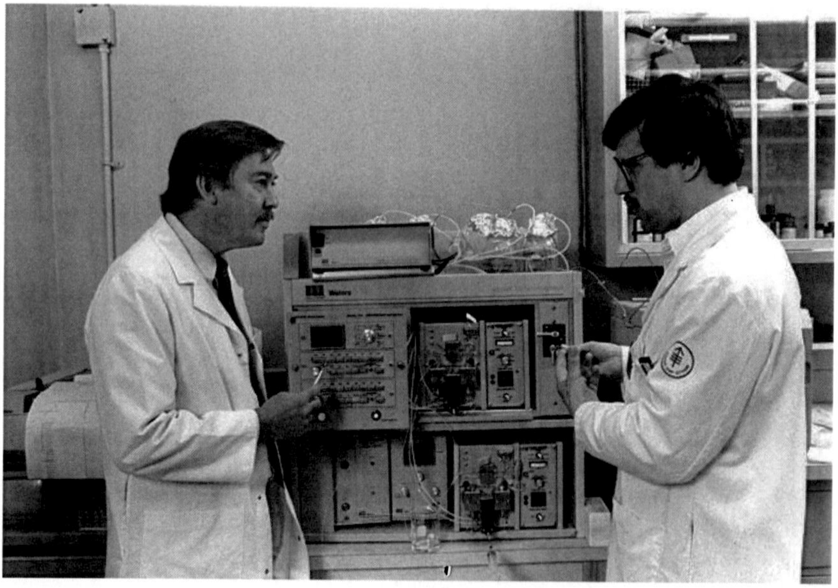

Fig. 1 Karl Welte (right) and Malcolm A.S. Moore at the HPLC equipment in the Memorial Sloan-Kettering Cancer Center, New York, NY, USA

purifying IL-2 [4] and Ihle and colleagues had recently successfully purified murine IL-3 [5]. These findings were one of the reasons why we called our CSF the first pluripotent hematopoietic CSF [3]. Starting with 40 L 5637-conditioned medium, we succeeded in purifying G-CSF to homogeneity by ion-exchange chromatography, gel filtration, and reversed-phase high-pressure liquid chromatography (HPLC), and produced 5 μg of pure G-CSF as judged by silver staining in polyacrylamide-gel electrophoresis. G-CSF is *O*-glycosylated and has a molecular weight of 19,600 Da. Because the amino acid sequence of G-CSF was not known, we asked Por Li and Lawrence Souza at Amgen, Thousand Oaks, CA, to help us to get the initial *N*-terminal amino acid sequence. Deduced from the amino acid sequence, molecular cloning of the cDNA for G-CSF and the first expression in *Escherichia coli* were achieved in 1986 by Souza and Boone at Amgen in cooperation with our laboratory [6]. The recombinant G-CSF (rG-CSF) was capable of supporting myeloid proliferation and differentiation in granulocyte-macrophage progenitor cell (CFU-GM) assays. We were also able to identify G-CSF receptors (G-CSFR) on myeloid leukemia cells by binding studies with ^{125}I-labeled G-CSF [6]. Later, in 1986, a Japanese group purified G-CSF from the cell line CHU-2 [7], and Nagata et al. [8] cloned the cDNA encoding G-CSF from the same cell line by using probes derived the partial amino acid sequence of the purified G-CSF from CHU-2 cells. However, they described a protein with 177 amino acids, which might be a splice variant of the 174-amino acid G-CSF we obtained from the 5637 cell line. In 1990, the specific G-CSFR was characterized and cloned [9] and was shown to be a homodimer. It is interesting to note that a few hundred receptors per cell can induce optimal responses to G-CSF. The binding of G-CSF to its receptor activates many signaling pathways such as phosphorylation of JAK2, STAT3, and STAT5, and the transcription factors LEF-1 and C/EBPα [10]. An interesting study demonstrated that G-CSF leads to upregulation of the expression of nicotinamide phosphoribosyltransferase (Nampt), which induces an increase in NAD^+ and sirtuins [11].

3 Biology of G-CSF

A range of actions have been documented for G-CSF. In a CFU-GM assay, G-CSF is capable of inducing growth of mainly neutrophiilic granulocyte colonies. It also acts on the function of mature neutrophils, such as enhancement of chemotactic peptide *N*-formyl-methionyl-leucyl-phenylalanine (fMLP) binding on mature neutrophils [12]. Granulocyte colony formation can be stimulated with GM-CSF, M-CSF, IL-3, and stem cell factor (SCF). An interesting study compared native G-CSF and rG-CSF and showed that they had identical biologic activities [13]. The study showed that the removal of monocytes and T-lymphocytes from the bone marrow abrogated the growth of erythroid progenitor cells (BFU-E) and granulocyte–erythrocyte–monocyte–megakaryocyte progenitor cells (CFU-GEMM) when native G-CSF was used as a stimulator. The initial term "pluripotent CSF" (PPO) was created because native purified G-CSF stimulated also BFU-E and CFU-GEMM in unseparated bone marrow due to stimulation of monocytes and T-lymphocytes to release GM-CSF [13]. The solution of the specificity of the single CSF came from gene deletion studies in mice. G-CSF knock-out studies revealed that it is clearly responsible for proliferation and differentiation of granulocyte progenitor and precursor cells [14]. Mice lacking endogenous G-CSF have chronic neutropenia and impaired neutrophil mobilization, indicating that G-CSF is indispensable for maintaining the normal balance of neutrophil production during steady-state myelopoiesis.

4 Nonclinical In Vivo Studies with G-CSF

Soon after the availability of rG-CSF, its potential clinical use was investigated. We investigated the effects of rG-CSF in nonhuman primates (cynomolgus monkeys) [15]. After subcutaneous administration, the white blood cell counts increased in a rG-CSF dose-dependent manner 24 h after initiation of treatment, and with daily treatment reached a plateau by day 6. The increase in the white blood cell count was mainly due to the increase in neutrophil counts. For example, at 10 μg/kg/day, the absolute neutrophil count (ANC) increased to approximately 50×10^9/L and at a dose of 100 μg/kg/day, the count increased to approximately 100×10^9/L.

The neutrophil counts could be maintained for the duration of a 4-week treatment period and returned to normal values 3 days after treatment ended. The absolute numbers of lymphocytes increased approximately twofold, but the number of monocytes, eosinophils, reticulocytes, and platelets did not change significantly. In the spleen, foci of extramedullary hematopoiesis were observed. No other organs demonstrated evidence of hematopoietic activities. The neutrophils functioned normally in standard tests for chemotaxis and there was evidence of enhanced ability to kill phagocytized bacteria [15].

We further investigated whether or not rG-CSF might be of benefit in chemotherapy-induced neutropenia and after bone marrow transplantation. rG-CSF was able to shorten the period of chemotherapy-induced neutropenia after high-dose cyclophosphamide and busulfan, or total body irradiation followed by autologous bone marrow transplantation [15, 16].

These results demonstrated that G-CSF is a potent granulopoietic growth and differentiation factor in vivo and opened the avenue of the use of rHuG-CSF in the treatment of patients not only with chemotherapy-induced neutropenias but also with other clinical situations associated with chronic neutropenias.

5 Clinical Uses of rHuG-CSF

5.1 Phase 1 and Phase 2 Studies

The first clinical use of recombinant human G-CSF (rHuG-CSF) was performed at the Memorial Sloan-Kettering Cancer Center, New York, in patients with transitional cell carcinoma of the urothelium in 1987 [17]. The study was designed as an open-label phase 1/2 study of both the safety and efficacy of five dosages of rHuG-CSF, with each dosage evaluated in three to five patients receiving the M-VAC chemotherapy (methotrexate, vinblastine, doxorubicin, and cisplatin). The dosages were 1, 3, 10, 30, and 60 μg/kg/day subcutaneously, given before (day 12 to day 7) or during the first cycle on day 4 through day 11 of treatment with M-VAC, or both. Treatment with rHuG-CSF before chemotherapy produced a dose-dependent increase in neutrophil counts, and the use of rHuG-CSF after M-VAC chemotherapy significantly reduced the number of days per patient on which the ANC was $\leq 1.0 \times 10^9$/L, reduced the number of days on which antibiotics were used to treat febrile neutropenia, and significantly increased the percentage of patients qualified to receive planned chemotherapy. These findings demonstrated that rHuG-CSF can reduce both the hematopoietic and oral toxicity of chemotherapy [17].

A second phase 1/2 study used rHuG-CSF in 12 patients who were receiving intensive chemotherapy for small-cell lung cancer [18]. Patients were treated by continuous infusion of rHuG-CSF at 1, 5, 10, 20, or 40 μg/kg/day for 5 days before chemotherapy and 14 days after adriamycin, ifosfamide, and etoposide chemotherapy on alternative cycles. As in the first study [17], use of rHuG-CSF reduced the period of neutropenia considerably and no infectious episodes were observed in the cycles when patients received rHuG-CSF [18].

5.2 *Phase 3 Studies*

Two randomized multicenter studies with rHuG-CSF were performed in patients with small-cell lung cancer in the USA [19] and in Europe [20]. These studies were designed to test whether or not rHuG-CSF could decrease the incidence of infections as manifested by febrile neutropenia, and whether or not the use of rHuG-CSF would lead to a reduction in the incidence of intravenous antibiotic use, hospitalization, and culture-confirmed infections. Both studies used chemotherapy with cyclophosphamide, doxorubicin, and etoposide. The reduction in febrile neutropenia episodes was similar in both studies, and no statistically significant differences were detected in tumor response rates or overall survival.

5.3 *Use of rHuG-CSF in Hematologic Malignancies*

The use of rHuG-CSF in patients with acute myeloid leukemia (AML) remained controversial because of the in vitro observation that leukemic cells express G-CSFR [6] and that G-CSF could stimulate leukemic cell growth [21]. A prospective, randomized study was done to determine the efficacy and safety of rHuG-CSF after standard intensive chemotherapy in 108 patients with relapsed or refractory acute leukemia [22]. Treatment with rHuG-CSF was shown to accelerate the recovery of neutrophils and reduce the incidence of infections. No difference was observed in remission rates between the two treatment groups. The researchers further explored whether or not the administration of rHuG-CSF before and during chemotherapy would result in enhanced killing of leukemic cells. No difference was seen between the groups in event-free survival or in disease-free survival in patients who did achieve a complete remission. Another group investigated the use of rHuG-CSF before and during chemotherapy in AML patients and did not see any effect on complete remission [23]. The ability of rHuG-CSF to induce remission by stimulating residual normal donor cells in patients after allogeneic stem cell transplantation was studied [24]. The investigators reported that rHuG-CSF might be effective in selected patients with early relapse after allogeneic bone marrow transplantation.

We conducted a randomized study with rHuG-CSF in children with high-risk acute lymphoblastic leukemia (ALL) treated according to the ALL-BFM chemotherapy protocol [25]. Children were randomly assigned to receive nine alternating cycles of chemotherapy alone or followed by rHuG-CSF at 5 μg/kg/day. In both groups, the planned interval between chemotherapy courses was 21 days. Of the 34 patients analyzed, the incidence of febrile neutropenia, the number of culture-confirmed infections, and the total duration of intravenous antibiotic treatment were reduced significantly; however, with a median follow-up of 15 years, no difference was observed with regard to the estimated event-free survival [25].

5.4 Use of rHuG-CSF in Stem Cell Transplantation

Myeloablative chemotherapy requires cellular reconstitution, and rHuG-CSF alone is not sufficient to produce recovery. Because prolonged neutropenia occurs after stem cell transplantation, this setting was one of the first in which rHuG-CSF was studied. An early study administered rHuG-CSF by 30-min bolus infusion at a dose of 60 μg/kg/day beginning 24 h after autologous marrow infusion in 18 patients with Hodgkin‘s disease [26]. Recovery to neutrophil counts of 1×10^9/L occurred 14 days earlier in rHuG-CSF-treated patients compared with that in control patients [26]. Other researchers administered rHuG-CSF as a continuous subcutaneous infusion (20 μg/kg/day) after autologous bone marrow transplantation in patients with relapsed Hodgkin‘s disease, non-Hodgkin’s lymphoma, ALL, and germ cell tumors [27]. The median time to an ANC $>0.5 \times 10^9$/L was 11 days compared to 20 days in the historical control patients.

5.5 Mobilization of Peripheral Blood Progenitor Cells

During the first clinical trials of rHuG-CSF in patients with cancer, an unexpected observation was made: The patients developed a 100-fold increase in the frequency of colony-forming progenitor cells in the peripheral blood. Dührsen et al. [28] were among the first to report the increases in various clonogenic hematopoietic progenitor cells on day 5 of rHuG-CSF treatment in patients with cancer. These results have been confirmed in subsequent studies [29, 30].

Sheridan et al. examined the ability of rHuG-CSF-mobilized peripheral blood progenitor cells (PBPC) to reconstitute hematopoiesis in 17 patients with poor-prognosis nonmyeloid malignancies. Of the 17 patients, 14 received high-dose chemotherapy, and the cryopreserved apheresis product was infused on day 0, followed by rHuG-CSF at 24 μg/kg/day starting on day 1. Platelet recovery was significantly faster in the rHuG-CSF-mobilized PBPC-treated patients than in historical controls [29]. Bensinger et al. studied the feasibility of using rHuG-CSF to mobilize granulocytes in normal donors and showed that rHuG-CSF was safe to administer in healthy adults [31]. In many subsequent studies, rHu G-CSF-mobilized PBPC were used instead of bone marrow stem cells in allogeneic stem cell transplantation and led to a paradigm change in stem cell transplantation. rHuG-CSF-mobilized PBPC from healthy donors have become the dominant cell populations used in transplantations to patients with leukemia and cancer. More information about the use of rHuG-CSF in the setting of PBPC transplantation is provided in the chapter "Use of rHuG-CSF in Peripheral Blood Progenitor Cell Transplantation" by Beligaswatte et al.

5.6 Use of rHuG-CSF in Patients with Severe Chronic Neutropenia

One of the first clinical studies with rHuG-CSF was performed in the late 1980 in patients with congenital neutropenia at the Memorial Sloan-Kettering Cancer Center, New York, by Bonilla et al. [32]. Children born with severe congenital neutropenia are diagnosed with neutropenia at birth or shortly thereafter, which is generally accompanied by frequent and often life-threatening infections. Initially, five patients were treated with rHuG-CSF in an attempt to reduce the morbidity and mortality associated with this disorder. Patients were treated with rHuG-CSF dosages between 3 and 60 μg/kg/day administered as a continuous subcutaneous infusion. In all five patients, an increase in the number of neutrophils was reached 1 week to 9 days after the initiation of the effective dose of rHuG-CSF. All patients had sustained neutrophil counts [32]. In subsequent phase 1/2 studies with rHuG-CSF in the setting of congenital neutropenia, hundreds of patients were enrolled worldwide and a Severe Chronic Neutropenia International Registry (SCNIR) was established [33–36]. More than 90% of the patients with severe chronic neutropenia (SCN) responded to the treatment with rHuG-CSF with an increase in ANC $>1.0 \times 10^9$/L, and in most patients, the bacterial infections and the requirement for intravenous antibiotic use were significantly reduced. The subcutaneous dose of rHuG-CSF necessary to reach and maintain this ANC varied from patient to patient, and ranged between 1 and 80 μg/kg/day. The initial patients have been treated for >20 years daily without exhaustion of myelopoiesis and without generation of anti-G-CSF antibodies. Both the prognosis and the quality of life of patients with congenital neutropenia improved dramatically after the introduction of rHuG-CSF therapy in 1987. Since the establishment of the SCNIR in 1994, data on >1,000 patients have been collected worldwide to monitor the clinical course, treatment, and disease outcomes. Current knowledge of the underlying pathomechanisms suggests that congenital neutropenia is a heterogeneous multigene disorder of myelopoiesis. Genetic analyses revealed mutations in the genes for ELANE, HAX1, G6PC3, and many others. It is now well accepted that congenital neutropenia is a preleukemic syndrome and an approximately 20% of these patient develop leukemia. Independent of the genetic subtype, conversion to leukemia in congenital neutropenia is associated with acquired genetic somatic aberrations, such as G-CSF receptor mutations, monosomy 7, and *ras*-mutations [37]. More information concerning the development and use of rHuG-SCF in the setting of SCN is provided in the chapter "rHuG-CSF for the Treatment of Severe Chronic Neutropenia" by Dale and Boulyard.

6 Summary and Conclusion

The identification, purification, and molecular cloning of rHuG-CSF in the 1980s; the nonclinical studies in the mid-1980s; and the subsequent development of rHuG-CSF as a therapeutic agent in the late 1980s and 1990s have had a major influence on the treatment of many diseases. rHuG-CSF has been the topic of many clinical researches and publications in the last 20 years. rHuG-CSF has specific and selective actions due to the restricted expression of G-CSFR on myelopoiesis. It stimulates proliferation, differentiation, and activation of cells of the neutrophil lineage. In the clinical setting, rHuG-CSF is of benefit to patients receiving chemotherapy or myeloablative treatment. It has been shown to reduce morbidity in many patient populations. Stem cell transplantation using rHuG-CSF-mobilized PBPC revolutionized stem cell transplantation, making it simpler, more efficient, and more widely applicable in the clinic. However, clinical research is still necessary to improve the use of rHuG-CSF in patients with cancer and other diseases.

References

1. Metcalf D (2010) The colony-stimulating factors and cancer. Nat Rev Cancer 10:425–434
2. Nicola NA, Metcalf D, Matsumoto M, Johnson GR (1983) Purification of a factor inducing differentiation in murine myelomonocytic leukemia cells. J Biol Chem 258:9017–9023
3. Welte K, Platzer E, Lu I et al (1985) Purification and biochemical characterization of human pluripotent hematopoietic colony-stimulating factor. Proc Natl Acad Sci U S A 82:1526–1530
4. Welte K, Wang CY, Mertelsmann R, Venuta S, Feldman SP, Moore MA (1982) Purification of human interleukin 2 to apparent homogeneity and its molecular heterogeneity. J Exp Med 156:454–464
5. Ihle JN, Keller J, Henderson L, Klein F, Palaszynski E (1982) Procedures for the purification of interleukin 3 to homogeneity. J Immunol 129:2431–2436
6. Souza LM, Boone TC, Gabrilove J et al (1986) Recombinant human granulocyte colony-stimulating factor: effects on normal and leukemic myeloid cells. Science 232:61–65
7. Nomura H, Imazeki I, Oheda M et al (1986) Purification and characterization of human granulocyte colony-stimulating factor (G-CSF). EMBO J 5:871–876
8. Nagata S, Tsuchiya M, Asano S et al (1986) Molecular cloning and expression of cDNA for human granulocyte colony-stimulating factor. Nature 319:415–418
9. Fukunaga R, Ishizaka-Ikeda E, Nagata S (1990) Purification and characterization of the receptor for murine granulocyte-colony-stimulating factor. J Biol Chem 265:14008–14015
10. Skokowa J, Cario G, Uenalan M et al (2006) LEF-1 is crucial for neutrophil granulocytopoiesis and its expression is severely reduced in congenital neutropenia. Nat Med 12:1191–1197
11. Skokowa LD, Thakur BK et al (2009) NAMPT is essential for the G-CSF induced myeloid differentiation via NAD^+-sirtuin-1-dependent pathway. Nat Med 15:151–158
12. Platzer E, Welte K, Gabrilove JL et al (1985) Biological activities of a human pluripotent hematopoietic colony stimulating factor on normal and leukemic cells. J Exp Med 162:1788–1801
13. Strife A, Lambek C, Wisniewski D et al (1987) Activities of four purified growth factors on highly enriched human hematopoietic progenitor cells. Blood 69:1508–1523

14. Lieschke GJ, Grail D, Hodgson G et al (1994) Mice lacking granulocyte colony-stimulating factor have chronic neutropenia, granulocyte and macrophage progenitor cell deficiency, and impaired neutrophil mobilization. Blood 84:1737–1746
15. Welte K, Bonilla MA, Gillio AP et al (1987) Recombinant human granulocyte colony-stimulating factor. Effects on hematopoiesis in normal and cyclophosphamide-treated primates. J Exp Med 165:941–948
16. Gillio AP, Bonilla MA, Potter GK et al (1987) Effects of recombinant human granulocyte colony stimulating factor on hematopoietic reconstitution after autologous bone marrow transplantation in primates. Transplant Proc 19:153
17. Gabrilove JL, Jakubowski A, Scher H et al (1988) Effect of granulocyte colony-stimulating factor on neutropenia and associated morbidity due to chemotherapy for transitional-cell carcinoma of the urothelium. N Engl J Med 318:1414–1422
18. Bronchud MH, Scarfe JH, Thatcher N et al (1987) Phase I/II study of recombinant human granulocyte colony-stimulating factor in patients receiving intensive chemotherapy for small cell lung cancer. Br J Cancer 56:809–813
19. Crawford J, Ozer H, Stoller R et al (1991) Reduction by granulocyte colony-stimulating factor of fever and neutropenia induced by chemotherapy in patients with small-cell lung cancer. N Engl J Med 325:164–170
20. Trillet-Lenoir V, Green J, Manegold C et al (1993) Recombinant granulocyte colony stimulating factor reduces the infectious complications of cytotoxic chemotherapy. Eur J Cancer 29A:319–324
21. Lowenberg B, Touw IT (1993) Hematopoietic growth factors and their receptors in acute leukemia. Blood 81:281–292
22. Ohno R, Naoe T, Kanamaru A et al (1994) A double-blind controlled study of granulocyte colony-stimulating factor started two days before induction chemotherapy in refractory acute myeloid leukemia. Blood 83:2015–2019
23. Estey E, Thall P, Andreeff M et al (1994) Use of granulocyte colony-stimulating factor before, during and after fludarabine plus cytarabine induction therapy of newly diagnosed acute myelogenous leukemia or myelodysplastic syndromes: comparison with fludarabine plus cytarabine without granulocyte colony-stimulating factor. J Clin Oncol 12:671–678
24. Giralt S, Escudier S, Kantarjian H et al (1993) Preliminary results of treatment with filgrastim for relapse of leukemia and myelodysplasia after allogeneic bone marrow transplantation. N Engl J Med 329:757–761
25. Welte K, Reiter A, Mempel K et al (1996) A randomized phase III study of the efficacy of granulocyte colony-stimulating factor in children with high-risk acute lymphoblastic leukemia. Blood 87:3143–3150
26. Taylor KM, Jagganath S, Spitzer G et al (1989) Recombinant human granulocyte colony-stimulating factor hastens granulocyte recovery after high-dose chemotherapy and autologous bone marrow transplantation in Hodgkin's disease. J Clin Oncol 7:1791–1799
27. Sheridan WP, Morstyn G, Wolf M et al (1989) Granulocyte colony stimulating factor and neutrophil recovery after high-dose chemotherapy and autologous bone marrow transplantation. Lancet 2:891–895
28. Dührsen U, Villeval JL, Boyd J, Kannourakis G, Morstyn G, Metcalf D (1988) Effects of recombinant human granulocyte colony-stimulating factor on hematopoietic progenitor cells in cancer patients. Blood 72:2074–2081
29. Sheridan WP, Begley CG, Juttner CA et al (1992) Effect of peripheral-blood progenitor cells mobilised by filgrastim (G-CSF) on platelet recovery after high-dose chemotherapy. Lancet 339:640–644
30. De Luca E, Sheridan WP, Watson D, Szer J, Begley CG (1992) Prior chemotherapy does not prevent effective mobilisation by G-CSF of peripheral blood progenitor cells. Br J Cancer 66:893–899
31. Bensinger W, Singer J, Appelbaum F et al (1993) Autologous transplantation with peripheral blood progenitor mononuclear cells collected after administration of recombinant granulocyte stimulating factor. Blood 81:3158–3163

32. Bonilla MA, Gillio AP, Ruggeiro M et al (1989) Effects of recombinant human granulocyte colony-stimulating factor on neutropenia in patients with congenital agranulocytosis. N Engl J Med 320:1574–1580
33. Welte K, Zeidler C, Reiter A et al (1990) Differential effects of granulocyte-macrophage colony stimulating factor and granulocyte colony-stimulating factor in children with severe congenital neutropenia. Blood 75:1056–1063
34. Dale DC, Bonilla MA, Davis MS et al (1993) A randomized controlled phase III trial of recombinant human granulocyte colony-stimulating factor (filgrastim) for treatment of severe chronic neutropenia. Blood 81:2496–2502
35. Bonilla MA, Dale D, Zeidler C et al (1994) Long-term safety of treatment with recombinant human granulocyte colony stimulating factor (r-metHuG-CSF) in patients with severe congenital neutropenias. Br J Haematol 88:723–730
36. Welte K, Zeidler C, Dale DC (2006) Severe congenital neutropenia. Semin Hematol 43:189–195
37. Zeidler C, Germeshausen M, Klein C, Welte K (2009) Clinical implications of ELA2-, HAX1-, and G-CSF-receptor (CSF3R) mutations in severe congenital neutropenia. Br J Haematol 144:459–467

Mouse Models of G-CSF Signaling in Hematopoiesis

Daniel C Link

1 G-CSF is the Principal Cytokine Regulating Granulopoiesis

To begin to define the role of granulocyte colony-stimulating factor (G-CSF) in the regulation of hematopoiesis, Lieschke and colleagues generated G-CSF$^{-/-}$ (knockout) mice [1]. Knockout mice are genetically engineered to contain a complete loss-of-function (null) mutation of the gene of interest. They are generated by homologous recombination in murine embryonic stem cells in which the normal gene (in this case *Csf3* encoding G-CSF) is replaced with a mutated gene. The targeted embryonic stem cells are implanted into pseudopregnant mice to generate a transgenic mouse line carrying the mutated gene. Importantly, the rest of the murine genome is intact, allowing investigators to examine the effect of the loss of that gene, in isolation, on a biologic process.

The G-CSF$^{-/-}$ and G-CSF receptor (G-CSFR)$^{-/-}$ mice have a similar phenotype and are discussed together [1, 2]. The major phenotype of these mice is chronic severe neutropenia. In both types of mice, the number of circulating neutrophils is reduced to approximately 20% that of wild-type mice. In the bone marrow, the major site of granulopoiesis, a uniform decrease in granulocytic precursors is present. The number of myeloid progenitors (e.g., colony-forming unit–granulocyte macrophage, CFU-GM) is either normal (G-CSFR$^{-/-}$ mice) or only modestly reduced (G-CSF$^{-/-}$ mice). Moreover, the number of mature neutrophils in the myeloid colonies is reduced, suggesting reduced proliferation of granulocytic precursors. Finally, neutrophils from G-CSFR$^{-/-}$ mice display increased susceptibility to apoptosis. Indeed, in vivo labeling studies of G-CSF$^{-/-}$ mice with the thymidine analog bromodeoxyuridine suggested that increased apoptosis of cells in

D.C. Link (✉)
Division of Oncology, Washington University, 660 S. Euclid Avenue, Campus Box 8007, St. Louis, MO 63110, USA
e-mail: dlink@dom.wustl.edu

G. Molineux et al. (eds.), *Twenty Years of G-CSF*, Milestones in Drug Therapy,
DOI 10.1007/978-3-0348-0218-5_3,

the granulocytic lineage is the primary mechanisms of neutropenia in these mice [3]. Collectively, these data suggest that:

- G-CSF is the principal cytokine regulating basal granulopoiesis.
- G-CSF signals are required for the normal proliferation and survival of granulocytic cells.
- G-CSF signals are not required for the production and/or maintenance of myeloid progenitor cells.

2 G-CSF-Independent Granulopoiesis is Mediated, in Part, by GM-CSF, IL-6, and Thrombopoietin

The presence of residual morphologically mature neutrophils in G-CSF$^{-/-}$ and G-CSFR$^{-/-}$ mice establishes that there must be G-CSF-independent pathways supporting granulopoiesis. Other hematopoietic cytokines that are known to stimulate neutrophil production are attractive candidates, most notably granulocyte-macrophage colony-stimulating factor (GM-CSF) and interleukin-6 (IL-6). Knockout mice for each of these cytokines have been generated. GM-CSF$^{-/-}$ mice have normal basal granulopoiesis but altered macrophage function resulting in impaired clearance of surfactant from the lung and a phenotype resembling human alveolar proteinosis [4]. IL-6$^{-/-}$ mice have normal steady-state granulopoiesis but impaired neutrophil responses to *Listeria monocytogenes* or *Candida albicans* infection [5, 6]. These studies show that GM-CSF and IL-6 are dispensable for steady-state granulopoiesis, but do not exclude a redundant role for these cytokines. To address this possibility, mice lacking G-CSF (or its receptor, G-CSFR) and either GM-CSF or IL-6 were generated. In adult G-CSF$^{-/-}$ × GM-CSF$^{-/-}$ (doubly deficient) mice, the degree of neutropenia was similar to that observed in G-CSF$^{-/-}$ mice [4]; however, in newborn G-CSF$^{-/-}$ × GM-CSF$^{-/-}$ mice, neutropenia was more severe than mice lacking only G-CSF. Walker and colleagues showed that conditioned media from G-CSF$^{-/-}$ × GM-CSF$^{-/-}$ bone marrow cells supported granulocytic cell proliferation and differentiation, and they provided evidence that this activity was dependent on IL-6 [3]. Consistent with the observation, the loss of IL-6 significantly worsened the neutropenia present in mice lacking the G-CSFR alone [7].

Somewhat surprisingly, thrombopoietin also appears to play an important role in granulopoiesis. Whereas thrombopoietin$^{-/-}$ mice have normal neutrophil numbers in the blood, thrombopoietin$^{-/-}$ × G-CSFR$^{-/-}$ (doubly deficient) mice display chronic severe neutropenia that is significantly worse than that seen in G-CSFR$^{-/-}$ mice [8].These data establish that GM-CSF, IL-6, and thrombopoietin provide, in part, the signals supporting basal granulopoiesis in the absence of G-CSF. It seems likely that other cytokines, such as stem cell factor (SCF), also support granulopoiesis in the absence of G-CSF signaling.

3 G-CSF is a Key Mediator of the Stress Granulopoiesis Response to Certain Infections

In response to infection and other environmental stresses, the number of neutrophils in the blood is markedly increased. The importance of G-CSF in regulating the stress granulopoiesis response, however, is controversial. G-CSF expression is often induced during infections, resulting in high levels both systemically (i.e., in the plasma) and locally in inflammatory fluids in both mice [9] and in humans [10]. G-CSF$^{-/-}$ mice infected intravenously with *C albicans* or intraperitoneally with *L monocytogenes* demonstrate a neutrophilia that matches that of wild-type littermates, suggesting a nonessential role for G-CSF in mediating stress granulopoiesis [11, 12]. In contrast, G-CSF$^{-/-}$ mice infected intravenously with *L. monocytogenes* have reduced neutrophil recruitment into the blood compared with wild-type littermates [1]. It is noteworthy that each of these models used large doses of infectious agents administered parenterally. Thus, their relevance to the more common local-regional infections seen in the clinical setting is unclear. To address this issue, Gregory et al. used a model of bronchopulmonary *Pseudomonas aeruginosa* infection induced by intratracheal injection with *P aeruginosa*-laden agarose beads [13]. In this model, they showed that the increase in circulating neutrophils is mainly due to enhanced neutrophil release from the bone marrow rather than increased production. Importantly, the increase in circulating neutrophils was severely blunted in G-CSFR$^{-/-}$ mice. Collectively, these data suggest that the contribution of G-CSF to the stress granulopoiesis response is variable and depends on both the type and magnitude of infection/inflammation.

4 G-CSF Signals Are Not Required for Terminal Granulocytic Differentiation

The fact that G-CSF provides key proliferative and survival signals to granulocytic cells is well established. Less clear is the role that G-CSF signals play in directing terminal differentiation of myeloid progenitor cells. Older studies using myeloid cell lines showed that the addition of recombinant G-CSF (rG-CSF) to the culture media is associated with granulocytic differentiation, suggesting that G-CSF signals actively direct granulocytic differentiation [14, 15]. On the contrary, in mice lacking G-CSF or G-CSFR, the number of myeloid progenitors is near normal and residual morphologically mature neutrophils are present [1, 2], indicating that G-CSF signals are not absolutely required for granulocytic differentiation.

To address this question, targeted ("knock-in") transgenic mice were generated in which the cytoplasmic (signaling) domain of the G-CSFR was replaced with cytoplasmic domain of the erythropoietin receptor (EpoR) [16], which was achieved by homologous recombination in embryonic stem cells in which exons

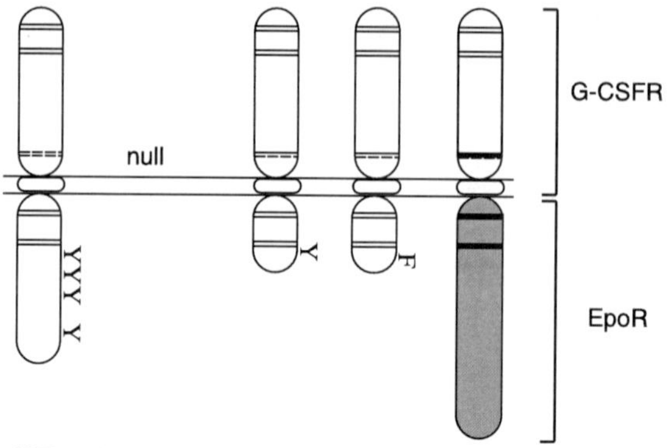

	WT	G-CSFR-/-	d715	d715F	GEpoR
PMN Production	++	+	++	+	++
PMN Function	++	Abnl	ND	ND	Abnl
HSPC mobilization	++	-	+++	+/-	+/-

Fig. 1 Knock-in mutations of the G-CSF receptor (G-CSFR).The wild-type G-CSFR has four cytoplasmic tyrosines. The d715 G-CSFR mutation deletes the distal 96 amino acids of the cytoplasmic domain of the G-CSFR and is typical of the mutations found in patients with severe chronic neutropenia (SCN) who have progressed to acute myeloid leukemia or myelodysplastic syndromes (AML/MDS). In the d715F G-CSFR mutation, the sole cytoplasmic tyrosine remaining in the d715 G-CSFR is mutated to phenylalanine. Note this receptor is unable to activate STAT3 or STAT5. Finally, the GEpoR is a chimeric receptor containing the extracellular (ligand-binding) and transmembrane domains of the G-CSFR and the cytoplasmic (signaling) domain on the erythropoietin receptor (EpoR). The GEpoR is predicted to bind G-CSF normally but transmit EpoR-specific signals. A summary of key phenotypic features for each transgenic mouse line is provided. *Abnl* abnormal; *ND* not done; *WT* wild-type

encoding the cytoplasmic domain of G-CSFR were replaced with an exon encoding EpoR (Fig. 1). Importantly, this replacement preserves the genomic organization of the G-CSFR locus and ensures that the resulting chimeric receptor (termed GEpoR) is expressed in the same cell types and at the proper stage of maturation as the wild-type G-CSFR. As expected, the GEpoR bound to G-CSF normally but transmitted EpoR-specific signals [16]. Although neutropenic at baseline, the number of morphologically mature neutrophils in the bone marrow is normal. Moreover, treatment with rG-CSF resulted in the robust production of mature neutrophils. These data show that signals generated by GEpoR are able to direct normal terminal granulocytic differentiation and support a model in which G-CSF provides "generic" signals that support the proliferation and survival of granulocytic precursors but does not provide specific signals that direct terminal granulocytic differentiation.

This model may extend to other hematopoietic lineages. Stoffel and colleagues generated knock-in mice in which the cytoplasmic domain of the thrombopoietin receptor (c-mpl) was replaced with that of G-CSFR [17]. This chimeric receptor bound thrombopoietin normally but transmitted signals through the cytoplasmic domain of the G-CSFR. These mice had normal platelet counts and megakaryopoeisis. Thus, G-CSFR signals were able to support terminal megakaryocyte differentiation. Two independent groups analyzed the effect of ectopic G-CSFR expression on hematopoietic differentiation. Ectopic expression of G-CSFR was achieved either by transduction of hematopoietic progenitors with a G-CSFR retrovirus [18] or in transgenic mice that ubiquitously express G-CSFR [19]. In each case, treatment with rG-CSF stimulated multilineage differentiation, including granulocytic, erythroid, and megakaryocytic differentiation.

5 G-CSF Signals Contribute to Commitment to the Myeloid Lineage

The signals that determine commitment of multipotential hematopoietic stem/progenitor cells (HSPC) to the myeloid or other hematopoietic lineages are incompletely understood. In particular, the role of hematopoietic growth factors in directing lineage commitment is controversial. Two general models exist for how hematopoietic growth factors lead to an expansion of a specific lineage (e.g., myeloid cell expansion after rG-CSF treatment). In the instructive model, cytokines transmit specific signals to multipotential HSPC directing their lineage commitment. In the stochastic model, lineage commitment is intrinsically determined with cytokines providing only growth and survival signals.

With respect to G-CSF, the number of myeloid progenitors is only modestly reduced in G-CSF$^{-/-}$ or G-CSFR$^{-/-}$mice, suggesting that G-CSF signals are not absolutely required for commitment [1, 2]; however, a potential confounding problem of loss-of-function mouse models is the induction of compensatory mechanisms that may mask an important phenotype. Specifically, compensatory mechanisms induced by the severe neutropenia present in G-CSF$^{-/-}$ or G-CSFR$^{-/-}$ mice may stimulate myeloid progenitor production. This possibility was addressed by analyzing bone marrow chimeras generated by mixing wild-type and G-CSFR$^{-/-}$bone marrow cells and transplanting them into irradiated recipient mice [20]. Since wild-type and G-CSFR$^{-/-}$cells share the same microenvironment in these chimeras, cell-intrinsic alterations in lineage commitment can be assessed. This study showed that the contribution of G-CSFR$^{-/-}$cells progressively decreased as cells became more committed to the myeloid lineage (Fig. 2). For example, whereas approximately 80% of B and T cells were derived from G-CSFR$^{-/-}$cells, only 24% of granulocyte–monocyte committed progenitors (GMP) were derived from G-CSFR$^{-/-}$cells. These data show that G-CSF signals, though not absolutely required, are directing commitment to the myeloid lineage. The signals transmitted by the G-CSFR that direct lineage commitment are currently not known.

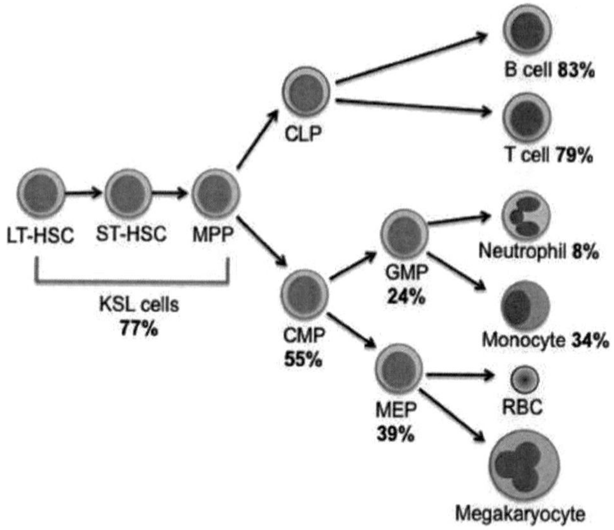

Fig. 2 Lineage commitment in G-CSFR$^{-/-}$bone marrow chimeras. Bone marrow chimeras were generated by transplanting G-CSFR$^{-/-}$ and wild-type bone marrow cells (in a 9:1 ratio) into irradiated mice. Shown is the percentage of G-CSFR$^{-/-}$ cells that contributed to each cell population. There is a progressive loss of myeloid-committed cells that derived from G-CSFR (G-CSF receptor) knock-out cells, indicating that G-CSF signals are directing commitment to the common myeloid lineage in a cell intrinsic fashion. *CLP* common lymphoid progenitor; *CMP* common myeloid progenitor; *GMP* granulocyte-macrophage progenitor; *KSL* Kit^{+} Sca^{+} lineage^{-} cells; *MEP* megakaryocyte-erythroid progenitor

6 G-CSF Signals Are Required for Normal Neutrophil Function

G-CSF concentrations are often elevated in the serum and at inflammatory sites in patients with infections [10, 21]. Thus, in addition to regulating neutrophil production, G-CSF signals may also regulate mature neutrophil function. Indeed, there is in vitro data showing that G-CSF can modulate neutrophil degranulation [22], phagocytosis [23], adhesion [24], chemokinesis [25], and superoxide burst [24]. In vivo studies with G-CSF$^{-/-}$ or G-CSFR$^{-/-}$ mice suggest a more modest, but still significant, role for G-CSF signals in regulating neutrophil function. Metcalf et al. showed that in G-CSF$^{-/-}$ mice, neutrophil recruitment into the peritoneal cavity in response to casein-induced inflammation was normal [26]. However, G-CSFR$^{-/-}$ neutrophils have selected defects in mature neutrophil function. Whereas neutrophil superoxide generation and degranulation are normal, chemotaxis and adhesion of G-CSFR$^{-/-}$ neutrophils are impaired, likely due to defective [2] integrin activation [27]. Likewise, neutrophil recruitment to an inflamed knee joint and the number of adherent neutrophils in synovial blood vessels in a collagen-induced arthritis model are reduced in G-CSF$^{-/-}$ and G-CSFR$^{-/-}$ mice [28]. These data show that G-CSF is providing nonredundant signals that modulate selected neutrophil functions. Of note, neutrophil chemotaxis also is impaired in GEpoR mice [16], suggesting that unlike generic proliferative/survival signals, the signals that regulate neutrophil function may be specific to the G-CSFR.

7 G-CSF Signals Contribute to the Regulation of Hematopoietic Stem Cell Function

The role of G-CSF signals in the regulation of hematopoietic stem cells is controversial. McKinstry et al. showed that the G-CSFR is expressed on the surface of Rhodamine-123^{lo} lineage-Sca^{+} c-Kit^{+} cells, a cell population highly enriched for hematopoietic stem cells (HSC) [29]. Moreover, G-CSF has been shown to stimulate the proliferation of long-term initiating cells in vitro [30]. In contrast, a recent report suggested that human stem cell activity was enriched in progenitors with low G-CSFR expression [31]. To assess the contribution of G-CSF signals to HSC maintenance in vivo, a competitive repopulation assay was performed with G-CSFR$^{-/-}$ hematopoietic cells [20]. The long-term repopulating activity of G-CSFR$^{-/-}$ HSC was markedly reduced, demonstrating that, at least in mice, constitutive G-CSF signals contribute to the regulation of HSC function.

8 Truncation Mutations of the G-CSFR in Severe Chronic Neutropenia

Severe chronic neutropenia (SCN) is a rare congenital bone marrow failure manifested by severe neutropenia from birth and a marked propensity to develop acute myeloid leukemia (AML) or myelodysplasic syndromes (MDS). The most recent update of the Severe Chronic Neutropenia International Registry (SCNIR) reported that cumulative incidence of AML/MDS in SCN is 21% after 10 years [32]. Mutations of G-CSFR are the most common type of mutation found in AML/MDS arising in the setting of SCN [33]. Interestingly, G-CSFR mutations are rarely seen in de novo AML [34, 35]. G-CSFR mutations are acquired, typically heterozygous, and nearly always introduce a premature stop codon resulting in the truncation of the distal cytoplasmic portion of the G-CSFR [36]. Importantly, these truncation mutations are strongly associated with the development of AML/MDS. The incidence of *CSF3R* mutations was reported to be 78% (18/23) in individuals with SCN and monosomy 7, MDS, or AML, compared with 34% (43/125) in patients without MDS or AML [37].

Two independent groups generated transgenic mice carrying knock-in mutations of their G-CSFR that reproduced the truncation mutations found in patients with SCN (Fig. 1, d715 G-CSFR) [38, 39]. These mice have mild neutropenia but otherwise normal granulopoiesis; however, they exhibit an exaggerated neutrophil response to rHuG-CSF treatment that is due to increased myeloid progenitor/precursor cell proliferation. The enhanced proliferation is secondary, in part, to sustained Stat5 activation that, in turn, is related to impaired receptor internalization and the failure of the truncated G-CSFR to recruit and activate SOCS3 and SHP2 [40, 41]. Of note, AML or MDS was not observed in d715 G-CSFR mice, despite chronic treatment with G-CSF, indicating that truncation mutations of the G-CSFR are not sufficient to induce AML/MDS [39].

With some exceptions, the frequency of circulating cells containing mutant G-CSFR increases over time in patients with SCN [37, 42], suggesting that expression of the mutant G-CSFR confers clonal dominance over time. Moreover, a study reported that the G-CSFR mutations were present in all blood lineages (including lymphocytes), suggesting that the G-CSFR mutations arose and were selected for in a HSC population [43]. To determine how cells expressing mutant G-CSFR gain clonal dominance, bone marrow chimeras were generated containing wild-type and d715 G-CSFR cells [44]. In the absence of exogenous G-CSF, the contribution of d715 G-CSFR cells to hematopoiesis was stable over time, indicating no clonal dominance; however, after short-term (3 weeks) treatment with rG-CSF, a marked and durable expansion of d715 G-CSFR-derived hematopoietic cells was observed. Further studies demonstrated that this clonal expansion occurred at the HSC level. Mechanistic studies suggested that accentuated activation of Stat5 by the d715 G-CSFR plays a key role in establishing HSC clonal dominance. Consistent with this observation, mice expressing a mutant G-CSFR (d715F, Fig. 1) engineered to have impaired Stat5 activation have impaired HSC function [45].

Collectively, these data suggest that truncation mutations of the G-CSFR do not contribute to the block in granulocytic differentiation present in patients with SCN. Although not sufficient to induce AML/MDS, these mutations appear to contribute to leukemic transformation by conferring a clonal advantage on HSC.

9 G-CSF Signals in Monocytic Cells Are Required for Hematopoietic Stem/Progenitor Mobilization

Under normal conditions, most HSPC reside in the bone marrow. The number of HSPC in the circulation can be markedly increased in response to a number of stimuli, including hematopoietic growth factors, myeloablative agents, and environmental stresses such as infection. The ability to "mobilize" HSPC from the bone marrow to the blood has been exploited clinically to obtain HSPC for stem cell transplantation. rHuG-CSF is the most commonly used mobilizing agent for stem cell transplantation. rHuG-CSF-mobilized HSC are associated with more rapid engraftment compared with unmanipulated bone marrow [46].

Studies of rHuG-CSF-induced HSPC mobilization in mice have greatly advanced our understanding of mechanisms of mobilization. One avenue of investigation is directed at defining target cell population(s) in the bone marrow that mediates HSPC mobilization by rHuG-CSF. In addition to mature neutrophils and monocytes, the G-CSFR is expressed on a broad range of hematopoietic progenitors, including HSC [47]. There are also reports of G-CSFR expression on endothelial cells [48]. To determine the cellular target(s) of G-CSF required for HSPC mobilization, a series of bone marrow chimeras was generated (Fig. 3) [49]. Wild-type mice reconstituted with G-CSFR$^{-/-}$ bone marrow cells failed to mobilize HSPC in response to rHuG-CSF. In contrast, G-CSFR$^{-/-}$ mice reconstituted with wild-type bone marrow displayed normal HSPC mobilization by rHuG-CSF. This experiment showed that G-CSFR signaling in hematopoietic cells but not stromal cells is required for HSPC mobilization by G-CSF. Since the G-CSFR is expressed on HSC, the simplest model suggests that G-CSF directly acts upon HSPC to induce their mobilization. However, strongly arguing against this model, in mixed bone marrow chimeras containing both wild-type and G-CSFR$^{-/-}$ HSPC, both types of cells were mobilized equally after rHuG-CSF treatment (Fig. 3). Collectively, these studies support a model in which G-CSF acts on a hematopoietic intermediary that, in turn, generates trans-acting signal(s) that lead to HSPC mobilization.

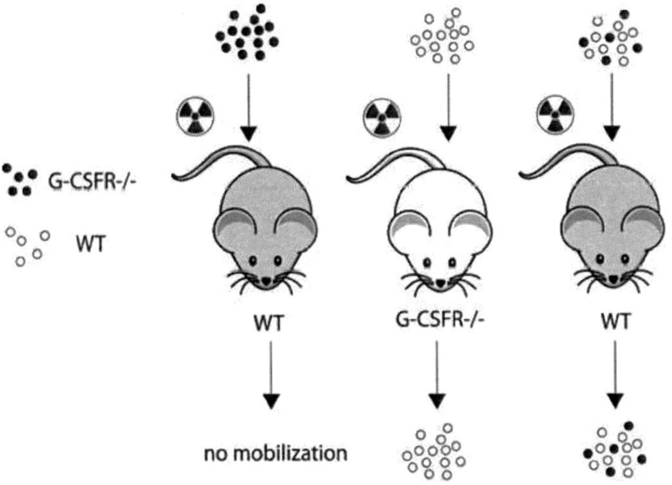

Fig. 3 Hematopoietic stem/progenitor cell (HSPC) mobilization by G-CSF in G-CSFR$^{-/-}$ bone marrow chimeras. Bone marrow chimeras were generated in which G-CSFR$^{-/-}$ bone marrow cells were transplanted into irradiated wild-type (WT) recipients (first cohort); conversely, WT bone marrow cells were transplanted into irradiated G-CSFR$^{-/-}$ mice (second cohort). G-CSF-induced hematopoietic stem/progenitor cell (HSPC) mobilization was only observed in mice reconstituted with WT bone marrow, demonstrating that G-CSF signaling in hematopoietic cells is required. Mixed bone marrow chimeras (last cohort) were generated by transplanting WT and G-CSFR$^{-/-}$ bone marrow cells into irradiated recipient mice. Importantly, both WT and G-CSFR$^{-/-}$ HSPC were mobilized equally in these mice, demonstrating that G-CSF acts in a noncell intrinsic fashion to induced HSPC mobilization. Both WT and G-CSFR$^{-/-}$ HSPC were mobilized equally in these mice, demonstrating that G-CSF acts in a noncell intrinsic fashion to induced HSPC mobilization

A series of studies strongly suggests that G-CSFR signals in cells in the monocyte lineage play a key role in initiating the mobilization cascade. Transgenic mice were generated that express the G-CSFR under control of the CD68 promoter, which directs expression only in cells in the monocyte lineage [50]. These CD68:G-CSFR mice were crossed with G-CSFR$^{-/-}$ mice to generate transgenic mice that express the G-CSFR only in monocytic cells. G-CSF treatment induced robust HSPC mobilization in these mice, suggesting that G-CSFR signals in monocytic cells were sufficient to initiate the HSPC mobilization cascade. Of note, G-CSF also resulted in a decrease in monocytes in the bone marrow in both CD68:G-CSFR and wild-type mice. This observation agrees with observations by two independent groups showing that monocyte ablation results in marked hematopoietic progenitor mobilization [51, 52]. Together, these data suggest a model in which monocytic cells provide key signals that are required for the retention of HSPC in the bone marrow. G-CSF acts on monocytes to suppress these signals, ultimately leading to HSPC mobilization.

The exact monocytic cell population that mediates HSPC mobilization is unknown. In the bone marrow, there are at least four distinct monocytic cell populations: inflammatory monocytes/macrophages, resident monocytes/macrophages, myeloid dendritic cells, and osteoclasts. There is considerable (though conflicting) data on the role of osteoclasts in HSPC mobilization. Activation of osteoclasts by injection of RANK ligand (RANKL) was associated with moderate HSPC mobilization, and inhibition of osteoclasts, either genetically by knocking out *PTPε* or by injecting mice with calcitonin, blunts the mobilization response to G-CSF [53]. Of note, osteoclasts produce the protease cathepsin K, which can cleave CXCL12 in vitro [53]. On the contrary, other studies indicate that osteoclasts may actually inhibit mobilization, as mice that were given pamidronate, an osteoclast-inhibiting bisphosphonate, exhibit increased mobilization in response to G-CSF [52, 54]. Ablation of CD169$^+$ macrophages resulted in constitutive HSPC mobilization, suggesting that CD169$^+$ macrophages may be the target cell population for G-CSF in the mobilization pathway [51].

10 rHuG-CSF Treatment is Associated with Disruption of the Stem Cell Niche

The bone marrow microenvironment plays a critical role in the maintenance of HSC, which preferentially localize in the bone marrow either to a perivascular region [55, 56] or near the endosteum [57–60]. There is a strong evidence that osteoblast lineage cells are required to maintain the "endosteal stem cell niche." Expansion of osteoblast lineage cells by genetic or pharmacologic means results in a concurrent expansion of HSC [61, 62]. Conversely, ablation of mature osteoblasts using a suicide gene results in a loss of HSC [63]. CXCL12 (also known as stromal-derived factor-1, SDF1) is key gene that is expressed by osteoblast lineage cells that regulates HSC function and trafficking. CXCL12 is constitutively expressed at high levels in the bone marrow and is a potent chemoattractant for HSPC [64, 65]. In mice lacking CXCL12 [66] or CXCR4 (the major receptor for CXCL12) [67], there is a failure of the migration of HSPC from the fetal liver to the bone marrow, and CXCR4$^{-/-}$ bone marrow chimeras exhibit constitutive mobilization [68, 69]. The importance of the CXCL12/CXCR4 axis is shown by the success of the CXCR4 inhibitor plerixafor (also known as AMD3100) to rapidly mobilize HSC in humans and mice [70, 71].

rHuG-CSF treatment is associated with a marked decrease in CXCL12 expression in the bone marrow. Bone marrow and results in a profound suppression of mature osteoblast number and function [16, 52, 72–74]. HSPC mobilization and osteoblast suppression were assessed in transgenic mice carrying the knock-in mutations depicted in Fig. 1. Although able to support normal neutrophil production, rHuG-CSF-induced HSPC mobilization and CXCL12 suppression were impaired in mice expressing the GEpoR, suggesting that the distinct signals are regulating these processes. Osteoblast suppression is a common feature of HSPC mobilization by other cytokines, including Flt3 ligand and stem cell factor [69]. Although alterations in osteoblast lineage cells are likely to be a key step in rHuG-CSF-induced mobilization, the mechanisms by which this leads to HSPC egress from the bone marrow remain to be elucidated.

11 Summary and Future Directions

Animal models are crucial to the study of complex biologic processes such as hematopoiesis. In particular, the study of transgenic mice carrying mutations of G-CSF or the G-CSFR have greatly improved our understanding of granulopoiesis and HSC trafficking. These studies show that G-CSF plays a central role in the regulation of both basal and stress granulopoiesis. G-CSF signals regulate granulopoiesis by: (1) stimulating the growth and survival of granulocytic precursors; (2) regulating the release of neutrophils from the bone marrow; and (3) directing commitment to the myeloid lineage. G-CSF signals also contribute to HSC maintenance, in part, through regulation of the bone marrow microenvironment. rHuG-CSF treatment leads to changes in the bone marrow microenvironment (osteoblast suppression and decreased CXCL12 expression) that profoundly affect the trafficking and function of HSC. This later property of G-CSF has stimulated studies to determine whether disruption of the stem cell niche might render leukemic cells more sensitive to chemotherapy [75]. Many important questions remain to be addressed. In particular, what are the molecular mechanisms by which G-CSF regulates the release of neutrophils and HSPC from the bone marrow, and how does G-CSF regulate osteoblast lineage cells? Collectively, these studies may lead to better strategies to treat patients with neutropenia or with hematologic malignancies.

Acknowledgments DCL is supported by grants from the NIH (RO1 CA136617; RC2 CA1455073; and RO1 HL60772) and by the Leukemia & Lymphoma Society TRA 6030-10). The author thanks Mahil Rao for the generation of Fig. 3.

References

1. Lieschke GJ, Grail D, Hodgson G et al (1994) Mice lacking granulocyte colony-stimulating factor have chronic neutropenia, granulocyte and macrophage progenitor cell deficiency, and impaired neutrophil mobilization. Blood 84:1737–1746
2. Liu F, Wu HY, Wesselschmidt R, Kornaga T, Link DC (1996) Impaired production and increased apoptosis of neutrophils in granulocyte colony-stimulating factor receptor-deficient mice. Immunity 5:491–501
3. Walker F, Zhang HH, Matthews V et al (2008) IL6/sIL6R complex contributes to emergency granulopoietic responses in G-CSF- and GM-CSF-deficient mice. Blood 111:3978–3985
4. Seymour JF, Lieschke GJ, Grail D, Quilici C, Hodgson G, Dunn AR (1997) Mice lacking both granulocyte colony-stimulating factor (CSF) and granulocyte-macrophage CSF have impaired reproductive capacity, perturbed neonatal granulopoiesis, lung disease, amyloidosis, and reduced long-term survival. Blood 90:3037–3049
5. Dalrymple SA, Lucian LA, Slattery R et al (1995) Interleukin-6-deficient mice are highly susceptible to *Listeria monocytogenes* infection: correlation with inefficient neutrophilia. Infect Immun 63:2262–2268
6. Romani L, Mencacci A, Cenci E et al (1996) Impaired neutrophil response and CD4+ T helper cell 1 development in interleukin 6-deficient mice infected with *Candida albicans*. J Exp Med 183:1345–1355
7. Liu F, Poursine-Laurent J, Wu HY, Link DC (1997) Interleukin-6 and the granulocyte colony-stimulating factor receptor are major independent regulators of granulopoiesis in vivo but are not required for lineage commitment or terminal differentiation. Blood 90:2583–2590
8. Kaushansky K, Fox N, Lin NL, Liles WC (2002) Lineage-specific growth factors can compensate for stem and progenitor cell deficiencies at the postprogenitor cell level: an analysis of doubly TPO- and G-CSF receptor-deficient mice. Blood 99:3573–3578
9. Quinton LJ, Nelson S, Boe DM et al (2002) The granulocyte colony-stimulating factor response after intrapulmonary and systemic bacterial challenges. J Infect Dis 185:1476–1482
10. Kawakami M, Tsutsumi H, Kumakawa T et al (1990) Levels of serum granulocyte colony-stimulating factor in patients with infections. Blood 76:1962–1964
11. Basu S, Hodgson G, Zhang HH, Katz M, Quilici C, Dunn AR (2000) "Emergency" granulopoiesis in G-CSF-deficient mice in response to *Candida albicans* infection. Blood 95:3725–3733
12. Zhan Y, Lieschke GJ, Grail D, Dunn AR, Cheers C (1998) Essential roles for granulocyte-macrophage colony-stimulating factor (GM-CSF) and G-CSF in the sustained hematopoietic response of *Listeria monocytogenes*-infected mice. Blood 91:863–869
13. Gregory AD, Hogue LA, Ferkol TW, Link DC (2007) Regulation of systemic and local neutrophil responses by G-CSF during pulmonary *Pseudomonas aeruginosa* infection. Blood 109:3235–3243
14. Dong F, van Buitenen C, Pouwels K, Hoefsloot LH, Lowenberg B, Touw IP (1993) Distinct cytoplasmic regions of the human granulocyte colony-stimulating factor receptor involved in induction of proliferation and maturation. Mol Cell Biol 13:7774–7781

15. Yoshikawa A, Murakami H, Nagata S (1995) Distinct signal transduction through the tyrosine containing domains of the granulocyte colony-stimulating factor receptor. EMBO 14:5288–5296
16. Semerad CL, Christopher MJ, Liu F et al (2005) G-CSF potently inhibits osteoblast activity and CXCL12 mRNA expression in the bone marrow. Blood 106:3020–3027
17. Stoffel R, Ziegler S, Ghilardi N, Ledermann B, de Sauvage FJ, Skoda RC (1999) Permissive role of thrombopoietin and granulocyte colony-stimulating factor receptors in hematopoietic cell fate decisions in vivo. Proc Natl Acad Sci U S A 96:698–702
18. Jacob J, Haug JS, Raptis S, Link DC (1998) Specific signals generated by the cytoplasmic domain of the granulocyte colony-stimulating factor (G-CSF) receptor are not required for G-CSF-dependent granulocytic differentiation. Blood 92:353–361
19. Yang FC, Tsuji K, Oda A et al (1999) Differential effects of human granulocyte colony-stimulating factor (hG-CSF) and thrombopoietin on megakaryopoiesis and platelet function in hG-CSF receptor-transgenic mice. Blood 94:950–958
20. Richards MK, Liu F, Link DC (2001) G-CSF receptor signals may play a pivotal role in directing commitment to the common myeloid lineage. Blood 98:795a
21. O'Grady NP, Preas HL, Pugin J et al (2001) Local inflammatory responses following bronchial endotoxin instillation in humans. Am J Respir Crit Care Med 163:1591–1598
22. de Haas M, Kerst JM, van der Schoot CE et al (1994) Granulocyte colony-stimulating factor administration to healthy volunteers: analysis of the immediate activating effects on circulating neutrophils. Blood 84:3885–3894
23. Hoglund M, Hakansson L, Venge P (1987) Effects of in vivo administration of G-CSF on neutrophil functions in healthy volunteers. Eur J Haematol 58:195–202
24. Yuo A, Kitagawa S, Ohsaka A et al (1989) Recombinant human granulocyte colony-stimulating factor as an activator of human granulocytes: potentiation of responses triggered by receptor-mediated agonists and stimulation of C3bi receptor expression and adherence. Blood 74:2144–2149
25. Yong KL (1996) Granulocyte colony-stimulating factor (G-CSF) increases neutrophil migration across vascular endothelium independent of an effect on adhesion: comparison with granulocyte-macrophage colony-stimulating factor (GM-CSF). Br J Haematol 94:40–47
26. Metcalf D, Robb L, Dunn AR, Mifsud S, Di Rago L (1996) Role of granulocyte-macrophage colony-stimulating factor and granulocyte colony-stimulating factor in the development of an acute neutrophil inflammatory response in mice. Blood 88:3755–3764
27. Betsuyaku T, Liu F, Senior RM et al (1999) A functional granulocyte colony stimulating factor receptor is required for normal chemoattractant-induced neutrophil activation. J Clin Invest 103:825–832
28. Eyles JL, Hickey MJ, Norman MU et al (2008) A key role for G-CSF-induced neutrophil production and trafficking during inflammatory arthritis. Blood 112:5193–5201
29. McKinstry WJ, Li CL, Rasko JE, Nicola NA, Johnson GR, Metcalf D (1997) Cytokine receptor expression on hematopoietic stem and progenitor cells. Blood 89:65–71
30. Zandstra PW, Conneally E, Petzer AL, Piret JM, Eaves CJ (1997) Cytokine manipulation of primitive human hematopoietic cell self-renewal. Proc Natl Acad Sci U S A 94:4698–4703
31. Gibbs KD Jr, Gilbert PM, Sachs K et al (2011) Single-cell phospho-specific flow cytometric analysis demonstrates biochemical and functional heterogeneity in human hematopoietic stem and progenitor compartments. Blood 117(16):4226–4233
32. Rosenberg PS, Alter BP, Link DC et al (2008) Neutrophil elastase mutations and risk of leukaemia in severe congenital neutropenia. Br J Haematol 140:210–213
33. Link DC, Kunter G, Kasai Y et al (2007) Distinct patterns of mutations occurring in de novo AML versus AML arising in the setting of severe congenital neutropenia. Blood 110:1648–1655
34. Bernard T, Gale RE, Linch DC (1996) Analysis of granulocyte colony stimulating factor receptor isoforms, polymorphisms and mutations in normal haemopoietic cells and acute myeloid leukaemia blasts. Br J Haematol 93:527–533

35. Carapeti M, Soede-Bobok A, Hochhaus A et al (1997) Rarity of dominant-negative mutations of the G-CSF receptor in patients with blast crisis of chronic myeloid leukemia or de novo acute leukemia. Leukemia 11:1005–1008
36. Dong F, Brynes RK, Tidow N, Welte K, Lowenberg B, Touw IP (1995) Mutations in the gene for the granulocyte colony-stimulating-factor receptor in patients with acute myeloid leukemia preceded by severe congenital neutropenia. N Engl J Med 333:487–493
37. Germeshausen M, Ballmaier M, Welte K (2007) Incidence of CSF3R mutations in severe congenital neutropenia and relevance for leukemogenesis: results of a long-term survey. Blood 109:93–99
38. Hermans MH, Ward AC, Antonissen C, Karis A, Lowenberg B, Touw IP (1999) Perturbed granulopoiesis in mice with a targeted mutation in the granulocyte colony-stimulating factor receptor gene associated with severe chronic neutropenia. Blood 92:32–39
39. McLemore ML, Poursine-Laurent J, Link DC (1998) Increased granulocyte colony-stimulating factor responsiveness but normal resting granulopoiesis in mice carrying a targeted granulocyte colony-stimulating factor receptor mutation derived from a patient with severe congenital neutropenia. J Clin Invest 102:483–492
40. van de Geijn GJ, Gits J, Aarts LH, Heijmans-Antonissen C, Touw IP (2004) G-CSF receptor truncations found in SCN/AML relieve SOCS3-controlled inhibition of STAT5 but leave suppression of STAT3 intact. Blood 104:667–674
41. Hermans MH, Antonissen C, Ward AC, Mayen AE, Ploemacher RE, Touw IP (1999) Sustained receptor activation and hyperproliferation in response to granulocyte colony-stimulating factor (G-CSF) in mice with a severe congenital neutropenia/acute myeloid leukemia-derived mutation in the G-CSF receptor gene. J Exp Med 189:683–692
42. Ancliff PJ, Gale RE, Liesner R, Hann I, Linch DC (2003) Long-term follow-up of granulocyte colony-stimulating factor receptor mutations in patients with severe congenital neutropenia: implications for leukaemogenesis and therapy. Br J Haematol 120:685–690
43. Germeshausen M, Ballmaier M, Welte K (2009) In vivo growth advantage of cells expressing acquired CSF3R mutations in patients with severe congenital neutropenia. Blood 113:668–670
44. Liu F, Kunter G, Krem MM et al (2008) Csf3r mutations in mice confer a strong clonal HSC advantage via activation of Stat5. J Clin Invest 118:946–955
45. McLemore ML, Grewal S, Liu F et al (2001) STAT-3 activation is required for normal G-CSF-dependent proliferation and granulocytic differentiation. Immunity 14:193–204
46. Stem Cell Trialists' Collaborative Group (2005) Allogeneic peripheral blood stem-cell compared with bone marrow transplantation in the management of hematologic malignancies: an individual patient data meta-analysis of nine randomized trials. J Clin Oncol 23:5074–5087
47. Demetri GD, Griffin JD (1991) Granulocyte colony-stimulating factor and its receptor. Blood 78:2791–2808
48. Bussolino F, Ziche M, Wang JM et al (1991) In vitro and in vivo activation of endothelial cells by colony-stimulating factors. J Clin Invest 87:986–995
49. Liu F, Poursine-Laurent J, Link DC (1997) The granulocyte colony-stimulating factor receptor is required for the mobilization of murine hematopoietic progenitors into peripheral blood by cyclophosphamide or interleukin-8 but not flt-3 ligand. Blood 90:2522–2528
50. Christopher MJ, Rao M, Liu F, Woloszynek JR, Link DC (2011) Expression of the G-CSF receptor in monocytic cells is sufficient to mediate hematopoietic progenitor mobilization by G-CSF in mice. J Exp Med 208:251–260
51. Chow A, Lucas D, Hidalgo A et al (2011) Bone marrow CD169+ macrophages promote the retention of hematopoietic stem and progenitor cells in the mesenchymal stem cell niche. J Exp Med 208:261–271
52. Winkler IG, Sims NA, Pettit AR et al (2010) Bone marrow macrophages maintain hematopoietic stem cell (HSC) niches and their depletion mobilizes HSCs. Blood 116:4815–4828
53. Kollet O, Dar A, Shivtiel S et al (2006) Osteoclasts degrade endosteal components and promote mobilization of hematopoietic progenitor cells. Nat Med 12:657–664

54. Takamatsu Y, Simmons PJ, Moore RJ, Morris HA, To LB, Levesque J-P (1998) Osteoclast-mediated bone resorption is itimulated during short-term administration of granulocyte colony-stimulating factor but is not responsible for hematopoietic progenitor cell mobilization. Blood 92:3465–3473
55. Kiel MJ, Yilmaz OH, Iwashita T, Terhorst C, Morrison SJ (2005) SLAM family receptors distinguish hematopoietic stem and progenitor cells and reveal endothelial niches for stem cells. Cell 121:1109–1121
56. Sugiyama T, Kohara H, Noda M, Nagasawa T (2006) Maintenance of the hematopoietic stem cell pool by CXCL12-CXCR4 chemokine signaling in bone marrow stromal cell niches. Immunity 25:977–988
57. Yoshimoto M, Shinohara T, Heike T, Shiota M, Kanatsu-Shinohara M, Nakahata T (2003) Direct visualization of transplanted hematopoietic cell reconstitution in intact mouse organs indicates the presence of a niche. Exp Hematol 31:733–740
58. Nilsson SK, Johnston HM, Coverdale JA (2001) Spatial localization of transplanted hemopoietic stem cells: inferences for the localization of stem cell niches. Blood 97:2293–2299
59. Lo Celso C, Fleming HE, Wu JW et al (2009) Live-animal tracking of individual haematopoietic stem/progenitor cells in their niche. Nature 457:92–96
60. Xie Y, Yin T, Wiegraebe W et al (2009) Detection of functional haematopoietic stem cell niche using real-time imaging. Nature 457:97–101
61. Calvi LM, Adams GB, Weibrecht KW et al (2003) Osteoblastic cells regulate the haematopoietic stem cell niche. Nature 425:841–846
62. Zhang J, Niu C, Ye L et al (2003) Identification of the haematopoietic stem cell niche and control of the niche size. Nature 425:836–841
63. Visnjic D, Kalajzic Z, Rowe DW, Katavic V, Lorenzo J, Aguila HL (2004) Hematopoiesis is severely altered in mice with an induced osteoblast deficiency. Blood 103:3258–3264
64. Aiuti A, Webb IJ, Bleul C, Springer T, Gutierrez-Ramos JC (1997) The chemokine SDF-1 is a chemoattractant for human CD34+ hematopoietic progenitor cells and provides a new mechanism to explain the mobilization of CD34+ progenitors to peripheral blood. J Exp Med 185:111–120
65. Wright DE, Bowman EP, Wagers AJ, Butcher EC, Weissman IL (2002) Hematopoietic stem cells are uniquely selective in their migratory response to chemokines. J Exp Med 195:1145–1154
66. Nagasawa T, Hirota S, Tachibana K et al (1996) Defects of B-cell lymphopoiesis and bone-marrow myelopoiesis in mice lacking the CXC chemokine PBSF/SDF 1. Nature 382:635–638
67. Zou YR, Kottmann AH, Kuroda M, Taniuchi I, Littman DR (1998) Function of the chemokine receptor CXCR4 in haematopoiesis and in cerebellar development. Nature 393:595–599
68. Ma Q, Jones D, Springer TA (1999) The chemokine receptor CXCR4 is required for the retention of B lineage and granulocytic precursors within the bone marrow microenvironment. Immunity 10:463–471
69. Christopher MJ, Liu F, Hilton MJ, Long F, Link DC (2009) Suppression of CXCL12 production by bone marrow osteoblasts is a common and critical pathway for cytokine-induced mobilization. Blood 114:1331–1339
70. Broxmeyer HE, Kohli L, Kim CH et al (2003) Stromal cell-derived factor-1/CXCL12 directly enhances survival/antiapoptosis of myeloid progenitor cells through CXCR4 and G(alpha)i proteins and enhances engraftment of competitive, repopulating stem cells. J Leukoc Biol 73:630–638
71. Liles WC, Broxmeyer HE, Rodger E et al (2003) Mobilization of hematopoietic progenitor cells in healthy volunteers by AMD3100, a CXCR4 antagonist. Blood 102:2728–2730
72. Froberg MK, Garg UC, Stroncek DF, Geis M, McCullough J, Brown DM (1999) Changes in serum osteocalcin and bone-specific alkaline phosphatase are associated with bone pain in donors receiving granulocyte-colony-stimulating factor for peripheral blood stem and progenitor cell collection. Transfusion 39:410–414

73. Katayama Y, Battista M, Kao WM et al (2006) Signals from the sympathetic nervous system regulate hematopoietic stem cell egress from bone marrow. Cell 124:407–421
74. Mayack SR, Wagers AJ (2008) Osteolineage niche cells initiate hematopoietic stem cell mobilization. Blood 112:519–531
75. Nervi B, Ramirez P, Rettig MP et al (2009) Chemosensitization of acute myeloid leukemia (AML) following mobilization by the CXCR4 antagonist AMD3100. Blood 113:6206–6214

The Clinical Pharmacology of Filgrastim and Pegfilgrastim

Lorin K. Roskos

1 Introduction

Granulocyte colony-stimulating factor (G-CSF), an endogenous hematopoietic growth factor, selectively stimulates granulopoietic cells of the neutrophil lineage [1]. G-CSF increases proliferation and differentiation of neutrophil progenitor cells, induces maturation of the progenitor cells, and enhances survival and function of mature neutrophils [2, 3]. Filgrastim is a recombinant methionyl form of human G-CSF (r-metHuG-CSF) produced in *Escherichia coli* and has been approved for the treatment of neutropenia in cancer patients receiving myelosuppressive chemotherapy, patients with acute myeloid leukemia (AML) receiving induction or consolidation chemotherapy, cancer patients receiving myeloablative chemotherapy followed by bone marrow transplantation, and patients with severe chronic neutropenia; filgrastim is also approved for mobilization of hematopoietic progenitor cells in patients undergoing peripheral blood progenitor cell (PBPC) collection and therapy [2, 3]. Filgrastim requires daily administration to maintain its therapeutic effects because of its short circulating half-life (approximately 3.5 h) [4, 5].

Pegfilgrastim is a sustained-duration form of filgrastim, produced by covalently binding a 20-kilodalton (kDa) polyethylene glycol molecule to the *N*-terminal methionine residue of filgrastim [6]. Results from in-vitro proliferation, receptor binding, and neutrophil function studies demonstrate that filgrastim and pegfilgrastim have the same mechanism of action [7]; however, pegfilgrastim in vivo has sustained effects on granulopoiesis compared with filgrastim [8]. Filgrastim is eliminated predominantly by neutrophil G-CSF receptor (G-CSFR)-mediated clearance [9] and by renal clearance [10]. During chemotherapy-induced neutropenia, nonreceptor-mediated clearance is the primary route for filgrastim and pegfilgrastim elimination. The sustained effects of pegfilgrastim have been attributed to decreased renal clearance of the

L.K. Roskos (✉)
MedImmune, 24500 Clawiter Road, Hayward, CA 94545, USA
e-mail: RoskosL@MedImmune.com

G. Molineux et al. (eds.), *Twenty Years of G-CSF*, Milestones in Drug Therapy,
DOI 10.1007/978-3-0348-0218-5_4, © Springer Basel AG 2012

pegylated molecule with a higher hydrodynamic radius [10, 11]. The negligible contribution of renal clearance creates a greater dependency of the clearance of pegfilgrastim on G-CSFR-mediated clearance, which causes a "self-regulated" pharmacokinetic profile: after a single dose, pegfilgrastim concentrations are sustained during periods of neutropenia, and are washed out quickly during neutrophil recovery.

This review summarizes the pharmacologic properties of G-CSF and the clinical pharmacology of filgrastim and pegfilgrastim in healthy volunteers and patients with chemotherapy-induced neutropenia. Pharmacokinetics of filgrastim and pegfilgrastim are reviewed, and pharmacokinetic–pharmacodynamic modeling results are emphasized to characterize the exposure-response relationships.

2 Pharmacology of G-CSF

G-CSF stimulates selective effects on the neutrophil lineage. G-CSF regulates the number of circulating neutrophils by binding to G-CSFR, which results in the proliferation, differentiation, and accelerated maturation of neutrophil precursor cells, and enhances the function of mature neutrophils. Binding of G-CSF to G-CSFR promotes homodimerization of the receptor, resulting in the formation of a tetrameric complex containing two G-CSF and two receptor molecules [12, 13]. The high-affinity dissociation constant (K_D) of G-CSF to the complex is approximately 120–360 pM (2–7 ng/mL). The G-CSFR has been found on all cells of the neutrophil lineage as well as on monocytes and monocyte precursors. Receptors for G-CSF have not been found on eosinophils, basophils, or their progeny [14]. The presence of functional G-CSFR on platelets has been reported [15]. G-CSF inducible expression on T cells in vivo and in vitro in healthy donors has been described [16].

Healthy individuals express low amounts of G-CSF in serum (<30–163 pg/mL). Infection causes an increase in the amount of G-CSF in humans (30–3,199 pg/mL) [17–19]. The increase in of G-CSF causes the bone marrow to respond to the physiologic demand of infection through "emergency granulopoiesis." [20] Even during infection, however, the amounts of G-CSF are generally below the G-CSFR affinity constant and the EC_{50} for granulopoietic effects. By comparison, the pharmacologic doses of filgrastim and pegfilgrastim approach the EC_{90} (described in subsequent sections), which allows the therapeutic effects of filgrastim and pegfilgrastim to be maximized in conditions such as chemotherapy-induced neutropenia.

2.1 *Pharmacologic Effects on Neutrophilic Granulopoiesis and Neutrophil Function*

G-CSF exerts multiple effects on the neutrophil lineage that promote granulopoiesis and a rapid increase in the numbers of circulating neutrophils. A schematic of the production and kinetics of neutrophils in humans is shown in Fig. 1 [21]. G-CSF

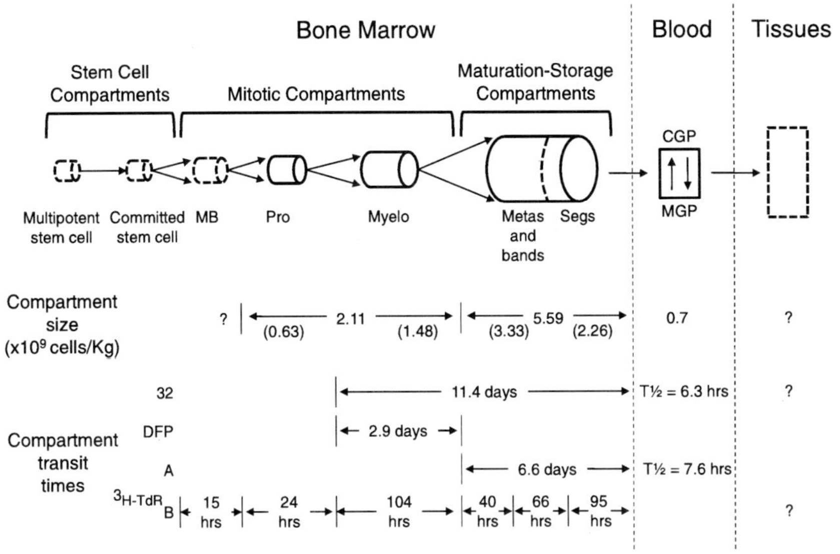

Fig. 1 Schematic representation of the kinetics of normal human neutrophilic granulopoiesis [21]

causes rapid mobilization of band cells and mature segmented neutrophils from a postmitotic storage pool within marrow; this storage pool is several-fold greater than the circulating neutrophil pool, and represents a rapidly available source of neutrophils during emergency granulopoiesis. Because of the release of both band cells and mature segmented neutrophils from the marrow after an increase in endogenous G-CSF, a neutrophil "left shift," or increase in the percentage of band cells in blood, is observed.

A sustained increase in the amount of G-CSF, as occurs during infection or during therapeutic dosing, promotes the proliferation and differentiation of the mitotic precursor cells: myeloblasts, promyelocytes, and myelocytes. These committed precursor cells show increased rates of proliferation during exposure to rG-CSF in vivo [22, 23]. The mitotic expansion occurs primarily at the promyelocyte and myelocyte stages [24]. The mitotic expansion drives and increase in the total number of G-CSF-bearing cells; this expansion acts as a negative feedback mechanism for granulopoiesis by increasing the receptor-mediated clearance of G-CSF [5].

G-CSF also shortens the maturation time of the postmitotic neutrophil precursor cells in marrow. As shown in Fig. 1, the normal transit time from the last mitotic step of neutrophil production (metamyelocyte) until the emergence of the mature-banded neutrophils in blood is approximately 5.6–7.1 days. A pulse-labeling study in healthy volunteers showed neutrophil precursor maturation time of 6.4 days in controls. By contrast, filgrastim produced a dose-dependent decrease in maturation time: 4.3 days after a 30-μg/kg dose and 2.9 days after a 300-μg/kg dose of filgrastim [25]. The accelerated maturation of neutrophil precursors stimulated by

G-CSF likely contributes to neutrophil recovery after therapeutic administration for chemotherapy-induced neutropenia.

G-CSF also affects neutrophil phenotype and function [26]. Treatment with rG-CSF causes enhancement of functions such as phagocytosis, superoxide anion generation, chemiluminescence, bacterial killing, and antibody-dependent cell-mediated cytotoxicity (ADCC). G-CSF causes rapid and direct activation of circulating neutrophils and delayed effects that include increased surface expression of effector molecules such as CD14, CD32, and CD64. Enhancement of phagocytic function includes respiratory burst metabolism, surface CD11b/CD18 expression, and cellular elastase activity. These functional changes might have beneficial clinical consequences in patients who are at risk of infection.

2.2 *Other Hematologic Effects*

Therapeutic doses of rHuG-CSF mobilize hematopoietic progenitor cells into the blood, which increases the yield of $CD34^+$ cells collected by leukopheresis for use in PBPC transplantation [27, 28]. Mobilization of PBPC is mediated, in part, by metalloproteases released from myeloid cells after rHuG-CSF administration [29].

G-CSF exerts selective effects on the neutrophil lineage, but some effects on other blood cell lineages have been described and reviewed [5, 30, 31]. In PBPC donors, a 5-day course of rHuG-CSF causes an average threefold increase in blood monocytes with associated monocyte activation and modulation of effector molecules on the monocytes [32, 33]. No consistent effect on hemoglobin or hematocrit has been noted, and measurements have remained in the normal range [34]. Filgrastim administration led to a twofold, dose-dependent increase in the number of circulating lymphocytes and a slight increase in the number of monocytes and the myeloid-to-erythroid ratio in the bone marrow. The latter effect could be due to a change in the proportion of early myeloid cells in marrow [34].

2.3 *Clearance Mechanisms and Homeostatic Regulation of G-CSF Levels*

Proteins can be cleared by multiple, parallel elimination pathways [35]. The elimination pathways include renal clearance, elimination by the reticuloendothelial system (RES), receptor-mediated endocytosis, and nonspecific proteolysis and degradation. Renal clearance is an important elimination pathway for proteins that are small enough to undergo glomerular filtration. The ability of the protein to be filtered depends on physical factors such as molecular weight, structure, charge, and water of hydration that contribute to the overall hydrodynamic radius of the molecule. Proteins >70 kDa are not filtered; but depending on the biophysical characteristics of the protein, smaller molecules can have negligible renal clearance. Generally some impediments to renal filtration exist for molecules larger than 7 kDa [36].

For G-CSF, the receptor-mediated clearance pathway is an important homeostatic mechanism regulating granulopoiesis. During neutropenia, G-CSF concentrations are increased because of decreased G-CSFR, which stimulates neutrophil mobilization and production. During neutrophilia, the increased number of G-CSFR-bearing cells increases the clearance of G-CSF, which decreases the amount of G-CSF and neutrophil production. The importance of G-CSFR to clearance has been demonstrated: G-CSFR$^{-/-}$ mice show greatly reduced clearance of filgrastim and pegfilgrastim relative to wild-type controls [37]. As discussed in subsequent sections, the relationship between filgrastim and pegfilgrastim clearance and absolute neutrophil count (ANC) has been well established.

A difference in renal clearance is the predominant characteristic that differentiates filgrastim and pegfilgrastim. The contribution of different clearance pathways to the elimination of filgrastim and pegfilgrastim was assessed by modeling the pharmacokinetics (PK) in rats receiving a bilateral nephrectomy compared to sham-operated controls [10]. Filgratim and pegfilgrastim were dosed at 5 μg/kg and 100 μg/kg. The contribution of renal clearance was evaluated by simultaneously modeling the PK collected from the nephrectomized rats and the sham controls. The contribution of renal clearance to filgrastim (MW: 18.8 kDa) elimination when receptor-mediated elimination is saturated by the high dose of filgrastim is clearly observed by the lower area-under-the-curve (AUC) in sham controls relative to nephrectomized rats. The higher hydrodynamic size of pegfilgrastim (MW: 38.8 kDa) prevented renal clearance, indicated by overlapping PK curves for the nephrectomized rats and sham controls after the high dose. Modeling of the data indicated that in absence of receptor-mediated clearance (as would occur in patients with chemotherapy-induced neutropenia), renal clearance accounts for >75% of filgrastim clearance. The role of the liver in the elimination of filgrastim appears to be insignificant, as pharmacokinetics of filgrastim after subcutaneous administration was similar between healthy volunteers and patients with hepatic impairment [38].

Filgrastim, which is used to treat chemotherapy-induced neutropenia, must be dosed subcutaneously daily to maintain effective concentrations because of high renal clearance; by contrast, pegfilgrastim can be dosed once per cycle of chemotherapy as a single subcutaneous injection because of the negligible renal clearance pathway and reduced G-CSFR-mediated clearance during neutropenia. This difference is an example of rational engineering of a protein drug: prospective, nonclinical modeling and simulation predicted that pegfilgrastim levels, after a single dose, would be sustained during neutropenia and would wash out rapidly after neutrophil recovery [39]. The PK–PD (pharmacodynamic) properties of pegfilgrastim that were predicted by modeling and simulation were confirmed in a well-designed phase 1 study by a head-to-head comparison of filgrastim and pegfilgrastim in cancer patients before and after chemotherapy [8].

3 Pharmacokinetics of Filgrastim

3.1 Pharmacokinetics of Filgrastim in Healthy Volunteers

The PK of filgrastim has been studied in healthy volunteers after single intravenous and subcutaneous doses and multiple subcutaneous doses [5]. A summary of the PK in healthy volunteers is listed in Table 1. After a single intravenous infusion in

Table 1 Summary of clinical pharmacokinetic studies in healthy volunteers

No. of volunteers	Dose and route	Results		
$N = 5$	3.45 μg/kg IV, single 30-min infusion	Elimination by first-order kinetics, $t_{1/2}\beta = 2.7$ h; $V_d = 162$ mL/kg; clearance = 0.6 mL/min/kg		
$N = 16$	4-period crossover absolute bioavailability study: each volunteer received 375 and 750 μg IV and SC	Dose-dependent PK		
		Filgrastim clearance was mediated by parallel Michaelis–Menten and linear clearance processes		
		Michaelis–Menten clearance (attributed to G-CSFR-mediated clearance) accounted for approximately 80% of total clearance		
		Absorption kinetics described by parallel absorption processes		
		The absolute bioavailability of filgrastim was 60% after 375 μg SC and 69% after 750 μg SC		
$N = 30$	75, 150, 300, or 600 μg per individual, SC, multiple dosing for 10 days	Decrease in serum levels upon multiple dosing		
		C_{max} (ng/mL)		
			Day 1	Day 10
		75 μg	1.65 ± 0.80	0.99 ± 0.28
		150 μg	4.60 ± 3.78	1.84 ± 1.29
		300 μg	14.79 ± 6.55	3.64 ± 1.63
		600 μg	16.28 ± 7.14	3.46 ± 0.52
		t_{max} (h)		
			Day 1	Day 10
		75 μg	5.5 ± 1.8	3.5 ± 0.9
		150 μg	5.0 ± 2.0	4.5 ± 1.0
		300 μg	4.0 ± 0.0	3.5 ± 1.0
		600 μg	5.8 ± 1.3	3.7 ± 0.8
		$AUC_{0–24}$ (ng h/mL)		
			Day 1	Day 10
		75 μg	14.3 ± 4.3	5.7 ± 1.6
		150 μg	33.2 ± 16.9	10.2 ± 4.5
		300 μg	119.0 ± 41.7	20.3 ± 8.8
		600 μg	209.4 ± 107.3	29.1 ± 5.3

AUC area under the curve; *G-CSFR* G-CSF receptor; *IV* intravenous; *PK* pharmacokinetics; *SC* subcutaneous; t_{max} time to peak concentration; V_d volume of distribution normal volunteers, serum filgrastim concentrations declined with a half-life of 2.7 h. Clearance averaged 0.6 mL/min/kg, and volume of distribution averaged 162 mL/kg. After single subcutaneous injections, peak serum concentrations occurred at approximately 4–6 h. A single 5-μg/kg dose resulted in mean peak concentrations of 20–30 ng/mL.

In an absolute bioavailability study conducted in normal volunteers at doses of 375 and 750 μg intravenous and subcutaneous, the clearance of filgrastim was dose dependent [40]. Filgrastim clearance was mediated by parallel Michaelis–Menten and linear clearance processes. The nonlinear clearance process, which was attributed to G-CSFR-mediated clearance, accounted for about 80% of total clearance. The linear clearance was 0.296 L/h (4.23 mL/h/kg for a 70-kg volunteer). The subcutaneous absorption kinetics were modeled by two parallel, first-order absorption processes with different absorption rates. The parallel absorption process probably reflects lymphatic and direct vascular absorption processes, as previously described for other subcutaneously administered proteins [35]. The vascular absorption was modeled as a first-order absorption process. The lymphatic absorption was modeled as first-order absorption with a lag time before absorption. The fraction of dose absorbed through the delayed absorption route was 57–67% for filgrastim. The absolute bioavailability of filgrastim (extent of absorption from the subcutaneous injection site) was dose dependent, ranging from 60 to 70%.

The effect of multiple dosing on the PK of filgrastim also was investigated [41]. Serum concentrations of filgrastim on day 1 were found to be substantially higher than on day 10, as shown by the higher peak concentrations (C_{max}) and AUC values (Table 1). The difference in the serum concentration profiles of filgrastim with multiple dosing is mostly related to increasing neutrophil counts, which in turn leads to an increase in cellular clearance of filgrastim. The C_{max} occurred 4–6 h after the first dose and 3.5–4.5 h after the tenth dose. Terminal half-life was 3–5 h.

3.2 Pharmacokinetics of Filgrastim in Neutropenic Adults

The PK of filgrastim has been studied in adult cancer patients with chemotherapy-induced neutropenia. Dose-linear clearance has been observed after high doses of filgrastim and during severe neutropenia. Nonlinear PK has been observed after low intravenous doses administered before chemotherapy, however. Clearance of intravenous filgrastim was more rapid after an infusion of 1.73 μg/kg than after 5.75–34.5 μg/kg. Filgrastim clearance appeared linear at higher doses, presumably due to saturation of receptor-mediated clearance [5]. When cellular clearance is saturated by high filgrastim concentrations or is diminished by neutropenia, the linear clearance pathway predominates.

Homeostatic regulation of filgrastim clearance was first described in patients receiving filgrastim after chemotherapy [9]. In patients receiving daily doses of filgrastim, steady-state concentrations were rapidly attained and were maintained until onset of hematopoietic recovery (as measured by increasing ANC). As neutrophil counts recovered toward normal levels, serum concentrations of filgrastim declined rapidly, consistent with increased clearance by neutrophils.

In a comparison of the PK and PD of filgrastim and pegfilgrastim before and after chemotherapy in patients with nonsmall-cell lung cancer (NSCLC), the PK of filgrastim was evaluated after five daily subcutaneous doses before chemotherapy, and then after standard daily therapeutic dosing after chemotherapy until neutrophil recovery [8]. Before chemotherapy, each consecutive dose of filgrastim produced progressively lower serum concentrations; these serum concentrations were inversely related to the ANC at the time of dosing. After daily doses of filgrastim after chemotherapy, drug levels accumulated during the period of neutropenia; after the ANC nadir, filgrastim concentrations decreased rapidly upon discontinuation of dosing.

3.3 Pharmacokinetics of Filgrastim in Neutropenic Pediatric Patients

The PK of filgrastim in pediatric cancer patients is similar, when comparing weight-normalized doses, to the PK observed in adult cancer patients. In both adults and children, ANC is the major factor affecting between-patient and within-patient PK variability.

The PK of filgrastim in 15 children (aged 1.2–9.4 years) administered daily (5, 10, and 15 μg/kg/day subcutaneously) for 10 days after chemotherapy has been reported [42]. Similar to adults, apparent clearance (CL/F) decreased with increasing dose. CL/F was lower and C_{max} was higher on day 1 (immediately after chemotherapy) relative to day 10 (after hematopoietic recovery). The PK on day 10 (normal ANC) was comparable to the PK in healthy adults after the first subcutaneous dose of filgrastim.

The PK of filgrastim was reported in 11 pediatric patients (aged 6–18 years) receiving ten daily intravenous doses (5 or 10 μg/kg/day; 30-min infusion) after chemotherapy [43]. A significant correlation between filgrastim clearance and ANC was noted, supporting that homeostatic regulation of filgrastim clearance occurs in pediatric populations.

Filgrastim PK also has been studied in children with severe chronic neutropenia (SCN) [44]. Eleven children were given 6–48 μg/kg filgrastim twice daily by subcutaneous injection. Peak serum concentrations occurred 2–8 h after dosing. A strong relationship between ANC and filgrastim clearance was described using a sigmoid model. At low neutrophil counts ($<0.1 \times 10^9$/L), clearance approached a minimum value of 0.29 mL/min/kg, which reflects the linear clearance rate in the absence of receptor-mediated clearance. Maximum clearance approached 2 mL/

min/kg at high values of ANC (>17.0 × 10^9/L). Mean half-life was 4.7 h at low ANC and <2 h at ANC >17.0 × 10^9/L.

4 Pharmacokinetics of Pegfilgrastim

4.1 Pharmacokinetics of Pegfilgrastim in Healthy Volunteers

Pegfilgrastim has been studied after single intravenous and subcutaneous doses in healthy volunteers. After subcutaneous administration of pegfilgrastim, a nonlinear increase in AUC is observed and the time to maximum concentration increases (from 8 to 24 h) with increasing dose (30–300 μg/kg), consistent with progressive saturation of G-CSFR-mediated clearance by higher concentrations of pegfilgrastim [45]. A 10-fold increase in the dose resulted in an approximately 25-fold increase in the C_{max} and a 75-fold increase in the AUC.

In a study comparing intravenous and subcutaneous dosing in healthy volunteers, the absolute bioavailability of pegfilgrastim after subcutaneous administration (defined as the fraction of dose absorbed from the site of subcutaneous administration) was estimated to be in the range of 20–30% using PK/PD modeling (Amgen data on file). The C_{max} and AUC values of pegfilgrastim after subcutaneous administration were approximately 3–4%, of the exposure obtained after intravenous administration. However, the duration of ANC elevation was approximately 3–4 days longer after subcutaneous administration than after intravenous administration. This observation suggests that efficacious concentrations after subcutaneous administration were sustained longer than those after intravenous administration.

The effect of renal function of the PK and PD of pegfilgrastim was evaluated across various renal function groups in a phase 1 clinical study. The PK and ANC profiles were similar in all groups, supporting that the kidney contributes negligibly to the elimination of pegfilgrastim [46]. This observation is consistent with renal clearance studies conducted in nonclinical nephrectomy models [10]. Although the hepatic clearance of pegfilgrastim has not been investigated, PK–PD modeling (described below) suggests that neutrophil-mediated clearance is the predominant pathway in eliminating pegfilgrastim, which allows pegfilgrastim to have a very efficient self-regulating clearance mechanism.

4.2 Pharmacokinetics of Pegfilgrastim in Neutropenic Patients

In patients receiving myelosuppressive chemotherapy, the PK of pegfilgrastim is highly dependent on dose and the ANC. The PK of pegfilgrastim was nonlinear in patients with NSCLC after a single dose before and after chemotherapy (Table 2) [8]. After chemotherapy, pegfilgrastim reached a C_{max} similar to that before chemotherapy; however, drug levels were sustained for a longer period of time and began to decline

Table 2 Pegfilgrastim pharmacokinetic parameter values before and after chemotherapy in patients with nonsmall-cell lung cancer [8]

Dose (μg/kg)	No. of patients	C_{max} (ng/mL)	T_{max} (h)	AUC$_{(0-\infty)}$ (ng* h/mL)	CL/F (mL/h/kg)
Before chemotherapy					
30	3	13.2 (2.64)	8 (8–24)	522 (208)	63.6 (23.3)
100	3	118 (61.8)	24 (24–36)	5,280 (2620)	23.6 (14.5)
300	4	937 (564)	48 (36–48)	64,400 (44,500)	7.06 (5.40)
After chemotherapy					
30	3	9.43 (5.15)	120 (120–120)	777 (242)	41.2 (12.4)
100	3	125 (73.1)	72 (24–96)	12,500 (10,000)	11.3 (6.27)
300	3	1,030 (278)	72 (48–120)	130,000 (39,000)	2.47 (0.83)

$AUC_{(0-\infty)}$ area under the concentration time curve from time 0 to infinity; *CL/F* time-averaged clearance after subcutaneous administration; C_{max} peak serum concentration; T_{max} time to C_{max}. Data represent mean (SD) values except for T_{max} data, which represent median (range) values.

rapidly at a time corresponding closely to the ANC nadir. The exposure to pegfilgrastim (AUC) after chemotherapy was higher than that before chemotherapy, supporting the hypothesis that neutrophil and neutrophil precursors are important contributors to the elimination of pegfilgrastim. The temporal correspondence between the rapid decline in pegfilgrastim beginning at the ANC nadir is consistent with a recovery of G-CSFR-expressing neutrophils and precursor cells in blood and marrow.

The relationship between PK and ANC was further supported by the PK–PD profiles of pegfilgrastim during multiple cycles of myelosuppressive chemotherapy. In a dose-finding study, women with high-risk stage II, III, or IV breast cancer received a single subcutaneous dose of pegfilgrastim at 30, 60, or 100 μg/kg 24 h after completion of doxorubicin/docetaxel chemotherapy. Treatment with pegfilgrastim was repeated every 21 days for up to four cycles of chemotherapy [47, 48]. The exposure to pegfilgrastim was lower in chemotherapy cycle 3 than in cycle 1, while an improvement in ANC nadir and a decrease in duration of neutropenia occurred in cycle 2 and subsequent cycles of treatment, suggesting that an expansion of neutrophil and neutrophil precursor mass in the later cycles resulted in an increase in drug clearance.

4.3 *Pharmacokinetics of Pegfilgrastim in Neutropenic Pediatric Patients*

The efficacy, safety, and PK of a single 100-μg/kg subcutaneous dose of pegfilgrastim have been evaluated and compared to 5 μg/kg subcutaneous daily filgrastim in pediatric patients receiving myelosuppressive chemotherapy for sarcoma [49]. The maximum median pegfilgrastim concentration was achieved approximately 1–2 days after pegfilgrastim administration and was sustained until the ANC nadir was reached. Median drug exposures in cycle 3 were lower than in cycle 1 for each age cohort. The youngest cohort (0–5 years) had a higher exposure to pegfilgrastim than the other two cohorts (6–11 years and 12–21 years), likely because the younger patients experienced a longer duration of neutropenia. Filgrastim PK was not evaluated. Single-dose pegfilgrastim exhibited comparable safety and efficacy as daily filgrastim. Pharmacokinetic profiles were consistent with those reported previously for adults.

5 PK–PD Modeling of Filgrastim and Pegfilgrastim

G-CSF has multiple effects on circulating neutrophils and neutrophil precursors. Stimulation of precursor cell proliferation leads to an increase in the total mass of G-CSFR-expressing cells, which serves as a negative regulator of G-CSF levels through accelerated clearance. Administration of cytotoxic chemotherapy decreases the number of mitotic precursor cells and subsequently the number of postmitotic precursors and circulating neutrophils, which decreases the clearance of G-CSF. A systems pharmacology approach can be used to describe the complex PK–PD relationships of filgrastim and pegfilgrastim.

A PK–PD model built on the biology of granulopoiesis (Fig. 1) has been developed to describe the PK, PD, and homeostatic regulation pegfilgrastim clearance after administration to healthy volunteers [45] (Fig. 2). In the PK model, the

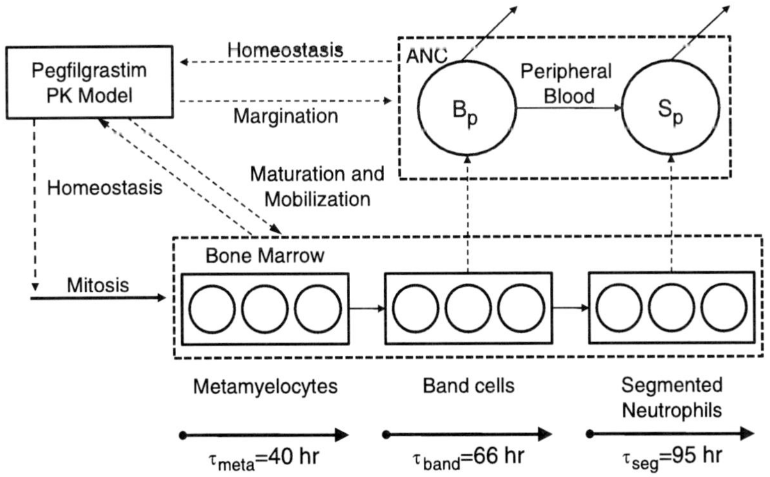

Fig. 2 Pharmacodynamic model describing the granulopoietic effects of pegfilgrastim. Serum concentrations of pegfilgrastim stimulate mitosis and mobilization of band cells and segmented neutrophils in bone marrow, decrease maturation times for postmitotic cells in marrow, and affect margination of the peripheral blood band cell (Bp) and segmented neutrophil (Sp) populations, the sum of which is the total absolute neutrophil count (ANC). Changes in neutrophil counts in peripheral blood provide feedback regulation of pegfilgrastim clearance [45]

V_{max} of the receptor-mediated clearance pathway is assumed to be proportional to ANC. In the Michaelis–Menten approximation of target-mediated drug disposition, the V_{max} is equal to the product of the total receptor concentration and the internalization rate constant of the drug–receptor complex [50]. The starting point of the PD model was the production of metamyelocytes from the last maturational step of mitotic precursors. The most distal effect of the drug, expansion of the mitotic promyelocyte and myelocyte pools, was empirically described by a Hill equation. Serum concentrations of pegfilgrastim were assumed to stimulate mitosis and mobilization of band cells and segmented neutrophils in bone marrow and increase adhesion of peripheral blood band cells (B_p) and segmented neutrophils (S_p) to blood vessels (margination), causing a change in the volume of distribution of the circulating neutrophils. Pegfilgrastim was also assumed to decrease the maturation time of metamyelocytes, band cells, and segmented neutrophils in marrow. All effects were assumed to have the same EC_{50}, since the effects are mediated by G-CSFR.

As described subsequently, the modeling approach can be used to describe single- or multiple-dose effects of filgrastim and pegfilgrastim, and the effects of chemotherapy on the production of neutrophils can also be incorporated.

5.1 *Pharmacokinetic–Pharmacodynamic Modeling of Filgrastim in Healthy Volunteers*

The PK–PD of filgrastim after single intravenous and subcutaneous doses of 375 and 750 μg were simultaneously modeled [40]. Because the half-life of filgrastim is very short, the upstream effects of filgrastim on the mitotic precursor pool do not need to be incorporated into the model following a single dose. Consequently, the clearance of filgrastim does not need to be adjusted by changes in ANC over this short time frame; the change in ANC reflects a change in neutrophil distribution rather than a change in total neutrophil and precursor mass. Effects on ANC after a single dose can be modeled as a simple indirect response model, where filgrastim promotes the rapid influx of neutrophils from the marrow storage pool. Early margination effects are accounted for by a filgrastim-induced increase in the volume of distribution of neutrophils in blood.

The fitting of the subcutaneous PK and ANC data are illustrated in Fig. 3. An excellent fit was obtained that described the dose-dependent PK and the effects of filgrastim on margination (early decrease in ANC) and mobilization of neutrophils from marrow (subsequent increase in ANC) and the return of ANC to baseline values. In the model, the EC_{50} was 4.72 ng/mL, which is similar to the high-affinity dissociation constant of filgrastim for G-CSFR.

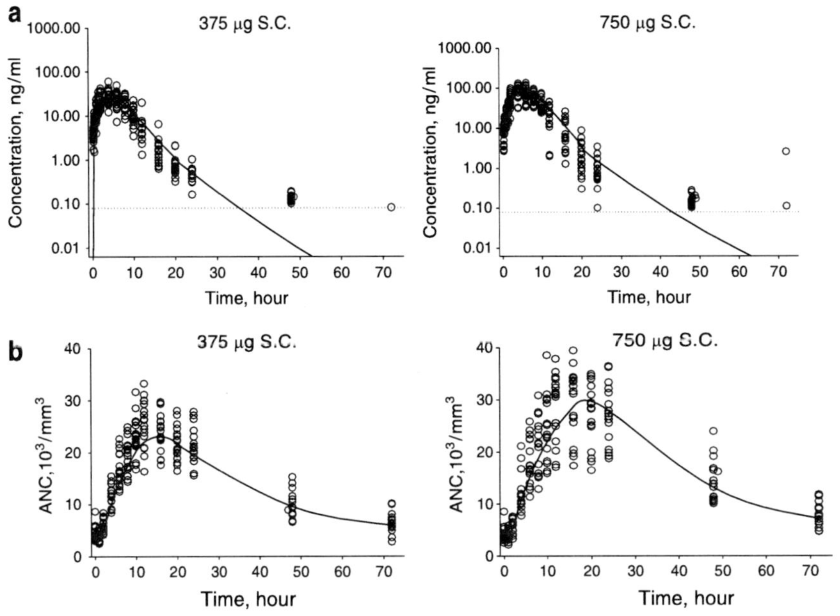

Fig. 3 PK–PD modeling of filgrastim after single 375 and 750 μg SC doses to healthy volunteers. (**a**) Pharmacokinetic profiles and model fit. (**b**) ANC profiles and model fit. *Symbols* represent the pooled data (PK and ANC) from all volunteers from each cohort. *Solid lines* represent the fit of population PK–PD model to the data [40]

5.2 Pharmacokinetic–Pharmacodynamic Modeling of Pegfilgrastim in Healthy Volunteers

Because pegfilgrastim has a long half-life relative to filgrastim, single doses are able to sustain G-CSF levels sufficiently long to promote proliferation of mitotic precursor cells and increase in the total neutrophil and neutrophil precursor mass. Therefore, the mechanistic PK–PD model illustrated in Fig. 2 was applied to the PK–PD profiles after a single subcutaneous dose of pegfilgrastim administered to healthy volunteers at doses of 30, 60, 100, and 300 μg/kg [45].

Modeling of the PK and ANC profiles are illustrated in Fig. 4. In healthy volunteers, the PK of pegfilgrastim were nonlinear in a dose range of 30–300 μg/kg; the clearance of pegfilgrastim decreased with increasing dose, which is attributed to the neutrophil G-CSFR-mediated pathway. The terminal half-life was independent of dose, suggesting that pegfilgrastim serum concentrations at the terminal phase decreased below levels saturating G-CSFR. PK modeling required that the V_{max} change in proportion to peripheral neutrophil counts, consistent with a G-CSFR-mediated clearance mechanism. The model accurately

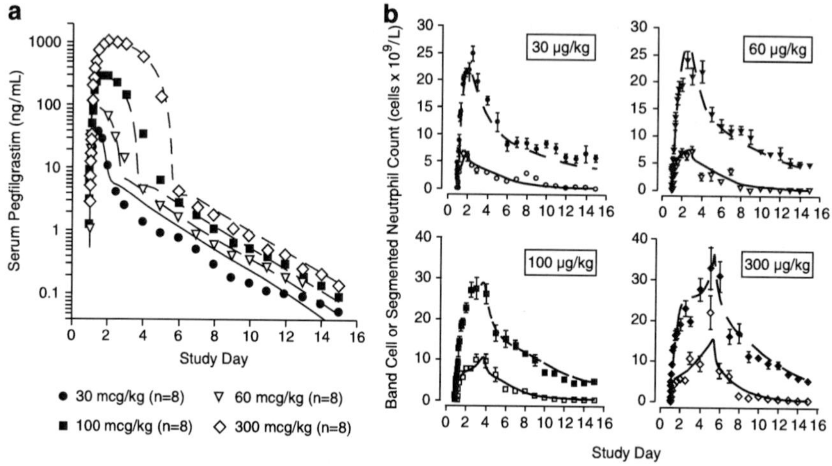

Fig. 4 PK–PD modeling of the dose-ranging effects of pegfilgrastim on neutrophil counts in the blood of healthy volunteers, with homeostatic regulation of pegfilgrastim clearance by changing neutrophil and precursor cell mass in blood and marrow. (**a**) Modeled (*lines*) and observed (*symbols*) pegfilgrastim serum concentrations after single SC doses; (**b**) Modeled (*lines*) and observed mean band cell (*open symbols*) and segmented neutrophil (*closed symbols*) counts in peripheral blood [45]

described the dose-dependent and neutrophil-dependent PK. The low rate of pegfilgrastim clearance by a neutrophil-independent, linear pathway (CL/F approximately 1.2 mL/h/kg) might explain the prolonged exposure to drug and sustained effects during chemotherapy-induced neutropenia [8]. During neutropenia, the linear clearance pathway is expected to be the predominant elimination pathway for pegfilgrastim because of the decreased numbers of neutrophils and neutrophil precursors in blood and marrow.

After administration of pegfilgrastim, a transient margination followed by a rapid, dose-dependent increase in ANC was observed. A dose-dependent increase in the percentage of band cells in peripheral blood was observed. The appearance of the younger, band cell population in peripheral blood – also observed after dosing with filgrastim – suggests that pegfilgrastim stimulates early release of neutrophils from marrow.

The model of granulopoiesis successfully described the relationship between pegfilgrastim serum concentrations and neutrophil counts in peripheral blood. Modeling predicted that pegfilgrastim could elicit a maximum fourfold increase in the rate of metamyelocyte production. Consistent with the modeling results, neutrophil kinetic studies of filgrastim in healthy volunteers have shown that filgrastim exerts mitotic effects primarily on promyelocytes and myelocytes, with a twofold to threefold amplification of mitosis and a minimal delay before influx of metamyelocytes [24, 25].

Differential effects of pegfilgrastim on blood neutrophil populations were modeled as a pegfilgrastim-dependent flux of younger neutrophils, including band cells and segmented neutrophils through early release from marrow and a decrease in maturation time. As suggested by the model, this decreased transit time may be due, in part, to the early release of band cells and segmented neutrophils from bone marrow. The early release of neutrophils and accelerated precursor maturation may be important contributors to accelerated ANC recovery after myelosuppressive chemotherapy. The EC_{50} for the granulopoietic effects was 9.86 ng/mL, which is reasonably similar to the EC_{50} reported for filgrastim in healthy volunteers. The results suggest that the PD differences between filgrastim and pegfilgrastim are predominantly driven by the PK differences and decreased linear clearance rate of pegfilgrastim.

5.3 Pharmacokinetic–Pharmacodynamic Modeling of Filgrastim and Pegfilgrastim in Cancer Patients Before and After Chemotherapy

As discussed, data from repeated dosing of filgrastim or single dosing of pegfilgrastim must be modeled using a complete PK–PD model of granulopoiesis that incorporates effects on progenitor cell proliferation and homeostatic regulation of drug clearance by changing neutrophil mass. In the setting of chemotherapy, the cytotoxic effects on neutrophil progenitors and the PK–PD relationship must be incorporated. To illustrate the effect of chemotherapy on the PK–PD profile, data from filgrastim and pegfilgrastim collected before and after chemotherapy [8] were modeled using a similar model to that of pegfilgrastim after administration to healthy volunteers [45]. The major difference was that an effect of chemotherapy on metamyelocyte production rate was introduced. This effect was incorporated as a step function, where chemotherapy halted proliferation of mitotic precursor cells and the effects of G-CSF on metamyelocyte production for period of time:

$$\frac{dM_1}{dt} = \left(S_0 + \frac{E_{\text{mit}}C}{EC_{50} + C}\right)X(t) - \frac{n_{\text{m}}}{\tau_{\text{meta}}}M_1, \tag{1}$$

where

$$X(t) = 1 \quad t < \text{Day } 15, \tag{2}$$

$$X(t) = \frac{t^{120}}{\tau_{\text{rec}}^{120} + t^{120}} \quad t \geq \text{Day } 15. \tag{3}$$

Equation (1) represents the rate of change of metamyelocytes emerging from the myelocyte compartment and the effect of G-CSF concentration (C) on the production rate of metamyelocytes. After chemotherapy is given on day 15, the production of metamyelocytes is halted for a time, τ_{rec}, which is the recovery time of the myelocyte pool from the cytotoxic effects of chemotherapy.

The model fit to the PK–PD model for filgrastim and pegfilgrastim are shown in Fig. 5. Notable differences in the PK profiles before and after chemotherapy were successfully described for both molecules. Before chemotherapy, filgrastim levels decreased after each dose, consistent with the increase in ANC. After chemotherapy, filgrastim levels accumulated on daily dosing during the period of neutropenia. Before chemotherapy, pegfilgrastim levels decreased quickly; after chemotherapy, levels after a single dose were sustained until onset of ANC recovery. For both molecules, the effect on ANC before and after chemotherapy was well predicted. The modeling estimates of EC_{50} were similar for filgrastim and pegfilgrastim, supporting that the PK–PD differences are primarily determined by PK.

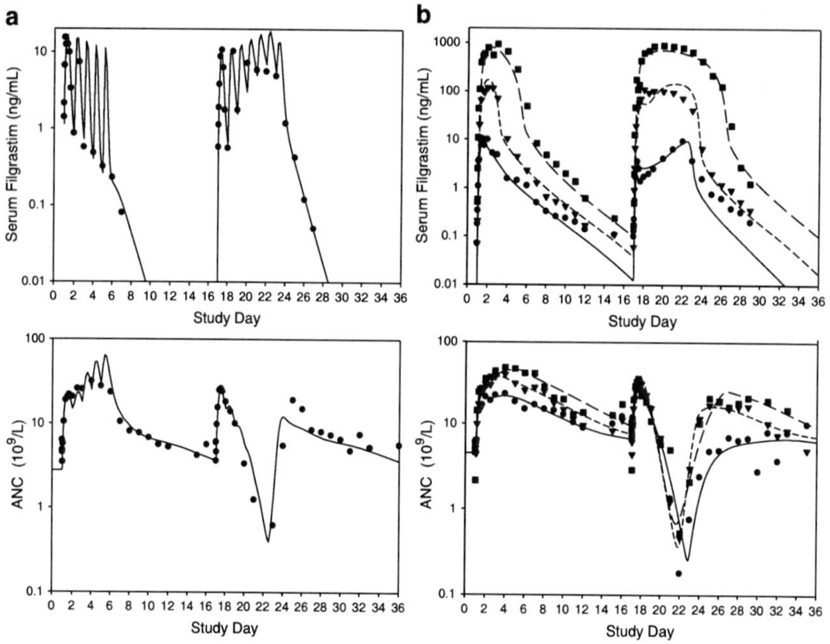

Fig. 5 PK–PD modeling of the effects of pegfilgrastim and filgrastim on neutrophil counts in the blood of patients with nonsmall-cell lung cancer (NSCLC) receiving myelosuppressive chemotherapy, with homeostatic regulation of pegfilgrastim clearance by changing neutrophil and precursor cell mass in blood and marrow. (**a**) PK–PD modeling of 5 μg/kg SC filgrastim given daily for five doses before chemotherapy and daily until ANC recovery after chemotherapy. (**b**) PK–PD modeling of single-dose pegfilgrastim given before and after chemotherapy at doses of 30, 100, and 300 μg/kg SC. *Symbols* represent median values for each cohort and *solid lines* are the model fit to the data [8]

6 Conclusions

Filgrastim and pegfilgrastim exhibit nonlinear PK related to clearance by G-CSFR expressed on neutrophils and neutrophil precursors. PK is subject to homeostatic regulation during conditions of neutropenia or neutrophilia. PD effects of filgrastim and pegfilgrastim are exerted in a similar manner, through stimulation of progenitor cell proliferation, decreased maturation time of postmitotic precursor cells, and mobilization of neutrophil storage pools from marrow. The complex relationships between PK and changes in neutrophil counts related to rHuG-CSF treatment and chemotherapy-induced neutropenia can be successfully described by mechanistic PK–PD models based on the biology of granulopoiesis. Modeling supports that the major differences between filgrastim and pegfilgrastim are related to the PK properties. Pegfilgrastim, due to a high molecular weight and hydrodynamic radius, has a negligible renal clearance; therefore, the clearance of pegfilgrastim is more dependent on neutrophil counts than filgrastim, which has a substantial renal clearance pathway. These properties allow pegfilgrastim to have a "self-regulating" PK profile that supports dosing once per cycle of chemotherapy.

References

1. Demetri GD, Griffin JD (1991) Granulocyte colony-stimulating factor and its receptor. Blood 78:2791–2808
2. Welte K, Gabrilove J, Bronchud MH, Platzer E, Morstyn G (1996) Filgrastim (r-metHuG-CSF): the first 10 years. Blood 88:1907–1929
3. Neupogen Prescribing Information (2011)
4. Kuwabara T, Kobayashi S, Sugiyama Y (1996) Pharmacokinetics and pharmacodynamics of a recombinant human granulocyte colony-stimulating factor. Drug Metab Rev 28:625–658
5. Roskos LK, Cheung EN, Vincent M, Foote M (1998) Pharmacology of filgrastim (r-metHuG-CSF). In: Morstyn G, Dexter TM, Foote M (eds) Filgrastim (r-metHuG-CSF) in clinical practice. Marcel Dekker, New York, pp 51–71
6. Kinstler O, Molineux G, Treuheit M, Ladd D, Gegg C (2002) Mono-N-terminal poly(ethylene glycol)-protein conjugates. Adv Drug Deliv Rev 54:477–485
7. Lord BI, Woolford LB, Molineux G (2001) Kinetics of neutrophil production in normal and neutropenic animals during the response to filgrastim (r-metHu G-CSF) or filgrastim SD/01 (PEG-r-metHu G-CSF). Clin Cancer Res 7:2085–2090
8. Johnston E, Crawford J, Blackwell S et al (2000) Randomized, dose-escalation study of SD/01 compared with daily filgrastim in patients receiving chemotherapy. J Clin Oncol 18:2522–2528
9. Layton JE, Hockman H, Sheridan WP, Morstyn G (1989) Evidence for a novel in vivo control mechanism of granulopoiesis: mature cell-related control of a regulatory growth factor. Blood 74:1303–1307
10. Yang BB, Lum PK, Hayashi MM, Roskos LK (2004) Polyethylene glycol modification of filgrastim results in decreased renal clearance of the protein in rats. J Pharm Sci 93:1367–1373
11. Tanaka H, Tokiwa T (1990) Influence of renal and hepatic failure on the pharmacokinetics of recombinant human granulocyte colony-stimulating factor (KRN8601) in the rat. Cancer Res 50:6615–6619

12. Aritomi M, Kunishima N, Okamoto T, Kuroki R, Ota Y, Morikawa K (1999) Atomic structure of the GCSF-receptor complex showing a new cytokine-receptor recognition scheme. Nature 401:713–717
13. Barreda DR, Hanington PC, Belosevic M (2004) Regulation of myeloid development and function by colony stimulating factors. Dev Comp Immunol 28:509–554
14. Nicola NA, Peterson L, Hilton DJ, Metcalf D (1988) Cellular processing of murine colony-stimulating factor (Multi-CSF, GM-CSF, G-CSF) receptors by normal hemopoietic cells and cell lines. Growth Factors 1:41–49
15. Shimoda K, Okamura S, Harada N, Kondo S, Okamura T, Niho Y (1993) Identification of a functional receptor for granulocyte colony-stimulating factor on platelets. J Clin Invest 91:1310–1313
16. Franzke A, Piao W, Lauber J et al (2003) G-CSF as immune regulator in T cells expressing the G-CSF receptor: implications for transplantation and autoimmune diseases. Blood 102:734–739
17. Cheers C, Haigh AM, Kelso A, Metcalf D, Stanley ER, Young AM (1988) Production of colony-stimulating factors (CSFs) during infection: separate determinations of macrophage-, granulocyte-, granulocyte-macrophage-, and multi-CSFs. Infect Immun 56:247–251
18. Watari K, Asano S, Shirafuji N et al (1999) Serum granulocyte colony-stimulating factor levels in healthy volunteers and patients with various disorders as estimated by enzyme immunoassay. Blood 73:117–122
19. Kawakami M, Tsutsumi H, Kumakawa T et al (1990) Levels of serum granulocyte colony-stimulating factor in patients with infections. Blood 76:1962–1964
20. Panopoulos AD, Watowich SS (2008) Granulocyte colony-stimulating factor: molecular mechanisms of action during steady state and 'emergency' hematopoiesis. Cytokine 42:277–288
21. Skubitz KM (1999) Neutrophilic leukocytes. In: Lee GR, Foerster J, Lukens J, Paraskevas F, Greer J, Rodgers GM (eds) Wintrobe's clinical hematology. Lippincott Williams & WIlkins, Baltimore, pp 300–350
22. Lord BI, Bronchud MH, Owens S et al (1989) The kinetics of human granulopoiesis following treatment with granulocyte colony-stimulating factor in vivo. Proc Natl Acad Sci U S A 86:9499–9503
23. Lord BI, Molineux G, Pojda Z, Souza LM, Mermod JJ, Dexter TM (1991) Myeloid cell kinetics in mice treated with recombinant interleukin-3, granulocyte colony-stimulating factor (CSF), or granulocyte-macrophage CSF in vivo. Blood 77:2154–2159
24. Chatta GS, Price TH, Allen RC, Dale DC (1994) Effects of in vivo recombinant methionyl human granulocyte colony-stimulating factor on the neutrophil response and peripheral blood colony-forming cells in healthy young and elderly adult volunteers. Blood 84:2923–2929
25. Price TH, Chatta GS, Dale DC (1996) Effect of recombinant granulocyte colony-stimulating factor on neutrophil kinetics in normal young and elderly humans. Blood 88:335–340
26. Carulli G (1997) Effects of recombinant human granulocyte colony-stimulating factor administration on neutrophil phenotype and functions. Haematologica 82:606–616
27. Sheridan WP, Morstyn G, Wolf M et al (1989) Granulocyte colony-stimulating factor and neutrophil recovery after high-dose chemotherapy and autologous bone marrow transplantation. Lancet 2:891–895
28. Chao NJ, Schriber JR, Grimes K et al (1993) Granulocyte colony-stimulating factor "mobilized" peripheral blood progenitor cells accelerate granulocyte and platelet recovery after high-dose chemotherapy. Blood 81:2031–2035
29. Cashen AF, Lazarus HM, Devine SM (2007) Mobilizing stem cells from normal donors: is it possible to improve upon G-CSF? Bone Marrow Transplant 39:577–588
30. Anderlini P, Champlin RE (2006) Biologic and molecular effects of granulocyte colony-stimulating factor in healthy individuals: recent findings and current challenges. Blood 111:1767–1772

31. Anderlini P (2009) Effects and safety of granulocyte colony-stimulating factor in healthy volunteers. Curr Opin Hematol 16:35–40
32. Ohsaka A, Saionji K, Endo K et al (1995) Alterations of effector cell molecule expression on neutrophils in granulocyte colony-stimulating factor-producing tumour. Br J Haematol 91:571–574
33. Ohsaka A, Saionji K, Kuwaki T, Takeshima T, Igari J (1995) Granulocyte colony-stimulating factor administration modulates the surface expression of effector cell molecules on human monocytes. Br J Haematol 89:465–472
34. Gabrilove JL, Jakubowski A, Fain K et al (1998) Phase I study of granulocyte colony-stimulating factor in patients with transitional cell carcinoma of the urothelium. J Clin Invest 82:1454–1461
35. Roskos LK, Ren S, Robbie G (2010) Application of modeling and simulation in the development of protein drugs. In: Kimko HHC, Peck CC (eds) Clinical trial simulations. Springer Science Businessz Media, LLC, New York, NY, pp 361–396
36. Lote C (2000) The loop of Henle, distal tubule and collecting duct. In: Lote C (ed) Principles of renal physiology, 4th edn. Kluwer Academic Publishers, Dordrecht
37. Kotto-Kome AC, Fox SE, Lu W, Yang BB, Christensen RD, Calhoun DA (2004) Evidence that the granulocyte colony-stimulating factor (G-CSF) receptor plays a role in the pharmacokinetics of G-CSF and PegG-CSF using a G-CSF-R KO model. Pharmacol Res 50:55–58
38. Lau D, Pilz D, Schwab G (1996) Phase I pharmacokinetic and pharmacodynamic studies of G-CSF (filgrastim) in patients with renal or liver impairment compared to healthy volunteers. Br J Haematol 93:277
39. Galluppi GR, Rogge MC, Roskos LK et al (2001) Integration of pharmacokinetic and pharmacodynamic studies in the discovery, development, and review of protein therapeutic agents: a conference report. Clin Pharmacol Ther 69:387–399
40. Wang B, Ludden TM, Cheung EN, Schwab GG, Roskos LK (2001) Population pharmacokinetic-pharmacodynamic modeling of filgrastim (r-metHuG-CSF) in healthy volunteers. J Pharmacokinet Pharmacodyn 28:321–342
41. Borleffs JC, Bosschaert M, Vrehen HM et al (1998) Effect of escalating doses of recombinant human granulocyte colony-stimulating factor (filgrastim) on circulating neutrophils in healthy subjects. Clin Ther 20:722–736
42. Stute N, Furman WL, Schell M, Evans WE (1995) Pharmacokinetics of recombinant human granulocyte-macrophage colony-stimulating factor in children after intravenous and subcutaneous administration. J Pharm Sci 84:824–828
43. Sturgill MG, Huhn RD, Drachtman RA, Ettinger AG, Ettinger LJ (1997) Pharmacokinetics of intravenous recombinant human granulocyte colony-stimulating factor (rhG-CSF) in children receiving myelosuppressive cancer chemotherapy: clearance increases in relation to absolute neutrophil count with repeated dosing. Am J Hematol 54:124–130
44. Kearns CM, Wang WC, Stute N, Ihle JN, Evans WE (1993) Disposition of recombinant human granulocyte colony-stimulating factor in children with severe chronic neutropenia. J Pediatr 123:471–479
45. Roskos LK, Lum P, Lockbaum P, Schwab G, Yang BB (2006) Pharmacokinetic/pharmacodynamic modeling of pegfilgrastim in healthy subjects. J Clin Pharmacol 46:747–757
46. Yang BB, Kido A, Salfi M, Swan S, Sullivan JT (2008) Pharmacokinetics and pharmacodynamics of pegfilgrastim in subjects with various degrees of renal function. J Clin Pharmacol 48:1025–1031
47. Holmes FA, Jones SE, O'Shaughnessy J et al (2002) Comparable efficacy and safety profiles of once-per-cycle pegfilgrastim and daily injection filgrastim in chemotherapy-induced neutropenia: a multicenter dose-finding study in women with breast cancer. Ann Oncol 13:903–909
48. Holmes FA, O'Shaughnessy JA, Vukelja S et al (2002) Blinded, randomized, multicenter study to evaluate single administration pegfilgrastim once per cycle versus daily filgrastim as an adjunct to chemotherapy in patients with high-risk stage II or stage III/IV breast cancer. J Clin Oncol 20:727–731

49. Spunt SL, Irving H, Frost J et al (2010) Phase II, randomized, open-label study of pegfilgrastim-supported VDC/IE chemotherapy in pediatric sarcoma patients. J Clin Oncol 28:1329–1336
50. Gibiansky L, Gibiansky E, Kakkar T, Ma P (2008) Approximations of the target-mediated drug disposition model and identifiability of model parameters. J Pharmacokinet Pharmacodyn 35:573–591

Structural Biology of G-CSF and Its Receptor

Tara L. Arvedson and Mike J. Giffin

1 Introduction

Two structures of the helical cytokine granulocyte colony-stimulating factor (G-CSF) and portions of the G-CSF receptor (G-CSFR) have been reported, and each suggests a different mechanism of interaction. Interpretation of the biologic relevance of the two structures has been facilitated by comparison of the structures with the structures of other helical cytokine family members, their receptors, and modes of binding. Although little sequence identity is shared within the helical cytokine family or within the helical cytokine receptor family, helical cytokines are structurally similar, as are helical cytokine receptors. Correspondingly, although the different signaling complexes contain a varying number of cytokine and receptor chains, they are each reported to use similar modes of binding to initiate a signaling response. By considering the two G-CSF/G-CSFR structures within the context of the greater helical cytokine/cytokine receptor family, a model of the G-CSF/G-CSFR signaling complex can be proposed.

2 G-CSF Cloning and Sequence Similarity with Other G-CSF Orthologs

The amino acid sequence of human G-CSF was determined by Souza et al. using protein purified from a human bladder carcinoma cell line [1]. The gene for human G-CSF encodes a 204-amino acid protein, which includes both a signal sequence and a 174-amino acid mature protein. To date, several G-CSF orthologs have been

T.L. Arvedson (✉) • M.J. Giffin
Amgen Inc, Hematology Research, One Amgen Center Drive, M/S 15-2-A, Thousand Oaks, CA, 91320, USA
e-mail: taraa@amgen.com; giffin@amgen.com

G. Molineux et al. (eds.), *Twenty Years of G-CSF*, Milestones in Drug Therapy,
DOI 10.1007/978-3-0348-0218-5_5,

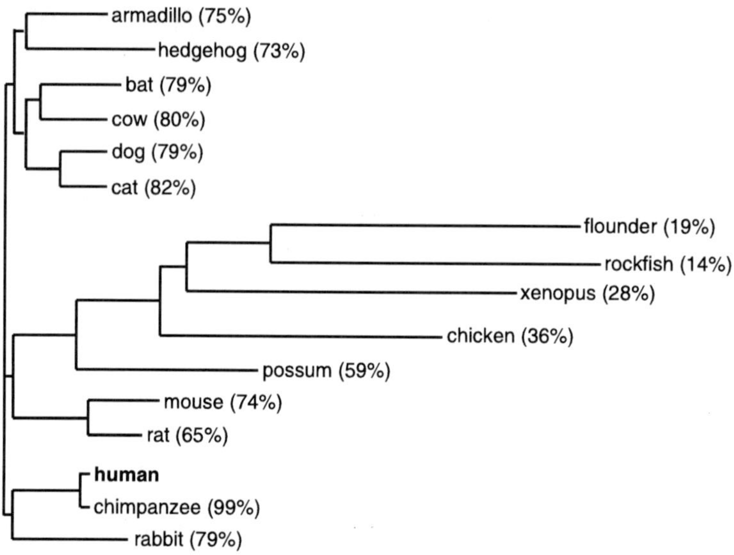

Fig. 1 Dendrogram illustrating the inferred evolutionary relationship between G-CSF orthologs. The number in parentheses is the percent identity shared with human G-CSF. Evaluation done using Vector NTI (Vector NTI Advance, Version 11 [software]; December 15, 2008. Invitrogen Corporation)

identified in mammals, birds, and fish [2] (Fig. 1). A high degree of sequence identity is shared between human G-CSF and the mammalian sequences (range: 99% identity with chimpanzee G-CSF to 59% identity with possum G-CSF), and a lower degree of sequence identity with chicken (36% identity) and fish (approximately 15% identity) G-CSF. In comparison to the degree of sequence identity shared between human G-CSF and other mammalian orthologs, very little sequence identity is shared between human G-CSF and other human helical cytokines. The closest relative is the p19 subunit of interleukin (IL)-23 (approximately 26% identity) (Fig. 2).

3 Three-Dimensional Structure of Helical Cytokines

The central structural feature shared by all helical cytokines is four alpha-helices (labeled A–D; Fig. 3) arranged in an up-up-down-down configuration (i.e., helices A and B both extend up, while helices C and D both extend down). Because helices A and B both extend in the same direction, the peptide linking them must span the length of the protein. This peptide region is called the "crossover region" as it either crosses over or behind helix D. The localization of the peptide is important as it is one of the features that enables classification of the helical cytokine family into either the short-chain or long-chain subfamilies. Additional characteristics defining

	G-CSF	IL-23	IL-6	GH	Prolactin	OSM	CNTF	Leptin	IL-12	LIF
G-CSF		25	19	17	16	20	19	17	20	15
IL-23	41		22	18	18	18	16	19	19	21
IL-6	38	34		15	17	16	14	17	16	17
GH	33	32	32		24	16	16	17	19	19
Prolactin	33	30	36	49		17	13	17	18	17
OSM	29	30	30	32	33		18	16	15	18
CNTF	34	35	33	31	30	33		15	14	18
Leptin	32	32	32	29	32	29	29		15	16
IL-12	38	33	35	34	37	30	33	28		17
LIF	32	40	33	34	31	29	33	34	31	

Fig. 2 Percent identity (*above gray boxes*) and similarity (*below gray boxes*) shared between human G-CSF and a panel of long-chain cytokines. Values generated using MatGAT [47]. *CNTF* ciliary neurotrophic factor; *GH* growth hormone; *IL* interleukin; *LIF* leukemia inhibitory factor; *OSM* oncostatin M

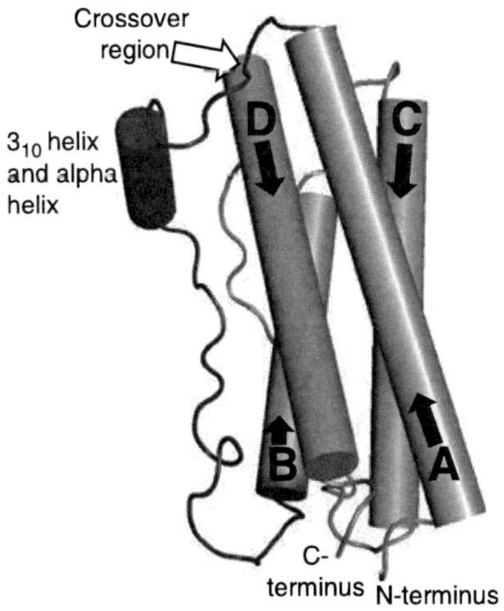

Fig. 3 Cartoon representation illustrating the helical bundle topology of G-CSF. G-CSF contains four alpha-helices shown as cylinders (labeled A–D). These helices are arranged in an up-up-down-down orientation. Both helices A and B extend from bottom to top, while helices C and D extend from *top* to *bottom*. The directionality of the helices is shown with an *arrow*. The peptide sequence between helices A and B is the crossover region (shown in *dark gray* and highlighted with an *arrow*). In G-CSF, this region passes in front of helix D and contains a 4-residue 3_{10}-helix and a 6-residue alpha-helix, shown here as an additional cylinder in the loop between helices A and B. Figure generated using Pymol (Pymol Molecular Graphics System, Version 1.3 [software]. Schrodinger, LLC) and G-CSF structure (PDB ID 1RHG [5])

	Long chain cytokine subfamily (Representative: G-CSF) 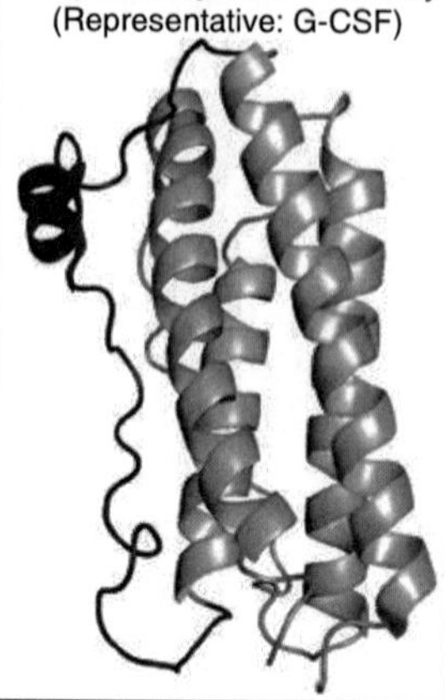	Short chain cytokine subfamily (Representative: GM-CSF) 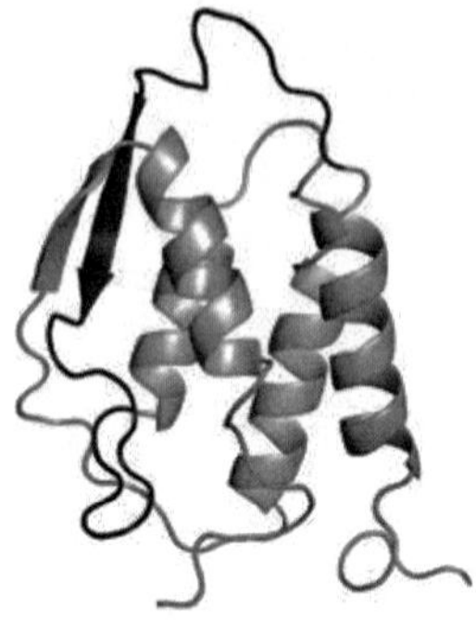
Crossover region	In front of helix D	Behind helix D
Helix packing	Tight	Loose
Range of amino acids/helix	20-31	10-16
Length of protein (amino acid)	204	144

Fig. 4 Schematic illustrating two representative helical cytokine structures, GM-CSF from the short-chain subfamily (PDB ID 1CSG [29]) and G-CSF from the long-chain subfamily (PDB ID 1RHG [5]). The differences between the subfamilies include the orientatioh of the crossover region, the packing and length of alpha-helices A–D, and the total length of the protein. *G-CSF* granulocyte colony-stimulating factor; *GM-CSF* granulocyte–macrophage colony-stimulating factor; *PDB* protein database

membership in one subfamily or the other are based on the total length of the protein, the average length of the alpha-helices A–D, and the packing of these alpha-helices (Fig. 4).

Examples of the short-chain subfamily include granulocyte–macrophage colony-stimulating factor (GM-CSF), stem cell factor (SCF), and several IL family members (e.g., IL-2, -4, -5, -13, -15, and -21). The average protein length is about 150 amino acids, although there are at least two members, thrombopoietin (TPO) and macrophage colony-stimulating factor (M-CSF) that contain >350 amino acids. The helices are generally 10–20 residues long and they are typically loosely packed against one another. The crossover region passes behind helix D and many of the crossover regions contain one strand of a β-sheet. Table 1 provides a list describing each of these characteristics for the short-chain cytokines for which structures have been determined.

Examples of the long-chain subfamily include G-CSF, growth hormone (GH), leptin, prolactin (PRL), IL-6, IL-12, ciliary neurotrophic factor (CNTF), leukemia inhibitory factor (LIF), and oncostatin M (OSM). The average protein length is approximately 260 amino acids and the helices are generally 20–30 residues long. The helices are typically well aligned with one another, and the crossover region

Table 1 Members and attributes of the short-chain subfamily for which structures have been determined

Protein	Protein length (no. of amino acids)	Length of helices (A–D)	Crossover region location relative to helix D	Secondary structure within crossover region	PDB (references)
GM-CSF	144	A: 16 B: 10 C: 16 D: 12	Behind	Alpha-helix, β-strand	1CSG [29]
IL-2	153	A: 23 B: 11 C: 16 D: 16	Behind	Alpha-helix	1M47 [30]
IL-4	153	A: 13 B: 18 C: 25 D: 19	Behind	β strand, alpha-helix	2INT [31]
IL-5	134	A: 14 B: 12 C: 20 D: 17	Behind	β-strands, alpha-helices	1HUL [32]
IL-13	146	A: 14 B: 5 C: 7 D: 18	Behind	Alpha-helix	3L5X [33]
IL-15	162	A: 16 B: 18 C: 15 D: 15	Behind	β-strand	2Z3Q [34]
IL-21	155	A: 18 B: 4 C: 11 D: 20	Behind	Alpha-helix	2OQP (NMR) [35]
M-CSF	554	A: 10 B: 9 C: 10 D: 13	Behind	β strand, alpha helix	1HMC [36]
Flt3 ligand	235	A: 9 B: 17 C: 14 D: 15	Behind	β-strand	1ETE [37]
SCF	273	A: 9 B: 16 C: 21 D: 16	Behind	β-strand, alpha-helix	1SCF [38]
EPO	193	A: 17 B: 27 C: 21 D: 13	Behind	β-strand, alpha-helix	1EER [39]
TPO	353	A:13 B: 22 C: 17 D: 10	Behind	β-strand, alpha-helix	1V7N [40]

EPO erythropoietin; *GM-CSF* granulocyte–macrophage colony-stimulating factor; *IL* interleukin; *M-CSF* macrophage colony-stimulating factor; *NMR* nuclear magnetic resonance; *PDB* protein database; *SCF* stem cell factor; *TPO* thrombopoietin passes in front of helix D. It is common for the crossover region to contain one or two short helices. Table 2 provides a list describing each of these characteristics for the long-chain cytokines for which structures have been determined, and Fig. 5 shows several structures from the long-chain cytokine family, illustrating the common fold and shared structure among members.

Erythropoietin (EPO) and TPO have features that make them difficult to classify as either short- or long-chain cytokines, as both are longer than the average

Table 2 Members and attributes of the long-chain subfamily for which structures have been determined

Protein	Protein length (no. of amino acids)	Length of helices (A–D)	Crossover region location relative to helix D	Secondary structure within crossover region	PDB (references)
G-CSF	204	A: 31 B: 20 C: 24 D: 29	In front	3_{10} and alpha-helix	1RHG [5]
GH	217	A: 28 B: 15 C: 21 D: 29	In front	Alpha-helix	1HGU [13]
Leptin	167	A: 20 B: 16 C: 23 D: 19	In front	None	1AX8 [41]
PRL	227	A: 31 B: 14 C: 28 D: 33	In front	β-strands, alpha-helices	1RW5 [42]
CNTF	200	A: 29 B: 27 C: 26 D: 29	In front	None	1CNT [43]
LIF	202	A: 27 B: 29 C: 27 D: 23	In front	Alpha-helix	1LKI [44]
IL-6	212	A: 27 B: 24 C: 21 D: 26	In front	Alpha-helix	1ALU [45]
OSM	252	A: 15 B: 24 C: 27 D: 27	In front	Alpha-helices	1EVS [46]

CNTF ciliary neurotrophic factor; *G-CSF* granulocyte colony-stimulating factor; *GH* growth hormone; *IL* interleukin; *LIF* leukemia inhibitory factor; *OSM* oncostatin M; *PDB* protein database; *PRL* prolactin

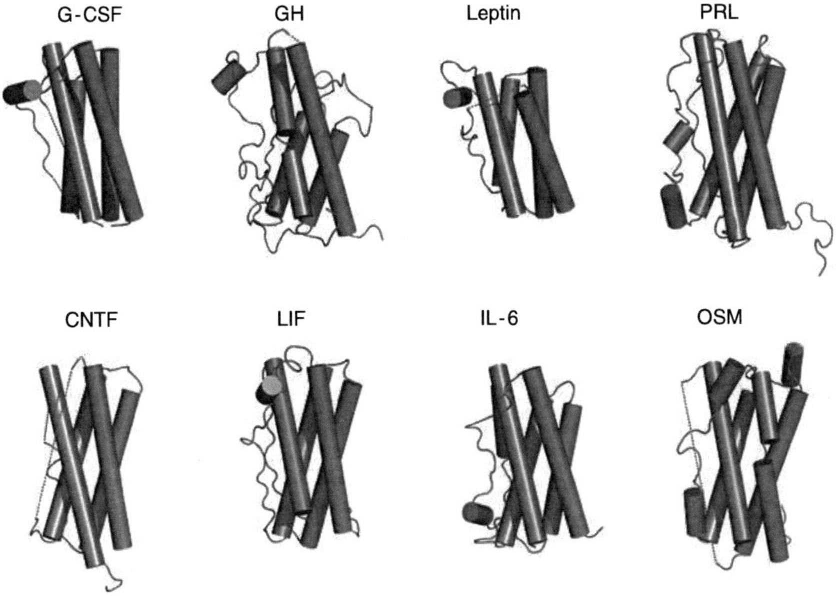

Fig. 5 Schematic illustrating the structures of several long-chain cytokines for which the structures have been determined. Figure generated using Pymol (Pymol Molecular Graphics System, Version 1.3 [software]. Schrodinger, LLC) and G-CSF structure (PDB ID 1RHG [5]), GH (PDB ID 1HGU [48]), leptin (PDB ID 1AX8 [41]), PRL (PDB ID [42]), CNTF (PDB ID 1CNT [43]), LIF (PDB ID 1LKI [44]), IL-6 (PDB ID 1ALU [14]), and OSM (PDB ID 1EVS [46]). Missing regions of the models are represented by dashed lines. *CNTF* ciliary neurotrophic factor; *G-CSF* granulocyte colony-stimulating factor; *GH* growth hormone; *IL* interleukin; *LIF* leukemia inhibitory factor; *OSM* oncostatin M; *PDB* protein database; *PRL* prolactin

short chain cytokine (EPO: 193 amino acids, TPO: 353 amino acids) and have longer helices (up to 27 or 22 amino acids, respectively), yet their topology matches that of the short-chain cytokines as their crossover regions pass behind helix D and each contain β strands. While β strands in the crossover region are not unique to the short-chain cytokines, they are more common than in the long-chain cytokines (see Tables 1 and 2). In light of the crossover region topology, both EPO and TPO are grouped with the short-chain cytokines in Table 1.

4 Structure of G-CSF

The structure of recombinant G-CSF has been determined by both nuclear magnetic resonance (NMR) and crystallography [3–5]. The structure contains four alpha helices: helix A consists of 29 amino acids (residues 11–39), helix B consists of 21 amino acids (residues 71–91), helix C consists of 24 amino acids (residues 100–123), and helix D consists of 30 amino acids (residues 143–172).

The crossover region passes in front of helix D and contains a 4-residue 3_{10}-helix (residues 44–47) followed by a 6-residue alpha-helix (residues 48–53). Both 3_{10}- and alpha-helices are right-handed helices; the difference between the two is that the 3_{10}-helix has three residues per turn, whereas the alpha-helix has 4 (3.6) residues per turn. Both the length of the protein and the structural features of G-CSF place it within the long-chain cytokine subfamily.

G-CSF has five cysteine residues. Four of these cysteines form disulfide bonds (Cys36–Cys42 and Cys64–Cys74). The remaining cysteine at position 17 is free, but is only partially accessible to solvent, so it is unlikely to be able to react with other proteins [6]. No *N*-linked glycosylation sites are in G-CSF; however, there is an *O*-linked glycan on residue threonine 133 in material expressed in mammalian cells [7, 8]. Glycosylation is not required for biologic activity as demonstrated by filgrastim, which is expressed in bacterial cells and is not glycosylated but retains potent biologic activity [9].

5 Structural Similarity Between Long-Chain Cytokine Receptors

Helical cytokine receptors, like helical cytokines, lack significant sequence identity. In addition, they vary in terms of length, domain number, and domain organization. Despite these differences, the receptor structures and the means of ligand binding have been reported to be conserved. In terms of domain usage, all helical cytokine receptors comprise mixtures of fibronectin type III (FN III) and immunoglobulin (Ig) domains. These domains are similar as each contains seven β-strands (A–G) oriented in two sheets forming a β-sandwich. For the FN III domains, one sheet comprises strands A, B, and E, while the other sheet comprises strands C, D, F, and G (Fig. 6a). For the Ig domain, strand D is switched from one sheet to the other, such that 1 β-sheet comprises strands A, B, D, and E, while the other β-sheet comprises strands C, F, and G (Fig. 6b). A common feature shared by all cytokine

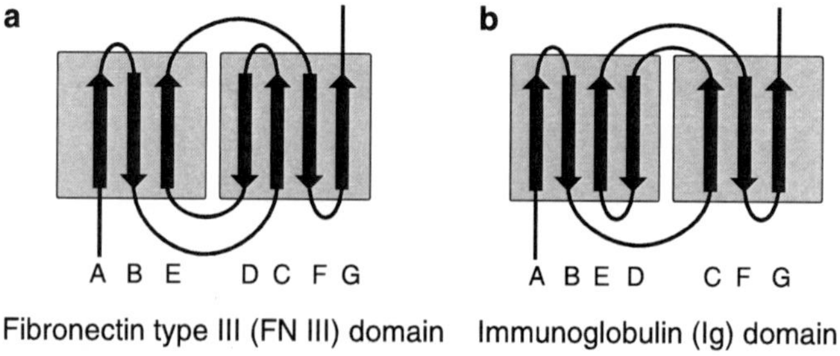

Fig. 6 β-strand topology map illustrating the folding for (**a**) fibronectin type III (FN III) and (**b**) immunoglobulin (Ig) domains. Both domains contain 7 β-strands that form a β-sandwich. Each *arrow* depicts an individual strand and the *gray* shading indicates strands that cluster together on each face of the β-sandwich

receptors is the cytokine-binding homology region (CHR). This module contains approximately 200 amino acids and is reported to contain several conserved elements, including the domain structure, the location of 6 conserved cysteine residues, and the presence of a WSXWS (Trp-Ser-X-Trp-Ser) sequence motif (Fig. 7a, b) [10]. In terms of the domain structure, the CHR consists of two Fibronectin type III (FN III) domain connected by a proline-rich linker. Because of the cyclic nature of the proline side chain, proline linkers are relatively rigid and contribute to the two domains being held at a fixed angle to one another. Analysis of the known structures suggests that this angle varies from 70 to 110° (defined as the angle between the first residue of the *N*-terminal FN III domain and the last residue of the *C*-terminal FN III domain). Another conserved element within the CHR is positionally conserved cysteines. Within the CHR *N*-terminal FN III domain are four positionally conserved cysteine residues that stabilize the β-sandwich structure by forming two disulfide bonds between the faces of the sandwich. Within the CHR *C*-terminal domain, two positionally conserved cysteine residues form one disulfide bond. An additional feature of cytokine receptors is the conserved WSXWS motif, where X can be any amino acid (Fig. 7a, b). Although there are few exceptions to

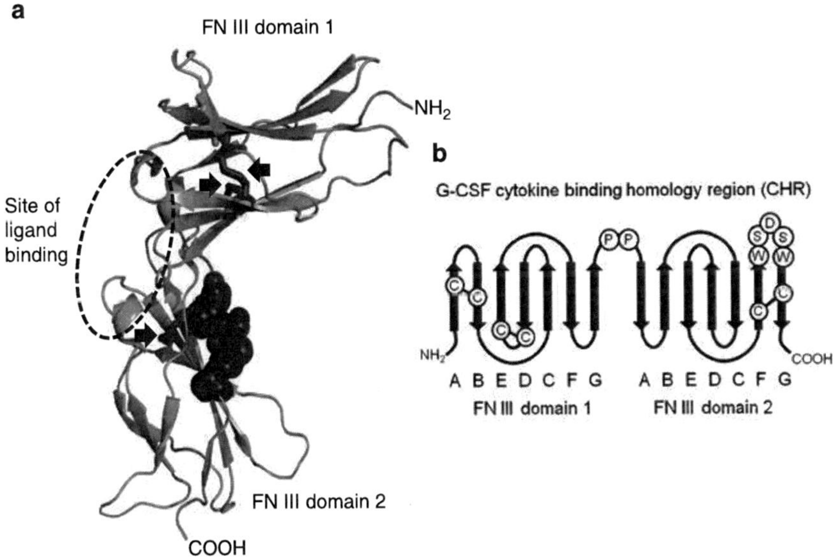

Fig. 7 Conserved features reported for the cytokine-binding homology region (CHR). CHR comprises two fibronectin type III (FN III) domains. Conserved features are reported to include three positionally conserved cysteines forming disulfide bonds, a WSXWS motif, and a proline-rich linker. (**a**) The structure of the G-CSF receptor (G-CSFR) (PDB ID 2D9Q [23]) CHR is used to illustrate the disulfide bonds formed by positionally conserved cysteines (*black arrows*), the WSXWS motif (*space-filling spheres*), and the site of ligand binding in the elbow region of the CHR (*dotted circled area*). Structure generated using Pymol (Pymol Molecular Graphics System, Version 1.3 [software]. Schrodinger, LLC). (**b**) The conserved features are illustrated on a β-strand topology map of the G-CSFR CHR

this sequence (e.g., YGEFS in growth hormone receptor and WQPWS in the IL-23 receptor), most receptors contain the WSXWS amino acid sequence. While there are six different DNA codons that encode serine, most receptor genes use only two different ones (AGC and ACT) to encode this motif [11]. The function of the WSXWS motif is not known, although it is thought to contribute to efficient receptor folding. This conclusion is supported by work done by Hilton et al. in which each residue in the EPO receptor (EPOR) WSXWS sequence was individually mutated to every other amino acid [11]. In all, a series of 100 different EPOR point mutants were generated. The mutated protein was then expressed in mammalian cells and tested for cell surface expression and the ability to bind EPO. While the mutant proteins appeared to be comparably expressed, most were retained in the endoplasmic reticulum, particularly if they contained mutations in the tryptophan or serine residues. Of the mutants that were able to reach the cell surface, all retained the ability to bind EPO. From this work, the authors concluded that the WSXWS sequence was not required for ligand binding but did appear to be required for protein folding.

Within the Cytokine-binding homology region (CHR), the residues involved in ligand binding are reported to be located in the "elbow" region between the two Fibronectin type III (FN III) domain (Fig. 7a). This region comprises the loops linking the β-strands to one another. Because the loops are structurally flexible, significant sequence diversity can be tolerated within them without affecting the overall structure of the domain. As a result, the same structural scaffold is maintained by all cytokine receptors, but sequence variation within the loops enables receptors to have different ligand specificities. The structures of several long-chain cytokine receptors have been determined and are shown schematically in Fig. 8.

6 Receptor Signaling and Stoichiometry of Cytokine/Receptor Complexes

Receptor signaling is reported to be induced by cytokine-mediated receptor oligomerization and juxtaposition of associated intracellular kinases [12]. Unlike growth factor receptors that contain integrated kinase domains, cytokine receptors do not have intrinsic kinase activity and must associate with intracellular kinases (e.g., members of the JAK or TYK family). Receptor oligomerization is reported to bring these intracellular kinases into an appropriate proximity to induce phosphorylation of themselves and the associated cytokine receptor chain. This initial phosphorylation event is suggested to trigger recruitment and phosphorylation of a secondary group of proteins (e.g., STAT proteins) that are then able to translocate to the nucleus to modify gene expression.

The first crystal structure to be determined between a cytokine and its receptor was GH with GH receptor (GHR) [13]. This complex is reported to comprise one ligand and two receptors (Fig. 9a). The published structure suggests that GH is able

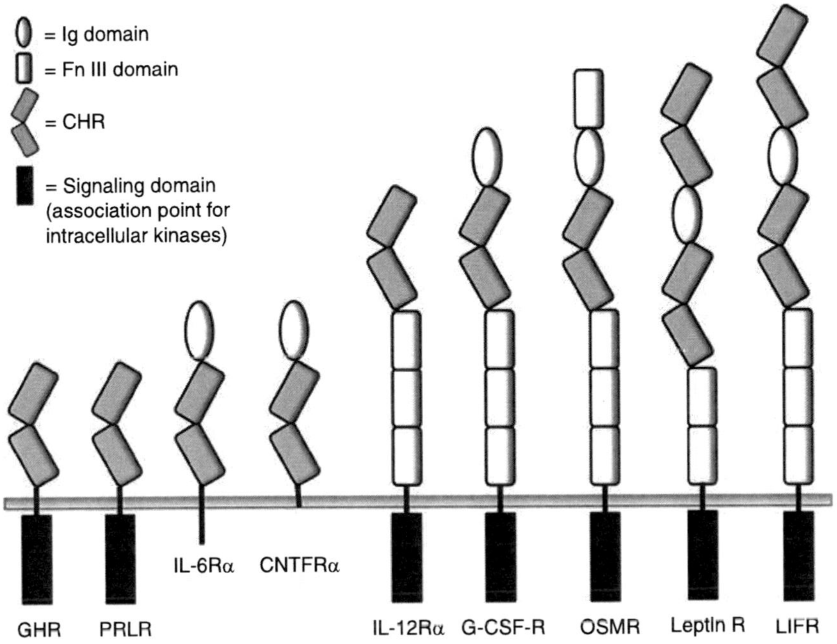

Fig. 8 Schematic illustrating the domain organization of receptors for several long-chain cytokines. While the receptors for growth hormone (GH) and prolactin (PRL) contain only cytokine homology regions (CHR), the remaining receptors contain additional Fibronectin type III (FN III) domain or immunoglobulin (Ig) domains. Receptors for both leptin and leukemia inhibitory factor (LIF) contain two CHR. Many of the receptors contain intracellular domains that are capable of associating with intracellular kinases; exceptions are the alpha receptors for both interleukin (IL)-6 (IL-6Rα) and ciliary neutrophic factor (CNTF) (CNTFRα). IL-6Rα contains a short cytoplasmic domain that cannot associate with intracellular kinases, and CNTFRα is linked to the membrane through a glycophosphatidylinositol (gpi) anchor

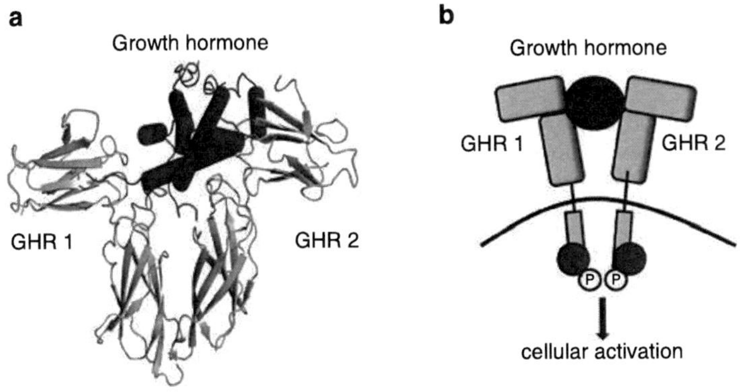

Fig. 9 (**a**) Structure reported for the growth hormone (GH) with its receptor (GHR). A single growth hormone molecule associates with two GHR. Figure generated using Pymol (Pymol Molecular Graphics System, Version 1.3 [software]. Schrodinger, LLC) and PDB ID 3HHR [13]. (**b**) Schematic illustrating the signaling complex formed by GH with GHR

to use two disparate surfaces to interact with similar regions on each of the two associated receptors. A similar 1:2 (ligand:receptor) stoichiometry has been observed between PRL and the PRL receptor (PRLR) [14] and has been predicted to occur with leptin and the leptin receptor [15]. In terms of initiating a signaling response, the intracellular domain of GH, PRL, and leptin receptors are reported to be able to associate with intracellular kinases and to propagate a signaling cascade upon dimerization [12] (Fig. 9b). It is not the case that all receptors within the long-chain cytokine family associate with intracellular kinases and can initiate a signaling response upon dimerization. For example, the IL-6 receptor alpha (IL-6Rα) chain and the CNTF receptor alpha (CNTFRα) chain lack the ability to bind intracellular kinases and cannot propagate a signal independently. To make up for the lack of signaling capability, additional receptors that are signal competent are reported to be recruited to the ligand/ligand receptor complex. One receptor that is frequently reported to be recruited is gp130.

The extracellular domain of gp130 contains six domains: D1 is an Cytokine-binding homology region (CHR) and D2–D6 are Fibronectin type III (FN III) domain. Domains 2 and 3 form the Cytokine-binding homology region (CHR). The intracellular domain of gp130 can associate with intracellular kinases and can, therefore, initiate a signaling response when associated with another signal-competent receptor (Fig. 10). While gp130 has little affinity for cytokines alone, it can form high-affinity interactions with complexes containing both a ligand and an accompanying ligand-specific receptor [16]. Long-chain cytokines that form complexes with gp130 include LIF, OSM, CNTF, and IL-6. The heteromeric complexes vary in the number and type of individual components required to form a cytokine-specific, signal-competent complex (i.e., a complex must contain at least two receptors capable of associating with intracellular kinases). The LIF signaling complex is reported to be a trimer containing one LIF molecule, one cytokine-specific receptor (LIFR), and one gp130 molecule (Fig. 11a) [17, 18].

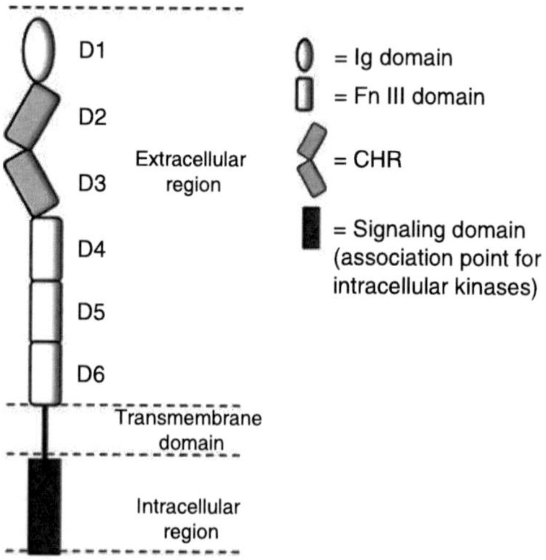

Fig. 10 The domains of gp130. The extracellular region contains six domains: D1 is an Ig domain and D2–D6 are fibronectin (FN) III domains. Domains 2 and 3 form the cytokine-binding homology region (CHR). The intracellular domain of gp130 can associate with intracellular kinases and can, therefore, initiate a signaling response when associated with another signal-competent receptor

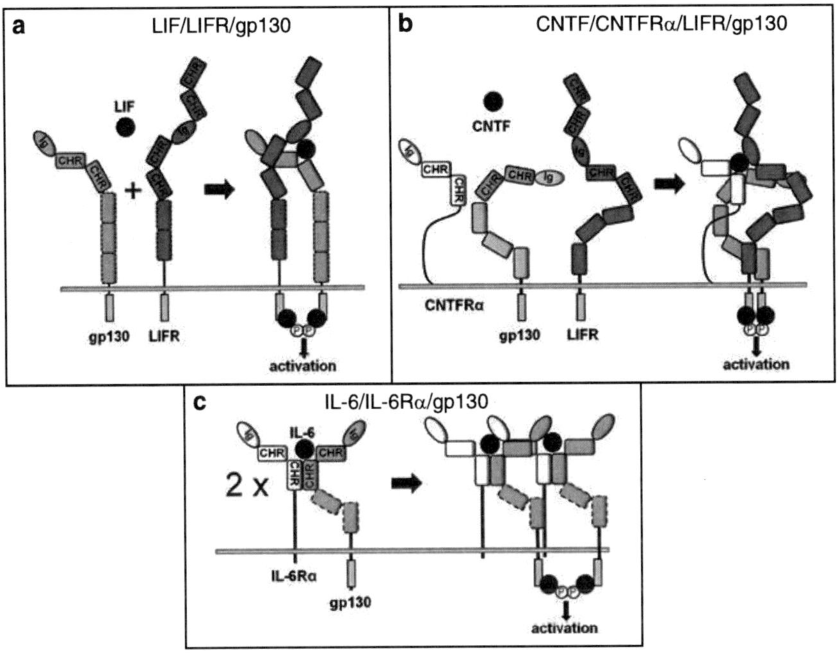

Fig. 11 Signaling complexes reported to be formed by leukemia inhibitory factor (LIF) [17, 19], ciliary neutrophic factor (CNTF) [19], and interleukin (IL)-6 [20]. Initiation of a signaling response requires dimerization of two receptors that can associate with intracellular kinases

Within this complex, both the LIFR and gp130 can associate with intracellular kinases and are capable of generating a signaling response upon dimerization. The CNTF signaling complex is reported to be a tetramer containing one molecule, one cytokine-specific receptor (CNTFRα), one auxiliary receptor (LIFR), and one gp130 molecule (Fig. 11b) [19]. Because the CNTFRα cannot associate with intracellular kinases, additional signal-competent receptors are required. This requirement is met by the LIFR and gp130. The IL-6 signaling complex is reported to be a hexamer containing two IL-6 molecules, two cytokine-specific receptors (IL-6Rα), and two gp130 molecules (Fig. 11c) [20]. In this complex, only the gp130 molecules associate with intracellular kinases. To initiate a signaling cascade, two gp130 molecules are brought together by the association of two IL-6:IL-6Rα:gp130 complexes.

7 Conserved Mechanisms of Cytokine/Receptor Binding

Analysis of cytokine/receptor structures suggests that although the various cytokines interact with different types and numbers of receptors, they do so using similar surfaces. On the cytokine, only three different regions (sites 1, 2, and 3) appear to be used for receptor binding (Fig. 12). Published reports suggest that site 1

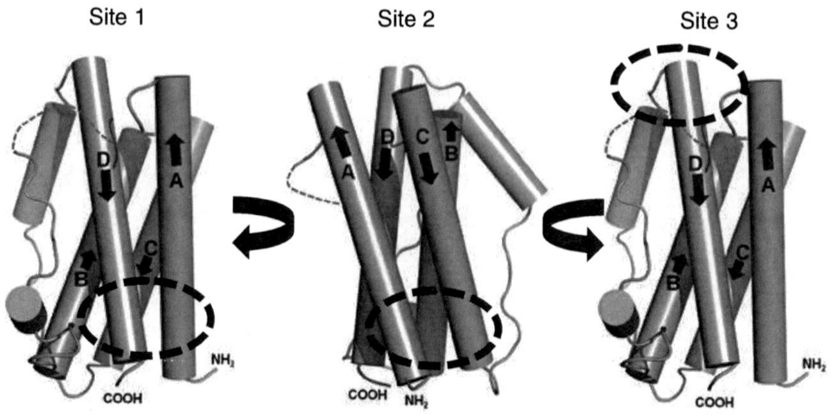

Fig. 12 Receptor-binding sites are reported to be conserved on the surface of cytokines. Cytokines interact with receptors using only three possible surfaces: site 1, site 2, and site 3. The localization of these sites is conserved. Site 1 localizes to helices A and D, site 2 localizes to helices A and C, and site 3 is located in the *N*-terminal region of helix D with contributions from residues in the crossover region (A–B loop). Sites 1 and 2 bind receptor cytokine-binding homology region (CHR) and site 3 binds receptor immunoglobulin (Ig) domain

incorporates residues from helices A and D, site 2 incorporates residues from helices A and C, and site 3 is located in the *N*-terminal region of helix D with contributions from residues in the crossover region (A–B loop). In all structures reported to date, sites 1 and 2 bind receptor Cytokine-binding homology region (CHR) and site 3 binds receptor Immunoglobulin (Ig) domain. While these are the possible interaction sites that may be used, they are not all used in every structure. In the GH/GHR and PRL/PRLR complexes, only sites 1 and 2 appear to be used (Fig. 13a). In the LIF/LIFR/gp130 structure, only sites 2 and 3 appear to be used: LIF binds gp130 using site 2 and the LIFR using site 3 (Fig. 13b). In the IL-6/IL-6Rα/gp130 complex, all three sites appear to be used: IL-6 site 1 contacts the IL-6Rα, site 2 contacts gp130, and site 3 contacts another gp130 molecule (Fig. 13c). All three sites also appear to be used in the CNTF/CNTFRα/LIFR/gp130 structure, with the CNTF site 1 contacting CNTFRα, site 2 contacting gp130, and site 3 contacting the LIFR [19]. Given the conservation of contact points between a cytokine and a receptor within the helical cytokine family, it would be expected that G-CSF would use a similar surface to contact G-CSFR.

8 Cloning and Domain Structure of G-CSF Receptor

The G-CSFR was cloned in 1990 by Fukunaga et al. [21]. It was found to be an 836-amino acid type I receptor (type I = the amino terminus is extracellular) with a 604-amino acid extracellular domain, a 23-amino acid transmembrane domain, and

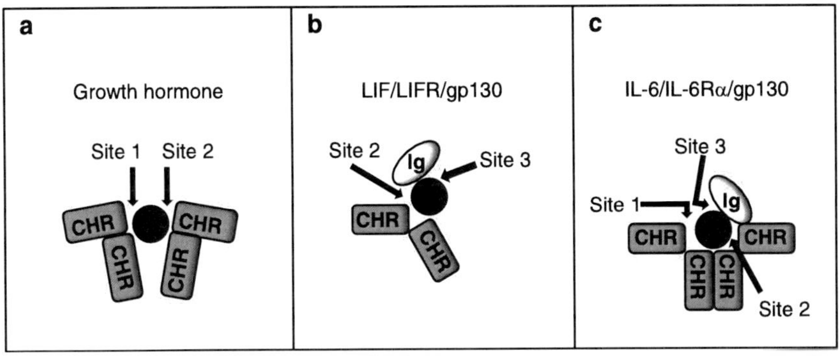

Fig. 13 The cytokine/cytokine receptor-binding sites are reported to be conserved. Despite many different cytokine/cytokine receptor combinations, there are only three observed sites of interaction. Sites 1 and 2 occur between the cytokine and receptor cytokine-binding homology region (CHR), and site 3 occurs between the cytokine and the receptor immunoglobulin (Ig) domain. While three sites of interaction are possible, not all are always used. For example, sites 1 and 2 are used by growth factor and its receptor (GH/GHR) [13]; sites 2 and 3 are used by leukemia inhibitory factor and its receptor (LIF/LIFR)/gp130 [17, 19], and all three sites are used by interleukin (IL)-6/IL-6Rα/gp130 [20]

a 183-amino acid intracellular domain. The extracellular domain structure of G-CSFR is identical to that of gp130 in that it contains six domains, D1 is an Ig domain and domains D2–D6 are FN III like.

9 Structural Characterization of G-CSF/G-CSFR Interaction

Two crystallography structures between G-CSF and portions of G-CSFR have been reported. While both structures contain a 2:2 G-CSF:G-CSFR complex and have two sites of interaction, they differ in the location of those sites of interaction.

The first structure, by Aritomi et al., included human G-CSF and the CHR (domains D2 and D3) from mouse G-CSFR [22]. In this structure, the primary contact region is reported to correspond to the canonical site 2, with G-CSF interacting with the G-CSFR CHR (Fig. 14). At this contact site, there appears to be good shape complementarity resulting in a large buried surface area (1,672 $Å^2$) and extensive contacts between G-CSF and the elbow region of the receptor CHR (10 polar interactions between 6 residues on G-CSF and 7 residues on G-CSFR, and 28 van der Waals contacts between 15 residues on G-CSF and G-CSFR, respectively). The second contact site appears to comprise residues in the amino terminus of G-CSF and D3 of the CHR. This binding site does not appear to correspond to any of the canonical sites (1, 2, or 3) seen in other cytokine/receptor systems. This interaction contains fewer polar and van der Waals interactions, making it the weaker of the two observed binding sites. The biologic relevance of the second interaction site is also called into question by mutagenesis studies that reported that

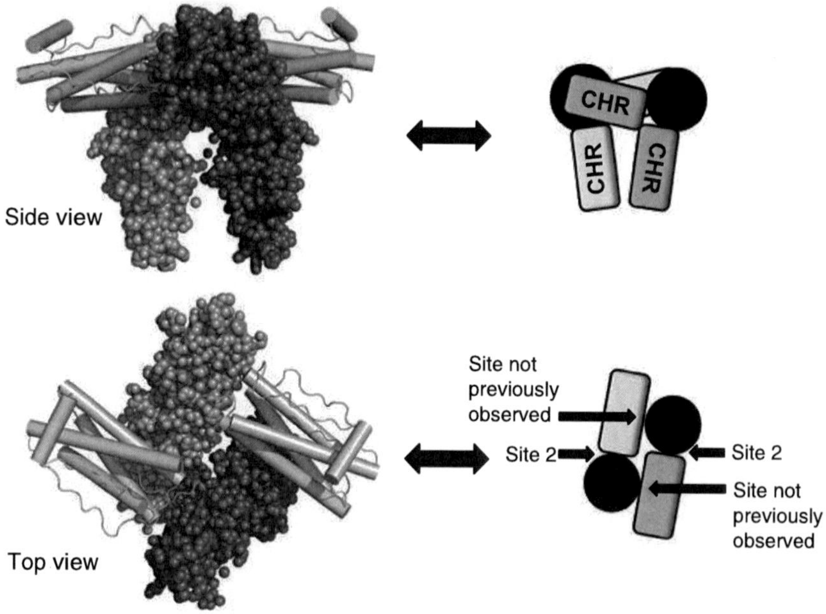

Fig. 14 Structure reported of mouse G-CSF and human G-CSFR cytokine-binding homology region (CHR) as determined by Aritomi et al. [22]. The 2 G-CSF molecules are shown using a wire backbone, the 2 G-CSFR molecules are shown using *spheres*. Schematics of the structures are shown to the *right*. The binding sites observed in the structure are indicated with *arrows*. While one binding site corresponds to the canonical site 2, the other binding site has not been previously observed in any other cytokine/cytokine receptor interaction. Figure generated using Pymol (Pymol Molecular Graphics System, Version 1.3 [software]. Schrodinger, LLC) and PDB ID 1CD9

the amino terminus of G-CSF could be deleted without affecting receptor binding. It is, therefore, hypothesized that this second site of interaction is an artifact of crystallizing a fragment of the protein.

The second structure, by Tamada et al., included human G-CSF and the Immunoglobulin (Ig) domain (D1) and CHR (D2 and D3) from human G-CSFR (Fig. 15) [23]. In this tetrameric structure, each G-CSF ligand appears to bind two different receptor molecules and each receptor molecule appears to bind two different G-CSF ligands. On the G-CSF molecule, sites 2 and 3 appear to be used. The receptor regions recognized by sites 2 and 3 are reported to be in agreement with what is seen with other cytokines (i.e., site 2 binding the CHR of 1 G-CSFR and site 3 binding the Ig domain of another G-CSFR). Both the site 2 and 3 interactions appear to contribute similarly to the strength of the binding event, with site 2 contributing 1,375 Å^2 of buried surface area and site 3 contributing 1,144 Å^2. The site 2 interaction appears to be similar in both the Tamada et al. and Aritomi et al. structures. When viewed from the side, the Tamada et al. structure resembles a table with two legs. The top of the table is formed by the two G-CSF molecules, the Ig domain (D1) and the first Fibronectin type III (FN III) domain (D2). The legs of the table are formed by the G-CSFR D3 domains.

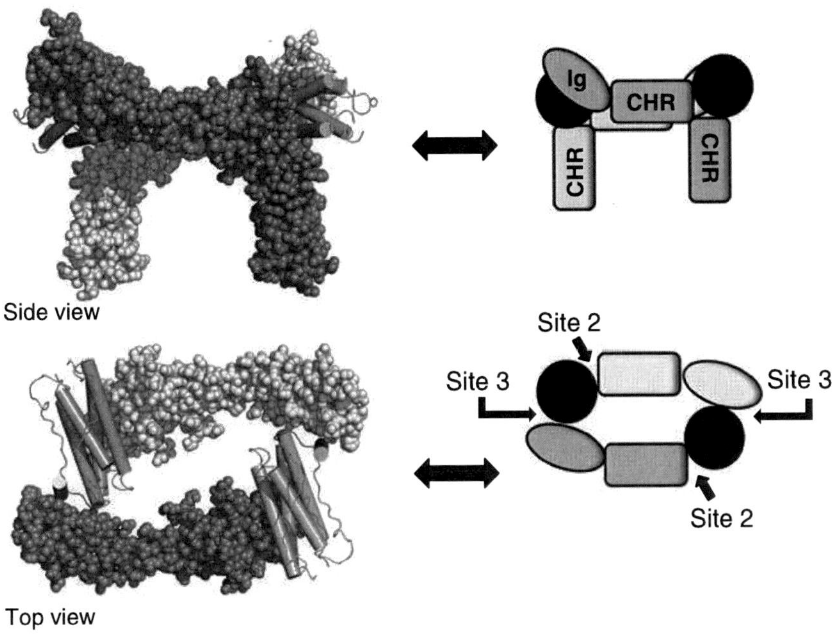

Fig. 15 Structure reported of human Granulocyte colony stimulating factor (G-CSF), human G-CSFR Immunoglobulin (Ig) domain, and cytokine-binding homology region (CHR) as determined by Tamada et al. [23]. The 2 G-CSF molecules are shown using a wire backbone, the 2 G-CSFR molecules are shown using *spheres*. Schematics of the structures are shown to the *right*. The binding sites observed in the structure are indicated with *arrows*. Figure generated using Pymol (Pymol Molecular Graphics System, Version 1.3 [software]. Schrodinger, LLC) and PDB ID 2D9Q

The Tamada et al. structure is supported by its similarity to the viral IL-6/gp130 structure (Fig. 16). Viral IL-6 (vIL-6) is an IL-6 homolog from Kaposi's sarcoma-associated herpesvirus (HHV8). Unlike IL-6 that requires an IL-6-specific receptor (IL-6Rα) to enable IL-6 binding to gp130, vIL-6 can directly bind and activate gp130 without an α-receptor [24]. Biochemical studies reported that vIL-6 formed a 2:2 complex with gp130, and this stoichiometry was supported by the crystal structure of vIL-6 with gp130 [25, 26]. Comparison of the vIL-6/gp130 and G-CSF/G-CSFR structures suggests that both vIL-6 and G-CSF contact their respective receptors through sites 2 and 3. The two receptors assume comparable conformations, and the overall size and symmetry of the two complexes are reported to be similar.

10 Modeling the G-CSF/G-CSFR Signaling Complex

Given the similarities between gp130 and G-CSFR, it has been suggested that it may be possible to use the full-length gp130 structure to model a G-CSF/G-CSFR signaling complex. The structures of several gp130-containing complexes have

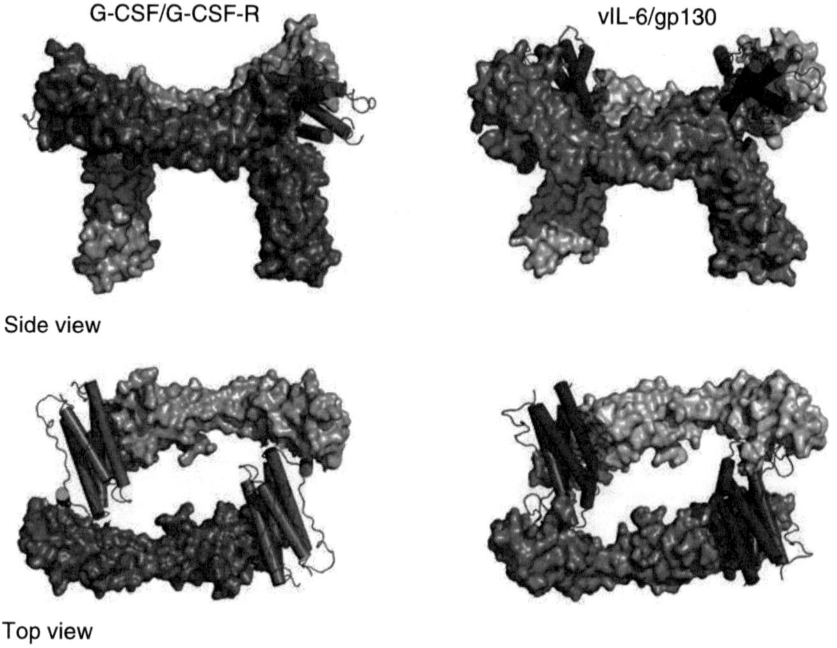

Fig. 16 Structural comparison of G-CSF/G-CSFR [23] and viral interleukin (vIL)-6/gp130 [26]. Both the G-CSF and vIL-6 ligands are shown using a wire backbone, and the 2 G-CSFR molecules and gp130 molecules are shown using space-filling spheres. Figure generated using Pymol (Pymol Molecular Graphics System, Version 1.3 [software]. Schrodinger, LLC) and PDB ID 2D9Q (G-CSF/G-CSF-R) and PDB ID 1L1R (vIL-6/gp130)

been characterized by low-resolution electron microscopy (EM) [19, 27], and each of these structures suggests that gp130 assumes a C-shaped structure that is proposed to bring the *C*-terminal, Jak-associated regions within sufficient proximity to initiate a signaling response. Jak molecules are large, containing approximately 1,100 residues across four globular domains, and estimates from EM structures suggest that a separation of 30–65 Å between the *C*-terminal domains would be sufficiently close to enable transactivation of two Jak molecules [28]. The crystal structure of full-length, unliganded gp130 was reported and it suggested that the C-shaped structure (Fig. 17) [28] was due in part to a sharp bend introduced by domains D4 and D5. These domains are held at an angle of about 80° relative to one another through the interaction between Trp-311, Lys-330, and Trp-448 with residues on the interfacing domain. All the key residues responsible for this interaction are conserved in G-CSFR D4 and D5, suggesting that G-CSFR will be oriented similarly. In terms of aligning gp130 with G-CSFR, the orientation of G-CSFR domains D1, D2, and D3 from the Tamada et al. structure resembles the orientation of gp130 domains D1, D2, and D3 observed in other structures. The conformational similarity enables the construction of a model of G-CSF with

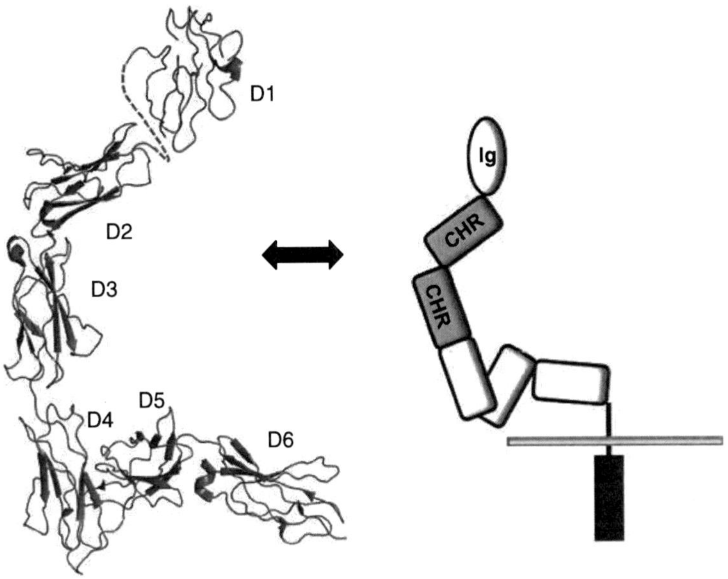

Fig. 17 Reported structure of gp130. Structure of gp130 determined by X-ray crystallography [28] and an illustration of the domain orientation. Figure generated using Pymol (Pymol Molecular Graphics System, Version 1.3 [software]. Schrodinger, LLC) and PDB ID 3L5H

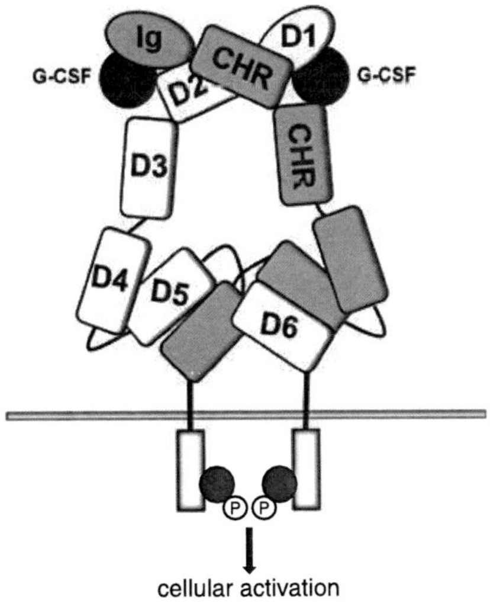

Fig. 18 Suggested model of the G-CSF/G-CSFR signaling complex. The orientation of G-CSF and domains D1–D3 of G-CSFR is based on the structure determined by Tamada et al. [23]. The orientation of G-CSFR domains D4–D6 based on the gp130 structure [28]

G-CSFR where there is a crossover between domains D5 and D6, resulting in a conformation where the end of the extracellular domains would be positioned such that the intracellular signaling region is sufficiently close to initiate a signaling response (Fig. 18).

11 Summary

G-CSF and its receptor appear to be structurally similar to reports of other ligand/receptor pairs within the helical cytokine/receptor family. Although there have been two co-crystal structures determined of G-CSF and G-CSF-R, only one appears to be structurally consistent with the other cytokine/receptor structures. Analysis of both structures within the context of the rest of the family suggests that only one structure is biologically relevant. By further using the similarities between G-CSFR and the common receptor gp130, a model has been proposed of the full-length signaling complex.

References

1. Souza L, Boone TC, Gabrilove J et al (1986) Recombinant human granulocyte colony-stimulating factor: effects on normal and leukemic myeloid cells. Science 232:61–65
2. Huising MO, Kruiswijk CP, Flik G (2006) Phylogeny and evolution of class-I helical cytokines. J Endocrinol 189:1–25
3. Zink T, Ross A, Lüers K, Cieslar C, Rudolph R, Holak TA (1994) Structure and dynamics of the human granulocyte colony-stimulating factor determined by NMR spectroscopy. Loop mobility in a four-helix-bundle protein. Biochemistry 33:8453–8463
4. Werner JM, Breeze AL, Kara B et al (1994) Secondary structure and backbone dynamics of human granulocyte colony-stimulating factor in solution. Biochemistry 33:7184–7192
5. Hill CP, Osslund TD, Eisenberg D (1993) The structure of granulocyte-colony-stimulating factor and its relationship to other growth factors. Proc Natl Acad Sci U S A 90:5167–5171
6. Arakawa T, Prestrelski SJ, Narhi LO, Boone TC, Kenney WC (1993) Cysteine 17 of recombinant human granulocyte-colony stimulating factor is partially solvent-exposed. J Protein Chem 12:525–531
7. Zsebo KM, Cohen AM, Murdock DC et al (1986) Recombinant human granulocyte colony stimulating factor: molecular and biological characterization. Immunobiology 172:175–184
8. Kubota N, Orita T, Hattori K, Oh-eda M, Ochi N, Yamazaki T (1990) Structural characterization of natural and recombinant human granulocyte colony-stimulating factors. J Biochem 107:486–492
9. Dale DC, Bonilla MA, Davis MW et al (1993) A randomized controlled phase III trial of recombinant human granulocyte colony-stimulating factor (filgrastim) for treatment of severe chronic neutropenia. Blood 81:2496–2502
10. Bazan JF (1990) Structural design and molecular evolution of a cytokine receptor superfamily. Proc Natl Acad Sci U S A 87:6934–6938
11. Hilton DJ, Watowich SS, Katz L, Lodish HF (1996) Saturation mutagenesis of the WSXWS motif of the erythropoietin receptor. J Biol Chem 27:4699–4708
12. Touw IP, De Koning JP, Ward AC, Hermans MH (2000) Signaling mechanisms of cytokine receptors and their perturbances in disease. Mol Cell Endocrinol 160:1–9
13. de Vos A, Ultsch M, Kossiakoff A (1992) Human growth hormone and extracellular domain of its receptor: crystal structure of the complex. Science 255:306–312
14. Somers W, Ultsch M, De Vos AM, Kossiakoff AA (1994) The X-ray structure of a growth hormone–prolactin receptor complex. Nature 372:478–481
15. Hiroike T, Higo J, Jingami H, Toh H (2000) Homology modeling of human leptin/leptin receptor complex. Biochem Biophys Res Comm 275:154–158
16. Bravo J, Heath JK (2000) Receptor recognition by gp130 cytokines. EMBO J 19:2399–2411

17. Huyton T, Zhang JG, Luo CS et al (2007) An unusual cytokine:Ig-domain interaction revealed in the crystal structure of leukemia inhibitory factor (LIF) in complex with the LIF receptor. Proc Natl Acad Sci U S A 104:12737–12742
18. Boulanger MJ, Bankovitch AJ, Kortemme T, Baker D, Garcia KC (2003) Convergent mechanisms for recognition of divergent cytokines by the shared signaling receptor gp130. Mol Cell 12:577–589
19. Skiniotis G, Lupardus PJ, Martick M, Walz T, Garcia KC (2008) Structural organization of a full-length gp130/LIF-R cytokine receptor transmembrane complex. Mol Cell 31:737–748
20. Boulanger MJ, Chow DC, Brevnova EE, Garcia KC (2003) Hexameric structure and assembly of the interleukin-6/IL-6 α-receptor/gp130 complex. Science 300:2101–2104
21. Fukunaga R, Ishizaka-Ikeda E, Pan CX, Seto Y, Nagata S (1991) Functional domains of the granulocyte colony-stimulating factor receptor. EMBO J 10:2855–65
22. Aritomi M, Kunishima N, Okamoto T, Kuroki R, Ota Y, Morikawa K (1999) Atomic structure of the GCSF-receptor complex showing a new cytokine receptor recognition scheme. Nature 401:713–717
23. Tamada T, Honjo E, Maeda Y et al (2006) Homodimeric cross-over structure of the human granulocyte colony-stimulating factor (GCSF) receptor signaling complex. Proc Natl Acad Sci U S A 103:3135–3140
24. Hoischen SH, Vollmer P, Marz P et al (2000) Human herpes virus 8 interleukin-6 homologue triggers gp130 on neuronal and hematopoietic cells. Eur J Biochem 267:3604–3612
25. Li H, Nicholas J (2002) Identification of amino acid residues of gp130 signal transducer and gp80 alpha receptor subunit that are involved in ligand binding and signaling by human herpesvirus 8-encoded interleukin-6. J Virol 76:5627–36
26. Chow DC, He X, Snow AL, Rose-John S, Garcia KC (2001) Structure of an extracellular gp130 cytokine receptor signaling complex. Science 291:2150–2155
27. Matadeen R, Hon WC, Heath JK, Jones EY, Fuller S (2007) The dynamics of signal triggering in a gp130-receptor complex. Structure 15:441–448
28. Xu Y, Kershaw NJ, Luo CS et al (2010) Crystal structure of the entire ectodomain of gp130. J Biol Chem 285:21214–21218
29. Walter MR, Cook WJ, Ealick SE, Nagabhushan TL, Trotta PP, Bugg CE (1992) Three-dimensional structure of recombinant human granulocyte-macrophage colony-stimulating factor. J Mol Biol 224:1075–1085
30. Arkin MR, Randal M, DeLano WL et al (2003) Binding of small molecules to an adaptive protein-protein interface. Proc Natl Acad Sci U S A 100:1603–1608
31. Walter MR, Cook WJ, Zhao BG et al (1992) Crystal structure of recombinant human interleukin-4. J Biol Chem 267:20371–20376
32. Milburn MV, Hassell AM, Lamber MH et al (1993) A novel dimer configuration revealed by the crystal structure at 2.4 A resolution of human interleukin-5. Nature 363:172–176
33. Fransson J, Teplyakov A, Raghunathan G et al (2010) Human framework adaptation of a mouse anti-human IL-13 antibody. J Mol Biol 398:214–231
34. Chirifu M, Hayashi C, Nakamura T et al (2007) Crystal structure of the IL-15-IL-15Rα complex, a cytokine-receptor unit presented in trans. Nat Immunol 8:1001–1007
35. Bondensgaard K, Breinholt J, Madsen D et al (2007) The existence of multiple conformers of interleukin-21 directs engineering of a superpotent analogue. J Biol Chem 282:23326–23336
36. Pandit J, Bohm A, Jancarik J, Hatenbeck R, Koths K, Kim SH (1992) Three-dimensional structure of dimeric human recombinant macrophage colony-stimulating factor. Science 258:1358–1362
37. Savvides SN, Boone T, Karplus PA (2000) Flt3 ligand structure and unexpected commonalities of helical bundles and cystine knots. Nat Struct Mol Biol 7:486–491
38. Jiang X, Gurel O, Mendiaz EA et al (2000) Structure of the active core of human stem cell factor and analysis of binding to its receptor Kit. EMBO J 19:3192–3203
39. Syed RS, Reid SW, Li C et al (1998) Efficiency of signalling through cytokine receptors depends critically on receptor orientation. Nature 395:511–516

40. Feese MD, Tamada T, Kato Y et al (2004) Structure of the receptor-binding domain of human thrombopoietin determined by complexation with a neutralizing antibody fragment. Proc Natl Acad Sci U S A 101:1816–1821
41. Zhang F, Basinski MB, Beals JM et al (1997) Crystal structure of the obese protein Ieptin-E100. Nature 387:206–209
42. Teilum K, Hoch JC, Goffin V, Kinet S, Martial JA, Kragelund BB (2005) Solution structure of human prolactin. J Mol Biol 351:810–823
43. McDonald N, Panayotatos N, Hendrickson WA (1995) Crystal structure of dimeric human ciliary neurotrophic factor determined by MAD phasing. EMBO J 14:2689–2699
44. Robinson RC, Grey LM, Stauton D et al (1994) The crystal structure and biological function of leukemia inhibitory factor: Implications for receptor binding. Cell 77:1101–1116
45. Somers W, Stahl M, Seehra JS (1997) 1.9 A crystal structure of interleukin 6: implications for a novel mode of receptor dimerization and signaling. EMBO J 16:989–997
46. Deller MC, Hudson KR, Ikemizu S, Bravo J, Jones EY, Heath JK (2000) Crystal structure and functional dissection of the cytostatic cytokine oncostatin M. Structure 8:863–874
47. Campanella JJ, Bitincka L, Smalley J (2003) MatGAT: an application that generates similarity/identity matrices using protein or DNA sequences. BMC Bioinformatics 4:29
48. Chantalat L, Jones N, Korber F, Navaza J, Pavlovsky AG (1995) The crystal-structure of wild-type growth-hormone at 2.5 angstrom resolution. Protein Pept Lett 2:333–340

G-CSF Receptor Structure, Function, and Intracellular Signal Transduction

Hoainam T. Nguyen-Jackson, Huiyuan Zhang, and Stephanie S. Watowich

1 Characterization of the G-CSF Receptor

1.1 G-CSF Receptor Molecular Structure

The G-CSF receptor (G-CSFR), like other members of the type-I cytokine receptor family, is a single transmembrane-spanning protein lacking intrinsic kinase activity, and is composed of an extracellular cytokine receptor homologous (CRH) domain containing four conserved cysteine residues and a Trp-Ser-X-Trp-Ser (WSXWS, where X is a nonconserved amino acid) motif, and shared elements in the intracellular region denoted as Box 1 and Box 2 (Fig. 1). This receptor also comprises an immunoglobulin (Ig)-like domain and 3 fibronectin type III (FNIII)-like domains in the extracellular portion, as well as a cytoplasmic region containing a conserved sequence termed Box 3 [1, 2]. By forming intramolecular disulfide bonds, the extracellular cysteine residues are important for maintaining the 3-dimensional structural integrity of the CRH domain, along with the WSXWS region; moreover, these motifs together with the Ig-like domain mediate intermolecular interactions with the ligand G-CSF, as revealed by X-ray crystallography studies of the ligand:receptor complex [3].

G-CSFR is similar to the receptors for erythropoietin (EPO) and growth hormone, as it forms homodimers upon binding of a ligand; however, the G-CSFR

H.T. Nguyen-Jackson
Department of Immunology, Unit 902, MD Anderson Cancer Center, PO Box 301402, Houston, Texas 77030, USA

The Graduate School of Biomedical Sciences, University of Texas, Houston, TX, USA

H. Zhang • S.S. Watowich (✉)
Department of Immunology, Unit 902, MD Anderson Cancer Center, PO Box 301402, Houston, Texas 77030, USA
e-mail: swatowic@mdanderson.org

G. Molineux et al. (eds.), *Twenty Years of G-CSF*, Milestones in Drug Therapy,
DOI 10.1007/978-3-0348-0218-5_6,

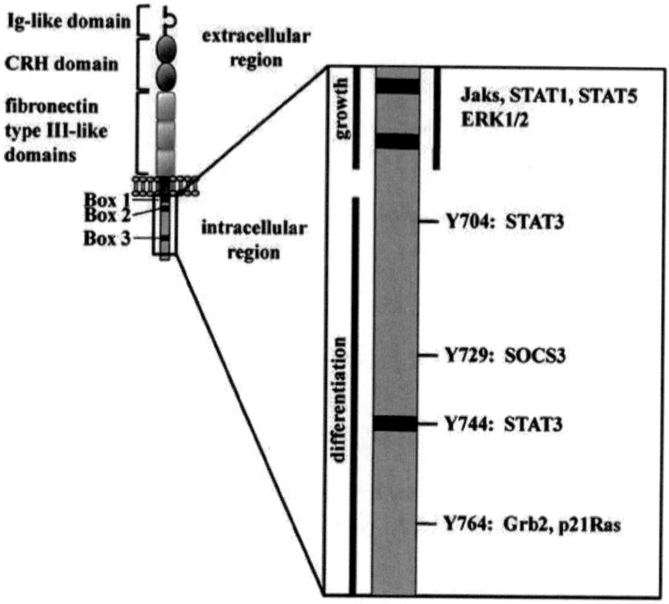

Fig. 1 G-CSFR structure and intracellular signaling pathways. The extracellular region of G-CSFR contains an immunoglobulin-like (Ig-like) domain, a cytokine receptor homologous (CRH) domain with four conserved cysteine residues and a conserved WSXWS motif, and three fibronectin type III (FNIII)-like domains. The intracellular region, which mediates downstream signal transduction, contains conserved Box 1, Box 2, and Box 3 motifs, and four tyrosine (Y) residues. The membrane-proximal region of the G-CSFR cytoplasmic domain has been linked to activation of Jaks, STAT1, STAT5, and ERK1/2. Y704 and Y744 recruit STAT3, Y729 is a docking site for SOCS3, and Y764 associates with Grb2, and is reported to activate p21Ras, as indicated

structure more closely resembles gp130, a single transmembrane protein that heterodimerizes with ligand-specific alpha subunits, and acts as the signaling chain for the receptors for interleukin-6 (IL-6), leukemia inhibitory factor (LIF), and oncostatin M (OSM) [1]. G-CSFR and gp130 share 46.3% amino acid homology, and, like G-CSFR, gp130 contains an Ig-like domain, FNIII-like domains, and Box 3 [4]. The G-CSFR molecular weight can range from 100,000 to 150,000 kDa due to posttranslational modification by glycosylation [2, 5]. Ligand binding occurs with high affinity ($K_d \approx 700$ pM), initiating and/or stabilizing homodimerization of the G-CSFR through interactions between the Ig-like domain of the first receptor subunit and the CRH domain of the second receptor subunit, resulting in a 2:2 G-CSF:G-CSFR crossover complex [1, 6] (Fig. 2). The 3-dimensional structure of the G-CSFR cytoplasmic region, like other type-I cytokine receptors, remains unresolved, although mutational analyses have indicated important roles in mediating interactions with signal transduction molecules.

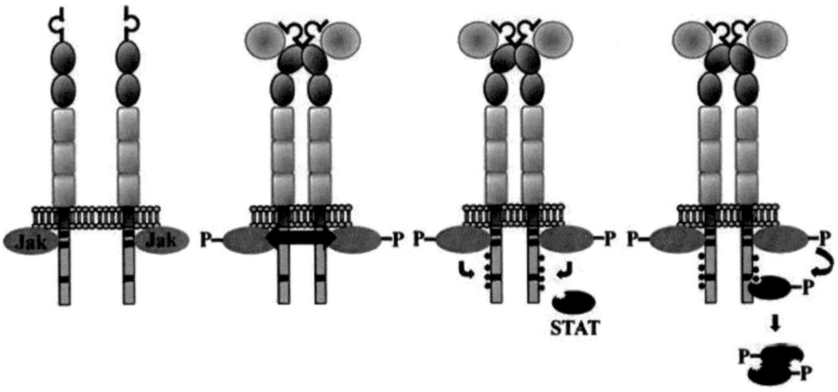

Fig. 2 Activation of Jak-STAT signal transduction. From *left* to *right*: Jaks are constitutively associated with G-CSFR in the absence of ligand. Ligand binding occurs at a 2:2 ligand:receptor subunit stoichiometry, resulting in a crossover configuration of the receptor subunits, as indicated. The resulting proximity of the Jak kinases enables their trans-phosphorylation and activation. The Jaks phosphorylate intracellular tyrosine residues of G-CSFR. STAT proteins interact with the phosphotyrosines on the G-CSFR through their SH2 domains, and become phosphorylated by the Jak proteins. STAT dimers are formed, accumulate in the nucleus, and regulate transcription of cytokine-responsive genes

1.2 Expression and Function of the G-CSFR

Expression of G-CSFR mRNA and/or protein has been detected in hematopoietic and nonhematopoietic cells. We delineate cell types in which protein versus mRNA expression has been established, an important issue considering that sensitive methods to detect mRNA expression do not always correlate with protein production or functional responses. Hematopoietic stem cells (HSC), common myeloid progenitors, granulocyte/macrophage progenitors, and cells of the granulocytic lineage are the key cell types to express this receptor, with G-CSFR protein and mRNA expression increasing during granulocytic maturation [7–9]. G-CSFR signaling chiefly mediates the proliferation, differentiation, survival, and mobilization of neutrophils and granulocytic progenitor cells, as demonstrated by impairments in these functions in G-CSFR-deficient mice (i.e., *Csf3r*$^{-/-}$) [10]. Surprisingly, while both G-CSFR- and G-CSF-deficient mice exhibit severe neutropenia, residual neutrophil production is detected in these animals, indicating that G-CSF-independent mechanisms play a role in the generation of neutrophils under homeostatic conditions and/or serve a compensatory function in directing granulopoiesis in the absence of G-CSF signaling [10, 11].

Additionally, G-CSFR protein expression has been detected in other hematopoietic cells, such as monocytes, which appear to respond to G-CSF by attenuating their secretion of proinflammatory cytokines [8, 12]; lymphocytes, with constitutive G-CSFR expression in B cells and inducible expression in T cells [13]; and platelets, in which pre-incubation with recombinant G-CSF boosts aggregation responses stimulated by adenosine diphosphate [14]). Administration of G-CSF affects the expression of stromal cell-derived factor (SDF)-1 by osteoblasts within the bone marrow stromal microenvironment, a process involved in hematopoietic cell retention and release from the marrow [15]; however, it is not known which stromal cell type responds to G-CSF in this pathway, and accordingly, G-CSFR expression on hematopoietic progenitor cells is dispensable for G-CSF-responsive mobilization [15, 16]. Multiple reports have demonstrated that leukemic cells express G-CSFR protein, including those of acute myeloid leukemia (AML) [17] and B-cell chronic lymphocytic leukemia (B-CLL) [18]. Initially, G-CSF treatment was considered as a method to induce leukemic cell differentiation; however, oncogenic mutations causing differentiation arrest preclude this approach. Currently, G-CSF is proposed as an agent to stimulate cell cycle entry of leukemic stem cells, enhancing their susceptibility to cytotoxic therapy (e.g., chemotherapeutics) [19].

Even nonhematopoietic cells have been demonstrated to express G-CSFR mRNA, with abundant amounts found in placental and fetal tissues [1, 20, 21]. Examples of nonhematopoietic cells expressing G-CSFR protein are cells of the central nervous system, where G-CSFR signaling was reported to promote neurogenesis [22]; cardiomyocytes, which respond to G-CSF by enhanced survival [23]; and endothelial cells, in which G-CSF can induce proliferation and migration [24]. Although there have been reports of G-CSFR signaling affecting multiple cell types, the profound defects in the granulocytic lineage in G-CSFR-deficient mice indicate that G-CSFR functions predominantly in myeloid cells.

1.3 Transcriptional Regulation of the G-CSFR Gene

The genes encoding human and murine G-CSFR, denoted as *CSF3R* and *Csf3r*, respectively, exhibit sequence homology and are regulated by comparable transcriptional mechanisms. Both consist of 17 exons and share 72% nucleotide sequence homology. Their gene products are of similar length, as *CSF3R* encodes 813 amino acids and *Csf3r* encodes 812 amino acids. Comparison of their amino acid sequences reveals 62.5% homology, indicating a high degree of conservation in structural and likely functional attributes of the receptor [21]. While *CSF3R* is located on chromosome 1, its murine counterpart is found on chromosome 4; the similarity between these two chromosomal regions further indicates evolutionary conservation of the murine and human genes [2, 25]. Expression of *CSF3R* is controlled by the myeloid transcription factors C/EBPα and PU.1 [26]; PU.1 also has a critical role in regulating murine *Csf3r* expression [27]. Furthermore, the upstream promoter regions of the human and murine genes share high sequence homology. For example, located approximately 100 base pairs upstream of the transcriptional start site of the human and murine genes is an 18-nucleotide sequence that is homologous to a region found in the promoters of the human myeloperoxidase and elastase genes, encoding granule components that are critical for neutrophil effector function. These data suggest the involvement of additional myeloid transcription factors in the regulation of *CSF3R* and *Csf3r* expression [25, 26]. Moreover, the similarities between *CSF3R* and *Csf3r* may contribute to the ability of recombinant human G-CSF to be used in mice, and for recombinant mouse G-CSF to stimulate the activation of the human G-CSFR; however, it is important to note that X-ray crystallography studies revealed that the 3-dimensional structure of the human G-CSF/human G-CSFR complex differs significantly from the human G-CSF/mouse G-CSFR complex, although both receptor:ligand complexes are capable of signaling [6, 28].

Five distinct transcripts of human *CSF3R* were uncovered through cDNA library screens; these transcripts exhibit differences in regions encoding the receptor transmembrane and intracellular domains. In one such transcript, the transmembrane-coding sequence is deleted and, therefore, is predicted to encode a soluble receptor [21]. These distinct transcripts have been proposed as alternative splice variants, since *CSF3R* is present in single copy, although the function of their protein products has not been revealed to date. Moreover, spontaneous mutations in the G-CSFR have been identified in human disease, indicating further variation of *CSF3R*-coding regions.

2 Intracellular Signaling Pathways Activated by G-CSFR

2.1 *Activation of Jak-STAT Signal Transduction by G-CSFR*

As with other cytokine receptors, G-CSFR activation requires phosphorylation of Janus protein tyrosine kinases (Jak) that are constitutively associated with the membrane-proximal intracellular region of the receptor (Fig. 2). Ligand binding brings receptor subunits into proximity, resulting in transphosphorylation of associated Jak proteins. G-CSF ligation strongly activates Jak1 and can also activate Jak family members Jak2 and Tyk2 [29–31]. Using Jak-deficient cell lines, it was determined that loss of any single Jak protein did not preclude activation of other Jaks by G-CSF. Interestingly, only the loss of Jak1 led to reduced activation of downstream signaling molecules in Jak-deficient cell lines, suggesting its non-redundancy [31]. Data from Jak1- or Jak2-deficient mice demonstrated that G-CSFR function was not strongly affected in either case, indicating that Jak kinases may have redundant roles in G-CSFR signaling in vivo [32, 33].

To elucidate the function of the intracellular domain of the G-CSFR, deletion mutants of the receptor were generated and overexpressed in cell lines. The membrane-proximal 55–57 amino acids of the cytoplasmic region of G-CSFR were found to be critical for stimulating cellular proliferation, whereas amino acids located between 57 and 96 of the intracellular domain were required for maximal proliferation and expression of candidate growth regulatory genes in cell culture [34–36]. The membrane-proximal intracellular region contains the conserved Box 1 and Box 2 domains, which are necessary for mediating G-CSF-responsive proliferation in cell lines, independent of G-CSFR tyrosine residues [36, 37]. It was later discovered that Box 1 includes a proline-rich sequence that is thought to serve as a docking site for Jak proteins [38] (Fig. 1). These results, as well as the proliferative activity of the receptor in the absence of tyrosine residues, suggest that signals elicited through G-CSFR and associated Jak molecules are sufficient for G-CSF-responsive myeloid cell growth.

By contrast, the membrane distal portion of the G-CSFR intracellular region, containing approximately 100 amino acid residues, mediates growth inhibitory activities in myeloid tissue culture lines and controls genes encoding surface markers and molecules that regulate neutrophil effector functions such as myeloperoxidase and elastase [34–36]. This region is essential for eliciting differentiation signals, which may be linked with growth inhibition [34–36]. Four tyrosine residues are located in the intracellular portion of G-CSFR, corresponding to Y704, Y729, Y744, and Y764 in the human G-CSFR, and Y703, Y728, Y743, and Y763 in the mouse receptor, which are phosphorylated by Jak proteins [39, 40] (Fig. 1). Phosphorylation of G-CSFR tyrosines leads to the recruitment of intracellular effector molecules, through interactions between phosphotyrosine-binding domains of the signaling proteins and phosphorylated tyrosines of the receptor. Canonically, Jak activation stimulates the recruitment of signal transducers and activators of transcription (STAT) proteins (reviewed in [41]). STAT proteins are predominantly localized in the cytoplasm in the absence of cytokine signals; after cytokine binding and receptor/Jak activation, the STAT proteins interact with phosphotyrosines on G-CSFR through the STAT SH2 domain, rendering their physical proximity to the G-CSFR–Jak complex. This close association between STAT proteins and the G-CSFR–Jak complex enables Jak-dependent phosphorylation of a critical *C*-terminal tyrosine residue in STAT proteins, a step that is necessary for STAT release from the receptor and subsequent STAT dimerization (Fig. 2). The STAT SH2 domains are reported to have higher affinity for STAT phosphotyrosine residues versus cytokine receptor phosphotyrosines, providing an elegant mechanism by which STAT proteins dissociate from the receptor and preferentially bind with other tyrosine-phosphorylated STAT family members in proximity. Homo- or heterodimerized STAT proteins accumulate in the nucleus, where they regulate cytokine-responsive transcription.

G-CSFR signaling has been shown to stimulate phosphorylation of STAT1, STAT3, and STAT5 in cell-line systems. Activation of STAT1 and STAT5 can occur through the membrane-proximal region of the cytoplasmic domain, indicating that these pathways may be independent of receptor tyrosines [42]. By contrast, STAT3 is recruited to Y704 and Y744 of G-CSFR, resulting in its robust, rapid, and sustained activation [30, 43] (Fig. 1). G-CSFR tyrosine-dependent STAT3 activation occurs at low doses of G-CSF, while at saturating doses, STAT3 activation can also proceed through tyrosine-independent mechanisms [44], possibly by direct recruitment and activation through G-CSFR-associated Jak proteins. Maximal activation of STAT3 requires residues 96–183 of the G-CSFR intracellular domain; however, STAT3 is still able to bind when G-CSFR is truncated 96 amino acids from the transmembrane domain, which leaves only Y704 intact among the receptor tyrosines [37]. Activation of STAT3 appears to drive neutrophil differentiation in myeloid cell culture systems, as judged by induction of neutrophil marker genes and morphologic changes associated with maturation; however, STAT3 has no effect on expression of myeloperoxidase [45, 46]. Experiments with receptor mutants expressed from the endogenous *Csf3r* locus in vivo (i.e., knock-in mouse models) confirmed that tyrosine-independent activation of STAT1 may occur, and demonstrated that while Y704 can facilitate STAT5 activation, this residue is chiefly involved in stimulating STAT3 and is important for regulating neutrophil progenitor proliferation, differentiation, and mobilization [47]. Collectively, the tissue culture and mouse studies have placed a particular focus on STAT3 as an important mediator of G-CSFR signaling.

2.2 Additional Signal Transduction Pathways Elicited by G-CSFR Activation

Several members of the mitogen-activated protein kinase (MAPK) pathway, which has well-established roles in controlling gene expression, differentiation, and proliferation, are stimulated by G-CSFR signaling; however, their precise mechanism of activation remains unclear. Phosphorylation of ERK1/2 in cell lines requires the membrane-proximal 57 amino acids of the cytoplasmic domain of G-CSFR, correlating with the Jak-binding domain and its role in inducing growth signals, therefore suggesting that the G-CSFR–Jak pathway activates MAPK to regulate cellular proliferation [37] (Fig. 1). G-CSFR has also been shown to stimulate p21Ras, a GTPase that can transduce signals upstream of ERK1/2. Activation of p21Ras has been attributed to Y764 of the G-CSFR (Fig. 1), and p21Ras is thought to affect proliferation by inducing cell cycle progression [48]. The adaptor proteins SHP-2, Grb2, and Shc, which can mediate activation of p21Ras and ERK1/2, are phosphorylated upon ligation of G-CSFR. While Grb2 has been reported to interact directly with G-CSFR Y764 (Fig. 1), SHP-2 can associate with Y704 in vitro [48–50]. Activation of SHP-2 and Shc requires G-CSFR Y764 (Fig. 1), which induces their association with Grb2, and through Grb2 association with the guanine exchange factor Son of Sevenless (SOS) results in stimulation of p21Ras signaling [51]. Because SHP-2, Grb2, and Shc have been previously associated with cellular proliferation and differentiation, they, too, are thought to transduce important signals through G-CSFR to regulate granulopoiesis. Thus, MAPK, particularly ERK1/2, may be activated by multiple mechanisms in the G-CSFR pathway and may participate in regulation of G-CSF-responsive cell growth and differentiation; however, the role for these molecules must be examined using appropriate conditional deletion experiments in primary cells in vivo.

The ligand-activated G-CSFR also appears to interact with Src and Syk tyrosine kinases [52, 53]. The Src-family kinases expressed in myeloid cells, Lyn and Hck, have been shown to be associated with and stimulated by G-CSFR in vitro [52, 53]. The activity of Src-family kinases is mediated by the membrane-proximal 55 amino acids of the G-CSFR cytoplasmic domain in cell lines and is negatively regulated by the *C*-terminal 98 amino acids [54]. While Lyn was reported to stimulate G-CSF-dependent proliferation in tissue culture systems [52], Hck is thought to be involved in eliciting differentiation signals and regulating the function of mature granulocytic cells [53]. Moreover, Src kinases can activate the serine/threonine kinase Akt, which is thought to control cell survival signals independently of Jak [54]. In vivo studies suggest that Src family members may function as negative regulators of granulopoiesis, with Lyn inhibiting the production of myeloid progenitors and Hck suppressing the proliferation of granulocytic precursors [55]. By contrast, activation of Syk may occur by association with Lyn or by direct binding to G-CSFR [52]; however, less is known about the role Syk plays in granulopoiesis in vivo.

2.3 Suppression of G-CSFR Signaling

Cytokine-receptor signaling is negatively regulated by several mechanisms, with a key response mediated through suppressor of cytokine signaling (SOCS) family members [56]. SOCS proteins directly interact with phosphotyrosines of cytokine receptors or Jak through their SH2 domains and thereby serve as competitive inhibitors of receptor-associated signal transduction molecules. Moreover, SOCS proteins are able to bind and inhibit Jak, and suppress kinase activity through the SOCS kinase inhibitory domain. SOCS proteins also target signaling molecules for ubiquitination and subsequent degradation by complexing with E3-ubiquitin ligases through their *C*-terminal SOCS box motif [56]. G-CSFR signaling is suppressed by SOCS3, which interacts directly with phosphorylated Y729 of human G-CSFR or Y728 of murine G-CSFR [57]. SOCS3 can also bind lysine residue 632 (K632) of G-CSFR, influencing receptor ubiquitination and routing to the lysosome, as mutation of this residue to arginine (K632R) blocks trafficking of G-CSFR, causing its accumulation in the early endosome [58]. While deletion of *Socs3* in mice results in embryonic lethality [59], hematopoietic-specific *Socs3* ablation causes G-CSF hyperresponsiveness in myeloid progenitor cells and granulocytes, and even G-CSF toxicity in vivo [60]. Interestingly, myeloid cells isolated from SOCS3-deficient mice demonstrate a modest shift in differentiation potential, favoring the production of macrophages versus neutrophils [60]. This phenotype is similar to the effects of mutation of G-CSFR Y729 to phenylalanine when human G-CSFR was overexpressed in murine bone marrow cells [40], suggesting that G-CSFR-dependent recruitment of SOCS3 may be involved in fine-tuning receptor signals that influence lineage specification events in myelopoiesis. The *Socs* genes are regulated transcriptionally through STATs, with *Socs3* directly activated by STAT3 [61]. The neutrophilia of aged hematopoietic-specific SOCS3-deficient mice is reminiscent of the phenotype of mice with conditional *Stat3* deletion in bone marrow, suggesting that STAT3 and SOCS3 act as components of a common pathway that suppress neutrophil production and/or accumulation in peripheral blood [60, 62].

Mice with spontaneous mutation in *Ptpn6*, the gene encoding Src homology region 2 domain-containing phosphatase-1, or SHP-1 (previously known as hematopoietic cell phosphatase [HCP]), are termed *motheaten*, based on their patchy skin appearance due to focal abscesses [63–65]. *Motheaten* mice exhibit severe autoimmunity and immunodeficiency, accompanied by an increased number of immature granulocytes, which demonstrate a hyperproliferative response to G-CSF [63–65]. These data implicate SHP-1 as a negative regulator of G-CSFR signaling, similar to its role in the erythropoietin receptor (EpoR) signal transduction cascade [63]. Although SHP-1 is not thought to interact physically with G-CSFR [64], it represses G-CSF-responsive STAT3 activation, an activity that requires the *C*-terminus of G-CSFR [65]. Hence, these results suggest an indirect association between SHP-1 and G-CSFR, although the molecular mechanisms involved remain unclear.

G-CSFR downregulation through ligand-induced internalization also terminates receptor signaling. The G-CSFR internalization mechanism is thought to be mediated by a dileucine motif in Box 3 [66]. Consequently, a mutant of G-CSFR (D716) lacking the dileucine motif demonstrates impaired internalization, and results in enhanced G-CSF-responsive proliferative signaling [66]. Collectively, numerous mechanisms operate to dampen G-CSFR signal transduction, indicating that a precise balance of activating and inhibitory signaling pathways is important in mediating the response to G-CSF in vivo.

3 Engineered Mutations of G-CSFR and Relationship to Human Disease

In the past 20 years, valuable mouse models have been generated through gene-targeting strategies that have introduced mutations in the endogenous *Csf3r* locus (Fig. 3). These *Csf3r* "knock-out" and "knock-in" mice have provided important and informative tools for understanding the precise role of G-CSFR in physiologic conditions. We highlight some of the mouse models generated and their relationship to human diseases.

3.1 G-CSFR-Deficient Mice

Insights into the role of G-CSFR in regulating granulopoiesis and multipotential hematopoietic progenitors have been provided by analysis of mice carrying a homozygous null mutation in *Csf3r* (referred to herein as *Csf3r*$^{-/-}$). A genomic fragment encompassing exons 3–8 of *Csf3r* was deleted, removing critical regions

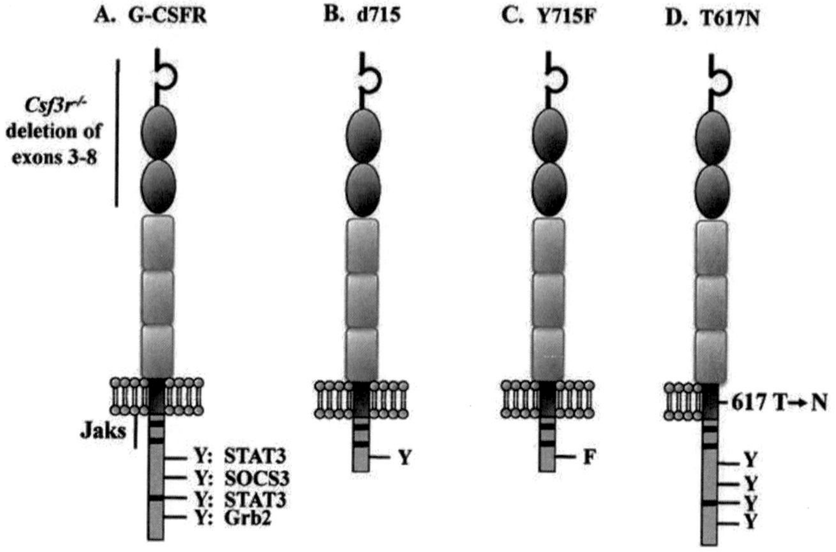

Fig. 3 Engineered G-CSFR mutations used in mouse models. Panel (**a**) *Csf3r*$^{-/-}$ mice were generated by deletion of exons 3–8, resulting in removal of the signal peptide, Ig-like, and CRH domains. Panel (**b**) The d715 mutation was generated by introduction of a stop codon in exon 17, truncating the *C*-terminal 96 amino acids. One tyrosine residue remains (Y704). Panel (**c**) d715F was generated by mutating Y704 of d715 to phenylalanine (F). Panel (**d**) The T617N mutation in the transmembrane domain of G-CSFR results in constitutive signaling

including the G-CSF-binding domains and signal peptide (Fig. 3a); importantly, no functional G-CSFR protein is detectable in hematopoietic cells from *Csf3r*$^{-/-}$ mice, e.g., *Csf3r*$^{-/-}$ bone marrow cells fail to respond to G-CSF in colony-forming assays ex vivo [10]. *Csf3r*$^{-/-}$ mice are severely neutropenic, with approximately 12% circulating neutrophil amounts compared with that in wild-type animals [10]. Late-stage granulocytes, such as metamyelocytes and bands, are reduced in the bone marrow of *Csf3r*$^{-/-}$ mice to approximately 50% of the amount in wild-type mice; however, no significant changes were detected in the amount of precursor cells such as myeloblasts or promyelocytes, or early myeloid progenitors including granulocyte colony-forming units (CFU-G) [10]. These results indicate that G-CSFR signaling is essential for driving the late stages of granulopoiesis in the bone marrow, as well as neutrophil efflux to the blood and/or survival in the peripheral circulation; however, G-CSFR may be dispensable or compensated by alternative pathways in early myeloid progenitors.

The role of G-CSFR in granulopoiesis was further investigated by competitive bone marrow repopulation assays, in which *Csf3r*$^{-/-}$ bone marrow cells were introduced into lethally irradiated wild-type mice in the presence of different amounts of wild-type competitor bone marrow. One purpose of this approach was to eliminate the effects of compensatory mechanisms driving granulopoiesis in the absence of functional G-CSFR [67]. These experiments revealed defective contribution of *Csf3r*$^{-/-}$ cells to the granulocytic and monocytic lineages, accompanied by increased contribution to lymphocyte subsets. Analysis of progenitor responses revealed that G-CSFR signals are required for promoting effective granulopoiesis from myeloid progenitors as early as the common myeloid progenitor stage [67]. Therefore, compensatory mechanisms support granulopoiesis, albeit at reduced amounts, in the absence of functional signals from G-CSFR, and may involve signals from other cytokines, such as IL-6 and granulocyte–macrophage colony-stimulating factor (GM-CSF) [67]. Indeed, loss of IL-6 function further suppresses the neutropenic phenotype observed in *Csf3r*$^{-/-}$ animals [68]. Moreover, the residual neutrophils found in *Csf3r*$^{-/-}$ mice exhibit severely impaired chemotaxis to IL-8, *N*-formyl-methionyl-leucyl-phenylalanine (fMLP), zymosan-activated serum, or macrophage inflammatory protein-2 (MIP-2) [69], indicating that signals from G-CSFR regulate critical neutrophil functions such as chemotactic migration. Consistent with this finding, it has been shown that STAT3 controls multiple pathways involved in neutrophil migration [70, 71], therefore suggesting that G-CSFR–STAT3 signaling has an important role in directing the generation of fully functional neutrophils.

3.2 Truncation of G-CSFR, Severe Congenital Neutropenia, and Aberrant G-CSFR Signaling

Severe congenital neutropenia (SCN), also known as Kostmann's disease, is a hematologic disorder characterized by significantly decreased amounts of circulating neutrophils (absolute neutrophil counts [ANC] $<0.5 \times 10^9$/L, as opposed to 3.5–4.5 $\times$ 10^9/L in healthy individuals) and early onset of acute, life-threatening bacterial infections ([72, 73]; see chapter "rHuG-CSF for the Treatment of Severe Chronic Neutropenia" by Dale and Boulyard). SCN is considered a premalignant condition, as it has been reported that approximately 21% of SCN patients eventually develop myelodysplastic syndrome (MDS) or acute myelogenous leukemia (AML) [74]. Before the advent of recombinant human G-CSF, SCN was nearly often fatal; however, clinical G-CSF treatments enhance circulating neutrophil amounts and suppress bacterial infections in patients with SCN. The underlying cause of SCN was attributed initially to point mutations introducing premature stop codons in G-CSFR, but recent studies suggest that these mutations are secondary to those that induce maturation arrest of myeloid progenitors, including mutations in the myeloid transcriptional regulator LEF [75]; mutations in *HAX-1*, which encodes a mitochondrial protein [76]; or mutations in *ELA2*, encoding neutrophil elastase [77]. Approximately 25% of patients have mutation in *CSF3R*, resulting in deletion of roughly 100 amino acids from the cytoplasmic domain of the G-CSFR protein, which is the region responsible for eliciting differentiation signals in vitro [78–80] (Fig. 1). It is thought that G-CSFR truncation mutations are acquired in these patients, perhaps as a mechanism to offset the lack of mature neutrophils by offering a proliferative advantage to myeloid progenitor cells harboring the mutant G-CSFR. *CSF3R* mutations and the use of G-CSF in patients with SCN have been investigated for possible associations with leukemic transformation. The available evidence to date suggests that the severity of SCN may differ among individuals, and those with more aggressive disease (e.g., more severe neutropenia) are more likely to be treated with higher doses of G-CSF and progress to leukemia; however, the use of G-CSF and leukemic transformation are not necessarily causally linked [74]. In other words, the severity of SCN would require high dose G-CSF, which may be independent of the events related to malignant progression.

A mouse model was generated to reproduce the mutation associated with SCN, in which introduction of a stop codon into exon 17 of *Csf3r* caused truncation of the *C*-terminal 96 amino acids (termed d715; Fig. 3b) [81]. While the tyrosine residues Y729, Y744, and Y764 were deleted, the STAT3 docking site Y704 remained intact [47, 81]. (Although the deletion occurred in murine *Csf3r*, the nomenclature for tyrosine residues on the human G-CSFR will be used, to maintain consistency with the respective publications.) Mice harboring the d715 mutation showed normal granulopoiesis, as judged by neutrophil numbers in blood and bone marrow at steady state, however progenitor proliferative responses to G-CSF were enhanced [47, 81]. Furthermore, G-CSF treatment in vivo induced sustained neutrophilia and increased granulocyte progenitor amounts in d715 mutant mice [81, 82]. This change was attributed to prolonged activation of STAT complexes in d715 bone marrow cells, e.g., 60 min after G-CSF treatment, increased amounts of STAT5:STAT1 and STAT3:STAT1 DNA-binding complexes were detected, although the latter to a lesser extent [82]. Activation of STAT3 homodimers appeared reduced relative to that in wild type, most likely due to loss of one STAT3 recruitment site (Y744) [82]. The contribution of G-CSF-responsive hyperactivation of STAT5 to enhanced clonal expansion of d715 hematopoietic progenitor cells was examined in STAT5A/B-deficient cells expressing d715 [83]. Deletion of STAT5A/B abrogated the proliferative advantage of d715 hematopoietic progenitor cells, establishing a critical role for STAT5 in G-CSF-driven proliferation through d715. G-CSF-driven granulopoiesis, however, is normal in STAT5A/B-deficient mice, as measured by neutrophil survival, neutrophil production, and mobilization from bone marrow to blood, as well as myeloid colony-forming activity [84]. Altogether, the data suggest that STAT5 is dispensable for G-CSF signaling through the wild-type receptor; however, STAT5 is able to confer a growth advantage to hematopoietic progenitors within the context of G-CSFR truncation that is reminiscent of clonal expansion of leukemic cells. These results are consistent with data demonstrating that sustained activation of STAT5 is associated with hyperproliferation of hematopoietic progenitor cells and hematological malignancies, implicating STAT5 as an important mediator of leukemogenesis [58, 83, 85]; however, additional studies are required to determine the mechanisms of STAT5 action during normal and pathological hematopoiesis.

3.3 d715F, a G-CSFR Mutation Abrogating STAT3 Recruitment

To assess the role of STAT3 in the G-CSFR signaling pathway in vivo, the remaining STAT3 recruitment site, Y704, was mutated to phenylalanine (F) within the context of d715 to generate the d715F mutation (Fig. 3c) [47]. This mutation abolishes detectable STAT3 and STAT1 activation, and demonstrates attenuated STAT5 activation, in response to G-CSF. Furthermore, d715F mice exhibit severe defects in granulopoieisis [47]. While white blood cell counts are normal in d715F mice, circulating neutrophil counts are approximately 15% compared with that in wild-type animals [47]. In addition, d715F mice have an attenuated response to G-CSF treatment, as they are unable to upregulate circulating neutrophil numbers to the same extent as wild-type animals, and d715F hematopoietic progenitors fail to respond to G-CSF in colony-forming assays ex vivo [47]. These proliferative defects were partially restored upon introduction of a constitutively active STAT3 isoform (STAT3C) into d715F progenitor cells, which resulted in an increase in total number of CFU-G and mature neutrophils in in vitro assays [47]. By contrast, introduction of a dominant-negative STAT3 isoform into wild-type bone marrow progenitor cells inhibited colony-forming responses to G-CSF [47]. These studies agree with results from earlier in vitro work, which indicated the importance of STAT3 in mediating G-CSF-responsive proliferation and differentiation signals [30, 37, 42, 43, 45, 47].

3.4 T617N, a Mutation Affecting G-CSFR Dimerization

Certain myeloproliferative disorders have been attributed to autosomal-dominant mutations in growth factor receptors, such as the erythropoietin and thrombopoietin receptors (reviewed in [86]). Similarly, an autosomal-dominant point mutation in G-CSFR has been identified in a family in which 12 of 16 individuals across three generations were diagnosed with chronic neutrophilia [87]. This threonine-to-asparagine mutation (T617N) is located within the transmembrane domain of G-CSFR, and increases the stability of G-CSFR dimers, resulting in constitutive receptor activation [87] (Fig. 3d). Hematopoietic progenitor ($CD34^+$) cells from these patients demonstrated constitutive activation of Jak2, STAT3, STAT5, ERK1/2, and Akt. Moreover, in response to G-CSF, phosphorylation of these signaling proteins was further enhanced and sustained [87], indicating continued responsiveness to ligand stimulation in addition to constitutive activation. Hematopoietic progenitor cells expressing the T617N mutation (i.e., retrovirally transduced cells) or $CD34^+$ cells from patients with chronic neutrophilia were transplanted into the immunodeficient NOD/SCID/$\gamma_c^{-/-}$ mice to evaluate functional responses; these studies revealed that G-CSFR T617N contributes to the development of myeloproliferative-like disorders in vivo [87]. A very low occurrence of the T617N mutation has been detected in patients with AML, but this

G-CSFR mutation is thought to be acquired secondary to the leukemic transformation events [88]. Collectively, these observations suggest that overactivation of signals from G-CSFR can contribute to myeloproliferation associated with hematologic disease.

4 Roles of STAT3 in Murine Granulopoiesis and Human Hyperimmunoglobulin E Syndrome

4.1 The Role of STAT3 in Granulopoiesis

Previous in vitro and in vivo models indicated that G-CSFR-activated STAT3 is a major regulator of granulopoiesis [30, 37, 42, 43, 45, 47]; however, deletion of *Stat3* results in embryonic lethality [89], necessitating the use of conditional knockout mouse models to study STAT3 function in hematopoiesis. Moreover, while deletion of *Stat1* or *Stat5* has a minimal effect on granulopoiesis in vivo [90, 91], supporting the idea that alternative pathways (i.e., STAT3) are important, conditional deletion of *Stat3* in the hematopoietic system yielded initially surprising results [62, 71, 92, 93]. As opposed to the neutropenia observed in d715F mice, a model in which G-CSF-responsive STAT3 activation is suppressed, hematopoietic STAT3-deficient mice (termed herein STAT3 deficient) exhibit neutrophilia and generalized inflammation [47, 62, 71]. STAT3-deficient mice have increased amounts of mature neutrophils in the bone marrow, yet similar amounts of myeloid progenitor cells and immature granulocytes, compared with that in wild-type animals, suggesting that STAT3 is a negative regulator of terminal neutrophil differentiation [71]. Taken together with the observation of increased neutrophil counts in the peripheral circulation of STAT3-deficient mice, the results suggest that STAT3 is required to suppress neutrophil production in homeostatic conditions. This phenotype is similar to that of aged mice with hematopoietic deletion of *Socs3*, a STAT3 target gene that suppresses G-CSFR-responsive signal transduction [60, 61]. Collectively, therefore, these data indicate that STAT3 and SOCS3 are important negative regulators of steady-state granulopoiesis. Studies indicate, however, that the role of STAT3 in granulopoiesis may be more complex than initially considered.

Emergency granulopoiesis is a response that occurs upon infection or clinical administration of pharmacological doses of G-CSF. For example, bacterial infections induce an increase in circulating amounts of endogenous G-CSF, which provides a feedback signal to the bone marrow to enhance granulocytic progenitor cell cycle progression, increase neutrophil production, and stimulate the release of mature neutrophils into circulating blood [94, 95]. This response is critical for inducing the amount of peripheral neutrophils that target and suppress infectious microorganisms. After repeated exposure to G-CSF, STAT3-deficient mice (*Tg[Tek-cre]12Flv*, $Stat3^{fl/\Delta}$) fail to increase immature granulocyte amounts in bone marrow, resulting in a skewed ratio of immature:mature cells in bone marrow, blood, and spleen [71]. This effect was not observed in myeloid-specific SOCS3-deficient mice, therefore indicating that the role of STAT3 in G-CSF-driven emergency granulopoiesis is independent of SOCS3, and implicating additional STAT3 target genes in regulating the neutrophil response [71]. STAT3 was found to promote G-CSF-dependent cell cycle progression and differentiation of immature granulocytes [96]. Moreover, the proliferation of multipotent and committed myeloid progenitors in response to G-CSF or after infection by *Listeria monocytogenes*, a pathogen that is regulated by G-CSF signals in vivo, is dependent upon STAT3. These data indicate that STAT3 is an important regulator of hematopoietic progenitor proliferation during emergency granulopoiesis driven by G-CSF or bacterial infection [96] (Fig. 4).

The phenotype of impaired emergency granulopoiesis in STAT3-deficient mice is reminiscent to that observed in *Cebpb*$^{-/-}$ mice [97], suggesting the possibility that STAT3 and C/EBPβ operate as components of a pathway regulating myeloid progenitor proliferation. C/EBPβ is a member of the CCAAT/enhancer-binding (C/EBP) family, which consists of transcription factors that have crucial functions

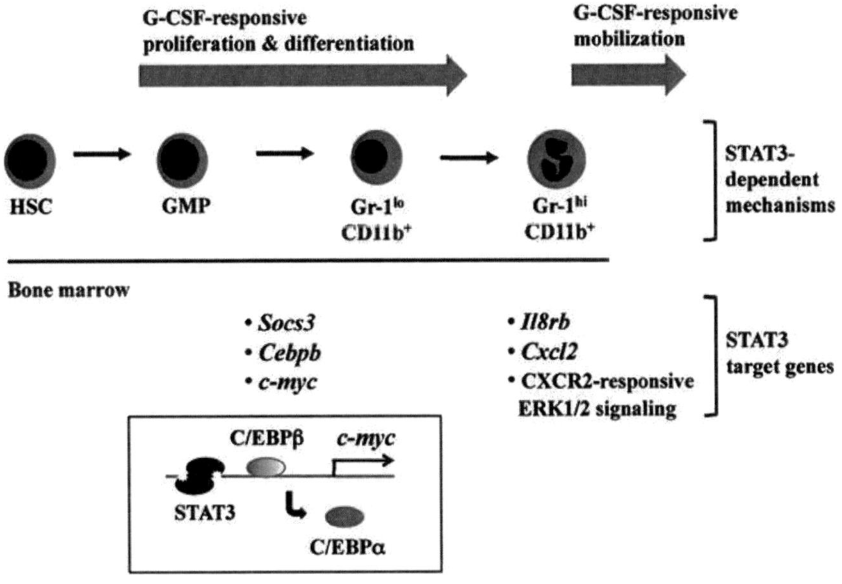

Fig. 4 Summary of the roles of STAT3 in neutrophil progenitor proliferation, differentiation, and mobilization. In the bone marrow, G-CSFR–STAT3 signaling induces proliferation and differentiation of myeloid progenitors, in part by inducing the STAT3 target genes *c-myc* and *Cebpb*. STAT3 also induces expression of *Socs3* to terminate signals from G-CSFR. G-CSFR–STAT3 signals contribute to neutrophil mobilization and migration by controlling the expression of genes encoding the chemokine MIP-2 (*Cxcl2*; H.N.-J. and S.S.W) and its receptor CXCR2 (*Il8rb*), in addition to MIP-2-responsive Raf/MEK/ERK signal transduction. G-CSFR–STAT3 induction of *Cebpb* results in the association of C/EBPβ with the *c-myc* proximal promoter, relieving suppression of *c-myc* by C/EBPα and resulting in sustained *c-myc* expression

in development including myelopoiesis. Specifically, C/EBPβ is an essential regulator of emergency granulopoiesis driven by G-CSF or infection [97], in contrast with C/EBPα, which is required for granulocyte–monocyte committed progenitor (GMP) development and granulocyte maturation in steady-state conditions. *Cebpb*$^{-/-}$ bone marrow progenitors exhibit reduced CFU-G formation, and knock down of *Cebpb* expression in *Cebpa*$^{-/-}$ cells impaired their G-CSF responsiveness, indicating that C/EBPβ is important for transmitting G-CSF-dependent proliferative signals [97]. Furthermore, G-CSF treatment induces expression of *Cebpb* in granulocyte/macrophage progenitors, suggesting that G-CSF-dependent signaling regulates *Cebpb* transcription. G-CSF-mediated induction of *Cebpb* was recently determined to require STAT3, which binds directly to the *Cebpb* proximal promoter to regulate *Cebpb* transcription [96]. C/EBPβ appears to act during emergency granulopoiesis in part by sustaining the expression of *c-myc* [97], thus providing a pathway by which G-CSF-responsive signals communicate with the cell cycle machinery. The upstream promoter region of *c-myc* contains consensus-binding sites for C/EBPα and C/EBPβ, which play antagonistic roles in the regulation of *c-myc* [96], with C/EBPα acting to suppress *c-myc* and C/EBPβ promoting expression. Induction of C/EBPβ expression by the G-CSFR–STAT3 pathway relieves *c-myc* promoter suppression by C/EBPα [96], thereby facilitating G-CSF- and STAT3-responsive *c-myc* transcription (Fig. 4). This pathway provides a molecular mechanism by which G-CSF signaling through STAT3 and C/EBPβ stimulates neutrophil progenitor proliferation during emergency granulopoiesis (Fig. 4).

4.2 STAT3 Controls Multiple Pathways Affecting Neutrophil Mobilization

Neutrophil mobilization, the release of neutrophils from bone marrow to blood, is necessary to increase circulating neutrophil amounts during infection. Little is known of this pathway, which is exploited clinically in the use of G-CSF to treat neutropenia. STAT3-deficient mice were found to have impaired acute neutrophil mobilization in response to G-CSF or the chemokine MIP-2 (CXCL2) [70, 71], indicating that STAT3 regulates important factors mediating neutrophil retention and/or release in the bone marrow. Studies show that STAT3 operates by controlling G-CSF-induced changes in the expression of chemokines within the bone marrow microenvironment. For example, SDF-1 (CXCL12), a chemokine that controls the retention of neutrophils and hematopoietic progenitors in the bone marrow, is downregulated upon administration of G-CSF in wild-type mice [15]. STAT3-deficient mice, however, fail to suppress *Cxcl12* and in fact show a slight induction in *Cxcl12* mRNA expression upon treatment with G-CSF [70]. Furthermore, expression of mRNA encoding the chemokines KC (CXCL1) and MIP-2, which are potent neutrophil chemoattractants, is upregulated in the bone marrow upon G-CSF treatment, in a STAT3-dependent manner [70]. Thus, STAT3 mediates the induction of chemoattractants that mobilize neutrophils and the suppression of retention signals (i.e., SDF-1). Moreover, STAT3 controls the expression of the shared KC and MIP-2 receptor, CXCR2, in neutrophil precursors, as well as MIP-2-stimulated neutrophil chemotaxis, actin polymerization, and Raf/MEK/ERK activation [70, 71]. Accordingly, STAT3-deficient mice exhibit impaired neutrophil mobilization in response to infection by *Listeria monocytogenes*, which correlates with increased bacterial burden in the liver and inability to resolve infection [70]. Taken together, these results demonstrate that STAT3 regulates responses in the bone marrow microenvironment, as well as neutrophil chemokine receptor expression and signal transduction, thereby influencing neutrophil release into the blood, which is an important aspect of antibacterial immunity. The target genes activated by G-CSFR–STAT3 signaling are summarized in Fig. 4.

While the neutrophilia of STAT3-deficient mice in homeostatic conditions has been attributed to STAT3-dependent control of SOCS3, circulating neutrophil amounts reflect net production from myeloid progenitors, survival of developing and mature neutrophils, and neutrophil lifetime in the circulation. Basal neutrophilia in STAT3-deficient mice is not likely to result from an increase in myeloid progenitors, as amounts of lin^{-} $Sca\text{-}1^{+}$ $c\text{-}kit^{+}$ cells, granulocyte/macrophage progenitors, and granulocyte precursors are similar in wild-type and STAT3-deficient mice in steady state [62, 96]. Furthermore, studies show that STAT3 has a fundamental role in controlling neutrophil mobilization and chemotaxis [70, 71]; this may affect the ability of neutrophils to marginate or enter peripheral tissues, and thus contribute to neutrophil accumulation in the blood of STAT3-deficient animals. Therefore, STAT3 appears to have distinct functions in granulopoiesis, including a cell-autonomous role in G-CSF-responsive progenitor proliferation [96], control of neutrophil migration [70, 71], and regulation of the negative feedback inhibitor SOCS3 [60, 62].

4.3 STAT3 Mutations in Hyperimmunoglobulin E Syndrome

STAT3 mutations have been implicated in hyperimmunoglobulin E syndrome (HIES), a primary immunodeficiency characterized by recurrent bacterial and fungal infections, skin lesions, and markedly increased amounts of immunoglobulin E (IgE) [98]. HIES has been attributed to mutations in the DNA binding or SH2 domains of STAT3, causing it to act as a dominant negative protein [99]. Consistent with the findings in STAT3-deficient neutrophils, patients with HIES demonstrate defects in neutrophil chemotaxis [98]. In support of the discovery that *Il8rb*, the gene encoding CXCR2, is a STAT3 target, neutrophils isolated from HIES patients express significantly reduced amounts of CXCR2, in addition to suppressed amounts of the fMLP receptor and CXCR1 [70, 100]. Because STAT3-deficient mice exhibit impaired neutrophil function and increased susceptibility to infection by *L. monocytogenes* [70], the neutrophil defects in HIES are likely to contribute to immunodeficiency, which is due at least in part to mutations in *Stat3*. Further studies are required to determine the contribution of impaired neutrophil responses to HIES and other human diseases.

5 Conclusions

G-CSFR has a fundamental role in regulating steady-state and emergency granulopoiesis. In addition, the discovery of G-CSFR expression on other cell types revealed potential ways to expand the clinical use of G-CSF. For example, expression of G-CSFR on hematopoietic progenitors may be exploited to treat leukemic conditions by activating dormant leukemic stem cells [19]. Furthermore, the expression of G-CSFR in the central nervous system and in cardiomyocytes mediates antiapoptotic signaling in both cell types after injury, suggesting that G-CSF may induce prosurvival signals outside the hematopoietic system [22, 23] (see chapter "Investigational Studies of rHuG-CSF to Promote the Regeneration of Nonhematopoietic Tissues" by Szilvassy; and chapter "rHuG-CSF-Induced Mobilization of Bone Marrow Stem Cells and Cardiac Repair" by Dawn et al.).

Acquired mutations in G-CSFR and aberrant activation of downstream signaling pathways have been linked to leukemogenesis, demonstrating that precise control of the G-CSFR pathway is required to maintain normal hematopoiesis versus pathologic function [66, 72, 74, 78–82, 85, 88, 98, 99]. As roles of G-CSFR and its signaling molecules are better understood, the advent of new therapeutic approaches for related hematologic disease or immunodeficiencies will be more feasible. Therefore, understanding the molecular aspects of G-CSFR signal transduction and function is important for generating new therapies for human diseases.

Acknowledgments HN-J is supported by an NIH predoctoral training grant in Cancer Immunology (T32-CA-09598-21). Related research in SSW's laboratory has been supported by grants from the NIH (AI073587, AR059010), a Preclinical Research Agreement with Amgen Inc., and a seed grant from the Center for Stem Cell and Developmental Biology at UT M D Anderson Cancer Center.

References

1. Larsen A, Davis T, Curtis BM et al (1990) Expression cloning of a human granulocyte colony-stimulating factor receptor: a structural mosaic of hematopoietin receptor, immunoglobulin, and fibronectin domains. J Exp Med 172:1559–1570
2. Fukunaga R, Ishizaka-Ikeda E, Seto Y, Nagata S (1990) Expression cloning of a receptor for murine granulocyte colony-stimulating factor. Cell 61:341–350
3. Yamasaki K, Naito S, Anaguchi H, Ohkubo T, Ota Y (1997) Solution structure of an extracellular domain containing the WSxWS motif of the granulocyte colony-stimulating factor receptor and its interaction with ligand. Nat Struct Biol 4:498–504

4. Fukunaga R, Ishizaka-Ikeda E, Pan CX, Seto Y, Nagata S (1991) Functional domains of the granulocyte colony-stimulating factor receptor. EMBO J 10:2855–2865
5. Nicola NA, Peterson L (1986) Identification of distinct receptors for two hemopoietic growth factors (granulocyte colony-stimulating factor and multipotential colony-stimulating factor) by chemical cross-linking. J Biol Chem 261:12384–12389
6. Tamada T, Honjo E, Maeda Y et al (2006) Homodimeric cross-over structure of the human granulocyte colony-stimulating factor (GCSF) receptor signaling complex. Proc Natl Acad Sci U S A 103:3135–3140
7. McKinstry WJ, Li CL, Rasko JE, Nicola NA, Johnson GR, Metcalf D (1997) Cytokine receptor expression on hematopoietic stem and progenitor cells. Blood 89:65–71
8. Nicola NA, Metcalf D (1985) Binding of 125I-labeled granulocyte colony-stimulating factor to normal murine hemopoietic cells. J Cell Physiol 124:313–321
9. Manz MG, Miyamoto T, Akashi K, Weissman IL (2002) Prospective isolation of human clonogenic common myeloid progenitors. Proc Natl Acad Sci U S A 99:11872–11877
10. Liu F, Wu HY, Wesselschmidt R, Kornaga T, Link DC (1996) Impaired production and increased apoptosis of neutrophils in granulocyte colony-stimulating factor receptor-deficient mice. Immunity 5:491–501
11. Lieschke GJ, Grail D, Hodgson G et al (1994) Mice lacking granulocyte colony-stimulating factor have chronic neutropenia, granulocyte and macrophage progenitor cell deficiency, and impaired neutrophil mobilization. Blood 84:1737–1746
12. Boneberg EM, Hareng L, Gantner F, Wendel A, Hartung T (2000) Human monocytes express functional receptors for granulocyte colony-stimulating factor that mediate suppression of monokines and interferon-gamma. Blood 95:270–276
13. Morikawa K, Morikawa S, Nakamura M, Miyawaki T (2002) Characterization of granulocyte colony-stimulating factor receptor expressed on human lymphocytes. Br J Haematol 118:296–304
14. Shimoda K, Okamura S, Harada N, Kondo S, Okamura T, Niho Y (1993) Identification of a functional receptor for granulocyte colony-stimulating factor on platelets. J Clin Invest 91:1310–1313
15. Semerad CL, Christopher MJ, Liu F et al (2005) G-CSF potently inhibits osteoblast activity and CXCL12 mRNA expression in the bone marrow. Blood 106:3020–3027
16. Liu F, Poursine-Laurent J, Link DC (2000) Expression of the G-CSF receptor on hematopoietic progenitor cells is not required for their mobilization by G-CSF. Blood 95:3025–3031
17. Budel LM, Touw IP, Delwel R, Lowenberg B (1989) Granulocyte colony-stimulating factor receptors in human acute myelocytic leukemia. Blood 74:2668–2673
18. Corcione A, Corrias MV, Daniele S, Zupo S, Spriano M, Pistoia V (1996) Expression of granulocyte colony-stimulating factor and granulocyte colony-stimulating factor receptor genes in partially overlapping monoclonal B-cell populations from chronic lymphocytic leukemia patients. Blood 87:2861–2869
19. Trumpp A, Essers M, Wilson A (2010) Awakening dormant haematopoietic stem cells. Nat Rev Immunol 10:201–209
20. Calhoun DA, Donnelly WH Jr, Du Y, Dame JB, Li Y, Christensen RD (1999) Distribution of granulocyte colony-stimulating factor (G-CSF) and G-CSF-receptor mRNA and protein in the human fetus. Pediatr Res 46:333–338
21. Fukunaga R, Seto Y, Mizushima S, Nagata S (1990) Three different mRNAs encoding human granulocyte colony-stimulating factor receptor. Proc Natl Acad Sci U S A 87:8702–8706
22. Schneider A, Kruger C, Steigleder T et al (2005) The hematopoietic factor G-CSF is a neuronal ligand that counteracts programmed cell death and drives neurogenesis. J Clin Invest 115:2083–2098
23. Harada M, Qin Y, Takano H et al (2005) G-CSF prevents cardiac remodeling after myocardial infarction by activating the Jak-Stat pathway in cardiomyocytes. Nat Med 11:305–311

24. Bussolino F, Wang JM, Defilippi P et al (1989) Granulocyte- and granulocyte-macrophage-colony stimulating factors induce human endothelial cells to migrate and proliferate. Nature 337:471–473
25. Seto Y, Fukunaga R, Nagata S (1992) Chromosomal gene organization of the human granulocyte colony-stimulating factor receptor. J Immunol 148:259–266
26. Smith LT, Hohaus S, Gonzalez DA, Dziennis SE, Tenen DG (1996) PU.1 (Spi-1) and C/EBP alpha regulate the granulocyte colony-stimulating factor receptor promoter in myeloid cells. Blood 88:1234–1247
27. Anderson KL, Smith KA, Conners K, McKercher SR, Maki RA, Torbett BE (1988) Myeloid development is selectively disrupted in PU.1 null mice. Blood 91:3702–3710
28. Aritomi M, Kunishima N, Okamoto T, Kuroki R, Ota Y, Morikawa K (1999) Atomic structure of the GCSF-receptor complex showing a new cytokine-receptor recognition scheme. Nature 401:713–717
29. Nicholson SE, Oates AC, Harpur AG, Ziemiecki A, Wilks AF, Layton JE (1994) Tyrosine kinase JAK1 is associated with the granulocyte-colony-stimulating factor receptor and both become tyrosine-phosphorylated after receptor activation. Proc Natl Acad Sci U S A 91:2985–2988
30. Tian SS, Lamb P, Seidel HM, Stein RB, Rosen J (1994) Rapid activation of the STAT3 transcription factor by granulocyte colony-stimulating factor. Blood 84:1760–1764
31. Shimoda K, Feng J, Murakami H et al (1997) Jak1 plays an essential role for receptor phosphorylation and Stat activation in response to granulocyte colony-stimulating factor. Blood 90:597–604
32. Rodig SJ, Meraz MA, White JM et al (1998) Disruption of the Jak1 gene demonstrates obligatory and nonredundant roles of the Jaks in cytokine-induced biologic responses. Cell 93:373–383
33. Parganas E, Wang D, Stravopodis D et al (1998) Jak2 is essential for signaling through a variety of cytokine receptors. Cell 93:385–395
34. Dong F, van Buitenen C, Pouwels K, Hoefsloot LH, Lowenberg B, Touw IP (1993) Distinct cytoplasmic regions of the human granulocyte colony-stimulating factor receptor involved in induction of proliferation and maturation. Mol Cell Biol 13:7774–7781
35. Ziegler SF, Bird TA, Morella KK, Mosley B, Gearing DP, Baumann H (1993) Distinct regions of the human granulocyte-colony-stimulating factor receptor cytoplasmic domain are required for proliferation and gene induction. Mol Cell Biol 13:2384–2390
36. Fukunaga R, Ishizaka-Ikeda E, Nagata S (1993) Growth and differentiation signals mediated by different regions in the cytoplasmic domain of granulocyte colony-stimulating factor receptor. Cell 74:1079–1087
37. Nicholson SE, Novak U, Ziegler SF, Layton JE (1995) Distinct regions of the granulocyte colony-stimulating factor receptor are required for tyrosine phosphorylation of the signaling molecules JAK2, Stat3, and p42, p44MAPK. Blood 86:3698–3704
38. Tanner JW, Chen W, Young RL, Longmore GD, Shaw AS (1995) The conserved box 1 motif of cytokine receptors is required for association with JAK kinases. J Biol Chem 270:6523–6530
39. Yoshikawa A, Murakami H, Nagata S (1995) Distinct signal transduction through the tyrosine-containing domains of the granulocyte colony-stimulating factor receptor. EMBO J 14:5288–5296
40. Akbarzadeh S, Ward AC, McPhee DO, Alexander WS, Lieschke GJ, Layton JE (2002) Tyrosine residues of the granulocyte colony-stimulating factor receptor transmit proliferation and differentiation signals in murine bone marrow cells. Blood 99:879–887
41. Aaronson DS, Horvath CM (2002) A road map for those who don't know JAK-STAT. Science 296:1653–1655
42. Tian SS, Tapley P, Sincich C, Stein RB, Rosen J, Lamb P (1996) Multiple signaling pathways induced by granulocyte colony-stimulating factor involving activation of JAKs, STAT5,

and/or STAT3 are required for regulation of three distinct classes of immediate early genes. Blood 88:4435–4444
43. Chakraborty A, Dyer KF, Cascio M, Mietzner TA, Tweardy DJ (1999) Identification of a novel Stat3 recruitment and activation motif within the granulocyte colony-stimulating factor receptor. Blood 93:15–24
44. Ward AC, Hermans MH, Smith L et al (1999) Tyrosine-dependent and -independent mechanisms of STAT3 activation by the human granulocyte colony-stimulating factor (G-CSF) receptor are differentially utilized depending on G-CSF concentration. Blood 93:113–124
45. Shimozaki K, Nakajima K, Hirano T, Nagata S (1997) Involvement of STAT3 in the granulocyte colony-stimulating factor-induced differentiation of myeloid cells. J Biol Chem 272:25184–25189
46. Panopoulos AD, Bartos D, Zhang L, Watowich SS (2002) Control of myeloid-specific integrin alpha Mbeta 2 (CD11b/CD18) expression by cytokines is regulated by Stat3-dependent activation of PU.1. J Biol Chem 277:19001–19007
47. McLemore ML, Grewal S, Liu F et al (2001) STAT-3 activation is required for normal G-CSF-dependent proliferation and granulocytic differentiation. Immunity 14:193–204
48. de Koning JP, Soede-Bobok AA, Schelen AM et al (1998) Proliferation signaling and activation of Shc, p21Ras, and Myc via tyrosine 764 of human granulocyte colony-stimulating factor receptor. Blood 91:1924–1933
49. de Koning JP, Schelen AM, Dong F et al (1996) Specific involvement of tyrosine 764 of human granulocyte colony-stimulating factor receptor in signal transduction mediated by p145/Shc/GRB2 or p90/GRB2 complexes. Blood 87:132–140
50. Ward AC, Smith L, de Koning JP, van Aesch Y, Touw IP (1999) Multiple signals mediate proliferation, differentiation, and survival from the granulocyte colony-stimulating factor receptor in myeloid 32D cells. J Biol Chem 274:14956–14962
51. Egan SE, Giddings BW, Brooks MW, Buday L, Sizeland AM, Weinberg RA (1993) Association of Sos Ras exchange protein with Grb2 is implicated in tyrosine kinase signal transduction and transformation. Nature 363:45–51
52. Corey SJ, Burkhardt AL, Bolen JB, Geahlen RL, Tkatch LS, Tweardy DJ (1994) Granulocyte colony-stimulating factor receptor signaling involves the formation of a three-component complex with Lyn and Syk protein-tyrosine kinases. Proc Natl Acad Sci U S A 91:4683–4687
53. Ward AC, Monkhouse JL, Csar XF, Touw IP, Bello PA (1998) The Src-like tyrosine kinase Hck is activated by granulocyte colony-stimulating factor (G CSF) and docks to the activated G-CSF receptor. Biochem Biophys Res Commun 251:117–123
54. Dong F, Larner AC (2000) Activation of Akt kinase by granulocyte colony-stimulating factor (G-CSF): evidence for the role of a tyrosine kinase activity distinct from the Janus kinases. Blood 95:1656–1662
55. Mermel CH, McLemore ML, Liu F et al (2006) Src family kinases are important negative regulators of G-CSF-dependent granulopoiesis. Blood 108:2562–2568
56. Wormald S, Hilton DJ (2004) Inhibitors of cytokine signal transduction. J Biol Chem 279:821–824
57. Hortner M, Nielsch U, Mayr LM, Johnston JA, Heinrich PC, Haan S (2002) Suppressor of cytokine signaling-3 is recruited to the activated granulocyte-colony stimulating factor receptor and modulates its signal transduction. J Immunol 169:1219–1227
58. Irandoust MI, Aarts LH, Roovers O, Gits J, Erkeland SJ, Touw IP (2007) Suppressor of cytokine signaling 3 controls lysosomal routing of G-CSF receptor. EMBO J 26:1782–1793
59. Roberts AW, Robb L, Rakar S et al (2001) Placental defects and embryonic lethality in mice lacking suppressor of cytokine signaling 3. Proc Natl Acad Sci U S A 98:9324–9329
60. Croker BA, Metcalf D, Robb L et al (2004) SOCS3 is a critical physiological negative regulator of G-CSF signaling and emergency granulopoiesis. Immunity 20:153–165
61. Auernhammer CJ, Bousquet C, Melmed S (1999) Autoregulation of pituitary corticotroph SOCS-3 expression: characterization of the murine SOCS-3 promoter. Proc Natl Acad Sci U S A 96:6964–6969

62. Lee CK, Raz R, Gimeno R et al (2002) STAT3 is a negative regulator of granulopoiesis but is not required for G-CSF-dependent differentiation. Immunity 17:63–72
63. Tapley P, Shevde NK, Schweitzer PA et al (1997) Increased G-CSF responsiveness of bone marrow cells from hematopoietic cell phosphatase deficient viable motheaten mice. Exp Hematol 25:122–131
64. Ward AC, Oomen SP, Smith L et al (2000) The SH2 domain-containing protein tyrosine phosphatase SHP-1 is induced by granulocyte colony-stimulating factor (G-CSF) and modulates signaling from the G-CSF receptor. Leukemia 14:1284–1291
65. Dong F, Qiu Y, Yi T, Touw IP, Larner AC (2001) The carboxyl terminus of the granulocyte colony-stimulating factor receptor, truncated in patients with severe congenital neutropenia/acute myeloid leukemia, is required for SH2-containing phosphatase-1 suppression of Stat activation. J Immunol 167:6447–6452
66. Hunter MG, Avalos BR (1999) Deletion of a critical internalization domain in the G-CSFR in acute myelogenous leukemia preceded by severe congenital neutropenia. Blood 93:440–446
67. Richards MK, Liu F, Iwasaki H, Akashi K, Link DC (2003) Pivotal role of granulocyte colony-stimulating factor in the development of progenitors in the common myeloid pathway. Blood 102:3562–3568
68. Liu F, Poursine-Laurent J, Wu HY, Link DC (1997) Interleukin-6 and the granulocyte colony-stimulating factor receptor are major independent regulators of granulopoiesis in vivo but are not required for lineage commitment or terminal differentiation. Blood 90:2583–2590
69. Betsuyaku T, Liu F, Senior RM et al (1999) A functional granulocyte colony-stimulating factor receptor is required for normal chemoattractant-induced neutrophil activation. J Clin Invest 103:825–832
70. Nguyen-Jackson H, Panopoulos AD, Zhang H, Li HS, Watowich SS (2010) STAT3 controls the neutrophil migratory response to CXCR2 ligands by direct activation of G-CSF-induced CXCR2 expression and via modulation of CXCR2 signal transduction. Blood 115:3354–3363
71. Panopoulos AD, Zhang L, Snow JW et al (2006) STAT3 governs distinct pathways in emergency granulopoiesis and mature neutrophils. Blood 108:3682–3690
72. Germeshausen M, Skokowa J, Ballmaier M, Zeidler C, Welte K (2008) G-CSF receptor mutations in patients with congenital neutropenia. Curr Opin Hematol 15:332–337
73. von Vietinghoff S, Ley K (2008) Homeostatic regulation of blood neutrophil counts. J Immunol 181:5183–5188
74. Rosenberg PS, Alter BP, Bolyard AA et al (2006) The incidence of leukemia and mortality from sepsis in patients with severe congenital neutropenia receiving long-term G-CSF therapy. Blood 107:4628–4635
75. Skokowa J, Cario G, Uenalan M et al (2006) LEF-1 is crucial for neutrophil granulocytopoiesis and its expression is severely reduced in congenital neutropenia. Nat Med 12:1191–1197
76. Klein C, Grudzien M, Appaswamy G et al (2007) HAX1 deficiency causes autosomal recessive severe congenital neutropenia (Kostmann disease). Nat Genet 39:86–92
77. Dale DC, Person RE, Bolyard AA et al (2000) Mutations in the gene encoding neutrophil elastase in congenital and cyclic neutropenia. Blood 96:2317–2322
78. Dong F, Brynes RK, Tidow N, Welte K, Lowenberg B, Touw IP (1995) Mutations in the gene for the granulocyte colony-stimulating-factor receptor in patients with acute myeloid leukemia preceded by severe congenital neutropenia. N Engl J Med 333:487–493
79. Dong F, Dale DC, Bonilla MA et al (1997) Mutations in the granulocyte colony-stimulating factor receptor gene in patients with severe congenital neutropenia. Leukemia 11:120–125
80. Germeshausen M, Ballmaier M, Welte K (2007) Incidence of CSF3R mutations in severe congenital neutropenia and relevance for leukemogenesis: results of a long-term survey. Blood 109:93–99
81. McLemore ML, Poursine-Laurent J, Link DC (1998) Increased granulocyte colony-stimulating factor responsiveness but normal resting granulopoiesis in mice carrying a

targeted granulocyte colony-stimulating factor receptor mutation derived from a patient with severe congenital neutropenia. J Clin Invest 102:483–492
82. Hermans MH, Antonissen C, Ward AC, Mayen AE, Ploemacher RE, Touw IP (1999) Sustained receptor activation and hyperproliferation in response to granulocyte colony-stimulating factor (G-CSF) in mice with a severe congenital neutropenia/acute myeloid leukemia-derived mutation in the G-CSF receptor gene. J Exp Med 189:683–692
83. Liu F, Kunter G, Krem MM et al (2008) Csf3r mutations in mice confer a strong clonal HSC advantage via activation of Stat5. J Clin Invest 118:946–955
84. Kimura A, Rieger MA, Simone JM et al (2009) The transcription factors STAT5A/B regulate GM-CSF-mediated granulopoiesis. Blood 114:4721–4728
85. Bunting KD (2007) STAT5 signaling in normal and pathologic hematopoiesis. Front Biosci 12:2807–2820
86. Silver RT, Tefferi A (eds) (2008) Myeloproliferative disorders: biology and management. Informa Healthcare USA, Inc., New York
87. Plo I, Zhang Y, Le Couedic JP et al (2009) An activating mutation in the CSF3R gene induces a hereditary chronic neutrophilia. J Exp Med 206:1701–1707
88. Forbes LV, Gale RE, Pizzey A, Pouwels K, Nathwani A, Linch DC (2002) An activating mutation in the transmembrane domain of the granulocyte colony-stimulating factor receptor in patients with acute myeloid leukemia. Oncogene 21:5981–5989
89. Takeda K, Noguchi K, Shi W et al (1997) Targeted disruption of the mouse Stat3 gene leads to early embryonic lethality. Proc Natl Acad Sci U S A 94:3801–3804
90. Durbin JE, Hackenmiller R, Simon MC, Levy DE (1996) Targeted disruption of the mouse Stat1 gene results in compromised innate immunity to viral disease. Cell 84:443–450
91. Meraz MA, White JM, Sheehan KC et al (1996) Targeted disruption of the Stat1 gene in mice reveals unexpected physiologic specificity in the JAK-STAT signaling pathway. Cell 84:431–442
92. Kamezaki K, Shimoda K, Numata A et al (2005) Roles of Stat3 and ERK in G-CSF signaling. Stem Cells 23:252–263
93. Welte T, Zhang SS, Wang T et al (2003) STAT3 deletion during hematopoiesis causes Crohn's disease-like pathogenesis and lethality: a critical role of STAT3 in innate immunity. Proc Natl Acad Sci U S A 100:1879–1884
94. Cheers C, Haigh AM, Kelso A, Metcalf D, Stanley ER, Young AM (1988) Production of colony-stimulating factors (CSFs) during infection: separate determinations of macrophage-, granulocyte-, granulocyte-macrophage-, and multi-CSFs. Infect Immun 56:247–251
95. Lord BI, Molineux G, Pojda Z, Souza LM, Mermod JJ, Dexter TM (1991) Myeloid cell kinetics in mice treated with recombinant interleukin-3, granulocyte colony-stimulating factor (CSF), or granulocyte-macrophage CSF in vivo. Blood 77:2154–2159
96. Zhang H, Nguyen-Jackson H, Panopoulos AD, Li HS, Murray PJ, Watowich SS (2010) STAT3 controls myeloid progenitor growth during emergency granulopoiesis. Blood 116:2462–2471
97. Hirai H, Zhang P, Dayaram T et al (2006) C/EBPbeta is required for 'emergency' granulopoiesis. Nat Immunol 7:732–739
98. Paslin D, Norman ME (1977) Atopic dermatitis and impaired neutrophil chemotaxis in Job's syndrome. Arch Dermatol 113:801–805
99. Holland SM, DeLeo FR, Elloumi HZ et al (2007) STAT3 mutations in the hyper-IgE syndrome. N Engl J Med 357:1608–1619
100. Mintz R, Garty BZ, Meshel T et al (2010) Reduced expression of chemoattractant receptors by polymorphonuclear leukocytes in Hyper IgE Syndrome patients. Immunol Lett 130:97–106

Part II
Use of rHuG-CSF in the Oncology Setting

Practice Guidelines for the Use of rHuG-CSF in an Oncology Setting

Santosh Saraf and Howard Ozer

1 Introduction

The United States Food and Drug Administration (US FDA) first approved recombinant human granulocyte colony-stimulating factor (rHuG-CSF) in 1991 for very broad indication to treat patients with cancer undergoing myelotoxic chemotherapy. Recombinant human granulocyte-macrophage colony-stimulating factor (rHuGM-CSF) was also approved for use in bone marrow and stem cell transplantation; however, the two drugs were often used interchangeably in the clinic and in subsequent guidelines developed by the American Society of Clinical Oncology (ASCO) and the National Comprehensive Cancer Network (NCCN).

Because the specific clinical indications for use of CSF were poorly defined by the FDA, concern grew regarding the significant costs of CSF solely as supportive care agents. For this reason, ASCO assembled an expert panel to review the clinical evidence and formulate guidelines for their use. The first ASCO guidelines were published in 1994 [1]. The goals were to provide evidence-based guidelines for clinical use of CSF to help direct practicing physicians on when and how to use them.

Febrile neutropenia is a significant cause of morbidity from infection, increases costs from hospitalization and treatment, and can potentially compromise effective treatment of the malignancy due to delays in administering chemotherapy or dose reductions. Prolonged and profound neutropenia may increase the risk of complications associated with infections [2]. As data emerged about the ability to reduce the incidence of febrile neutropenia using CSF therapy, concern grew regarding the high costs of the therapy.

S. Saraf • H. Ozer (✉)
Department of Medicine, Section of Hematology-Oncology, University of Illinois at Chicago, 840 South Wood Street, Suite #820-E CSB, M/C 713, Chicago, IL 60612, USA
e-mail: ozer@uic.edu

G. Molineux et al. (eds.), *Twenty Years of G-CSF*, Milestones in Drug Therapy,
DOI 10.1007/978-3-0348-0218-5_7,

As new data emerged, the guidelines were revisited with updated recommendations. Abbreviated updates were published in 1996 and 1997, and a more comprehensive guideline update was reported in 2000 [3–5]. With the advent of better tolerated chemotherapeutic agents and regimens, the guidelines in 2000 questioned the utility of CSF in primary prophylaxis. After publication of these guidelines, landmark trials emerged that helped change the paradigm for primary prophylaxis and dose intensification [6, 7]. New data demonstrated that chemotherapy regimens with a 20% risk of febrile neutropenia had significant reductions in the incidence of febrile neutropenia, hospitalizations, and antibiotic use with CSF primary prophylaxis, which permitted the advocacy for using CSF in more practical and commonly used chemotherapy regimens. Data also emerged suggesting that maintaining dose-dense therapy with CSF support improved outcomes in patients with breast cancer and aggressive lymphomas [8–10]. Until 2005, both NCCN and ASCO guidelines cited data that failed to show clinical outcome benefits by maintaining dose intensity with CSF support and also failed to show improvements in cost–benefit of CSF use in the more common chemotherapy regimens. In 2005, data began to emerge suggesting that CSF-supported, dose-dense treatment might increase response rates and survival, and subsequent studies examining primary prophylaxis confirmed a number of supportive-care benefits in regimens producing <20% febrile neutropenia. Based on new data, the NCCN initiated new guidelines in 2005, and the ASCO guidelines were revised soon thereafter [11, 12].

2 Primary Prophylaxis

Administration of chemotherapy is often limited by neutropenia that places patients already at an increased risk for infection at greater risk for more severe complications. The degree and length of neutropenia, important predictors for infection, vary based on the chemotherapy regimen, the patient's clinical condition, and the underlying disease [2, 13–15]. Although the infectious mortality from febrile neutropenia remains low, it is not zero and quality of life, cost of care, and timely delivery of chemotherapy can be affected.

Alternative approaches to primary prophylaxis with CSF include, when feasible, using an equally effective but less myelotoxic regimen, initial chemotherapy dose reduction, or prophylactic antibiotics; however, prophylactic antibiotic use in this setting is controversial. The Infectious Disease Society of America discourages this practice due to potential emergence of antibiotic-resistant microorganisms [16].

Primary prophylaxis with CSF had initially been recommended when the incidence of febrile neutropenia was >40% [1]. This recommendation was based on three prospective, randomized, placebo-controlled trials that showed that using rHuG-CSF in the initial cycle of treatment reduced the incidence of febrile neutropenia by approximately 50% with a quicker time to recovery of neutrophil counts when the incidence of febrile neutropenia from a chemotherapy regimen was >40% [17–19]. Two of the three trials evaluated cyclophosphamide, doxorubicin, and etoposide (CAE) in small-cell lung cancer (SCLC) and the third used vincristine, doxorubicin, prednisolone, etoposide, cyclophosphamide, and bleomycin (VAPEC-B) in non-Hodgkin's lymphoma (NHL). When used in the first cycle of chemotherapy, rates of neutropenic fever were reduced from 41–57% to 20–28% in the SCLC trials. In the trial including patients with NHL, rates were reduced in all cycles from 44% to 23%. These reductions translated into significant reductions in rates of hospitalization and antibiotic use in both of the SCLC trials but not in the case of the NHL trial. The rates of infectious mortality were low in both groups of these trials and none of the trials were able to demonstrate a reduction with rHuG-CSF. Data from a fourth trial, which included patients with various primary diseases and chemotherapy regimens, showed a reduction of febrile neutropenia from 32% to 12% with the use of rHuG-CSF versus placebo [20]. Although three of the four trials were able to show higher rates of chemotherapy intensity with rHuG-CSF support, improvements in disease-free or overall survival were not observed.

Initial data for rHuGM-CSF were less consistent. Only one randomized, placebo-controlled trial using molgramostim in patients with NHL found clinical benefit with reductions in episodes of febrile neutropenia, hospitalization, and faster neutrophil recovery [21]; however, these clinical benefits were only noted in the 72% of patients who were able to tolerate molgramostim. In contrast, no reduction in the incidence of febrile neutropenia was seen with the use of molgramostim in randomized, controlled trials in testicular or SCLC [22, 23].

With the advent of better tolerated chemotherapy regimens with lower rates of febrile neutropenia, the practicality of CSF in primary prophylaxis was questioned. The CAE chemotherapy regimen in SCLC had been superseded by less toxic regimens with expected rates of febrile neutropenia $\leq 10\%$ [24, 25]. Debate ensued whether or not dose intensification or maintaining the chemotherapy schedule with CSF support improved overall or disease-free survival. Data until the 2000 ASCO guidelines with dose intensity in patients with NHL or breast cancer did not show any advantage in disease-free or overall survival [26–29]. Similarly, using CSF to complete the timely administration of chemotherapy was not shown to have survival benefit in patients with AIDS-related tumors or in elderly patients with NHL, Hodgkin's disease, or testicular tumors [24, 25, 30].

It was not until 2003 that data were reported emerged showing survival benefits with dose-dense and dose-intense regimens. The first large randomized study showing a disease-free and overall survival benefit was in patients with node-positive breast cancer using rHuG-CSF support to maintain dose-dense chemotherapy [8]. In this phase 3 trial, a 2 × 2 factorial design was used to randomly assign 2,005 patients to one of four chemotherapy regimens:

1. Sequential doxorubicin × four doses followed by paclitaxel × four doses followed by cyclophosphamide × four doses every 3 weeks
2. Sequential doxorubicin × four doses followed by paclitaxel × four doses followed by cyclophosphamide × four doses every 2 weeks supported with filgrastim

3. Concurrent doxorubicin and cyclophosphamide × four doses followed by paclitaxel × four doses every 3 weeks or
4. Concurrent doxorubicin and cyclophosphamide × four doses followed by paclitaxel × four doses every 2 weeks supported with filgrastim

The dose-dense regimens using filgrastim for support (number 2 and number 4) had significantly improved disease-free survival (RR 0.74, $p = 0.010$) and overall survival (RR 0.69, $p = 0.013$). Furthermore, the dose-dense regimens were well tolerated and fewer patients had grade 4 neutropenia (6 vs. 33%, $p < 0.0001$). The following year, two studies in patients with NHL showed the benefit of dose-dense chemotherapy with rHuG-CSF support in both young and elderly patients using CSF support [9, 10]. The first study included 866 patients aged <60 years who were randomly assigned to one of four treatment groups: (1) CHOP-14 (chemotherapy with cyclophosphamide, doxorubicin, vincristine, and prednisone) plus rHuG-CSF every 2 weeks; (2) CHOP-21 every 3 weeks; (3) CHOEP-14 (CHOP with etoposide) plus rHuG-CSF every 2 weeks; or (4) CHOEP-21 every 3 weeks. In a multivariate analysis, reduction of the chemotherapy regimens from 3 to 2 weeks improved the rate of progressive disease and overall survival ($p = 0.032$ and $p = 0.044$, respectively). The second NHL study evaluated the benefit of dose density in 689 patients aged 60–75 years who were randomly assigned to the same four treatment groups. Although the regimens that contained etoposide were considered too toxic for this cohort, dose-dense CHOP-14 using rHuG-CSF support improved event-free (RR 0.66, $p = 0.003$) and overall survival (RR 0.58, $p < 0.001$).

In patients being treated with curative intent, the concern for dose-limiting neutropenia led to the investigation of CSF support in patients with early breast cancer, testicular cancer, and lymphoma [19, 31, 32]. Using CSF for primary prophylaxis in patients with NHL was shown to decrease the incidence of febrile neutropenia and likelihood of dose reduction. In patients with germ cell cancer who were randomly assigned to dose-intense chemotherapy plus rHuG-CSF support, a higher percentage of patients were able to tolerate at least six cycles of chemotherapy with higher dose intensities and reduced frequencies of febrile neutropenia and toxic deaths. Using a predictive model in patients with early stage breast cancer, CSF support in higher risk patients for neutropenia helped to maintain the chemotherapy regimen in a manner similar to lower risk patients not supported with CSF.

The threshold to recommend CSF in primary prophylaxis was reduced from 40% to 20% based on large randomized clinical trials showing a significant reduction of febrile neutropenia episodes with CSF when the risk for febrile neutropenia was ≥20%. In the first study of patients with breast cancer, pegfilgrastim reduced the incidence of febrile neutropenia and hospitalization for febrile neutropenia by >90% [7]. In this phase 3 study, 928 patients were randomly assigned to treatment with pegfilgrastim 6 mg or placebo to be administered the day after receiving docetaxel for a total of at least four cycles spaced 3 weeks apart. If the patients in the control group developed febrile neutropenia, they were treated with open-label pegfilgrastim. The incidence of febrile neutropenia was reduced for all cycles of treatment from 17% in the placebo group to 1% in the treatment group ($p < 0.001$). Clinical benefits were seen in the treatment group with reductions in the rates of hospitalization (14% vs. 1%, $p < 0.001$) and intravenous antibiotic use (10% vs. 2%, $p < 0.001$). Another phase 3 study that helped support the reduced threshold for primary prophylaxis was conducted in 175 patients with SCLC being treated with cyclophosphamide, doxorubicin, and etoposide and randomly assigned to prophylactic antibiotics alone or rHuG-CSF plus antibiotics [6]. The rates of febrile neutropenia in the antibiotic-alone group and rHuG-CSF plus antibiotic group were 24% and 10%, respectively ($p = 0.01$). Similarly, the overall rates of febrile neutropenia were 32% and 18%, respectively ($p = 0.01$). Although the rates of febrile neutropenia did decrease, there was no clinical benefit with regard to duration of hospitalization or intravenous antibiotic use.

Some chemotherapy regimens not typically associated with a ≥20% risk of febrile neutropenia may still warrant use of primary prophylaxis in select patient circumstances. This decision is often based on clinical judgment assessing the risk based on the patient's clinical status for febrile neutropenia and serious morbidity from infection. Although there are some data to suggest benefit, there is a dearth of prospective data to support this practice. Inferences have been made based on phase 1 data showing a reduction of prolonged neutropenia with rHuG-CSF support in patients who had received prior chemotherapy or irradiation, as well as the potential for a higher risk for chemotherapy-related infections in those patients who have a compromised bone marrow from prior therapy, poor performance status, or advanced cancer [2, 33–35]. Other suggestions for higher risk patients can be inferred from observations of quicker responses to CSF plus antibiotics versus antibiotics alone in patients with documented tissue infections or serious complications of febrile neutropenia being seen in patients with poor performance status, uncontrolled cancer, and an early onset of febrile neutropenia [14, 15, 36]. Using the current literature combined with clinical judgment, the guidelines have suggested considering CSF support in patients with preexisting neutropenia, heavy prior chemotherapy, previous irradiation to areas of active hematopoiesis such as the pelvis, a history of recurrent febrile neutropenia from prior chemotherapy regimens of equal or less myelotoxicity, poor performance status, advanced cancer, decreased immune function, open wounds, or currently active tissue infections [1].

The strongest evidence to date for a potential reduction in infection-related mortality and early death comes from a meta-analysis of 17 randomized controlled trials including 3,493 patients treated with rHuG-CSF in the primary prophylaxis setting [37]. This analysis included a wide range of malignancies and chemotherapy regimens. The relative risk reduction of infection-related mortality was 45% (RR 0.55, 95% CI, 0.33–0.90) and all-cause mortality during chemotherapy 40% (RR 0.60, 95% CI, 0.43–0.83). A similar sized meta-analysis of 14 studies with 1,569 patients found a marginally significant result for reduction of infection-related mortality (OR 0.51, 95% CI, 0.26–1.00) [38]; however, in a meta-analysis of 148 studies including 16,839 patients, a reduction in infection-related mortality was not observed (RR 0.82, 95% CI, 0.66–1.02) [39].

The NCCN presented a set of guidelines in 2005 based on the new data showing clinical benefit with primary prophylaxis when the risk of febrile neutropenia is 20% or higher and a survival advantage with dose-dense therapy with CSF support, both in breast cancer. The most current NCCN guidelines divide patients into high risk (defined as risk of febrile neutropenia >20%), intermediate risk (risk of febrile neutropenia 10–20%), and low risk (risk of febrile neutropenia <10%) [11]. Regardless of the treatment intent, all high-risk patients should be considered for routine CSF prophylaxis. In the intermediate-risk category, the NCCN panel recommended an individualized approach for CSF prophylaxis. The guidelines specified that if patient risk factors were the determinant for the risk for febrile neutropenia, then CSF prophylaxis is reasonable. Conversely, if the risk of febrile neutropenia is due exclusively to the chemotherapy regimen, an alternative regimen of similar efficacy should be considered. For patients at low risk for febrile neutropenia, the routine use of CSF was not considered cost effective and an alternative treatment regimen should be considered. The exceptions to the guidelines are that if the treatment is curative or for adjuvant treatment with a significant risk for morbidity or mortality from febrile neutropenia, then CSF can be considered. The following year, ASCO published its 2006 updated guidelines that modified their existing higher threshold to recommend primary prophylaxis to 20% as well [12].

3 Secondary Prophylaxis

Treating a patient with a CSF during subsequent cycles of chemotherapy can theoretically target the use of CSF support to a subset of patients who are at higher risk for neutropenia and complications thereof. Using a secondary prophylaxis approach allows for the use of CSF in patients who have presented with febrile neutropenia and are thus at high risk during subsequent cycles. It also may avoid the unnecessary use of CSF in patients who tolerated the first cycle well without febrile neutropenia and who are most likely low risk during subsequent cycles. The alternative approaches to secondary prophylaxis with CSF would be either using an alternative regimen with similar efficacy, dose reduction, or dose delay as long as clinical outcome is not negatively affected.

The initial recommendations for secondary prophylaxis with CSF arose from a randomized trial of rHuG-CSF in patients with SCLC treated with cyclophosphamide, doxorubicin, and etoposide [17]. Patients were initially randomly assigned to chemotherapy with or without primary CSF support. Patients in the control group who developed neutropenic fevers were permitted to be treated with the same dose of chemotherapy plus open-label CSF in subsequent cycles. The study found that using CSF in this setting reduced the days of neutropenia (6 days in cycle 1 vs. 2.5 days in cycle 2 with CSF) and reduced the rate of neutropenic fever (100% in cycle 1–23% in cycle 2). A phase 1/2 study using sargramostim in patients with documented neutropenia in a prior cycle of chemotherapy similarly found a reduction in the duration of grade 4 neutropenia from 6 days to 3 days [40].

Although no further prospective randomized trials have been reported, prediction models for CSF benefit in subsequent chemotherapy cycles have. One example is a study using a nadir neutrophil count of $<0.5 \times 10^9$/L to assign women with breast cancer receiving adjuvant chemotherapy to rHuG-CSF on subsequent cycles [31]. Fewer episodes of hospitalization for febrile neutropenia and greater dose intensity compared with historical controls not supported with CSF were seen, although cost analysis, survival, and quality of life were not reported.

4 Therapeutic Use of CSF

A lack of clinical benefit has been consistently seen in studies of CSF used to treat chemotherapy-induced neutropenia in afebrile patients. One of the earlier studies randomized CSF treatment in patients with non-small-cell lung cancer (NSCLC) being treated with cisplatin, vindesine, and mitomycin to either primary treatment or use when the absolute neutrophil count (ANC) was $<1.0 \times 10^9$/L. The investigators found neither a reduction in the duration of neutropenia nor any clinical benefit with the therapeutic use of CSF [41]. Similarly, no clinical benefit was seen when rHuGM-CSF was administered to afebrile patients with white blood cell counts $<2 \times 10^9$/L [42]. Both studies were underpowered making it difficult to draw definite conclusions regarding the lack of benefit for the therapeutic use of CSF for afebrile neutropenia.

Stronger evidence for the lack of clinical benefit from therapeutic CSF therapy in afebrile patients with chemotherapy-induced neutropenia later came from a larger randomized study [43]. In this trial, afebrile patients with solid tumors or lymphoma and severe neutropenia were randomly assigned to either rHuG-CSF or placebo. Neutrophil recovery was enhanced by 2 days in patients receiving rHuG-CSF (2 vs. 4 days) without any improvement in the rate of hospitalization, duration of hospitalization, duration of parenteral antibiotic therapy, or frequency of confirmed infections.

In contrast to therapeutic use in the afebrile setting, several randomized trials have examined therapeutic use in the febrile setting. The earlier guidelines included five randomized studies that had conflicting results. One study was a large, multicenter trial randomly assigning patients with febrile neutropenia being treated with intravenous antibiotics to either rHuG-CSF or placebo [44]. Although there was no benefit in major clinical outcomes such as fever duration or time on antibiotics, some patients with protracted neutrophil recovery may have benefited based on reduction in hospitalization beyond 11 days and a reduced need for empiric antifungal therapy. Another study comparing 50 episodes of febrile neutropenia treated with rHuGM-CSF plus antibiotics compared with antibiotics alone did not find improvements in duration of fever, neutropenia, or antibiotic therapy [36]; however, patients with pneumonia, cellulitis, an abscess, or sinusitis did have a higher response rate to rHuGM-CSF with antibiotics versus antibiotics alone (100% vs. 59%, respectively) providing evidence for the potential benefit of CSF in select situations of febrile neutropenia. A reduction in hospitalization was not seen in another smaller randomized trial of rHuGM-CSF used in addition to antibiotics in 15 patients with fever and grade 4 leucopenia [45]. In contrast, two studies did show reductions in hospitalization with CSF therapy in patients with febrile neutropenia, although biases may have been introduced by requiring specific endpoints for discontinuing treatment and discharge that may have confounded the results [46, 47].

The next series of randomized trials using CSF in febrile neutropenia also had varying clinical outcomes that were considered in developing the 2000 ASCO guidelines. The largest study of this period included 134 patients who were randomly assigned to rHuGM-CSF or placebo; no decrease in length of hospitalization with the potential risk of reducing quality-of-life indicators was seen [48]. Clinical benefit was seen in three other randomized trials. One of these included 121 patients who were randomly assigned to placebo, rHuG-CSF, or rHuGM-CSF [46]. In this study, a clinical benefit was seen in each CSF treatment group with a reduction in duration of neutropenia and length of hospitalization. When comparing the outcome between rHuG-CSF and rHuGM-CSF, a nonsignificant trend towards increased infection-related mortality was noted with rHuG-CSF treatment. A randomized study of 68 patients with neutropenic fever using rHuGM-CSF reduced median duration of neutropenia, days of antibiotic therapy, and duration of hospitalization [49]. A subset analysis found the greatest benefit in patients who were considered low risk for neutropenic fever. Similar positive findings were seen in a pediatric study of 186 children being treated for leukemia, lymphoma, or a solid tumor complicated by febrile neutropenia who were randomly assigned to either rHuG-CSF or placebo [50]. Median hospital stay and duration of antibiotics were reduced in patients receiving rHuG-CSF, with subset analysis finding benefit in patients with acute lymphocytic leukemia (ALL), non-dose-intensive alkylating agent-based regimens, and fever within 10 days after completing chemotherapy, lack of a focal source of infection, or without documented septicemia.

A multicenter trial conducted compared rHuG-CSF plus antibiotics to antibiotics alone in adult patients with solid tumors or lymphoma presenting with febrile neutropenia and at least one high-risk factor (defined as ANC $<0.1 \times 10^9$/L, short latency of <10 days from prior chemotherapy cycle, sepsis or clinically documented infection, severe comorbidity, or prior inpatient status, or ECOG 3–4) [51]. Patients who received rHuG-CSF had a reduced period of grade 4 neutropenia, antibiotic therapy, and duration of hospitalization although survival was similar between the two groups.

Two meta-analyses were included in the 2006 ASCO guidelines. The first included 962 patients and found no mortality advantage when using CSF for febrile neutropenia [52]. A Cochrane systematic review and meta-analysis included 1,518 patients from 13 randomized trials and observed less prolonged neutropenia, decreased length of hospitalization, and marginally less infection-related mortality with no difference in overall mortality between patients treated with CSF plus antibiotics versus antibiotics alone [38].

Cost analysis has been performed in three randomized trials for use of CSF in febrile neutropenia [46, 48, 50]. The results have been difficult to interpret because cost differences were not statistically different and were inconsistent among the three studies. One of the studies showed a trend toward cost savings with the use of CSF, while the other two found trends toward decreased costs for placebo recipients. A "Markov-type" economic model has been applied to assess cost comparisons with the use of CSF in patients with febrile neutropenia [53]. Savings were observed only when CSF was used in patients with a high risk of complicated infections.

Prediction models identified patients at high risk for developing complications from febrile neutropenia [15, 54]. Patients who were considered high risk included those who developed febrile neutropenia as an inpatient; or had hypotension, sepsis, cardiovascular or pulmonary disease, or a diagnosis of leukemia or lymphoma; were aged >65 years; or had prior fungal infection, uncontrolled malignancy, visceral organ involvement, organ dysfunction, and profound or prolonged neutropenia. A high mortality rate in septic patients with febrile neutropenia was seen in more recent studies as well. Mortality rates as high as 82% were reported in patients presenting with febrile neutropenia and shock [55]. A 30-day mortality of 54% was seen in patients with febrile neutropenia admitted to intensive care [56]. Since the 2006 ASCO guidelines, two randomized, double-blinded studies have been published with conflicting results. The larger of the two studies included 166 patients who were admitted to intensive care with septic shock and randomly assigned to either rHuG-CSF or placebo [57]. The in-hospital mortality was similar between the two groups at 27% in the rHuG-CSF group and 25% in the placebo group; however, the rHuG-CSF group did have a higher rate of liver toxicity (11% vs. 1%, $p = 0.007$). In the second randomized study, 38 patients with severe sepsis or septic shock and immunosuppression (defined as monocytic HLA-DR <8,000 monoclonal antibodies per cell) were randomly assigned to receive either rHuGM-CSF or placebo for 8 days [58]. Patients receiving rHuGM-CSF had shorter times on a mechanical ventilator (148 vs. 207 h, $p = 0.04$), an improved Acute Physiology and Chronic Health Evaluation-II Score ($p = 0.02$), and a nonsignificantly shorter length of in-hospital and intensive care stay compared with placebo. Given the conflicting nature of the results, further randomized studies will need to be performed to address the indication of CSF for treatment of sepsis in this setting.

5 CSF to Increase Chemotherapy Dose-Intensity and Dose-Density

From evidence that there is a dose–response curve for chemotherapeutic agents and from retrospective data suggesting higher doses may result in better efficacy, investigators have evaluated a number of dose-intense and dose-dense regimens to improve clinical outcomes [59–65]. Prospective studies that had not incorporated CSF support to achieve dose intensity beyond conventional doses showed promise for clinical benefit that led to trials incorporating CSF to try to maintain the dose intensity while potentially reducing toxicity [66–68]. Until the 2006 ASCO guidelines, however, advantages in treatment outcome or survival had not been shown in dose-intense or dose-dense regimens supported with CSF. Three large randomized studies in patients with node-positive breast cancer or NHL were available for review in the current guidelines showing improvement in disease-free and overall survival with dose-dense regimens using CSF to maintain the schedule [8–10].

Phase 1 data have shown that use of CSF can increase doses of chemotherapy by up to 30% when the primary toxicity is neutropenia [69]. Use of rHuG-CSF to maintain chemotherapy regimens in settings of SCLC and NHL improved delivery of chemotherapy with increases in dose intensity of 8% and 13% with CSF support, respectively [18, 19]. Previous to the most current guidelines, however, this finding did not translate into improvements in response rates or disease-free survival. Randomized studies using rHuGM-CSF had shown less consistent improvements in dose intensity [21–23]. When studies were able to show the ability of rHuGM-CSF to reduce hematologic toxicity thus allowing for dose intensification, enhanced clinical outcomes were again not demonstrated [70].

Randomized trials of dose intensification did begin to emerge before publication of the ASCO 2000 guidelines. At that time, the benefit of using CSF to achieve dose intensification was not clearly defined. In SCLC, patients were randomly assigned to ifosfamide, carboplatin, etoposide, and vincristine to be given every 3 versus 4 weeks with or without rHuGM-CSF support [71]. Although survival was improved in the dose-intensified group, no benefit was seen from the reduction of myelosuppression using rHuGM-CSF. A large randomized, multicenter clinical trial in SCLC did find benefit of rHuG-CSF to achieve dose-intense chemotherapy [72]. Using rHuG-CSF to support doxorubicin, cyclophosphamide, and etoposide every 2 weeks versus 3 weeks without support increased dose intensity with longer survival (HR 0.80, 95% CI, 0.65–0.99) with similar quality of life. In contrast, a number of trials showing dose intensification with CSF support failed to show improvements in response rates or survival [73–77].

The 2006 guidelines have had the addition of positive data from randomized clinical trials supporting the use of CSF to maintain dose density in patients with node-positive breast cancer or NHL. In patients with node-positive breast cancer, 2005 women were randomly assigned to receive either sequential or combined chemotherapy regimens of adriamycin and cyclophosphamide followed by paclitaxel in 3-week intervals or 2-week intervals supplemented with rHuG-CSF [8]. Both dose-dense regimens had a survival advantage over the standard 3-week regimen for disease-free (RR 0.74, $p = 0.01$) and overall survival (RR 0.69, $p = 0.013$). Despite these data, controversy remains as to the long-term benefits of dose-dense therapy in node-positive breast cancer. In diffuse aggressive NHL, benefit with CHOP chemotherapy every 14 days with CSF support showed clinical benefit with a reduction in the primary endpoint, event-free survival, in comparison to an every 21-day regimen in both in young (RR 0.61, $p = 0.048$) and elderly patients (RR 0.66, $p = 0.003$) [9, 10]. Furthermore, in the elderly cohort, an overall survival advantage was observed (RR 0.58, $p < 0.001$).

The benefit of dose-dense regimens with CSF support is not generalizable to all diseases, however. Data from a randomized study in patients with SCLC investigated dose-dense ifosfamide, carboplatin, and etoposide every 2 weeks with CSF support followed by autologous stem-cell rescue versus the standard arm of the same chemotherapy regimen administered every 4 weeks [78]. Although duration of treatment and incidence of febrile neutropenia were reduced in the dose-dense group receiving CSF, the study failed to show improvements in overall response rates, 1- or 2-year survival, or overall survival. Similarly, benefits have not been seen with dose-dense regimens using CSF support in advanced ovarian cancer or soft tissue sarcoma [79, 80]. Based on the current data, the 2006 ASCO guidelines advocate use of CSF to maintain dose intensity only in the settings of node-positive breast cancer and NHL, and recommend the use of dose-dense regimens only in clinical trials.

6 CSF as Adjuncts to Hematopoietic Stem-Cell Transplantation

Hematopoietic stem cell (HSC) transplantation is an important treatment modality for many hematologic and nonhematologic diseases. Peripherally mobilized HSC is the most frequent modality for hematopoietic transplantation due to the ease of obtaining the HSC and evidence of more rapid recovery in cell counts [81]. The success of engraftment strongly correlates to the number of $CD34^+$ cells transplanted, with a minimum of 2×10^6 $CD34^+$ cells being required [82]. Despite transplanting more than this threshold, complications with delayed engraftment do occur, increasing the potential for severe morbidity and mortality from infectious or bleeding events. For these reasons, CSF have been investigated to mobilize a sufficient number of $CD34^+$ cells and to accelerate engraftment.

In the 1994 guidelines, 13 randomized studies of CSF administered after bone marrow transplantation were evaluated to help formulate the guidelines. Eight were studies in autologous HSC transplantation, three in allogeneic HSC transplantation, and two in either autologous or allogeneic HSC transplantations [83–96]. Of the 13 studies, 8 incorporated rHuGM-CSF and 5 rHuG-CSF. The studies showed consistent benefits in reducing the duration of neutropenia using CSF after autologous bone marrow transplantation. Although one study also found a reduction in duration of thrombocytopenia, this result was not reproducible in the other studies. Improving the duration of neutropenia did translate into shorter lengths of hospitalization, antibiotic requirements, and frequency of infections, but not in decreasing infectious mortality. The duration of prolonged fevers was reduced in studies using rHuG-CSF. Cost-savings analysis did show benefit in using CSF after bone marrow transplantation. In the 1994 ASCO guidelines, only a single randomized study of peripherally mobilized autologous HSC was available for review. This study showed a statistically significant improvement in neutrophil recovery in patients supported with CSF although the clinical benefit was difficult to determine [97].

The data were less consistent with the use of CSF in the setting of allogeneic transplantation. No increase in complications of graft-versus-host disease (GVHD), rejection, or relapse of the underlying disease was noted, and most studies observed reduced durations for neutrophil engraftment. In contrast, the data were mixed on the effects of duration of fever and antibiotic use and did not show any significant improvement in length of hospitalization.

Level II and III evidence was available during the first ASCO guidelines showing that CSF could enhance the yield of peripherally mobilized HSC [98–109]. Having an adequate number of $CD34^+$ cells in circulation can reduce the number of apheresis procedures required, with one study showing a reduction in the cost of collecting HSC [104].

With the evolution of transplantation from bone marrow to peripherally mobilized HSC, more data became available in developing further guidelines. In the 1996 ASCO update, data from many trials were available supporting the use of CSF after peripherally mobilized HSC in autologous transplantation for a more rapid recovery of cell counts [91, 110, 111]. Use of CSF after allogeneic HSC transplantation was further supported by new data as well as the efficacy and safety of CSF to mobilize donor HSC for allogeneic transplantation [112–116]. In this update, a new potential application of CSF for patients with delayed or inadequate neutrophil engraftment was presented based on data from a randomized trial [117].

Further data were available in the development of the 2000 ASCO guidelines with respect to the use of CSF-induced peripherally mobilized HSC for both allogeneic and autologous transplantation. The administration of CSF after transplanting either autologous or allogeneic peripherally mobilized HSC continued to show faster neutrophil recovery with cost savings [118, 119]. Although comparisons between peripherally mobilized HSC versus bone marrow HSC showed more rapid hematopoietic recovery favoring peripherally mobilized HSC, there was concern regarding the potential for worsened chronic GVHD [120]. A large, multicenter, randomized study showed no statistically significant increased risks for GVHD with peripherally mobilized HSC [81]. With the available data, the 2000 ASCO guidelines continued to recommend use of CSF to mobilize HSC and for use after HSC infusion. The recommendations mentioned that the optimal dose may be higher than previously expected with up to 10 μg/kg/day rHuG-CSF possibly yielding a larger number of $CD34^+$ cells [110, 121].

A significant change in the 2006 ASCO guidelines was withdrawing the recommendation for the use of CSF after allogeneic transplantation. This decision was based on a study of 1,789 patients with acute leukemia that reported that rHuG-CSF use after allogeneic HSC infusion increased the incidence of GVHD and reduced survival without improving length of hospitalizations, antibiotic use, or cost savings [122]. Although neutrophil engraftment to a threshold of 0.5×10^9/L was accelerated ($p < 0.01$), platelet recovery to $>50 \times 10^9$/L was reduced in patients who had received rHuG-CSF after HSC infusion. Patients who received rHuG-CSF after bone marrow transplantation had a 50% rate of grade 2–4 GVHD versus 39% in patients who had not received rHuG-CSF (RR 1.33, $p = 0.007$). The effects on survival were concerning because reduced lower overall survival (RR 0.59,

$p < 0.0001$), leukemia-free survival (RR 0.645, $p = 0.0003$), and higher transplantation-related mortality (RR 1.73, $p = 0.00016$) were seen with rHuG-CSF. In contrast to the changes in the recommendation for CSF use after allogeneic HSC transplantation, data continued to show a benefit in neutrophil recovery and cost savings with the use of CSF after autologous HSC transplantation and for the use of rHuG-CSF to mobilized HSC into the peripheral blood [123, 124].

7 CSF in Acute Leukemias and Myelodysplastic Syndromes

Infections represent a significant cause of morbidity and mortality in patients with acute myeloid leukemia (AML) and myelodysplastic syndrome (MDS) [125, 126]. In AML, both the underlying disease and aggressive chemotherapy place patients at risk for infections that are often difficult to control. In certain situations, the antibiotic therapy itself often carries a significant cost and morbidity. Elderly patients are especially susceptible to infectious complications leading to a high rate of treatment failure [127]. Fungal infections are not uncommon complications during the course of treatment and often require administration of amphotericin-B that has many potential toxicities and may lead to treatment delays or prohibit more aggressive therapies. Using CSF can potentially reduce the duration of neutropenia, thus preventing complicated infections. Another hypothesized advantage to using CSF in patients with myeloid malignancies was based on in vitro data showing CSF can alter the cell cycle promoting sensitivity to S-phase specific agents, namely cytarabine [128–136]. Priming leukemic cells, however, may activate normal hematopoietic cells resulting in an exacerbation of the cytotoxicity. Another concern regarding CSF therapy is that most myeloid leukemic cells express CSF receptors [137] that may potentially lead to CSF-stimulating proliferation of leukemic cells, preventing chemotherapy-induced apoptosis of leukemic cells, or transforming MDS cells into leukemic cells.

Approaches to managing neutropenia after chemotherapy for treating leukemia include frequent monitoring of neutropenic patients and prompt empiric antibiotics for febrile neutropenia. The prophylactic use of antibiotics and antifungals in this setting is still controversial [138].

During development of the 1994 ASCO guidelines, three large randomized clinical trials evaluating the efficacy of CSF to reduce the duration of neutropenia after induction chemotherapy in AML were reviewed and incorporated. The first study randomly assigned patients aged 55–70 years with AML having completed induction chemotherapy to sargramostim or placebo after showing an aplastic bone marrow [139]. Preliminary reports showed a reduction in the duration of grade 4 neutropenia and grade 3 or 4 infections in the patients, with a trend towards higher complete remission rates and median survival that matured into a statistically significant difference during final analysis. This finding was in contrast to the results from a CALGB trial in newly diagnosed AML patients aged >60 years randomly assigned to treatment with molgramostim versus placebo [140]. Again, preliminary data were available during the 1994 ASCO guidelines showing that, although there was a statistically significant reduction in the duration of neutropenia, it was clinically insignificant because of a lack of benefits in rates of serious infections, duration of hospitalization, acute mortality, or response rates. The third trial evaluated rHuG-CSF in adults with acute leukemia (AML, ALL, or chronic myeloid leukemia [CML] in blast crisis, or acute leukemia in relapse) undergoing induction chemotherapy [141]. Patients were randomly assigned to rHuG-CSF or no cytokine support after demonstrating bone marrow hypoplasia, and the investigators found that the duration of severe neutropenia was reduced by 7 days with a decrease in documented infections with rHuG-CSF. Although antibiotic usage and duration of hospitalization were similar in the two groups, there were trends towards better leukemia responses with higher complete remission rates and disease-free survival with rHuG-CSF.

Data that were available during the first ASCO guidelines did not suggest benefit from CSF priming therapy and were in fact more concerning for harm. The only completed study involved a cohort of patients with AML given concurrent rHuGM-CSF and high-dose cytarabine therapy and compared with matched, historical controls [142]. The investigators were unable to show any clinically relevant benefits from CSF priming and, contrary to what they expected, found a lower complete remission rate with rHuGM-CSF. Other preliminary data that were available at that time were consistent with a detrimental effect from CSF priming with increased hematologic and nonhematologic toxicity.

Patients with MDS often are neutropenic due to ineffective hematopoiesis resulting in greater susceptibility to infection and death. The numbers of neutrophils increase after treatment with CSF in patients with MDS with improvement in infections [143, 144]. Because the neutrophil count declines once CSF therapy is discontinued, investigators evaluated whether continuous CSF treatment would have additional clinical benefit. Two randomized studies were available in abstract form addressing the utility of CSF therapy in MDS patients at the time of the 1994 ASCO guidelines. The first included 133 patients with refractory anemia (RA), refractory anemia with ringed sideroblasts (RARS), refractory anemia with excess blasts (RAEB), or chronic myelomonocytic leukemia (CMML) who were randomly assigned to either 90 days of molgramostim versus placebo [145]. Reductions in major infections were observed with molgramostim with concomitant increases in the neutrophil count without impact on survival or increased risk of transforming the MDS to overt leukemia. The second was a study of 102 patients in RAEB or refractory anemia with excess blasts in transformation (RAEB-T) randomized to rHuG-CSF versus placebo [146]. Neutrophil count in the rHuG-CSF–treated group significantly increased without differences in progression to leukemia or overall survival. Patients with RAEB, however, were found to have reduced median survival (10.4 months vs. 21.4 months) from nonleukemic deaths and these results may have been confounded by higher poor-risk factors in the rHuG-CSF group. Based on the limited data with no survival benefit, the 1994 ASCO guidelines could not recommend the routine use of rHuG-CSF in MDS but did find that it may be appropriate to consider in neutropenic MDS patients with infection.

In the 1996 update, newer data continued to show decreases in duration of neutropenia with administration of CSF in older patients with AML completing induction chemotherapy [147–149]. Less consistent were the effects of CSF administration after induction chemotherapy for AML on improving infectious complications, complete remission rates, and long-term outcomes. No adverse features were identified with CSF in this setting, and the 1996 guidelines continued to recommend the option of using CSF support for patients aged ≥55 years completing induction chemotherapy for AML.

For the 2000 ASCO guidelines, several new placebo-controlled, randomized trials were available to provide strong evidence for the recommendations of rHuG-CSF after induction chemotherapy in AML [150–153]. A summary of the current literature was that CSF support after induction chemotherapy for AML decreased neutrophil recovery time between 2 and 6 days with most data showing a reduction in duration of hospitalization and antibiotic use. It did not appear that CSF had consistent effects on response rates, patient survival, or platelet and red blood cell transfusion requirements. When comparing rHuG-CSF to rHuGM-CSF, both cytokines appeared to have similar outcomes when used in this indication. No differences were noted in outcome when the CSF was begun the day after completion of the chemotherapy or after a 2- to 3-day delay. In fact, using a delayed strategy may be more cost effective, reducing the number of required doses. Three of the studies reported cost analysis, with two studies showing savings of US$2230 and US$2310 and the third study showing an increase in costs of US$120.

Four larger randomized trials using rHuGM-CSF and one smaller trial using rHuG-CSF were available for the 2000 ASCO guidelines to assess for any benefit in trying to prime leukemic cells during induction chemotherapy. Three of the rHuGM-CSF studies were used in the initial induction and the fourth during induction for relapsed AML [149, 151, 152, 154]. The trial using rHuG-CSF started the therapy 2 days before induction chemotherapy for relapsed or refractory AML [155]. Priming effects were consistently seen to not have any impact on response rates, disease-free survival, or overall survival when CSF were used either before or during induction chemotherapy and the 2000 ASCO guidelines did not recommend the use in this setting.

Due to the reduction in neutropenia resulting in shorter durations of hospitalization and lowered requirement for antibiotic use seen when using CSF after induction chemotherapy, investigators began considering the utility of CSF after consolidation chemotherapy. Two new studies using rHuG-CSF after completing consolidation chemotherapy were included in the 2000 ASCO guidelines [148, 156]. Patients with AML in complete remission were randomly assigned to rHuG-CSF or no rHuG-CSF therapy after completing a course of standard-dose consolidation chemotherapy. Both studies showed reductions in duration of severe neutropenia with the benefit of decreasing rates of infection requiring antibiotic therapy. Overall patient survival and complete remission rates were not affected. In the one study, using more intensive consolidation chemotherapy followed by rHuG-CSF, a reduction in length of neutropenia was seen compared with historical controls [156]. Based on these two studies, the 2000 ASCO guidelines added the recommendation for CSF therapy after completing consolidation chemotherapy.

Another new area of investigation included in the 2000 ASCO guidelines was the use of CSF during induction and postremission therapy for ALL. An important distinction between AML and ALL is that for ALL, most protocols incorporate intermittent or continuous treatment with corticosteroid and antimetabolite chemotherapy after the initial induction therapy. Therefore, these trials often had to use CSF while simultaneously receiving chemotherapy. Six studies were available during the 2000 Update of which five were prospective, randomized trials [157–162]. A reduction in the length of neutropenia (ANC $>1 \times 10^9$/L) was seen in all six studies. The effects on incidence of severe infections, febrile neutropenia, maintaining timing of chemotherapy, and hospitalization were more variable. None of the randomized studies found any improvement in disease-free or overall survival with the use of rHuG-CSF. The largest of the trials in adults was conducted by CALGB, which included 198 patients who were randomly assigned to rHuG-CSF or placebo. A reduction in duration of neutropenia by 6 days and a trend towards a higher complete remission rate was noted in patients aged >60 years who had received rHuG-CSF. The largest pediatric trial included 164 children with ALL. In this trial, the length of neutropenia was reduced by 8 days that resulted in shorter median hospital durations and fewer documented infections. Rates of hospitalization were not improved and the costs were increased in patients who received rHuG-CSF by US$2497. Given the benefit of reducing the duration of neutropenia without any adverse consequences of administering the CSF concurrent with corticosteroid and/or antimetabolite chemotherapy, the 2000 Update endorsed the recommendation for use of rHuG-CSF after completing induction chemotherapy or the first postremission course in ALL.

The 2006 update maintained its recommendation for the use of CSF after initial induction or consolidation for AML and did not recommend CSF for chronic use in MDS based on no new major studies being available in these settings. Two randomized trials evaluated the possible priming effect of leukemia cells in AML, one in younger adults (18–60 years old) and the other in older adults (aged >60 years) [163, 164]. The first trial in younger adults administered rHuG-CSF during induction therapy and found a higher rate of disease-free survival in patients without any effects on complete remission or overall survival. This result was most pronounced in patients with "standard risk" cytogenetics and no benefit was seen in the more unfavorable risk category. Because the standard risk patients were deemed heterogeneous and did not have set defining criteria, the panel members felt that rHuG-CSF priming in this situation could not be recommended. In the latter study, patients were randomly assigned to either rHuG-CSF concomitant with chemotherapy, concomitant and postchemotherapy, or postchemotherapy alone. Patients in the first two groups who received rHuG-CSF concurrently with chemotherapy had higher complete remission rates (60% vs. 50%, $p = 0.01$) but this did not translate into either higher disease-free or overall survival. Based on lack of consistent clinical benefit with prior data showing the possibility of toxicity to normal hematopoietic cells, the ASCO guidelines in 2006 remained unchanged in this regard.

In ALL, one new large randomized study was reported in 287 children with high-risk ALL who were randomly assigned to receive rHuG-CSF during either induction or initial consolidation [165]. Although the median duration of neutropenia was reduced by 2.5 days, it did not result in any improvements in febrile neutropenia, serious infections, or hospitalization, and no benefit was found in event-free or overall survival. Based on data showing reductions in duration of neutropenia with variable effects on clinical outcome, the 2006 ASCO guidelines maintained its recommendation for the use of CSF after completing initial induction or first postremission chemotherapy in ALL. The new guidelines cautioned regarding the use of CSF in this setting based on data regarding the increased risk for therapy related myeloid leukemias and AML [166].

Areas that may need to be included in future updates include the use of CSF with erythropoietic-stimulating agents (ESA) in MDS and severe aplastic anemia. Based on retrospective data of cohorts of MDS patients treated with ESA and rHuG-CSF, there may be both reduced transfusion requirements and improved survival [167–169].

8 CSF Use in Concurrent Chemotherapy and Irradiation

Using chemotherapy agents to act as radiosensitizers can potentiate antitumor effects when given concomitantly with radiation therapy in certain diseases including head and neck, esophageal, lung, cervical, and rectal cancer [170–174]. Toxicities from concurrent multimodality therapy include myelosuppression, mucositis, and pneumonitis. Before the 1994 guidelines, data showed that CSF can hasten myeloid recovery and preliminary data showed CSF use might reduce the incidence of mucositis [175]. Based on these data, investigators postulated that there may be clinical benefit from using CSF support with concurrent chemoradiation therapy. Options for treating toxicity from concurrent chemoradiation therapy include holding either the chemotherapy or radiation for severe myelotoxicity, mucositis, or pneumonitis and supportive treatment with antibiotics, analgesics, steroids, and nutritional support with oral supplements or gastrointestinal tube feeds.

Data from two studies in lung cancer using CSF during concurrent chemoradiation therapy were evaluated during the first ASCO guidelines. The first study included seven patients with stage IIIB NSCLC who were treated with cisplatin, etoposide, and mitomycin with concurrent radiation during the initial cycles. rHuG-CSF was given on days 4 through 17 of each cycle [176]. The investigators found that the patients receiving rHuG-CSF had significantly more thrombocytopenia than the patients not receiving rHuG-CSF. The second study was a randomized Southwest Oncology Group (SWOG) trial including 213 patients with limited-stage SCLC treated with cisplatin and etoposide plus concurrent radiation during the first two cycles of chemotherapy [177]. The patients were randomly assigned to rHuGM-CSF (sargramostim) or placebo for days 4–14 of each cycle; a significantly higher percentage of patients receiving rHuGM-CSF were shown to have grade 3 or 4 thrombocytopenia (65% vs. 10%, $p < 0.001$). The data from this study were only available in abstract form during the 1994 ASCO guidelines and final data available during the ASCO guidelines in 2000 confirmed significant toxicity without any clinical benefit. Given these findings, the panel members agreed that administering CSF with mediastinal radiation increased the risk for adverse hematologic reactions and was not recommended. Trying to explain the toxicities seen with mediastinal radiation, two smaller studies were cited that did not show cytopenias with nonmediastinal radiation [178, 179]. Possible explanations for the increased toxicity with chest radiation are that the CSF mobilizes HSC into the periphery that are then damaged by the radiation when passing through the radiation portal involving the heart and great vessels.

At the time of the 2000 update, mature data from the SWOG study investigating rHuGM-CSF in limited-stage SCLC concurrently with chemoradiation was included [177]. This randomized study found higher rates of grade 3 or 4 thrombocytopenia (54% and 35%) with rHuGM-CSF versus placebo (12% and 6%, $p < 0.001$). Pulmonary toxicity was another major concern since death from toxicities was significantly higher with rHuGM-CSF and was attributed to pulmonary damage ($p < 0.01$). Although improvement in the neutrophil nadir was seen with rHuGM-CSF, there were higher rates of hospitalization and antibiotic usage. Smaller studies investigating CSF for craniospinal disease, Hodgkin's disease, and multiple myeloma treated with large-field irradiation alone have shown that there may be some improvements in neutropenia [179–181]. Based on these data, the 2000 ASCO guidelines continued to not recommend the use of CSF for concurrent chemoradiation but did provide the option for using CSF when radiation alone to large fields is complicated by neutropenia to reduce delays in therapy. No changes were made to these guidelines in 2006 based on no new available data from randomized studies.

9 CSF Use in Older Patients

Age is an important risk factor for development of neutropenia and its resulting complications after chemotherapy. Studies have shown that people aged >60 years are at higher risks for neutropenia after chemotherapy and that mortality from infection may be higher from neutropenic infections in older patients with lymphoma [182–187]. Convincing evidence for the need to investigate the use of CSF in the elderly comes from a retrospective study comparing the incidence of neutropenic fever during treatment for NHL [188]. For patients aged >65 years, 34% developed neutropenic fever compared with 21% in the younger cohort; most episodes occurred during the first cycle of therapy. Similarly, longer lengths of hospitalization were seen in patients aged >65 years at 12.1 days versus 8.2 days in the younger group. With one exception, prospective studies using CHOP-like regimens for patients with NHL aged >60 years have shown high rates of neutropenic infections [189–197]. Incidences ranged from 27 to 47%, with only one study showing a rate of 11–15%. Again, most of the complications of infection were during the first course of treatment with improvement in subsequent cycles from either dose reduction or widening the time intervals between cycles.

The alternative option to CSF therapy in elderly patients is dose reduction; however, in patients with lymphoma undergoing therapy with curative intent, this option is not preferred because dose reduction may reduce the response rate and survival [192, 194, 195, 198–200].

Decreases in the rates of febrile neutropenia and infection were consistently observed in all randomized and retrospective studies [190, 193, 196–198, 201, 202]. The impact on survival was less clear because of conflicting data. In one study, 455 patients aged >60 years with NHL being treated with either a CHOP or a CNOP (cyclophosphamide, mitoxantrone, vincristine, and prednisone) regimen were randomized to receive rHuG-CSF or no rHuG-CSF [193]. The percentage of patients able to undergo >90% of their recommended dose was significantly higher with rHuG-CSF ($p = 0.05$) and this translated into fewer deaths (62% vs. 45%) and improved projected 5-year survivals (45% vs. 60%) favoring rHuG-CSF ($p = 0.04$). In contrast, another study was unable to observe any improvement in projected 5-year survival despite an increase in dose intensity with rHuG-CSF [197]. Based on the consistent improvement in febrile neutropenia and reductions in infection, the 2006 update endorsed primary prophylaxis with CSF in older patients being treated for lymphoma with curative intent.

10 CSF Use in the Pediatric Population

Due to the greater curability of pediatric cancers, treatment regimens for pediatric malignancies are often more intensive. Additionally, most pediatric patients are enrolled in clinical trials. As expected, with more aggressive therapy, myelosuppression is typically more frequent and profound [203]. Febrile neutropenia and life-threatening infection are of great concern in the pediatric population, especially infants, because of an immature immune and hematopoietic system. Alternative options to CSF support include reducing the doses of chemotherapeutic agents, dose delay, and close patient monitoring with consideration of antibiotic support.

The pediatric trials that were included in the 1994 guidelines consistently found reduced lengths of neutropenia with less consistent effects on the incidence of febrile neutropenia or hospitalization rates [47, 180, 204–213]. These studies failed to show any benefit in tumor response or tumor-associated mortality in these pediatric trials. Toxicities such as bone pain and fever were similar to adults but there was concern for a higher rate of splenomegaly. CSF was also seen to be effective for mobilization of HSC for transplantation in pediatric patients [214]. The data showed that children may have a greater response and tolerate higher doses of rHuG-CSF and rHuGM-CSF with doses of 5–10 μg/kg/day and 500–750 μg/m^2/day, respectively [47, 204, 210, 213].

In the 2000 update, it was noted that most pediatric patients are enrolled in pediatric protocols and these protocols directed the use of CSF. A review of practices in the Pediatric Oncology Group was cited showing that primary prophylaxis with CSF occurred more frequently in children than adults, and use was based on the prediction of neutropenia >7 days [215]. The pediatric protocols did not have uniform indications for CSF prophylaxis. For secondary prophylaxis, pediatric oncologists preferred the use of CSF to dose reduction, but otherwise indications appeared similar to adults.

A meta-analysis was available for the 2006 update, incorporating data from 16 randomized clinical trials including 1,183 children; clinical benefits with the use of CSF were noted [216], including significant reductions in febrile neutropenia (RR 0.80, $p = 0.01$), infections (RR 0.78, $p = 0.02$), and length of hospitalization by an average of 1.9 days (95% CI, 1.1–2.7 days, $p < 0.00001$). A cost-savings analysis regarding the use of rHuG-CSF in pediatric ALL patients failed to show benefit in this setting [217]. Another investigation raised concern with respect to the use of CSF in pediatric ALL patients because of an increased risk of developing therapy-related myeloid leukemia or MDS in this setting [166]. This risk was most prominent for children who had been treated with concurrent irradiation, topoisomerase II inhibitors, or alkylating agents. Taking these data into account, the 2006 guidelines advocate use of CSF for primary and secondary prophylaxis in children at high risk for febrile neutropenia while cautioning regarding their use in children with ALL.

11 CSF Initiation, Duration, Dosing, and Route of Administration

In three large randomized clinical trials published before 1994, rHuG-CSF was given at 230 μg/m^2/day (equivalent to 5 μg/kg/day), and the results showed decreasing rate of febrile neutropenia [17–19]. Based on smaller trials, lower doses of rHuG-CSF may be effective in improving the length and severity of neutropenia. One small randomized trials comparing 1–4 μg/kg/day of rHuG-CSF found that the lower dose may reduce the duration of neutropenia better than higher doses [218]. In trials comparing efficacy of subcutaneous versus intravenous administration of rHuG-CSF, subcutaneous doses of rHuG-CSF as low as 50 μg/m^2/day were shown to reduce the duration of neutropenia [219–221]. These studies found the effective dose range of intravenous rHuG-CSF to be 100–200 μg/m^2/day (equivalent to 2.5–5.0 μg/kg/day). When comparing the intravenous to the subcutaneous route, the studies suggested that the subcutaneous route may have more activity. One exception is with lenograstim after bone marrow transplantation or high-dose chemotherapy for ALL [222, 223]. Intravenous lenograstim may enhance neutropenic recovery in this setting. Doses of filgrastim up to 30 μg/kg/day have been evaluated after bone marrow transplantation without any clinical advantages compared with lower doses [90, 91, 96]. In 1994, data were lacking that suggested a dose–response effect with rHuG-CSF for peripheral mobilization of HSC and these data indicated no improvement in HSC yield with doses of rHuG-CSF >3 μg/kg/day [100]. Two studies titrated doses of rHuG-CSF up to 24 μg/kg/day and were unable to show improvements in peripherally mobilized HSC yields [101, 224]. Neutrophil responses to filgrastim in patients with MDS have been observed in doses ranging from 0.3 to 10 μg/kg/day [144, 225].

Two forms of GM-CSF, sargramostim and molgramostim, were evaluated in 1994. An important difference between the two drugs is that sargramostim is glycosylated resulting in a greater molecular weight that may lead to lower specific activity, receptor-binding affinity, and distinct pharmacokinetics. Based on a phase 1 study of sargramostim after intensive chemotherapy, doses ranging from 250 to 750 μg/kg/day have been shown to have activity [226]. When given as a 2-h intravenous infusion after autologous bone marrow transplantation, a dose of 250 $\mu g/m^2$/day of sargramostim was shown to accelerate neutrophil recovery and result in decreased length of hospitalization and decreased antibiotic usage [83]. In MDS, neutrophil responses have been seen with doses of 30–250 $\mu g/m^2$/day [143, 227–229]. Two randomized studies compared different doses of molgramostim after chemotherapy. The first was in SCLC after CAE regimen. The doses of 5 or 10 μg/kg/day had similar efficacy with regards to neutropenia while 20 μg/kg/day was found to be too toxic [22]. The second study compared 5 versus 10 μg/kg/day of molgramostim and found a slight advantage with 10 μg/kg/day when used in later chemotherapy cycles without any improvement in infection or hospitalization rates [230]. In MDS, doses of molgramostim as low as 0.5 μg/kg/day were sufficient to resolve neutropenia [231].

In the 2000 update, new studies found that higher doses of rHuG-CSF or rHuGM-CSF had no additional clinical benefit with the exception of peripherally mobilizing HSC [232–237]. Using 10 μg/kg/day rHuG-CSF improved the leukapheresis yields compared with lower doses. Furthermore, new evidence showed that the scheduling of rHuG-CSF to 5 μg/kg/bid was more effective than 10 μg/kg/day for mobilization [238]. In this update, the preference of subcutaneously administered CSF was strengthened with new pharmacokinetic analyses [239, 240].

The 2006 update included a randomized study of 506 patients with early stage breast cancer being treated with epirubicin and cyclophosphamide [241]. Varying doses and schedules of subcutaneous rHuG-CSF were administered (480 μg/day on days 8–14; 480 μg subcutaneously on days 8, 10, 12, and 14; 300 μg/day on days 8–14; 300 μg/day on days 8, 10, 12, and 14; or 300 μg/day on days 8 and 12). Although the rates of grade 4 neutropenia were decreased from 41.6% in the control group to 5.4% in the rHuG-CSF groups, febrile neutropenia only occurred in 7% of non-CSF patients. When comparing the different rHuG-CSF regimens, the incidences of grade 3 or 4 neutropenia and the rates of delaying cycles of chemotherapy were similar among all five treatments. Low grades of toxicity including grade 1–3 bone pain (53% vs. 29%, $p = 0.01$) and grade 1–2 fever (24% vs. 8%, $p = 0.04$) were lower in the least intensive rHuG-CSF cohort of 300 μg/day on days

8 and 12 in comparison to the daily schedules. Although these data suggest that alternative dosing may have less low-grade toxicity and reduce costs by administering less frequent and lower doses without a negative impact on efficacy, the panel members identified flaws in the study. The rates of febrile neutropenia were too low to recommend CSF support in any of the existing guidelines and the small number of patients in each group led to the study being underpowered. The panel members suggested that this be an area of further investigation in a larger randomized clinical trial.

To maximize the clinical benefits while minimizing the cost of CSF therapy, it is important to understand the optimal time to initiate and discontinue CSF therapy. The package insert instructions for filgrastim states it should not be initiated any earlier than 24 h after completion of chemotherapy, within 24 h of starting the next chemotherapy cycle, and daily dosing continued until the ANC reaches at least 10×10^9/L from the nadir [242]. After bone marrow transplantation, once an ANC threshold of 1×10^9/L is achieved, the dose should be tapered from 10 to 5 μg/kg/day for an additional 3 days and then discontinued if the ANC is maintained $>1 \times 10^9$/L. Because sargramostim has a very specific indication for use after autologous bone marrow transplantation, the package insert only addresses their recommendations in this setting [243].

In 1994, the panel reviewed several studies to determine the validity of the package insert guidelines. There was concern about the risk of heightened progenitor HSC chemosensitivity if CSF was started too soon after chemotherapy [244]. Varying levels of data had shown that concurrent chemotherapy with rHuG-CSF support had higher degrees of neutropenia with delayed recovery [41, 226, 245–248]. Data supported the package insert recommendation for rHuG-CSF therapy to continue until an ANC of 10×10^9/L was achieved. There was concern that stopping the G-CSF prematurely at lower numbers may cause a decline in the neutrophil counts [249, 250]; however, data showed that tissue concentrations of neutrophils recover several days before the circulating numbers normalize and reinfection is uncommon at low but increasing ANC [251–255].

The rHuGM-CSF package insert was modified before the 2000 update. One of the discontinuation rules was changed such that an ANC $>1.5 \times 10^9$/L for three consecutive days should be the trigger to stop. The package insert modified the rule when the ANC reached 20×10^9/L to allow for the option of either a 50% dose reduction or discontinuing rHuGM-CSF. New data available during the 2000 update questioned the timing of CSF after peripherally mobilized HSC. The duration of neutropenia was shown to be reduced if CSF was administered within 1–5 days after HSC reinfusion [256, 257]. New recommendations were made regarding the use of CSF in graft failure or delays in engraftment with the suggested dose of rHuGM-CSF being 250 μg/m^2/day for 14 days followed by a 7-day break. The panel recommended up to three such courses with dose escalation to 500 μg/m^2/day in the third course.

A pegylated formulation of rHuG-CSF, pegfilgrastim, received FDA approval in 2002 and data were available for guideline recommendations in the 2006 update. Pegfilgrastim was developed to address the inconvenience of daily administration of CSF. One of the pivotal studies available for the 2006 update evaluated the efficacy of primary prophylaxis with pegfilgrastim in a phase 3, placebo-controlled study [7]. Pegfilgrastim reduced the incidence of febrile neutropenia from 17% to 1% ($p < 0.001$) and resulted in clinical benefit by reducing the rate of hospitalization and intravenous antibiotic use in a cohort of patients with breast cancer being treated with docetaxel-based chemotherapy.

Two studies were evaluated by the panel members regarding the efficacy of pegfilgrastim compared with filgrastim. One was a retrospective analysis that combined data from two randomized, double-blind trials of patients with breast cancer treated with myelosuppressive chemotherapy and receiving CSF support [258]. The combined analysis compared a single dose of 6 mg ($n = 77$) or 100 μg/kg ($n = 149$) of pegfilgrastim given on day 2 of each chemotherapy cycle to daily filgrastim (5 μg/kg/day; $n = 222$) administered until either the ANC was $>10 \times 10^9$/L or 14 days duration. Separately, the two randomized trials found that pegfilgrastim was equally effective at reducing the duration of severe neutropenia as daily dose of filgrastim. When combining the data to evaluate for clinical benefit, the risk of febrile neutropenia was observed to be significantly lower with pegfilgrastim (RR 0.56; 95% CI 0.35–0.89) compared with filgrastim. Although not statistically significant, there were trends for lower rates of hospitalization and intravenous antibiotic use with pegfilgrastim. The second trial reviewed compared pegfilgrastim to filgrastim in a randomized phase 2 study in 66 patients with lymphoma treated with etoposide, cisplatin, cytarabine, and methylprednisolone [259]. Both the incidence and duration of grade 4 neutropenia were similar for pegfilgrastim (69% and 2.8 days) and filgrastim (68% and 2.4 days). Based on these studies, the panel members included pegfilgrastim in the guideline for chemotherapy support. The panel members did not recommend pegfilgrastim for HSC mobilization, use in dose-dense therapy, or in the pediatric population based on a lack of strong data supporting efficacy in these situations. Furthermore, the committee presented caveats, importantly the need for long-term safety data.

Of interest, several studies since the 2006 ASCO guidelines have compared the cost savings of pegfilgrastim versus filgrastim [260–263]. Cost-effect analyses suggest that pegfilgrastim may be cost saving when compared to an 11-day course of filgrastim and also cost effective to a 6-day course of filgrastim.

12 Comparative Activity of G-CSF and GM-CSF

The 1994 guidelines were unable to definitively comment on any comparison between rHuG-CSF and rHuGM-CSF due to a dearth of large-scale, prospective clinical trials addressing this question. For the 2000 update, a randomized double-blind comparison study of CSF efficacy after chemotherapy was available for evaluation [264]. Patients were included if they had an ANC $<0.5 \times 10^9$/L and had received chemotherapy within 4 weeks of entry, were aged >17 years, and were afebrile. A total of 181 patients were enrolled of which 170 were available to compare efficacy. The study found no significant time difference to achieve an ANC of 0.5×10^9/L or incidences in fever or hospitalization between the two CSF. Statistically shorter times required to achieve an ANC of 1.0 and 1.5×10^9/L were observed favoring rHuG-CSF. Although the panel felt that the study had multiple design flaws making a comparative analysis of efficacy difficult, comparing the toxicity profiles did have some value. Most of the toxicities were similar between the two CSF with the only exception of a higher incidence of grade 2 fever 4 h after administering rHuG-CSF (10.7%) compared with rHuGM-CSF (3.8%).

Another randomized trial that was included in the 2000 ASCO guidelines compared rHuG-CSF and rHuGM-CSF or the sequential combination of both CSF in attempting to mobilize $CD34^+$ HSC in 156 patients with myeloma, lymphoma, or breast cancer [265]. After chemotherapy and HSC infusion, all patients in this study received rHuG-CSF. More fevers and unexplained increases in the incidence of anemia and red cell transfusion requirements were seen with sargramostim, but there were no differences in platelet nadirs or platelet transfusion requirements. Patients who received rHuG-CSF, either alone or in combination, demonstrated improvements in days to ANC of 0.5×10^9/L (11 vs. 14 days, $p = 0.0001$), which translated into clinical benefits with reduced incidence of fever (18% vs. 52%, $p = 0.001$), hospitalization, and antibiotic therapy. The improved rapidity in ANC recovery was attributed to improved $CD34^+$ cell yields with each apheresis in the patients receiving rHuG-CSF.

For the 2006 update, well-designed studies for support after chemotherapy were lacking. A systematic review of comparison trials between rHuG-CSF and rHuGM-CSF concluded that there was a higher incidence of fever in rHuGM-CSF-treated patients but efficacy data were lacking [266]. In contrast to the lack of randomized studies comparing CSF for support after chemotherapy, studies were published comparing rHuG-CSF to rHuGM-CSF in the posttransplant setting. The first study randomized 71 patients with either breast or ovarian cancer to rHuG-CSF or rHuGM-CSF until completion of the stem cell harvest [267]. This study found improved clinical outcomes with rHuG-CSF with a longer time to progression (61 months for rHuG-CSF vs. 25 months for rHuGM-CSF) and overall survival at a median follow-up of 40 months (75% for rHuG-CSF vs. 50% for rHuGM-CSF). Another study showed that rHuG-CSF consistently improved earlier mobilization of $CD34^+$ cells compared with rHuGM-CSF. The evidence to support better mobilization with rHuG-CSF was further supported by a crossover design study where rHuG-CSF alone was shown to mobilize more $CD34^+$ cells and monotherapy with rHuG-CSF was effective in mobilizing $CD34^+$ cells after failure by combination regimens [268]. Two studies failed to demonstrate any significant differences in mobilization ability with respect to rapidity of neutrophil engraftment [269, 270]. Based on limited available data, the panel was unable to make well-supported recommendations and suggested better comparative trials.

13 CSF as a Treatment for Radiation Injury

Exposure to radiation at doses of 3–10 Grays can lead to probable death from bone marrow failure without proper supportive care, CSF, and/or bone marrow transplantation [271–274]. Doses below this level are typically survivable and doses above this level are almost universally lethal from injury to other organs. This section was added to the 2006 update despite a lack of prospective, randomized trials based on data from the REACT/TS registry where neutrophil recovery was hastened in 25 of 28 cases with CSF support. This improvement was observed in animal models where survival was improved with prompt administration of CSF after exposure to lethal total body irradiation [275–279].

14 CSF Impacts on Quality of Life and HealthCare Costs

In the 2006 update, the emphasis of recommendation for CSF use shifted from cost savings to clinical benefit. New data showed reductions in the incidence of febrile neutropenia, hospitalization, and antibiotic use when the expected rate of febrile neutropenia was 20%. Accordingly, the panel made recommendations based on these clinical benefits and not on economic outcomes. The previous recommendation for primary prophylaxis at a cutoff of a 40% expected rate of febrile neutropenia was based on cost savings. In contrast, the new threshold of 20% is based on the reduction in clinical outcomes.

Understanding that the costs of CSF are substantial, the 2006 ASCO guidelines do recommend that when equivalent regimens are available that do not require CSF, these regimens should be preferred. Alternative dosing strategies will also need to be further investigated for potentially more cost savings.

The effects of CSF on quality of life also need to be further investigated. Data that are available typically have addressed quality of life only as a secondary endpoint and are thus underpowered to detect differences. For example, when using rHuG-CSF with CHOP regimens in elderly patients with NHL, quality of life was not found to be statistically different between the groups, although this endpoint was investigated in only a subgroup of the subjects [197]. Similarly in patients with SCLC, a dose-intense regimen plus CSF versus a standard regimen found equivalent palliation of symptoms and quality of life [72]. The panel advocates more studies focusing on quality of life as a primary measure as this is an important outcome to therapy.

References

1. American Society of Clinical Oncology (1994) American Society of Clinical Oncology recommendations for the use of hematopoietic colony-stimulating factors: evidence-based, clinical practice guidelines. J Clin Oncol 12:2471–2508

2. Bodey GP, Buckley M, Sathe YS, Freirich EJ (1966) Quantitative relationships between circulating leukocytes and infection in patients with acute leukemia. Ann Intern Med 64:328–340
3. American Society of Clinical Oncology (1996) Update of recommendations for the use of hematopoietic colony-stimulating factors: evidence-based clinical practice guidelines. J Clin Oncol 14:1957–1960
4. American Society of Clinical Oncology (1997) 1997 update of recommendations for the use of hematopoietic colony-stimulating factors: evidence-based clinical practice guidelines. J Clin Oncol 15:3288
5. Ozer H, Armitage J, Bennett C et al (2000) 2000 update of recommendations for the use of hematopoietic colony-stimulating factors: evidence-based clinical practice guidelines. J Clin Oncol 18:3558–3585
6. Timmer-Bonte J, de Boo T, Smit H et al (2005) Prevention of chemotherapy induced febrile neutropenia by prophylactic antibiotics plus or minus granulocyte colony-stimulating factor in small-cell lung cancer: a Dutch randomized phase III study. J Clin Oncol 23:7974–7984
7. Vogel CL, Wojtukiewicz M, Carroll R et al (2005) First and subsequent cycle use of pegfilgrastim prevents febrile neutropenia in patients with breast cancer: a multicenter, double-blind, placebo-controlled phase III study. J Clin Oncol 23:1178–1184
8. Citron ML, Berry DA, Cirrincione C et al (2003) Randomized trial of dose-dense versus conventionally scheduled and sequential versus concurrent combination chemotherapy as postoperative adjuvant treatment of node-positive primary breast cancer: first report of Intergroup Trial C9741/Cancer and Leukemia Group B Trial 9741. J Clin Oncol 21:1431–1439
9. Pfreundschuh M, Trumper L, Kloess M et al (2004) Two-weekly or 3-weekly CHOP chemotherapy with or without etoposide for the treatment of elderly patients with aggressive lymphomas: results of the NHL-B2 trial of the DSHNHL. Blood 104:634–641
10. Pfreundschuh M, Trümper L, Kloess M et al (2004) Two-weekly or 3-weekly CHOP chemotherapy with or without etoposide for the treatment of young patients with good-prognosis (normal LDH) aggressive lymphomas: results of the NHL-B1 trial of the DSHNHL. Blood 104:626–633
11. The National Comprehensive Cancer Network. Practice guidelines in oncology: myeloid growth factors. Available at http://www.nccn.org (accession date: 2010)
12. Smith TJ, Khatcheressian LG et al (2006) 2006 update of recommendations for the use of white blood cell growth factors: an evidence-based clinical practice guidelines. J Clin Oncol 24:3187–3205
13. Pizzo PA (1993) Management of fever in patients with cancer and treatment-induced neutropenia. N Engl J Med 328:1323–1332
14. Talcott JA, Finberg R, Mayer RJ, Goldman L (1988) The medical course of cancer patients with fever and neutropenia. Arch Intern Med 148:2561–2568
15. Talcott JA, Siegel RD, Finberg R, Goldman L (1992) Risk assessment in cancer patients with fever and neutropenia: a prospective, two center validation of a prediction rule. J Clin Oncol 10:316–322
16. Hughes WT, Armstrong D, Bodey GP et al (1997) 1997 guidelines for the use of antimicrobial agents in neutropenic patients with fever. Clin Infect Dis 25:551–573
17. Crawford J, Ozer H, Stoller R et al (1991) Reduction by granulocyte colony-stimulating factor of fever and neutropenia induced by chemotherapy in patients with small-cell lung cancer(r-metHUG-CSF). N Engl J Med 325:164–170
18. Trillet-Lenoir V, Green J, Manegold C et al (1993) Recombinant granulocyte colony stimulating factor reduces the infectious complications of cytotoxic chemotherapy. Eur J Cancer 29A:319–324
19. Pettengell R, Gurney H, Radford JA et al (1992) Granulocyte colony-stimulating factor to prevent dose-limiting neutropenia in non-Hodgkin's lymphoma: a randomized controlled trial. Blood 80:1430–1436

20. Gebbia V, Testa A, Valenza R et al (1993) A prospective evaluation of the activity of human granulocyte-colony stimulating factor on the prevention of chemotherapy-related neutropenia in patients with advanced carcinoma. J Chemother 5:186–190
21. Gerhartz HH, Engelhard M, Meusers P et al (1993) Randomized double-blind, placebo-controlled phase III study of recombinant human granulocyte-macrophage colony-stimulating factor as adjunct to induction treatment of high-grade malignant non-Hodgkin's lymphoma. Blood 82:2329–2339
22. Hamm JT, Schiller JH, Oken MM et al (1991) Granulocyte-macrophage colony-stimulating factor (GM-CSF) in small cell carcinoma of the lung (SCCL): preliminary analysis of a randomized controlled trial. Proc Am Soc Clin Oncol 10:255 (abstract)
23. Bajorin DF, Nichols CR, Schmoll HJ et al (1995) Recombinant human granulocyte-macrophage colony-stimulating factor as an adjunct to conventional-dose ifosfamide-based chemotherapy for patients with advanced or relapsed germ cell tumors: a randomized trial. J Clin Oncol 13:79–86
24. Savarese DM, Hsieh C, Stewart FM (1997) Clinical impact of chemotherapy dose escalation in patients with hematologic malignancies and solid cancers. J Clin Oncol 15:2981–2995
25. Phillips K, Tannock IF (1998) Design and interpretation of clinical trials that evaluate agents that may offer protection from the toxic effects of cancer chemotherapy. J Clin Oncol 16:3179–3190
26. Henderson IC, Berry DA, Demetri GD et al (2003) Improved outcomes from adding sequential Paclitaxel but not from escalating Doxorubicin dose in an adjunct chemotherapy regimen for patients with node-positive breast cancer. J Clin Oncol 15:976–983
27. Rodenhuis S, Richel DJ, van der Wall E et al (1998) Randomized trial of high-dose chemotherapy and haemopoietic progenitor-cell support in operable breast cancer with extensive axillary lymph-node involvement. Lancet 352:515–521
28. Hortobagyi GN, Buzdar AU, Theriault RL et al (2000) Randomized trial of high-dose chemotherapy for high-risk primary breast carcinoma. J Natl Cancer Inst 92:225–233
29. Stadtmauer EA, O'Neill A, Goldstein LJ et al (2000) Conventional dose chemotherapy compared with high-dose chemotherapy plus autologous hematopoietic stem-cell transplantation for metastatic breast cancer. N Engl J Med 342:1069–1076
30. Bokemeyer C, Kuczyk MA, Köhne H, Einsele H, Kynast B, Schmoll HJ (1996) Hematopoietic growth factors and treatment of testicular cancer: biological interactions, routine use and dose-intensive chemotherapy. Ann Hematol 72:1–9
31. Rivera E, Erder MH, Moore TD et al (2003) Targeted filgrastim support in patients with early-stage breast carcinoma: toward the implementation of a risk model. Cancer 98:222–228
32. Fossa SD, Kaye SB, Mead GM et al (1998) Filgrastim during combination chemotherapy of patients with poor-prognosis metastatic germ cell. J Clin Oncol 16:716–724
33. Figueredo AT, Hryniuk WM, Strautmanis I, Frank G, Rendell S (1985) Co-trimoxazole prophylaxis during high-dose chemotherapy of small-cell lung cancer. J Clin Oncol 3:54–64
34. Gabrilove JL, Jakuhowski A, Fain K et al (1988) Phase I study of granulocyte colony-stimulating factor in patients with transitional cell carcinoma of the urothelium. J Clin Invest 82:1454–1461
35. Tsakoma CP, Khwaja A, Goldstone AH (1993) Does treatment with haemopoietic growth factors affect the incidence of bacteraemia in adult lymphoma transplant patients? Bone Marrow Transplant 11:433–436
36. Anaissie E, Vartivarian S, Bodey GP et al (1996) Randomized comparison between antibiotics alone and antibiotics plus granulocyte-macrophage colony-stimulating factor (*Escherichia coli*-derived) in cancer patients with fever and neutropenia. Am J Med 100:17–23
37. Kuderer NM, Crawford J, Dale DC, Lyman GH (2007) Impact of primary prophylaxis with granulocyte colony-stimulating factor on febrile neutropenia and mortality in adult cancer patients receiving chemotherapy: a systematic review. J Clin Oncol 25:3158–3167

38. Clark O, Lyman GH, Castro AA, Clark LG, Djulbegovic B (2005) Colony-stimulating factors for chemotherapy-induced febrile neutropenia: a meta-analysis of randomized controlled trials. J Clin Oncol 23:4198–4214
39. Sung L, Nathan PC, Alibhai SM, Tomlinson GA, Beyene J (2007) Meta-analysis: effect of prophylactic hematopoietic colony-stimulating factors on mortality and outcomes of infection. Ann Intern Med 147:400–411
40. Vadhan-Raj S, Broxmeyer HE, Hittelman WN et al (1992) Abrogating chemotherapy-associated myelosuppression by recombinant granulocyte-macrophage colony-stimulating factor in patients with sarcoma: protection at the progenitor cell level. J Clin Oncol 10:1266–1277
41. Soda H, Oka M, Fukuda M et al (1996) Optimal schedule for administering granulocyte colony-stimulating factor in chemotherapy-induced neutropenia in non-small cell lung cancer. Cancer Chemother Pharmacol 38:9–12
42. Gerhartz HH, Stern AC, Wolf-Hornung B et al (1993) Intervention treatment of established neutropenia with human recombinant granulocyte-macrophage colony-stimulating factor (rhGM-CSF) in patients undergoing cancer chemotherapy. Leuk Res 17:175–185
43. Hartmann LC, Tschetter LK, Habermann TM et al (1997) Granulocyte-colony stimulating factor in severe chemotherapy-induced afebrile neutropenia. N Engl J Med 336:1776–1780
44. Maher DW, Lieschke GJ, Green M et al (1994) Filgrastim in patient with chemotherapy-induced febrile neutropenia: a double-blind, placebo-controlled trial. Ann Intern Med 121:492–501
45. Biesma B, de Vries EG, Willemse PH et al (1990) Efficacy and tolerability of recombinant human granulocyte-macrophage colony-stimulating factor in the patients with chemotherapy-induced leucopenia and fever. Eur J Cancer 26:932–936
46. Mayordomo JI, Rivera F, Díaz-Puente MT et al (1995) Improving treatment of chemotherapy-induced neutropenic fever administration of colony-stimulating factors. J Natl Cancer Inst 87:803–808
47. Riikonen P, Saarinen UM, Mäkipernaa A et al (1994) Recombinant human granulocyte-macrophage colony-stimulating factor in the treatment of febrile neutropenia: a double-blind, placebo-controlled study in children. Pediatr Infect Dis J 13:197–202
48. Vellenga E, Uyl-de Groot CA, de Wit R et al (1996) Randomized placebo-controlled trial of granulocyte-macrophage colony-stimulating factor in patients with chemotherapy-related febrile neutropenia. J Clin Oncol 14:619–627
49. Ravaud A, Chevreau C, Cany L et al (1998) Granulocyte-macrophage colony-stimulating factor in patients with neutropenic fever is potent after low-risk but not after high-risk neutropenic chemotherapy regimens: results of a randomized phase III trial. J Clin Oncol 16:2930–2936
50. Mitchell PL, Morland B, Stevens MC et al (1997) Granulocyte colony-stimulating factor in established febrile neutropenia: a randomized study of pediatric patients. J Clin Oncol 15:1163–1170
51. Garcia-Carbonero R, Mayordomo JI, Tornamira MV et al (2001) Granulocyte-colony stimulating factor in the treatment of high-risk febrile neutropenia: a multicenter randomized trial. J Natl Cancer Ins 93:31–38
52. Berghmans T, Paesmans M, Lafitte JJ et al (2002) Therapeutic use of granulocyte and granulocyte-macrophage colony-stimulating factors in febrile neutropenic cancer patients: a systematic review of the literature with meta-analysis. Support Care Cancer 10:181–188
53. Uyl-de Groot CA, Vellenga E, Rutten FF (1996) An economic model to assess the savings from a clinical application of haematopoietic growth factors. Eur J Cancer 32A:57–62
54. Klastersky J, Paesmans M, Rubenstein EB et al (2000) The Multinational Association for Supportive Care in Cancer risk index: a multinational scoring system for identifying low-risk febrile neutropenic cancer patients. J Clin Oncol 18:3038–3051

55. Malik I, Hussain M, Yousuf H (2001) Clinical characteristics and therapeutic outcome of patients with febrile neutropenia who present in shock: the need for better strategies. J Infect 42:120–125
56. Darmon M, Azoulay E, Alberti C et al (2002) Impact of neutropenia duration on short-term mortality in neutropenic critically cancer patients. Int Care Med 28:1775–1780
57. Stephens DP, Thomas JH, Higgins A et al (2008) Randomized, double-blind, placebo-controlled trial of granulocyte colony-stimulating factor in patients with septic shock. Crit Care Med 36:448–454
58. Meisel C, Schefold JC, Pschowski R et al (2009) Granulocyte-macrophage colony-stimulating factor to reverse sepsis-associated immunosuppression: a double-blind, randomized, placebo-controlled multicenter trial. Am J Respir Crit Care Med 180:640–648
59. Frei E, Blum RH, Pitman SW et al (1980) High dose methotrexate with leucovorin rescue: rationale and spectrum of antitumor activity. Am J Med 68:370–376
60. Bonadonna G, Valagussa P (1981) Dose-response effect of adjuvant chemotherapy in breast cancer. N Engl J Med 304:10–15
61. Carde P, MacKintosh FR, Rosenberg SA (1983) A dose and time response analysis of the treatment of Hodgkin's disease with MOPP chemotherapy. J Clin Oncol 1:146–153
62. Hryniuk W, Levine MN (1986) Analysis of dose intensity for adjuvant chemotherapy trials in stage II breast cancer. J Clin Oncol 4:1162–1170
63. Levin L, Hryniuk WM (1987) Dose intensity analysis of chemotherapy regimens in ovarian carcinoma. J Clin Oncol 5:756–767
64. Kwak LW, Halpern J, Olshen RA, Horning SJ (1990) Prognostic significance of actual dose intensity in diffuse large-cell lymphoma: results of a tree-structured survival analysis. J Clin Oncol 8:963–977
65. Meyer RM, Hryniuk WM, Goodyear MD (1991) The role of dose intensity in determining outcome in intermediate-grade non-Hodgkin's lymphoma. J Clin Oncol 9:339–347
66. Budman DR, Berry DA, Cirrincione CT et al (1998) Dose and dose intensity as determinants of outcome in the adjuvant treatment of breast cancer. The Cancer and Leukemia Group B. J Natl Cancer Inst 90:1205–1211
67. Mayer RJ, Davis RB, Schiffer CA, Berg DT, Sarno E, Frei E (1992) Intensive post-remission therapy with Ara-C in adults with acute myeloid leukemia: initial results of a CALGB phase III trial. Leukemia 6(Suppl 2):66–67
68. Kaye SB, Lewis CR, Paul J et al (1992) Randomised study of two doses of cisplatin with cyclophosphamide in epithelial ovarian cancer. Lancet 340:329–333
69. Miller LL (1993) Current status of G-CSF in support of chemotherapy and radiotherapy. Oncol 10:67–88
70. Clark DA, Neidhart JA (1992) Granulocyte-macrophage colony-stimulating factor with dose-intensified treatment of cancer. Semin Hematol 29(supp 3):27–32
71. Steward WP, von Pawel J, Gatzemeier U et al (1998) Effects of granulocyte-macrophage colony-stimulating factor and dose intensification of V-ICE chemotherapy in small-cell lung cancer: a prospective randomized study of 300 patients. J Clin Oncol 16:642–650
72. Thatcher N, Girling DJ, Hopwood P, Sambrook RJ, Qian W, Stephens RJ (2000) Improving survival without reducing quality of life in small-cell lung cancer patients by increasing the dose-intensity of chemotherapy with granulocyte colony-stimulating factor support: results of a British Research Medical Council Multicenter Randomized Trial. J Clin Oncol 18:395–404
73. Gordon LI, Young M, Weller E et al (1999) A phase II trial of 200% Pro-MACE-CytaBOM in patients with previously untreated aggressive lymphomas: analysis of response, toxicity, and dose intensity. Blood 94:3307–3314
74. Shipp MA, Neuberg D, Janicek M, Canellos GP, Shulman LN (1995) High-dose CHOP as initial therapy for patients with poor-prognosis aggressive non-Hodgkin's lymphoma: a dose finding pilot study. J Clin Oncol 13:2916–2923
75. Santoro A, Balzarotti M, Tondini C et al (1999) Dose-escalation of CHOP in non-Hodgkin's lymphoma. Ann Oncol 10:519–525

76. Talbot SM, Westerman DA, Grigg AP et al (1999) Phase I and subsequent phase II study of filgrastim (r-met-HuG-CSF) and dose intensified cyclophosphamide plus epirubicin in patients with non-Hodgkin's lymphoma and advanced solid tumors. Ann Oncol 10:907–914
77. Tanosaki R, Okamoto S, Akatsuka N et al (1994) Dose escalation of biweekly cyclophosphamide, doxorubicin, vincristine, and prednisolone using recombinant human granulocyte colony stimulating factor in non-Hodgkin's lymphoma. Cancer 74:1939–1944
78. Lorigan P, Woll PJ, O'Brien ME, Ashcroft LF, Sampson MR, Thatcher N (2005) Randomized phase III trial of dose-dense chemotherapy supported by whole-blood hematopoietic progenitors in better-prognosis small-cell lung cancer. J Natl Cancer Inst 97:666–674
79. Fayette J, Penel N, Chevreau C et al (2009) Phase III trial of standard versus dose-intensified doxorubicin, ifosfamide, and dacarbazine (MAID) in the first-line treatment of metastatic and locally advanced soft tissue sarcoma. Invest New Drugs 27:482–489
80. Ray-Coquard I, Paraiso D, Guastalla JP et al (2007) Intensified dose of cyclophosphamide with G-CSF support versus standard dose combined with platinum in first-line treatment of advanced ovarian cancer: a randomised study from the GINECO group. Br J Cancer 97:1200–1205
81. Bensinger WI, Martin PJ, Storer B et al (2001) Transplantation of bone marrow as compared with peripheral-blood cells from HLA-identical relatives in patients with hematologic cancers. N Engl J Med 344:175–181
82. Shpall E, Champlin R, Glaspy JA (1998) Effect of CD34+ peripheral blood progenitor cell dose on hematopoietic recovery. Biol Blood Marrow Transplant 4:84–92
83. Nemunaitis J, Rabinowe SN, Singer JW et al (1991) Recombinant granulocyte-macrophage colony-stimulating factor after autologous bone marrow transplantation for lymphoid cancer. N Engl J Med 324:1773–1778
84. Advani R, Chao NJ, Horning SJ et al (1992) Granulocyte-macrophage colony-stimulating factor (GM-CSF) as an adjunct to autologous hemopoietic stem cell transplantation for lymphoma. Ann Intern Med 116:183–189
85. Gulati S, Bennett C (1992) Granulocyte-macrophage colony-stimulating factor (GM-CSF) as an adjunct therapy in relapsed Hodgkin's disease. Ann Intern Med 116:177–182
86. Rabinowe SN, Neuberg D, Bierman PJ et al (1993) Long-term follow up of a phase III study of recombinant human granulocyte-macrophage colony-stimulating factor after autologous bone marrow transplantation for lymphoid malignancies. Blood 81:1903–1908
87. Gorin NC, Coiffier B, Hayat M et al (1992) Recombinant granulocyte-macrophage colony-stimulating factor after high-dose chemotherapy and autologous bone marrow transplantation with unpurged marrow in non-Hodgkin's lymphoma: a double-blind placebo-controlled trial. Blood 80:1149–1157
88. Link H, Boogaerts MA, Carella AM et al (1992) A controlled trial of recombinant human granulocyte-macrophage colony-stimulating factor after total body irradiation, high-dose chemotherapy, and autologous bone marrow transplantation for acute lymphoblastic leukemia or malignant lymphoma. Blood 80:2188–2195
89. Khwaja A, Linch DC, Goldstone AH et al (1992) Recombinant granulocyte-macrophage colony-stimulating factor after autologous bone marrow transplantation for malignant lymphoma: a British National Lymphoma Investigation double-blind, placebo-controlled trial. Br J Haematol 82:317–323
90. Stahel RA, Muller E, Pichert G et al (1992) Dose intensification with autologous marrow support in high-risk lymphoma: acceleration of hematopoietic recovery and reduction of days of hospitalization with granulocyte colony-stimulating factor (G-CSF) in a randomized open-label trial. Proc Am Soc Clin Oncol 11:331 (abstract)
91. Schmitz N, Dreger P, Zander AR et al (1995) Results of a randomised, controlled, multicenter study of recombinant human granulocyte colony stimulating factor (filgrastim) in patients with Hodgkin's disease and non-Hodgkin's lymphoma undergoing autologous bone marrow transplantation. Bone Marrow Transplant 15:261–266

92. Masaoka T, Moriyama Y, Kato S et al (1990) A randomized, placebo-controlled study of KRN8601 (recombinant human granulocyte colony-stimulating factor) in patients receiving allogeneic bone marrow transplantation. Jpn J Med 3:233–239
93. DeWitte T, Gratwohl A, Van Der Lely N (1992) Recombinant granulocyte-macrophage colony-stimulating factor accelerates neutrophil and monocyte recovery after allogeneic T-cell-depleted bone marrow transplantation. Blood 79:1359–1365
94. Powles R, Smith C, Milan S et al (1990) Human recombinant GM-CSF in allogeneic bone-marrow transplantation for leukaemia: double-blind, placebo-controlled trial. Lancet 336:1417–1420
95. Gupta P, Tiley C, Powles R, Treleaven J, Millar J, Catalano J (1992) No increase in relapse in patients with myeloid leukaemias receiving rhGM-CSF after allogeneic bone marrow transplantation. Bone Marrow Transplant 9:491–493
96. Blaise D, Vernant JP, Fiere D et al (1992) A randomised, controlled, multicenter trial of recombinant human granulocyte colony stimulating factor (filgrastim) in patients treated by bone marrow transplantation (BMT) with total body irradiation (TBI) for acute lymphoblastic leukemia (ALL) or lymphoblastic lymphoma (LL). Blood 80:982a (abstract)
97. Spitzer G, Adkins DR, Spencer V et al (1994) Randomized study of growth factors post-peripheral-blood stem-cell transplant: neutrophil recovery is improved with modest clinical benefit. J Clin Oncol 12:661–670
98. Schwartzberg LS, Birch R, Hazelton B et al (1993) Peripheral blood stem cell mobilization by chemotherapy with and without recombinant human granulocyte colony-stimulating factor. J Hematother 1:317–327
99. Kessinger A, Armitage JO (1991) The evolving role of autologous peripheral stem cell transplantation following high-dose therapy for malignancies. Blood 77:211–213
100. Duhrsen U, Villeval JL, Boyd J, Kannourakis G, Morstyn G, Metcalf D (1988) Effects of recombinant human granulocyte colony-stimulating factor on hematopoietic progenitor cells in cancer patients. Blood 72:2074–2081
101. Sheridan WP, Begley CG, Juttner CA et al (1992) Effect of peripheral-blood progenitor cells mobilized by filgrastim (G-CSF) on platelet recovery after high-dose chemotherapy. Lancet 339:640–644
102. Socinski MA, Cannistra SA, Elias A, Antman KH, Schnipper L, Griffin JD (1986) Granulocyte-macrophage colony-stimulating factor expands the circulating hematopoietic progenitor cell compartment in man. Lancet 1:1194–1198
103. Gianni AM, Siena S, Bregni M et al (1989) Granulocyte-macrophage colony stimulating factor to harvest circulating haemopoietic stem cells for autotransplantation. Lancet 2:580–585
104. Chao NJ, Schriber JR, Grimes K et al (1993) Granulocyte colony-stimulating factor "mobilized" peripheral blood progenitor cells accelerate granulocyte and platelet recovery after high-dose chemotherapy. Blood 81:2031–2035
105. Haas R, Hohaus S, Egerer G, Ehrhardt R, Witt B, Hunstein W (1992) Recombinant human granulocyte-macrophage colony-stimulating factor (rhGM-CSF) subsequent to chemotherapy improves collection of blood stem cells for autografting in patients not eligible for bone marrow harvest. Bone Marrow Transplant 9:459–465
106. Haas R, Ho AD, Bredthauer U et al (1990) Successful autologous transplantation of blood stem cells mobilized with recombinant human granulocyte-macrophage colony-stimulating factor. Exp Hematol 18:94–98
107. Tarella C, Ferrero D, Bregni M et al (1991) Peripheral blood expansion of early progenitor cells after high-dose cyclophosphamide and rhGM-CSF. Eur J Cancer 27:22–27
108. Teshima T, Harada M, Takamatsu Y et al (1993) Granulocyte colony-stimulating factor (G-CSF)-induced mobilization of circulating haemopoietic stem cells. Br J Haematol 84:570–573
109. Chao NJ, Long GD, Negrin RS et al (1992) G-CSF and peripheral blood progenitor cells. Lancet 339:1410–1411

110. Nademanee A, Sniecinski I, Schmidt GM et al (1994) High-dose chemotherapy followed by autologous peripheral-blood stem-cell transplantation for patients with Hodgkin's disease and non-Hodgkin's lymphoma using unprimed and granulocyte colony-stimulating factor mobilized peripheral-blood stem cells. J Clin Oncol 12:2176–2186
111. Klumpp TR, Mangan KF, Goldberg SL, Pearlman ES, Macdonald JS (1995) Granulocyte colony-stimulating factor accelerates neutrophil engraftment following peripheral-blood stem-cell transplantation: a prospective, randomized trial. J Clin Oncol 13:1323–1327
112. Nemunaitis J, Rosenfeld CS, Ash R et al (1995) Phase III randomized, double-blind placebo-controlled trial of rhGM-CSF following allogeneic bone marrow transplantation. Bone Marrow Transplant 15:949–954
113. Korbling M, Przepiorka D, Huh YO et al (1995) Allogeneic blood stem cell transplantation for refractory leukemia and lymphoma: potential advantage of blood over marrow allografts. Blood 85:1659–1665
114. Dreger P, Haferlach T, Eckstein V et al (1994) G-CSF-mobilized peripheral blood progenitor cells for allogeneic transplantation: safety, inetics of mobilization, and composition of the graft. Br J Haematol 87:609–613
115. Schmitz N, Dreger P, Suttorp M et al (1995) Primary transplantation of allogeneic peripheral blood progenitor cells mobilized by filgrastim (granulocyte colony-stimulating factor). Blood 85:1666–1672
116. Bensinger WI, Weaver CH, Appelbaum FR et al (1995) Transplantation of allogeneic peripheral blood stem cells mobilized by recombinant human granulocyte colony-stimulating factor. Blood 85:1655–1658
117. Weisdorf DJ, Verfaillie CM, Davies SM et al (1995) Hematopoietic growth factors for graft failure after bone marrow graft failure: a randomized trial of granulocyte macrophage colony stimulating factor (GM-CSF) versus sequential GM-CSF plus granulocyte-CSF. Blood 85:3452–3456
118. Linch DC, Milligan DW, Winfield DA et al (1997) G-CSF after peripheral blood stem cell transplantation in lymphoma patients significantly accelerated neutrophil recovery and shortened time in hospital: results of a randomized BLNI trial. Br J Haematol 99:933–938
119. Lee S, Weller E, Alyea E, Ritz J, Soiffer RJ (1998) Efficacy and costs of granulocyte colony-stimulating factor in allogeneic T-cell depleted bone marrow transplantation. Blood 92:2725–2729
120. Pavletic S, Bishop M, Tarantolo S et al (1997) Hematopoietic recovery after allogeneic blood stem-cell transplantation compared with bone marrow transplantation in patients with hematologic malignancies. J Clin Oncol 15:1608–1616
121. Snowden JA, Biggs JC, Milliken ST et al (1998) A randomized, blinded, placebo-controlled, dose escalation study of the tolerability and efficacy of filgrastim for hematopoietic stem cell mobilization in patients with severe rheumatoid arthritis. Bone Marrow Transplant 22:1035–1041
122. Ringden O, Labopin M, Gorin NC et al (2004) Treatment with granulocyte colony-stimulating factor after allogeneic bone marrow transplantation for acute leukemia increases the risk of graft-versus-host disease and death: a study from the Acute Leukemia Working Party of the European Group for Blood and Marrow Transplantation. J Clin Oncol 22:416–423
123. McQuaker IG, Hunter AE, Pacey S, Haynes AP, Iqbal A, Russell NH (1997) Low-dose filgrastim significantly enhances neutrophil recovery following autologous peripheral-blood stem-cell transplantation in patients with lymphoproliferative disorders: evidence for clinical and economic benefit. J Clin Oncol 15:451–457
124. Kroger N, Zander AR (2002) Dose and schedule effect of G-CSF for stem cell mobilization I healthy donors for allogeneic transplantation. Leuk Lymphoma 43:1391–1394
125. Schiffer CA, Wade JC (1987) Issues in the use of blood products and treatment of infection. Semin Oncol 14:454–467

126. Pomeroy C, Oken MM, Rydell RE, Filice GA (1991) Infection in the myelodysplastic syndromes. Am J Med 90:338–344
127. Glucksberg H, Cheever MA, Farewell VT, Fefer A, Sale GE, Thomas ED (1981) High-dose combination chemotherapy for acute nonlymphoblastic leukemia in adults. Cancer 48:1073–1081
128. Jakubowski A, Andreeff M, Tafuri A et al (1989) In vivo and in vitro studies of rhG-CSF in acute non-lymphocytic leukemia. Blood 74(Suppl 1):274a (abstract)
129. Tafuri A, Andreeff M (1990) Kinetic rationale for cytokine-induced recruitment of myeloblastic leukemia followed by cycle-specific chemotherapy in vitro. Leukemia 4:826–834
130. Aglietta M, Piacibello W, Sanavio F et al (1989) Kinetics of human hemopoietic cells after in vivo administration of granulocyte-macrophage colony-stimulating factor. J Clin Invest 83:551–557
131. Bhalla K, Birkhofer M, Arlin A, Grant S, Lutzky J, Graham G (1988) Effect of recombinant GM-CSF on the metabolism of cytosine arabinoside in normal and leukemic human bone marrow cells. Leukemia 2:810–813
132. Cannistra SA, Groshek P, Griffin JD (1989) Granulocyte-macrophage colony-stimulating factor enhances the cytotoxic effects of cytosine arabinoside in acute myeloblastic leukemia and in the myeloid blast crisis phase of chronic myeloid leukemia. Leukemia 3:328–334
133. Griffin JD, Young D, Gerrmann F, Wiper D, Wagner K, Sabbath KD (1986) Effects of recombinant human GM-CSF on proliferation of clonogenic cells in acute myeloblastic leukemia. Blood 67:1448–1453
134. Hiddemann W, Kiehl M, Zuhlsdorf M et al (1992) Granulocyte-macrophage colony-stimulating factor and interleukin-3 enhance the incorporation of cytosine arabinoside into the DNA of leukemic blasts and the cytotoxic effect on clonogenic cells from patients with acute myeloid leukemia. Semin Oncol 19:31–37
135. Miyauchi J, Kelleher CA, Wang C, Minkin S, McCulloch EA (1989) Growth factors influence the sensitivity of leukemic stem cells to cytosine arabinoside in culture. Blood 73:1272–1278
136. Van Der Lely N, De Witte T, Muus P, Raymakers R, Preijers F, Haanen C (1990) Growth factors enhance the toxicity of cytosine arabinoside towards leukemic clonogenic cells with self-renewal capacity. Exp Hematol 18:615 (abstract)
137. Motoji T, Watanabe M, Uzumaki H et al (1991) Granulocyte colony-stimulating factor (G-CSF) receptors on acute myeloblastic leukaemia cells and their relationship with the proliferative response to G-CSF in clonogenic assay. Br J Haematol 77:54–59
138. Donnelly JP (1993) Selective decontamination of the digestive tract and its role in antimicrobial prophylaxis. J Antimicrob Chemother 31:813–829
139. Rowe JM, Anderson J, Maza JJ et al (1995) Phase III randomized placebo-controlled study of yeast derived GM-CSF in adult patients (55–70 years) with AML: a study of the Eastern Cooperative Oncology Group (E1490). Blood 86:457–462
140. Stone R, George S, Berg D et al (1994) Granulocyte-macrophage colony-stimulating factor after initial chemotherapy for elderly patients with primary acute myelogenous leukemia. Cancer and Leukemia Group B. N Engl J Med 332:1671–1677
141. Ohno R, Tomonaga M, Kobayashi T (1990) Effect of granulocyte colony-stimulating factor after intensive induction therapy in relapsed or refractory acute leukemia. N Engl J Med 323:871–877
142. Estey E, Thall PF, Kantarjian H et al (1992) Treatment of newly diagnosed acute myelogenous leukemia with granulocyte-macrophage colony-stimulating factor (GM-CSF) before and during continuous-infusion high-dose ara-C + daunorubicin: comparion to patients treated without GM-CSF. Blood 79:2246–2255
143. Vadhan-Raj S, Keating M, LeMaistre A et al (1987) Effects of recombinant human granulocyte-macrophage colony-stimulating factor in patients with myelodysplastic syndromes. N Engl J Med 317:1545–1552

144. Negrin RS, Haeuber DH, Nagler A et al (1990) Maintenance treatment of patients with myelodysplastic syndromes using recombinant human granulocyte colony-stimulating factor. Blood 76:36–43
145. Schuster MW, Larson RA, Thompson JA et al (1990) Granulocyte-macrophage colony-stimulating factor (GM-CSF) for myelodysplastic syndrome (MDS): results of a multi-center, randomized, controlled trial. Blood 76:318a (abstract)
146. Greenberg P, Taylor K, Larson R et al (1993) Phase III randomized multicenter trial of G-CSF vs. observation for myelodysplastic syndromes (MDS. Blood 82:196a (abstract)
147. Dombret H, Chastang C, Fenaux P et al (1995) A controlled study of recombinant human granulocyte colony-stimulating factor in elderly patients after treatment for acute myeloid leukemia: AML Cooperative Study Group. N Engl J Med 332:1678–1683
148. Heil G, Hoelzer D, Sanz MA et al (1997) The International Acute Myeloid Leukemia Study Group: a randomized, double-blind, placebo-controlled phase III study of filgrastim in remission induction and consolidation therapy for adults with de novo acute myeloid leukemia. Blood 90:4710–4718
149. Zittoun R, Suciu S, Mandelli F et al (1996) Granulocyte-macrophage colony-stimulating factor associated with induction treatment of acute myelogenous leukemia: a randomized trial by the European Organization for Research and Treatment of Cancer Leukemia Cooperative Group. J Clin Oncol 14:2150–2159
150. Godwin JE, Kopecky KJ, Head DR et al (1998) A double-blind placebo-controlled trial of granulocyte colony-stimulating factor in elderly patients with previously untreated acute myeloid leukemia. Blood 91:3607–3615
151. Lowenberg B, Boogaerts MA, Daenen SM et al (1997) Value of different modalities of granulocyte-macrophage colony-stimulating factor applied during or after induction therapy of acute myeloid leukemia. J Clin Oncol 15:3496–3506
152. Lowenberg B, Suciu S, Archimbaud E et al (1997) Use of recombinant GM-CSF during and after remission induction chemotherapy in patients aged 61 years and older with acute myeloid leukemia: final report of AML-111, a phase III randomized study of the Leukemia Cooperative Group of European Organisation for the Research and Treatment of Cancer and the Dutch-Belgian Hemato-Oncology Cooperative Group. Blood 90:2952–2961
153. Witz F, Sadoun A, Perin MC et al (1998) A placebo-controlled study of recombinant human granulocyte-macrophage colony-stimulating factor administered during and after induction treatment for de novo acute myelogenous leukemia in elderly patients. Blood 91:2722–2730
154. Peterson BA, George SL, Bhalla K et al (1996) A phase III trial with or without GM-CSF administered before and during high dose cytarabine in patients with relapsed or refractory acute myelogenous leukemia: CALGB 9021. Proc Am Soc Clin Oncol 14:1749 (abstract 2016)
155. Ohno R, Naoe T, Kanamaru A et al (1994) A double-blind controlled study of granulocyte-stimulating factor started two days before induction chemotherapy in refractory acute myeloid leukemia. Blood 83:2086–2092
156. Harousseau JL, Witz B, Lioure B et al (2000) G-CSF after intensive consolidation chemotherapy in acute myeloid leukemia: results of a randomized trial of the Groupe Ouest-Est Leucemies Aigues Myeloblastiques. J Clin Oncol 18:780–787
157. Larson RA, Dodge RK, Linker CA et al (1998) A randomized controlled trial of filgrastim during remission induction and consolidation chemotherapy for adults with acute lymphoblastic leukemia. Blood 92:1556–1564
158. Pui C, Boyett JM, Hughes WT et al (1997) Human granulocyte colony-stimulating factor after induction chemotherapy in children with acute lymphoblastic leukemia. N Engl J Med 336:1781–1787
159. Ottmann OG, Hoelzer D, Gracien E et al (1995) Concomitant granulocyte colony-stimulating factor and induction chemoradiotherapy in adult acute lymphoblastic leukemia: a randomized phase III trial. Blood 86:444–450

160. Welte K, Reiter A, Mempel K et al (1996) A randomized phase III study of the efficacy of granulocyte colony-stimulating factor in children with high-risk acute lymphoblastic leukemia. Blood 87:3143–3150
161. Geissler K, Koller E, Hubmann E et al (1997) Granulocyte colony-stimulating factor as an adjunct to induction chemotherapy for adult lymphoblastic leukemia: a randomized phase III study. Blood 90:590–596
162. Scherrer R, Geissler K, Kyrle PA et al (1993) Granulocyte colony-stimulating factor (G-CSF) as an adjunct to induction chemotherapy of adult lymphoblastic leukemia (ALL). Ann Hematol 66:283–289
163. Lowenberg B, van Putten W, Theobald M et al (2003) Effect of priming with granulocyte colony-stimulating factor on the outcome of chemotherapy for acute myeloid leukemia. N Engl J Med 349:743–752
164. Amadori S, Suciu S, Jehn U et al (2005) Use of glycosylated recombinant human G-CSF during and/or after induction chemotherapy in elderly patients with acute myeloid leukemia: final results of AML-13, a randomized phase III study of the EORTC and GIMEMA Leukemia Groups. Blood 106:27–34
165. Heath JA, Steinherz PG, Altman A et al (2003) Human granulocyte colony-stimulating factor in children with high-risk acute lymphoblastic leukemia: a Children's Cancer Group Study. J Clin Oncol 21:1612–1617
166. Relling MV, Boyett JM, Blanco JG et al (2003) Granulocyte colony-stimulating factor and the risk of secondary myeloid malignancy after etoposide treatment. Blood 101:3862–3867
167. Park S, Grabar S, Kelaidi C et al (2008) Predictive factors of response and survival in myelodysplatic syndrome treated with erythropoietin and G-CSF: the GFM experience. Blood 111:574–582
168. Jadersten M, Malcovati L, Dybedal I et al (2008) Erythropoietin and granulocyte-colony stimulating factor treatment associated with improved survival in myelodysplastic syndrome. J Clin Oncol 26:3607–3613
169. Teramura M, Kimura A, Iwase S et al (2007) Treatment of severe aplastic anemia with antithymocyte globulin and cyclosporine A with or without G-CSF in adults: a multicenter randomized study in Japan. Blood 110:1756–1761
170. Robbins KT, Kumar P, Wong FS et al (2000) Targeted chemoradiation for advanced head and neck cancer: analysis of 213 patients. Head Neck 22:687–693
171. Bossett JF, Gignoux M, Triboulet JP et al (1997) Chemoradiotherapy followed by surgery compared with surgery alone in squamous cell cancer of the esophagus. N Engl J Med 337:661–667
172. Schaake-Koning C, van der Bogart W, Dalesio O et al (1992) Effects of concomitant cisplatin and radiotherapy on inoperable non-small cell lung cancer. N Engl J Med 326:524–530
173. Morris M, Eifel P, Lu J et al (1999) Pelvic radiation with concurrent chemotherapy compared with pelvic and para-aortic radiation for high-risk cervical cancer. N Engl J Med 340:1137–1143
174. Rich TA, Skibber JM, Ajani JA et al (1995) Preoperative infusional chemoradiation therapy for stage T3 rectal cancer. Int J Radiat Oncol Biol Phys 32:1025–1029
175. Gabrilove JL, Jakubowski A, Scher H et al (1988) Effect of granulocyte colony-stimulating factor on neutropenia and associated morbidity due to chemotherapy for transitional-cell carcinoma of the urothelium. N Engl J Med 318:1414–1422
176. Momin F, Kraut M, Lattin P et al (1992) Thrombocytopenia in patients receiving chemoradiotherapy and G-CSF for locally advanced non-small cell lung cancer (NSCLC). Proc Am Soc Clin Oncol 11:294 (abstract)
177. Bunn PA Jr, Crowley J, Kelly K et al (1995) Chemoradiotherapy with or without granulocyte-macrophage colony-stimulating factor in the treatment of limited-stage small-cell lung cancer: a prospective, phase III randomized study of the Southwest Oncology Group. J Clin Oncol 13:1632–1641

178. Fushiki M, Abe M (1992) Randomized double-blind controlled study of rhG-CSF in patients with neutropenia induced by radiation therapy. Proc Am Soc Clin Oncol 11:410 (abstract)
179. Knox SJ, Fowler S, Marquez C, Hoppe RT (1994) Effect of filgrastim (G-CSF) in Hodgkin's disease patients treated with radiation therapy. Int J Radiat Oncol Biol Phys 28:445–450
180. Marks LB, Friedman HS, Kurtzberg J, Oakes WJ, Hockenberger BM (1992) Reversal of radiation induced neutropenia by granulocyte colony-stimulating factor. Med Pediatr Oncol 20:240–242
181. Troussard X, Macro M, Vie B et al (1995) Human recombinant granulocyte-macrophage colony-stimulating factor (rhGM-CSF) improves double hemibody irradiation (DHBI) tolerance in patients with stage III multiple myeloma: a pilot study. Br J Haematol 89:191–195
182. Crivellari D, Bonetti M, Castiglione-Gertsch M et al (2000) Burdens and benefits of adjuvant cyclophosphamide, methotrexate, and fluorouracil and tamoxifen for elderly patients with breast cancer. J Clin Oncol 18:1412–1422
183. Dees E, O'Reilly S, Goodman SN et al (2000) A prospective pharmacologic evaluation of age-related toxicity of adjuvant chemotherapy in women with breast cancer. Cancer Invest 18:521–529
184. Gelman RS, Taylor SG (1984) Cyclophosphamide, methotrexate, and 5-fluorouracil chemotherapy in women more than 65 years old with advanced breast cancer: the elimination of age trends in toxicity by using doses based on creatinine clearance. J Clin Oncol 2:1404–1413
185. Armitage JO, Potter JF (1984) Aggressive chemotherapy for diffuse histiocytic lymphoma in the elderly: increased complications with advancing age. J Am Geriatr Soc 32:269–273
186. Doorduijn J, Van Der Holt B, Van Der Kem F et al (2000) Randomized trial of colony-stimulating factor (G-CSF) added to CHOP in elderly patients with aggressive non-Hodgkin's lymphoma. Blood 96:133a
187. Gomez H, Mas L, Casanova L et al (1998) Elderly patients with aggressive non-Hodgkin's lymphoma treated with CHOP chemotherapy plus granulocyte-macrophage colony-stimulating factor: identification of two age subgroups with differing hematologic toxicity. J Clin Oncol 16:2352–2358
188. Morrison VA, Picozzi V, Scott S et al (2001) The impact of age on delivered dose intensity and hospitalizations for febrile neutropenia in patients with intermediate-grade non-Hodgkin's lymphoma receiving initial CHOP chemotherapy: a risk factor analysis. Clin Lymphoma 2:47–56
189. Aviles A, Nambo MJ, Talavera A et al (1997) Epirubicin (CEOP-Bleo) versus idarubicin (CIOP-Bleo) in the treatment of elderly patients with aggressive non-Hodgkin's lymphoma: dose escalation studies. Anticancer Drugs 8:937–942
190. Bertini M, Freilone R, Vitolo U et al (1996) The treatment of elderly patients with aggressive non-Hodgkin's lymphomas: feasibility and efficacy of an intensive multidrug regimen. Leuk Lymphoma 22:483–493
191. Coiffier B, Lepage E, Briere J et al (2002) CHOP chemotherapy plus rituximab compared with CHOP alone in elderly patients with diffuse large-B-cell lymphoma. N Engl J Med 346:235–242
192. Gisselbrecht C, Haioun C, Lepage I et al (1997) Placebo-controlled phase III study of lenograstim (glycosylated recombinant human granulocyte colony-stimulating factor) in aggressive non-Hodgkin's lymphoma: factors influencing chemotherapy administration. Leuk Lymphoma 25:289–300
193. Osby E, Hagberg H, Kvaloy S et al (2003) CHOP is superior to CNOP in elderly patients with aggressive lymphoma while outcome is unaffected by filgrastim treatment: results of a Nordic Lymphoma Group randomized trial. Blood 101:3840–3848
194. Sonneveld P, de Ridder M, van der Lelie H et al (1995) Comparison of doxorubicin and mitoxantrone in the treatment of elderly patients with advanced diffuse non-Hodgkin's lymphoma using CHOP versus CNOP. J Clin Oncol 13:2530–2539
195. Tirelli U, Errante D, Van Glabbeke M et al (1998) CHOP is the standard regimen in patients > or = 70 years of age with intermediate-grade and high-grade non-Hodgkin's lymphoma:

results of a randomized study of the European Organization for Research and Treatment of Cancer Lymphoma Cooperative Study Group. J Clin Oncol 16:27–34
196. Zinzani PL, Storti S, Zaccaria A et al (1999) Elderly aggressive-histology non-Hodgkin's lymphoma: first-line VNCOP-B regimen experience on 350 patients. Blood 94:33–38
197. Doorduijn JK, van der Holt B, van Imhoff GM et al (2003) CHOP compared with CHOP plus granulocyte colony-stimulating factor in elderly patients with aggressive non-Hodgkin's lymphoma. J Clin Oncol 21:3041–3050
198. Bjorkholm M, Osby E, Hagberg H et al (1999) Randomized trial of R-methugranulocyte colony-stimulating factors as adjunct to CHOP or CNOP treatment of elderly patients with aggressive non-Hodgkin's lymphoma. Blood 94:599a
199. Dixon DO, Neilan B, Jones SE et al (1986) Effect of age on therapeutic outcome in advanced diffuse histiocytic lymphoma: the Southwest Oncology Group experience. J Clin Oncol 4:295–305
200. Meyer RM, Browman GP, Samosh ML et al (1995) Randomized phase II comparison of standard CHOP with weekly CHOP in elderly patients with non-Hodgkin's lymphoma. J Clin Oncol 12:2386–2393
201. Chrischilles E, Delgado DJ, Stolshek BS, Lawless G, Fridman M, Carter WB (2002) Impact of age and colony-stimulating factor use on hospital length of stay for febrile neutropenia in CHOP-treated non-Hodgkin's lymphoma. Cancer Control 9:203–211
202. Zagonel V, Babare R, Merola MC et al (1994) Cost-benefit of granulocyte colony-stimulating factor administration in older patients with non-Hodgkin's lymphoma treated with combination chemotherapy. Ann Oncol 5:127–132
203. Marsoni S, Ungerleider RS, Hurson SB, Simon RM, Hammershaimb LD (1985) Tolerance of antineoplastic agents in children and adults. Cancer Treat Rep 64:1263–1269
204. Furman W, Fairclough D, Huhn R et al (1991) Therapeutic effects and pharmacokinetics of recombinant human granulocyte-macrophage colony-stimulating factor in childhood cancer patients receiving myelosuppressive chemotherapy. J Clin Oncol 9:1022–1028
205. Blazar B, Kersey J, McGlave P et al (1989) In vivo administration of recombinant human granulocyte-macrophage colony-stimulating factor in acute lymphoblastic leukemia patients receiving purged autografts. Blood 73:849–857
206. Smith MA, Horowitz ME, Balis F et al (1991) Pediatric experience with granulocyte-macrophage colony-stimulating factor (GM-CSF): pharmacokinetics and cytokine induction. Proc Am Assoc Cancer Res 32:196 (abstract)
207. Stute N, Santana VM, Rodman JH, Schell MJ, Ihle JN, Evans WE (1992) Pharmacokinetics of subcutaneous recombinant human granulocyte colony-stimulating factor in children. Blood 79:2849–2854
208. Michon J, Hartmann O, Bouffet E et al (1992) Preliminary analysis of the first open label randomized phase III study of recombinant colony stimulating factor as an adjunct to combination induction chemotherapy in pediatric patients with metastatic neuroblastoma. Blood 10:248a (abstract)
209. Weinthal J, Gillan E, Hodder F et al (1992) G-CSF significantly reduces the nadir of neutropenia, hospitalizations and costs during intensive chemotherapy in children with solid tumors. Proc Am Soc Clin Oncol 11:362 (abstract)
210. Santana VM, Bowman L, Furman W et al (1990) Trial of chemotherapy plus recombinant G-CSF in children with advanced neuroblastoma. Med Ped Oncol 18:395 (abstract)
211. Stork L, Busselmaier C, Odom L et al (1993) IV G-CSF has similar efficacy to SQ G-CSF in pediatric patients. Proc Am Soc Clin Oncol 12:439 (abstract)
212. Shen WP, Gillan E, Tishler D et al (1993) A pilot study evaluating the toxicity and response rate of ifosfamide, carboplatin, and etoposide (ICE) chemotherapy with G-CSF support for recurrent pediatric solid tumors. Proc Am Soc Clin Oncol 12:426 (abstract)
213. Marina N, Shema S, Bowman LC et al (1994) Failure of granulocyte-macrophage colony-stimulating factor to reduce febrile neutropenia in children with recurrent solid tumors treated with ifosfamide, carboplatin, etoposide chemotherapy. Med Pediatr Oncol 23:328–334

214. Fukuda M, Kojima S, Matsumoto K, Matsuyama T (1992) Autotransplantation of peripheral blood stem cells mobilized by chemotherapy and recombinant human granulocyte colony-stimulating factor in childhood neuroblastoma and non-Hodgkin's lymphoma. Br J Haematol 80:327–331
215. Parsons SK, Mayer DK, Alexander SW, Xu R, Land V, Laver J (2000) Growth factor practice among pediatric oncologists: results of a 1998 Pediatric Oncology Group survey – Economic Evaluation Working Group of the Pediatric Oncology Group. J Pediatr Hematol Oncol 22:227–241
216. Sung L, Nathan PC, Lange B, Beyene J, Buchanan GR (2004) Prophylactic granulocyte colony-stimulating factor and granulocyte-macrophage colony-stimulating factor decrease febrile neutropenia after chemotherapy in children with cancer: a meta-analysis of randomized controlled trials. J Clin Oncol 22:3350–3356
217. Bennett CL, Stinson TJ, Laver JH, Bishop MR, Godwin JE, Tallman MS (2000) Cost analyses of adjunct colony stimulating factors for acute leukemia: can they improve clinical decision making. Leuk Lymphoma 37:65–70
218. Maugard-Louboutin C, Chastang C, Chevallier B et al (1993) Dose-effect relationship of granulocyte colony stimulating factor (G-CSF): PE 2601 in patients with advance breast carcinoma (ABC) treated by intensive chemotherapy. Proc Am Soc Clin Oncol 12:90 (abstract)
219. Eguchi K, Shinkai T, Sasaki Y et al (1990) Subcutaneous administration of recombinant human granulocyte colony-stimulating factor (KRN8601) in intensive chemotherapy for patients with advanced lung cancer. Jpn J Cancer Res 81:1168–1174
220. Eguchi K, Sasaki S, Tamura T (1989) Dose escalation study of recombinant human granulocyte colony-stimulating factor (KRN8601) in patients with advanced malignancy. Cancer Res 49:5221–5224
221. Kaneko T, Takaku F, Ogawa M (1991) Outline of clinical studies on recombinant human granulocyte colony-stimulating factor (KRN8601) in Japan. Tokai J Exp Clin Med 16:51–61
222. Linch DC, Scarffe H, Proctor S et al (1993) Randomised vehicle-controlled dose-finding study of glycosylated recombinant human granulocyte colony-stimulating factor after marrow transplantation. Bone Marrow Transplant 11:307–311
223. Ohno R, Tomonaga M, Ohshima T et al (1993) A randomized controlled study of granulocyte colony stimulating factor after intensive induction and consolidation therapy in patients with acute lymphoblastic leukemia. Japan Adult Leukemia Study Group. Int J Hematol 58:73–81
224. Sheridan W, Begley G, Juttner C et al (1992) The impact of r-metHuG-CSF (filgrastim) dose on the mobilization of mononuclear and progenitor cells in peripheral blood in patients with malignancy. Blood 80(Suppl 1):420a (abstract)
225. Negrin RS, Haeuber DH, Nagler A et al (1989) Treatment of myelodysplastic syndromes with recombinant human granulocyte colony-stimulating factor: a phase I-II trial. Ann Intern Med 110:976–984
226. Neidhart JA, Mangalik A, Stidley CA, Tebich SL et al (1992) Dosing regimen of granulocyte-macrophage colony-stimulating factor to support dose-intensive chemotherapy. J Clin Oncol 10:1460–1469
227. Ganser A, Volkers B, Greher J et al (1989) Recombinant human granulocyte-macrophage colony-stimulating factor in patients with myelodysplastic syndromes – a phase I/II trial. Blood 73:31–37
228. Rosenfeld CS, Sulecki M, Evans C, Shadduck RK (1991) Comparison of intravenous versus subcutaneous recombinant human granulocyte-macrophage colony-stimulating factor in patients with primary myelodysplasia. Exp Hematol 19:273–277
229. Antin J, Smith BR, Holmes W, Rosenthal RS (1988) Phase I/II study of recombinant human granulocyte-macrophage colony-stimulating factor in aplastic anemia and myelodysplastic syndrome. Blood 72:705–713

230. Bokemeyer C, Schmoll HJ, Metzner B et al (1993) Comparison of 5 versus 10 micrograms/kg per day of GM-CSF following dose-intensified chemotherapy with cisplatin, etoposide, and ifosfamide in patients with advanced testicular cancer. Ann Hematol 67:75–79
231. Rose C, Wattel E, Bastion Y et al (1994) Treatment with very low-dose GM-CSF in myelodysplastic syndromes with neutropenia: a report on 28 cases. Leukemia 8:1458–1462
232. Hamm J, Schiller JH, Cuffie C et al (1994) Dose-ranging study of recombinant human granulocyte-macrophage colony-stimulating factor in small-cell lung carcinoma. J Clin Oncol 12:2667–2676
233. Stahel RA, Jost LM, Honegger H, Betts E, Goebel ME, Nagler A (1997) Randomized trial showing equivalent efficacy of filgrastim 5 mg/kg/d and 10 mg/kg/d following high-dose chemotherapy and autologous bone marrow transplantation in high-risk lymphomas. J Clin Oncol 15:1730–1735
234. Grigg AP, Roberts AW, Raunow H et al (1995) Optimizing dose and scheduling of filgrastim (granulocyte colony-stimulating factor) for mobilization and collection of peripheral blood progenitor cells in normal volunteers. Blood 86:4437–4445
235. Weaver CH, Birch R, Greco FA et al (1998) Mobilization and harvesting of peripheral blood stem cells: randomized evaluation of different doses of filgrastim. Br J Haematol 100:338–347
236. Somlo G, Sniecinski I, Ahn C et al (1993) Priming with G-CSF 10 mg/kg is more effective than 5 mg/kg in patients receiving high dose chemotherapy followed by peripheral stem cell rescue. Proc Am Soc Hematol 82:642a (abstract 2551)
237. Erban J, Miler K, Berkman E et al (1985) Filgrastim priming of PBSC and hematopoietic reconstitution following high-dose chemotherapy for breast cancer: effect of dose on PBSC yield and engraftment. Proc Am Soc Clin Oncol 14:316 (abstract)
238. Somlo G, Sniecinski I, Odom-Maryon T et al (1997) Effect of CD34+ selection and various schedules of stem cell reinfusion and granulocyte colony stimulating factor priming on hematopoietic recovery after high-dose chemotherapy for breast cancer. Blood 89:1521–1528
239. Stute N, Furman WL, Schell M, Evans WE (1995) Pharmacokinetics of recombinant human granulocyte-macrophage colony-stimulating factor in children after intravenous and subcutaneous administration. J Pharm Sci 84:824–828
240. Honkoop AH, Hoekman K, Wagstaff I et al (1996) Continuous infusion or subcutaneous injection of granulocyte-macrophage colony-stimulating factor: increased efficacy and reduced toxicity when given subcutaneously. Br J Cancer 74:1132–1136
241. Papaldo P, Lopez M, Marolla P et al (2005) Impact of five prophylactic filgrastim schedules on hematologic toxicity in early breast cancer patients treated with epirubicin and cyclophosphamide. J Clin Oncol 23:6908–6918
242. Amgen Inc: Neupogen (filgrastim) (1994) Package insert. Amgen, Thousand Oaks, CA
243. Immunex Corporation: Leukine (sargramostim) (1991) Package Insert. Immunex, Seattle, WA
244. Broxmeyer HE, Benninger L, Patel SR, Benjamin RS, Vadhan-Raj S (1994) Kinetic response of human marrow myeloid progenitor cells to in vivo treatment of patients with granulocyte colony-stimulating factor is different from the response to treatment with granulocyte-macrophage colony-stimulating factor. Exp Hematol 22:100–102
245. Meropol NJ, Miller LL, Korn EL, Braitman LE, MacDermott ML, Schuchter LM (1992) Severe myelosuppression resulting from concurrent administration of granulocyte colony-stimulating factor and cytotoxic chemotherapy. J Natl Cancer Inst 84:1201–1203
246. Rowinsky E, Sartorius S, Grochow L et al (1992) Phase I and pharmacologic study of topotecan, an inhibitor of topoisomerase I, with granulocyte colony-stimulating factor (G-CSF): toxicologic differences between concurrent and post-treatment G-CSF administration. Proc Am Soc Clin Oncol 11:116 (abstract)

247. Shaffer DW, Smith LS, Burris HA et al (1993) A randomized phase I trial of chronic oral etoposide with or without granulocyte-macrophage colony-stimulating factor in patients with advanced malignancies. Cancer Res 53:5929–5933
248. Kaplan LD, Kahn JO, Crowe F et al (1991) Clinical and virologic effects of recombinant human granulocyte-macrophage colony-stimulating factor in patients receiving chemotherapy for human immunodeficiency virus-associated non-Hodgkin's lymphoma: results of a randomized trial. J Clin Oncol 9:929–940
249. Brandt SJ, Peters WP, Atwater DJ et al (1988) Effect of recombinant human granulocyte-macrophage colony-stimulating factor on hematopoietic reconstitution after high dose chemotherapy and autologous marrow transplantation. N Engl J Med 318:869–876
250. Sheridan W, Wolf M, Lusk J et al (1989) Granulocyte colony-stimulating factor and neutrophil recovery after high-dose chemotherapy and autologous bone marrow transplantation. Lancet 2:891–895
251. Lieshke GJ, Ramenchi U, O'Connor MP, Sheridan W, Szer J, Morstyn G (1992) Studies of oral neutrophil levels in patients receiving G-CSF after autologous marrow transplantation. Br J Haematol 82:589–595
252. Wright DG, Meierovics AI, Foxley JM (1986) Assessing the delivery of neutrophils to tissues in neutropenia. Blood 67:1023–1030
253. Cohen KJ, Leamer K, Odom L, Greffe B, Stork L (1995) Cessation of antibiotics regardless of ANC is safe in children with febrile neutropenia: a preliminary prospective trial. J Pediatr Hematol Oncol 17:325–330
254. Mullen CA, Buchanan GR (1990) Early hospital discharge of children with cancer treated for fever and neutropenia: identification and management of the low-risk patient. J Clin Oncol 8:1998–2004
255. Griffin TC, Buchanan GR (1992) Hematologic predictors of bone marrow recovery in neutropenic patients hospitalized for fever: implications for discontinuation of antibiotics and early discharge from the hospital. J Pediatr 121:28–33
256. Brugger W, Bross KJ, Glatt M, Weber F, Mertelsmann R, Kanz L (1994) Mobilization of tumor cells and hematopoietic progenitor cells into peripheral blood of patients with solid tumors. Blood 83:636–640
257. Pedrazzoli P, Battaglia M, Da Prada GA et al (1997) Role of tumor cells contaminating the graft in breast cancer recurrence after high-dose chemotherapy. Bone Marrow Transplant 20:167–169
258. Siena S, Piccart MJ, Holmes FA, Glaspy J, Hackett J, Renwick JJ (2003) A combined analysis of two pivotal randomized trials of a single dose of pegfilgrastim per chemotherapy cycle and daily filgrastim in patients with stage II-IV breast cancer. Oncol Rep 10:715–724
259. Vose JM, Crump M, Lazarus H et al (2003) Randomized, multicenter, open-label study of pegfilgrastim compared with daily filgrastim after chemotherapy for lymphoma. J Clin Oncol 21:514–519
260. Lyman G, Lalla A, Barron R, Dubois RW (2009) Cost-effectiveness of pegfilgrastim versus 6-day filgrastim primary prophylaxis in patients with non-Hodgkin's lymphoma receiving CHOP-21 in United States. Curr Med Res Opin 25:401–411
261. Ramsey SD, Liu Z, Boer R et al (2009) Cost-effectiveness of primary versus secondary prophylaxis with pegfilgrastim in women with early-stage breast cancer receiving chemotherapy. Value Health 12:217–225
262. Liu Z, Doan QV, Malin J, Leonard R (2009) The economic value of primary prophylaxis using pegfilgrastim compared with filgrastim in patients with breast cancer in the UK. Appl Health Econ Health Policy 7:193–205
263. Sehouli J, Goertz A, Steinle T et al (2010) Pegfilgrastim vs filgrastim in primary prophylaxis of febrile neutropenia in patients with breast cancer after chemotherapy: a cost-effective analysis for German. Dtsch Med Wochenschr 135:385–389
264. Beveridge RA, Miller JA, Kales AN et al (1998) A comparison of the efficacy of sargramostim (yeast-derived RhuGM-CSF) and filgrastim (bacteria-derived RhuG-CSF) in

the therapeutic setting of chemotherapy-induced myelosuppression. Cancer Invest 16:366–373
265. Weaver CH, Schulman KA, Wilson-Relyea B, Birch R, West W, Buckner CD (2000) Randomized trial of filgrastim, sargramostim, or sequential sargramostim and filgrastim after myelosuppressive chemotherapy for the harvesting of peripheral-blood stem cells. J Clin Oncol 18:43–53
266. Dubois RW, Pinto LA, Bernal M, Badamgarav E, Lyman GH (2004) Benefits of GM-CSF versus placebo or G-CSF in reducing chemotherapy-induced complications: a systematic review of the literature. Support Cancer Ther 2:34–41
267. Pierelli L, Perillo A, Ferrandina G et al (2001) The role of growth factor administration and T-cell recovery after peripheral blood progenitor cell transplantation in the treatment of solid tumors: results of a randomized comparison of G-CSF and GM-CSF. Transfusion 41:1577–1585
268. Koç ON, Gerson SL, Cooper BW et al (2000) Randomized cross-over trial of progenitor cell mobilization: high dose cyclophosphamide plus granulocyte colony-stimulating factor (G-CSF) versus granulocyte-macrophage colony-stimulating factor plus G-CSF. J Clin Oncol 18:1824–1830
269. Comenzo RL, Sanchorawala V, Fisher C et al (1999) Intermediate-dose intravenous melphalan and blood stem cells mobilized with sequential GM+G-CSF or G-CSF alone to treat AL (amyloid light chain) amyloidosis. Br J Haematol 104:553–559
270. Recchia F, Accorsi P, Bonfini T et al (2000) Randomized trial of sequential administration of G-CSF and GM-CSF vs G-CSF alone following peripheral blood progenitor cell autograft in solid tumors. J Interferon Cytokine Res 20:171–177
271. Waselenko JK, MacVittie TJ, Blakely WF et al (2004) Medical management of acute radiation syndrome. Ann Intern Med 140:1037–1051
272. US Department of Health and Human Services PHS, and agency for toxic substances and disease registry: toxicological profile for cesium. Available at http://www.atsdr.cdc.gov/toxprofiles/tp157.pdf
273. Hall EJ (2000) Acute effects of total-body irradiation, radiobiology for the radiologist. Lippincott Williams & Wilkins, Philadelphia, PA
274. Schull WJ (1998) The somatic effects of exposure to atomic radiation: the Japanese experience, 1947–1997. Proc Natl Acad Sci U S A 95:5437–5441
275. Farese AM, Case DB, Vigneulle RM et al (2001) A single dose of pegylated leridistim significantly improves neutrophil recovery in sublethally irradiated rhesus macaques. Stem Cells 19:514–521
276. Farese AM, Gunt P, Grab LB, MacVittie TJ (1996) Combined administration of recombinant human megakaryocyte growth and development factor and granulocyte colony-stimulating factor enhances multilineage hematopoietic reconstitution in nonhuman primates after radiation induced marrow aplasia. J Clin Invest 97:2145–2151
277. MacVittie TJ, Farese AM, Herodin F, Grab LB, Baum CM, McKearn JP (1996) Combination therapy for radiation-induced bone marrow aplasia in nonhuman primates using synthokine SC-55494 and recombinant human granulocyte colony-stimulating factor. Blood 87:4129–4135
278. Nash RA, Schuening FG, Seidel K et al (1994) Effect of recombinant caning granulocyte-macrophage colony-stimulating factor on hematopoietic recovery after otherwise lethal total body irradiation. Blood 83:1963–1970
279. Neelis KJ, Dubbelman YD, Wingliang L, Thomas GR, Eaton DL, Wagemaker G (1997) Simultaneous administration of TPO and G-CSF after cytoreductive treatment of rhesus monkeys prevents thrombocytopenia, accelerates platelet and red cell reconstitution, alleviates neutropenia, and promotes the recover of immature bone marrow cells. Exp Hematol 25:1084–1093

Use of r-metHuG-CSF to Enable Chemotherapy Delivery for Solid Tumors

Tara L. Arvedson and Graham Molineux

1 Introduction

Chemotherapy-induced neutropenia is a serious side effect of cancer treatment that can negatively impact patient outcomes by increasing the risk of serious infections or compromising the efficacy of chemotherapy treatment by necessitating chemotherapy dose reductions and/or delays. Managing neutropenia and its related complications has been improved by two forms of recombinant r-metHuG-CSF (filgrastim and a longer-lived form, pegfilgrastim) that have been shown to shorten the depth and duration of neutropenia, to reduce the incidence of infections, and to support delivery of chemotherapy treatment on time. As chemotherapy regimens have become more dose dense and dose intense, the time for neutrophil recovery has been further shortened. Both forms of r-metHuG-CSF have been shown to reduce neutropenia in these settings and to enable dose-dense and dose-intense chemotherapy regimens.

2 Neutropenia and Infection

A normal absolute neutrophil count (ANC) is $>1.5 \times 10^9$/L. Neutropenia reflects a reduction in the ANC and is graded according to the magnitude of the decrease (e.g., ANC $<1.0 \times 10^9$/L is considered mild [grade 3] neutropenia; $<0.5 \times 10^9$/L is moderate [grade 4]; and $<0.2 \times 10^9$/L is considered severe). Febrile neutropenia is the combination of neutropenia with a fever (single oral temperature of $\geq$38.3°C or $\geq$38.0°C for >1 h), and the ramifications of febrile neutropenia are typically

T.L. Arvedson (✉) • G. Molineux
Amgen Inc., One Amgen Center Drive, M/S 15-2-A, Thousand Oaks, CA, 91320, USA
e-mail: taraa@amgen.com; grahamm@amgen.com

G. Molineux et al. (eds.), *Twenty Years of G-CSF*, Milestones in Drug Therapy,
DOI 10.1007/978-3-0348-0218-5_8,

hospitalization for a median stay of 6 days (figure derived from tracking of >40,000 US patients) [1].

Early studies by Bodey et al. demonstrated that the risk of infection in neutropenic patients was directly related to the depth and duration of neutropenia [2]. In one study that followed 52 chemotherapy-treated leukemic patients (34 with acute lymphocytic leukemia [ALL] and 18 with acute myelogenous leukemia [AML]), over a total of 17,743 patient days, it was observed that 53% of those days were spent with infection if the patient had severe neutropenia. The percentage of days with infection decreased with increasing granulocyte counts until the range of 1–1.5 $\times$ 10^9/L, above which, there was no further protection from infection. The types of infections that were observed included urinary tract and disseminated fungal infections, pneumonia, and septicemia and the fatality rate overall was 41%. Treatment for infection at the time of these studies (1960s) involved antibiotic therapy and possibly leukocyte transfusions. Leukocyte transfusions were phased out as it was difficult to harvest sufficient numbers of neutrophils, and transfusion increased the risk of transmission of infectious agents, alloimmunization, and pulmonary complications [3, 4]. Antibiotics continue to be used to treat infections once they occur; however, prophylactic use of antibiotics is being debated given the concern that this practice will lead to the emergence of resistance [5, 6].

3 Neutropenia in Patients with Cancer

Neutropenia in patients with cancer may be the result of the disease itself, chemotherapy regimen, or patient-specific factors. An example of disease-induced neutropenia may be seen in AML, where there can be an expansion of abnormal progenitor cells that are unable to reach full myeloid development. Neutropenia may also be induced by the chemotherapy regimen, although the risk of this event varies between regimens. For example, the incidence of grade 3 or grade 4 neutropenia associated with docetaxel treatment has been reported to range between 65% in patients with metastatic or recurrent lung cancer and 86% in patients with metastatic or recurrent breast cancer (Taxotere package insert). The percentage of patients reported to develop febrile neutropenia during various combination treatment ranges from 0 to 2% for FEC (chemotherapy with fluorouracil, epirubicin, and cyclophosphamide) for the treatment of breast cancer [7] to 70% for VICE (chemotherapy with vincristine, ifosfamide, carboplatin, and etoposide) for the treatment of small-cell lung cancer (SCLC) [8]. Patient-specific factors may also increase the risk of chemotherapy-induced neutropenia. These factors include age ($\geq$65 years), the presence of comorbidities, type of cancer, advanced stage of disease, and a previous history of febrile neutropenia. Examples of comorbidities include congestive heart failure, cerebrovascular disease, lung disease, liver disease, renal disease, thrombosis, pulmonary embolism, and diabetes. The relationship between the risk of mortality and comorbidities in patients with febrile neutropenia was evaluated in a survey of 41,779 patients treated for febrile neutropenia at 115 US medical centers between 1995 and 2000 [1]. While the overall mortality was 9.5%, mortality for patients with no serious comorbidities was 2.6%, 10.3% in patients with 1 major comorbidity, and >21% for patients with >1 comorbidity. The average mortality rate in patients with febrile neutropenia varied according to the type of cancer, with patients with leukemia having the highest mortality rate among all cancer types at 14.3%, patients with lung cancers having the highest mortality rate among solid cancers at 13.4%, and patients with breast cancer having the lowest mortality rate at 3.6%. In addition to risk of infection and mortality associated with febrile neutropenia, another consequence that may impact patient outcome is the delay or reduction of subsequent chemotherapy dose. It is clinically undesirable to delay or reduce dose as data have shown that these actions can reduce tumor response and overall patient survival.

4 The Importance of Maintaining Dose in Cancer Chemotherapy

Failure of chemotherapy treatment has been attributed to an inability to successfully eliminate chemotherapy-sensitive cells (i.e., tumor-cell proliferation exceeds tumor-cell elimination) or to the outgrowth of chemotherapy-resistant cells. To address the tumor-cell proliferation versus elimination issue, the rates of tumor proliferation have been modeled and on the basis of these models, chemotherapy regimens have been proposed. Early studies of tumor growth kinetics in model systems of leukemia led to the proposal that cancer cells proliferate exponentially [9]. On the basis of exponential tumor cell growth, the "log kill" hypothesis was proposed, which states that a given dose of chemotherapy will eliminate a constant proportion of tumor cells rather than a constant number of tumor cells, meaning that if a given dose of a chemotherapeutic eliminates 10^3 cancer cells, it will reduce a tumor containing 10^{10} cells to 10^7 or a tumor containing 10^6 cells to 10^3. Because this relationship between chemotherapy and tumor-cell death is dependent on the tumor proliferating exponentially, it accurately modeled chemotherapeutic response of experimental leukemia, but failed when applied to human solid tumors.

The kinetics of solid tumor proliferation have been shown to follow a Gompertzian model of sigmoidal tumor growth [10]. This model suggests that while early tumor growth could follow an exponential function, growth slows and becomes nonlogarithmic as tumor mass increases. Based on the Gompertzian growth curve, Norton and Simon hypothesized that chemotherapy would result in a rate of regression of tumor that was proportional to the rate of growth for an unperturbed tumor of similar size [11]. In an attempt to minimize tumor regrowth, the most effective dose of chemotherapy should be given over the shortest period of time possible [12]. As a result, tumors would be given less time to recover between treatments and would be more effectively eradicated. This model gave rise to dose-dense chemotherapy regimens where the cumulative drug dose remains constant but the amount of drug is administered over a shorter period of time.

5 Dose Density

The benefit of dose-dense adjuvant chemotherapy has been evaluated in a cohort of node-positive breast cancer patients ($n = 2005$) in the Cancer and Leukemia Group B (CALGB)Trial 9741 [13]. In this trial, dose-dense, 2-week treatment intervals of doxorubicin, cyclophosphamide, and paclitaxel were compared to conventional 3-week intervals. At a median follow-up of 3 years, the dose-dense schedule was shown to be superior to the standard schedule with improvement in disease-free survival (risk ratio [RR] = 0.74; $p = 0.01$) and overall survival (RR = 0.69; $p = 0.013$) (Fig. 1). At a median follow-up of 4 years, the disease-free survival was 82% in the dose-dense group and 75% in the group treated with conventional 3-week intervals.

Additional studies have evaluated the value of dose-dense chemotherapy regimens and although each study has caveats (e.g., discrepancies in total dose between groups as in the ECOG 1199 [14] and AGO trials [15] or lower than expected event rates as in the GONO-MIG trial [16]), data suggest that increases in dose density have a beneficial impact on patient outcomes.

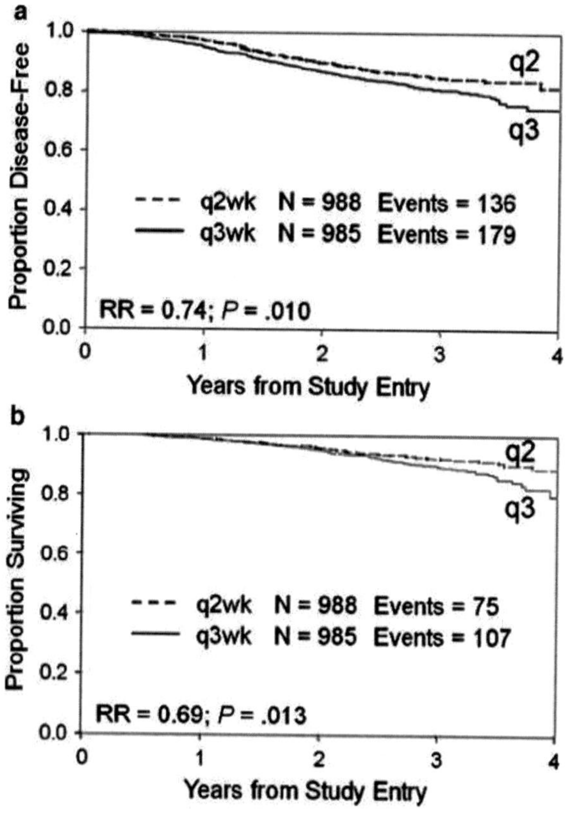

Fig. 1 (**a**) Disease-free and (**b**) overall survival by dose density from Cancer and Leukemia Group B Trial 9741 demonstrating the benefit of increased chemotherapy dose density compared to the standard schedule. Data from [13]

6 Dose Intensity

An additional reason for the failure of chemotherapy is the development and/or outgrowth of chemotherapy-resistant clones. The probability that a tumor population will contain resistant clones has been described by the Goldie–Coldman hypothesis to be a function of the size and inherent genetic instability of the tumor [17]. Consideration of this model suggests that combination chemotherapy would be preferable to single-agent therapies as it should be more difficult to develop resistance to multiple chemotherapeutics than to just one. Data by the same group suggested that increases in the dose intensity of the combination chemotherapy could reduce the likelihood that resistant cell lines would emerge and proliferate [17]. These models gave rise to dose-intense regimens where the drug dose/cycle is increased.

The association between increased dose intensity and improved patient outcome has been demonstrated in both retrospective and prospective clinical trials. One of the first retrospective studies evaluated the effect of dose on relapse-free survival and overall survival in a group of patients with node-positive breast cancer ($n = 386$, Milan Cancer Institute) 20 years after chemotherapy treatment (cyclophosphamide, methotrexate, fluorouracil) [18] (Fig. 2). The study groups were divided into patients who received ≥85%, 65% to 84%, and <65% of the optimal chemotherapeutic dose compared to control patients who did not receive chemotherapy. Patients treated with ≥85% of the optimal dose had the highest probability of both relapse-free survival and overall survival, whereas patients who received <65% of the optimal dose had relapse-free survival and overall survival probabilities that were comparable to control patients. In terms of survival, 52%

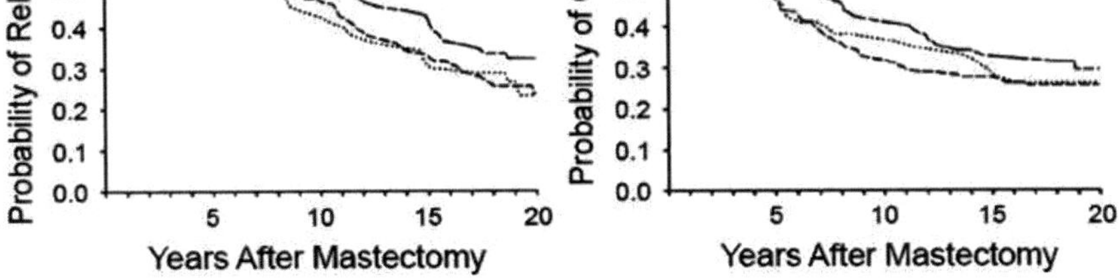

Fig. 2 (**a**) Relapse-free and (**b**) overall survival according to the percentage of the optimal dose received by patients ($n = 386$ node-positive breast cancer patients). Data generated by Bonadonna et al. [18] reflect a retrospective assessment of outcome 20 years after chemotherapy treatment

of the patients who received ≥85% of the optimal dose were alive after 20 years compared with 29% of patients who received <85%.

A similar relationship between dose intensity and response was observed in a retrospective analysis of patients with breast cancer ($n = 793$ stages I–IIIA, Valencia University Hospital) 10 years after treatment with anthracycline-based nontaxane chemotherapy [19]. This study compared disease-free and overall survival in patients who received ≥85% of the optimal chemotherapy dose to those who received <85%. The disease-free survival was 81% in patients who received ≥85% of the optimal dose compared with 67% of patients who received <85%. This study also reported that delayed cycles of treatment adversely affected disease-free survival (hazard ratio [CI 95%] = 1.61–2.67 for >2 delayed cycles) and overall survival (hazard ratio [CI 95%] = 1.24–2.33 for >2 delayed cycles), emphasizing the importance of maintaining chemotherapy treatment on schedule.

Prospective clinical trials have demonstrated a link between increased dose intensity and improved patient outcome. The CALGB 8541 study evaluated disease-free and overall survival in patients with node-positive breast cancer ($n = 1550$) who were treated with cyclophosphamide, doxorubicin, and fluorouracil at relative intensities of 1, 0.67, and 0.5 [20]. At a median follow-up of 9 years, the disease-free and overall survival for patients in the high- and moderate-dose groups were significantly better than that of patients in the low dose-intensity group (disease-free survival, $p = 0.0001$; overall survival, $p = 0.004$). Similarly, the French Adjuvant Study Group 05 compared two chemotherapy treatment regimens: FEC100 (fluorouracil 500 mg/m^2, cyclophosphamide 500 mg/m^2, and epirubicin 100 mg/m^2) and FEC50 (same fluorouracil and cyclophosphamide but epirubicin at 50 mg/m^2) in patients with early-stage breast cancer ($n = 565$) [21]. The results of this study demonstrated better disease-free and overall survival in the FEC100 group with a 10-year disease-free survival of 50.7% (95% CI, 47.3%–54.1%) in the FEC100 group and 45.3% (95% CI, 41.9%–48.7%) in the FEC50 group (Wilcoxon $p = 0.036$; log rank $p = 0.08$). The overall survival over this same time span was 54.8% (95% CI, 51.3%–58.3%) in the FEC100 group and 50.0% (95% CI, 46.7%–53.3%) in the FEC50 group (Wilcoxon $p = 0.039$; log rank $p = 0.05$).

7 Intense Dose Density

As treatment continues to evolve, studies have been done evaluating the effect of combining chemotherapy dose density with dose intensity in intensive dose-dense trials. Untch et al. compared the objective pathologic complete response rate and overall survival in patients with high-risk breast cancer ($n = 671$; patients with noninflammatory breast cancer with a tumor ≥3 cm or inflammatory breast cancer) treated preoperatively with epirubicin (90 mg/m^2) and paclitaxel (175 mg/m^2) every 3 weeks with an intense dose-density regimen of epirubicin (150 mg/m^2) followed by paclitaxel (250 mg/m^2) every 2 weeks [22]. Results demonstrated that intense dose-dense treatment significantly improved the pathologic complete response rate (18% vs. 10% for intense dose density vs. standard treatment; odds ratio = 1.89; $p = 0.008$), disease-free survival (hazard ratio [HR] = 0.71; $p = 0.011$) and overall survival (HR = 0.83; $p = 0.041$) (Fig. 3).

A benefit of intense dose-density treatment was seen in a separate trial described in patients with breast cancer ($n = 1284$) with stage II–IIIA disease and ≥4 positive axillary lymph nodes [23]. In this trial, patients were randomly assigned to receive intense dose-density treatment (three cycles each of epirubicin 150 mg/m^2, paclitaxel 225 mg/m^2, and cyclophosphamide 2,500 mg/m^2; all cycles separated by 2 weeks) or standard treatment (four cycles each of epirubicin/cyclophosphamide 90/600 mg/m^2, paclitaxel 175 mg/m^2; all cycles separated by 3 weeks). At a median of 62 months, the event-free survival rates were 70% in the intense dose-density group and 62% in the standard treatment group ($p < 0.001$), and overall survival was 82% in the intense dose-density group and 77% in the standard group ($p = 0.0285$).

Together these results demonstrate the value in pursuing more aggressive chemotherapy regimens. However, while increasing tumor cell killing, increased chemotherapy dose density and intensity also affect neutrophil counts and impair neutrophil repopulation. This issue has been largely addressed with the development of recombinant growth factors, r-metHuG-CSF, which can be administered in conjunction with chemotherapy and have been shown in clinical trials to mitigate

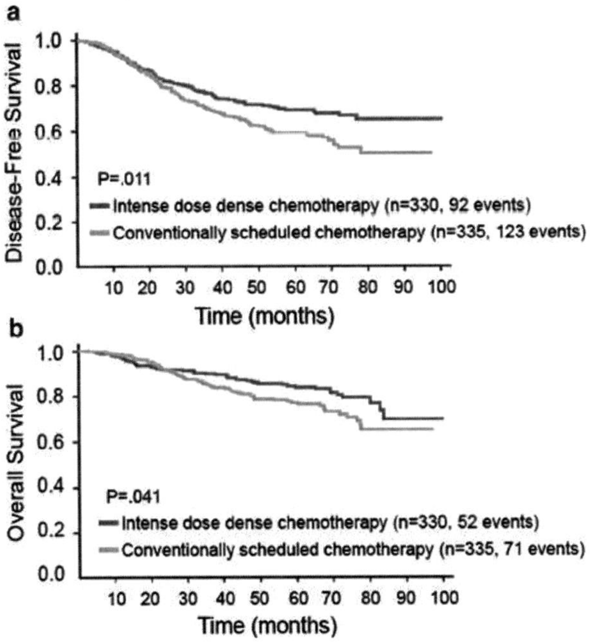

Fig. 3 Kaplan–Meier curve of (**a**) disease-free and (**b**) overall survival curves for breast cancer patients treated with intense dose dense or conventional chemotherapy regimens. Data generated by Untch et al. [22]

the risk of febrile neutropenia and to increase the likelihood of successfully administering chemotherapy dose on time.

8 Filgrastim and the Treatment of Chemotherapy-Induced Neutropenia

r-metHuG-CSF has been shown in various animal models and humans to cause a dose-dependent increase in peripheral blood neutrophil counts through decreased maturation time, increased cell divisions, and accelerated rates of neutrophil release from the bone marrow into the peripheral system [24–26]. Following on these observations, r-metHuG-CSF was used in clinical trials with the goal of preventing neutropenia and associated complications. The first form of r-metHuG-CSF to enter clinical trials was filgrastim. Studies to gain marketing approval were initiated in 1986 in the setting of solid tumors (Table 1). In the phase 1 studies, a rapid increase in peripheral neutrophils was observed after administration of filgrastim [27–31]. These neutrophils were shown to be functionally normal in terms of mobility (e.g., chemotaxis in response to chemotactic agent *N*-formyl-methionyleucyl phenylalanine), phagocytic activity (e.g., opsonization of zymosan), and neutrophil alkaline phosphatase activity, a marker of secondary granule formation and neutrophil maturation [28, 32]. In the phase 2 portions of these same studies, filgrastim was shown to hasten the recovery of neutrophils in chemotherapy-treated patients and to reduce the incidence of infection.

Table 1 Early experience with r-metHu-G-CSF (filgrastim) in the setting of solid tumors

Disease state	No of patients (treatment/control)	r-metHuG-CSF dosage	Chemotherapy	References
Phase 1/2				
Urothelial	18/6	1–60 μg/kg IV infusion; 10 days	Standard dose	[28]
Breast/ ovarian	17/4	10 μg/kg IV; 7 days 5 μg/kg IV; 3 days	High dose	[57]
SCLC	9/0	1–40 μg/kg IV infusion, 14 days	Standard dose	[27]
Various	15/0	1–60 μg/kg short IV; 8 days	Standard dose	[29]
Various	31/0	0.3–10 μg/kg SC; 5–16 days	Standard dose	[58]
Various	10/11	20–60 μg/kg IV; 20 days	High dose	[59]
Phase 2/3				
SCLC	20/17	50 μg/kg SC	High dose	[60]
Phase 3				
SCLC	95/104	230 μg/m^2 SC; 17 days max	Standard dose	[33]
SCLC	64/66	230 μg/m^2 SC; 4–17 days	Standard dose	[34]

IV intravenous; *SC* subcutaneous; *SCLC* small-cell lung cancer

These studies were followed by two pivotal phase 3 trials that evaluated the reduction in incidence of febrile neutropenia in patients with SCLC treated with standard-dose chemotherapy (cyclophosphamide, doxorubicin, and etoposide) [33, 34]. This chemotherapy regimen is associated with a 60% risk of febrile neutropenia. Both studies demonstrated that the use of filgrastim significantly decreased the incidence of febrile neutropenia by approximately 50% (Fig. 4). In the study by Crawford et al. [33], the incidence of febrile neutropenia was 77% in the placebo group and 40% in the filgrastim group ($p < 0.001$), and in the study by Trillet-Lenoir et al. [34], the incidence was 53% and 26% ($p < 0.002$) in the placebo and filgrastim groups, respectively. Filgrastim treatment also reduced the incidence of hospitalization and intravenous antibiotic use. The Trillet-Lenoire et al. trial further demonstrated that filgrastim significantly improved the ability to deliver chemotherapy on time. In this study, 61% of patients in the placebo group required at least one dose reduction compared with 29% in the filgrastim group ($p < 0.001$).

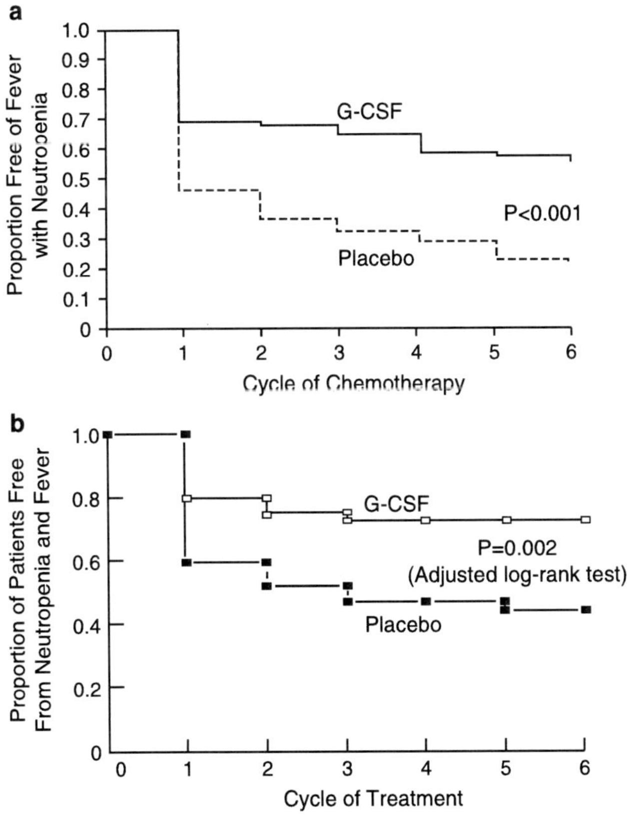

Fig. 4 Kaplan–Meier curve for the proportion of patients remaining free of fever with neutropenia according to treatment cycles (**a**) data from [33]; $n = 194$ patients. (**b**) Data from [34]; $n = 129$ patients

These studies supported FDA approval of filgrastim in 1991 for the prophylaxis of febrile neutropenia in patients with nonmyeloid malignancies treated with myelosuppressive chemotherapy. Initial guidelines recommended use when the risk threshold of febrile neutropenia was >40%. As additional clinical results have been generated, these guidelines have been refined. Usage guidelines are discussed in the chapter by Saraf and Ozer.

9 Pegfilgrastim and the Treatment of Chemotherapy-Induced Neutropenia

The half life of filgrastim is 3.5 h in humans and, as a result, it must be given daily until neutrophil recovery is achieved (as per Neupogen, filgrastim, prescribing information). This requirement for daily dosing is inconvenient for some patients and to address this problem, a longer-lived form of r-metHuG-CSF (pegfilgrastim) was generated through the addition of a 20 kiloDalton (kDa) polyethylene glycol (PEG) molecule to the amino terminus of r-metHuG-CSF. There are different chemistry methods to attach a PEG molecule to a protein, but typically these processes result in >1 PEG attached to the protein and/or attachment at different sites leading to unpredictable heterogeneity; however, the conditions of the coupling reaction have been optimized such that the PEG molecule is specifically added to the amino terminus [35]. Because the amino terminus is not involved in receptor binding [36], the addition of the PEG molecule does not inhibit the interaction between G-CSF and its receptor (G-CSFR) (see chapter by Arvedson and Giffin for further information).

The pharmacokinetic effect of the PEG addition is a significant reduction in the renal clearance of pegfilgrastim [37]. Renal clearance is highly influenced by the size of molecules passing through the kidney and proteins within the size range of r-metHuG-CSF (18.8 kDa) are thought to be filtered by the glomeruli, reabsorbed, and degraded [38]. Pegfilgrastim, in comparison, has an increased molecular weight (38.8 kDa) and an increased hydrodynamic size stemming from the fact that PEG molecules are highly hydrated. As a result, pegfilgrastim is not filtered by the glomeruli and instead is predominantly cleared through its interaction with G-CSFR. Consequently, the circulating amounts of pegfilgrastim persist until neutrophil recovery is achieved, meaning that only one dose is required per chemotherapy cycle (as per Neulasta, pegfilgrastim, prescribing information).

Nonclinical studies comparing filgrastim to pegfilgrastim (also referred to as SD/01 where SD = sustained duration) in mice demonstrated that the neutrophil response to a single bolus injection of filgrastim up to 2,500 μg/kg could only sustain increased neutrophil counts for 24–48 h, whereas a single bolus of 1,000 μg/kg pegfilgrastim could sustain increased neutrophil counts for up to 5 days [39] (Fig. 5). In chemotherapy-treated mice (150 mg/kg 5-fluorouracil), a single injection of pegfilgrastim at 1,000 μg/kg comparably reduced both the depth and duration of neutropenia as 10 days of daily filgrastim injections at 300 μg/kg (Fig. 6). In early evaluations in humans, pegfilgrastim was shown to increase and sustain neutrophils in a dose-dependent manner [39, 40]. In chemotherapy patients, the effects of a single dose of SD/01 (100 μg/kg) on ANC after chemotherapy were comparable or greater than those achieved with daily filgrastim (5 μg/kg) [40].

Pegfilgrastim was approved by the FDA in 2001 after the demonstration that pegfilgrastim was as effective as filgrastim in managing chemotherapy-induced

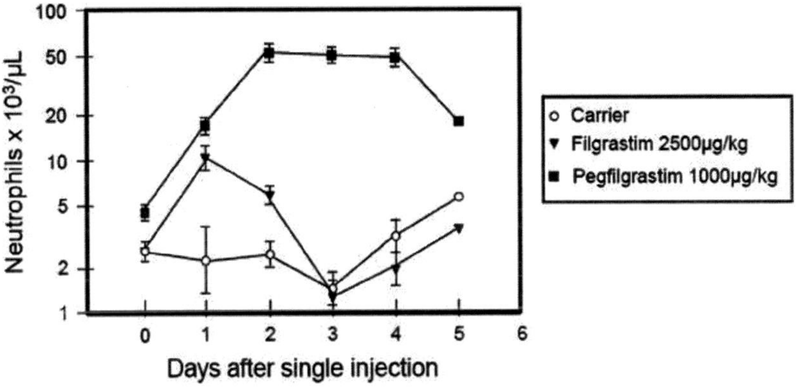

Fig. 5 Neutrophil response after a single injection of carrier, filgrastim or pegfilgrastim in splenectomized mice on day 0. Each point represents the mean of 5–10 mice. Data adapted from [39]

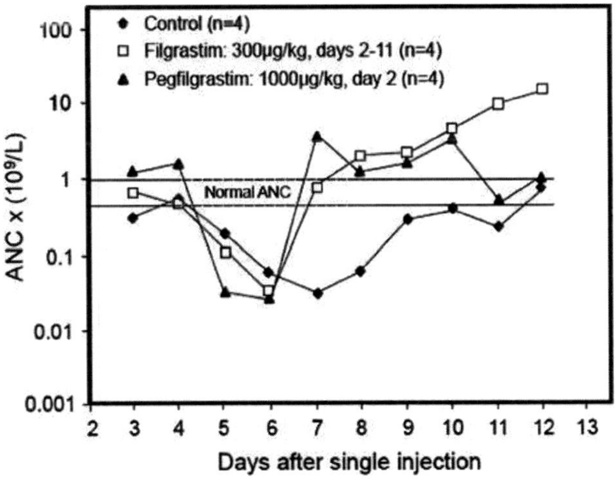

Fig. 6 Neutrophil response to carrier, filgrastim, or pegfilgrastim in chemotherapy-treated mice. Each point represents the mean of four mice. Data adapted from [39]

neutropenia. The clinical trials supporting this result [41, 42] were done in the setting of breast cancer with patients treated with doxorubicin and docetaxel, a chemotherapy regimen with a 38% febrile neutropenia rate in patients not treated with r-metHuG-CSF [43]. Both trials compared either a single injection of pegfilgrastim followed by daily injections of placebo or daily injections of filgrastim. Results from these trials demonstrated that a single dose of pegfilgrastim was as effective as daily filgrastim in preventing the incidence and limiting the duration of grade 4 neutropenia. The two treatments were also similar in terms of the depth of neutropenia and the time for neutrophil recovery. Although neither study was designed to detect a difference in febrile neutropenia incidence between the two groups, Green et al. [42] observed febrile neutropenia in 20% of patients ($n = 80$) treated with filgrastim versus 13% of patients ($n = 77$) treated with pegfilgrastim (difference was not significant) and Holmes et al. [41] observed febrile neutropenia in 18% of patients ($n = 156$) treated with filgrastim versus 9% of patients ($n = 154$) treated with pegfilgrastim ($p = 0.029$). Pegfilgrastim treatment enabled maintenance of dose schedules.

10 Febrile Neutropenia Timing and Effect on Mortality

Febrile neutropenia frequently occurs in the first cycle of chemotherapy. In an evaluation of 2,302 patients from 137 US community oncology centers, 50–75% of all neutropenic events were found to occur in the first cycle, and neutropenia was more likely to occur in the first cycle than in the subsequent three cycles combined [44]. Febrile neutropenia occurred most frequently in patients who were treated with a full dose of chemotherapy without supportive care, particularly if the chemotherapy regimen was taxane based. The reduced incidence of neutropenia in later cycles was reported to be due to chemotherapy delays, dose reductions, or the introduction of r-metHuG-CSF [45]. In the event that r-metHuG-CSF was not introduced, neutropenia was reported to persist across the chemotherapy cycles [46].

While the benefit of prophylactic r-metHuG-CSF to prevent febrile neutropenia was demonstrated in the registrational trials, an effect on mortality has also been observed. Data from a meta-analysis of 17 studies comparing prophylactic use of r-metHuG-CSF with placebo or no treatment in 3,593 patients being treated for solid tumors or lymphoma demonstrated a reduced risk in infection-related mortality of 45% (1.5% vs. 2.8%; RR = 0.55; $p = 0.018$) and all cause early mortality of 40% (3.4% vs. 5.7%; RR = 0.6; $p = 0.002$) [47]. Two other meta-analyses have been performed demonstrating no survival benefit [48, 49]; however, both of these analyses combined studies that used recombinant human granulocyte-macrophage colony-stimulating factor (GM-CSF) in addition to those that used rHuG-CSF, which may have affected the results.

11 r-metHuG-CSF and the Effects on Chemotherapy Dose Delivery

Studies have demonstrated that prophylactic r-metHuG-CSF can mitigate neutropenia-mediated effects of dose delivery. In a study by Pettengell et al., 80 patients with non-Hodgkin's lymphoma (NHL) were treated with intensive weekly chemotherapy (VAPEC-B: vincristine, doxorubicin, prednisolone, etoposide, cyclophosphamide, and bleomycin) alone or in combination with r-metHuG-CSF [50]. Chemotherapy dose reduction was decreased in r-metHuG-CSF-treated patients compared to controls (4 of 41 dose reductions in r-metHuG-CSF-treated patients vs. 13 of 39 dose reductions in controls; $p = 0.01$). Chemotherapy dose intensity was significantly increased in r-metHuG-CSF-treated patients, with patients receiving a median of 95% of the scheduled dose compared to controls, where patients received a median of 83% of the scheduled dose.

Similarly, in a trial of patients with non-small cell lung cancer (NSCLC) treated with etoposide and cisplatin alone or with filgrastim, chemotherapy treatment was intensified by >50% in the filgrastim-treated patients [51]. In a trial of patients with lymphoma, patients treated with CHOP (cyclophosphamide, doxorubicin, vincristine, and prednisone) either alone or with filgrastim, cyclophosphamide was intensified by 270% and doxorubicin was intensified by 87% in the filgrastim-treated group [51].

Similar to filgrastim, pegfilgrastim has been shown to be effective in supporting dose-dense chemotherapy regimens. As discussed previously, the CALGB 9741 trial demonstrated that dose-dense administration of doxorubicin/cyclophosphamide followed by paclitaxel improved disease-free and overall survival in women with node-positive breast cancer [13]. In that study, filgrastim was used to limit neutropenia-related complications, and the result was a febrile neutropenia rate of 2–3% for patients receiving filgrastim. This percentage was compared to a febrile neutropenia incidence of 3–5% in patients who received the standard (less dose-dense) treatment in the absence of filgrastim. Subsequently, a phase 2 trial including 135 women was conducted to evaluate pegfilgrastim in this same chemotherapy regimen [52]. In this study, 6 mg of pegfilgrastim was administered subcutaneously approximately 24 h after chemotherapy, and the resulting febrile neutropenia rate was 1.5% with two of the 135 patients developing febrile neutropenia.

Finally, a meta-analysis of 17 randomized controlled trials evaluating 3,493 patients being treated for solid tumors and malignant lymphoma found that patients being treated with prophylactic r-metHuG-CSF (one study with pegfilgrastim, the others with filgrastim or lenograstim) received an increased mean relative dose intensity compared to control patients [47]. The average relative dose intensity between control patients ranged from 71.0 to 95.0% with a mean (median) of 86.7% (88.5%). This result compares to the average relative dose intensity between r-metHuG-CSF patients, which ranged from 91.0 to 99.0% with a mean (median) of 95.1% (95.5).

12 Comparison of Filgrastim to Pegfilgrastim

While phase 3 trials comparing filgrastim to pegfilgrastim demonstrated the two agents to be comparable [41, 42], subsequent studies suggest that pegfilgrastim may be more effective than filgrastim. A retrospective analysis of data from two randomized, double-blind trials of patients with breast cancer ($n = 448$) compared a single dose pegfilgrastim to daily filgrastim and found that the risk of febrile neutropenia was significantly lower in patients receiving pegfilgrastim than for patients who were receiving filgrastim (11% vs. 19%; RR = 0.56; 95% CI = 0.35–0.89) [53]. A meta-analysis of five randomized controlled trials ($n = 617$ solid tumor and lymphoma patients) comparing the efficacy of pegfilgrastim to that of filgrastim when administered according to the package insert found that a single dose of pegfilgrastim more effectively reduced febrile neutropenia than a median of 10–14 days of filgrastim (RR = 0.64; 95% CI, 0.43–0.97) [54].

The apparent efficacy of filgrastim may be further reduced by current standard practice. While it is recommended that filgrastim be given daily until the ANC is restored to the normal range (typically 10 or 11 doses/chemotherapy cycle was required in clinical trials [29, 31, 33]), it is common that fewer doses are administered or treatment is delayed. In an observational study of patients treated in 99 community US oncology practices, Morrison et al. observed that while pegfilgrastim was initiated, on average 2.4 days (±3.2) after chemotherapy in the first cycle of use and 1.9 (±3.0) days in subsequent cycles of use, filgrastim was started on average 7.7 (±6.5) days and 4.9 (±4.6) days after chemotherapy in the first and subsequent cycles of use in 2001 [55]. In 2003, the delay in start time increased to 9.6 (±6.2) and 6.4 (±6.4) days for the first and subsequent cycles of chemotherapy. In terms of the length of use, filgrastim was only administered for an average of 5.2 (±3.5) days in 2001 and 3.7 (±2.8) days in 2003 ($p < 0.001$). Reduced courses of filgrastim treatment were also observed by Weycker et al. in an analysis of treatment records for patients with NHL, breast, and lung cancer [56]. It was further observed from this analysis that the risk of hospitalization for neutropenia decreased with each additional day of filgrastim administration (NHL patients OR $= 0.81, p = 0.003$; breast cancer patients OR $= 0.77, p = 0.001$; and lung cancer patients OR $= 0.91, p = 0.084$). These data emphasize the importance of administering r-metHuG-CSF according to recommendations in order to achieve maximum benefit.

13 Summary

Two forms of r-metHuG-CSF have been developed that address neutropenic complications associated with cancer treatment. Both are effective in reducing the risk of febrile neutropenia, and meta-analysis results further suggest that prophylactic use of r-metHuG-CSF improves infection-related and early mortality. In addition, both agents have been shown to support delivery of chemotherapy treatment on time and to enable dose-dense, dose-intense, and intense dose-dense chemotherapy regimens. While recent studies suggest that pegfilgrastim may be more effective than filgrastim, several of these studies may reflect reduced or delayed daily administration of filgrastim. Pegfilgrastim, by comparison must only be given once per cycle, resulting in increased patient compliance and improved outcomes.

Disclosure Drs Arvedson and Molineux are employees and shareholders in Amgen, the manufacturer of filgrastim and pegfilgrastim.

References

1. Kuderer NM, Dale DC, Crawford J, Cosler LE, Lyman GH (2006) Mortality, morbidity, and cost associated with febrile neutropenia in adult cancer patients. Cancer 106:2258–2266
2. Bodey G, Buckley M, Sathe Y, Freireich E (1966) Quantitative relationships between circulating leukocytes and infection in patients with acute leukemia. Ann Intern Med 64:328–340
3. Winston D, Ho W, Howell C et al (1980) Cytomegalovirus infections associated with leukocyte transfusions. Ann Intern Med 93:671–675
4. Schiffer C, Aisner J, Daly P, Schimpff S, Wiernik P (1979) Alloimmunization following prophylactic granulocyte transfusion. Blood 54:766–774
5. Cometta A, Calandra T, Bille J, Glauser MP (1994) *Escherichia coli* resistant to fluoroquinolones in patients with cancer and neutropenia. N Engl J Med 330:1240–1241
6. Wingard JR, Elmongy M (2009) Strategies for minimizing complications of neutropenia: prophylactic myeloid growth factors or antibiotics. Crit Rev Oncol Hematol 72:144–154
7. Zielinski C, Beslija S, Mrsic-Krmpotic Z et al (2005) Gemcitabine, epirubicin, and paclitaxel versus fluorouracil, epirubicin, and cyclophosphamide as first-line chemotherapy in metastatic breast cancer: a Central European Cooperative Oncology Group international, multicenter, prospective, randomized phase III trial. J Clin Oncol 23:1401–1408
8. Woll P, Hodgetts J, Lomax L, Bildet F, Cour-Chabernaud V, Thatcher N (1995) Can cytotoxic dose-intensity be increased by using granulocyte colony-stimulating factor? A randomized controlled trial of lenograstim in small-cell lung cancer. J Clin Oncol 13:652–659
9. Skipper H (1967) Criteria associated with destruction of leukemia and solid tumor cells in animals. Cancer Res 27:2636–2645
10. Laird A (1964) Dynamics of tumor growth. Br J Cancer 18:490–502
11. Simon R, Norton L (2006) The Norton-Simon hypothesis: designing more effective and less toxic chemotherapeutic regimens. Nat Clin Prac Oncol 3:406–407
12. Norton L, Simon R, Brereton H, Bogden A (1976) Predicting the course of Gompertzian growth. Nature 264:452–545
13. Citron ML, Berry DA, Cirrincione C et al (2003) Randomized trial of dose-dense versus conventionally scheduled and sequential versus concurrent combination chemotherapy as postoperative adjuvant treatment of node-positive primary breast cancer: first report of Intergroup Trial C9741/Cancer and Leukemia Group B Trial 9741. J Clin Oncol 21:1431–1439
14. Sparano J, Wang M, Martino S et al (2007) Phase III study of doxorubicin-cyclophosphamide followed by paclitaxel or docetaxel given every 3 weeks or weekly in operable breast cancer: results of Intergroup Trial E1199. J Clin Oncol 25:516

15. Moebus V, Lueck H, Thomssen C et al (2006) Dose-dense sequential chemotherapy with epirubicin (E), paclitaxel (T) and cyclophosphamide (C) (ETC) in comparison to conventional schedule chemotherapy in high-risk breast cancer patients (≥4+ LN). Mature results of an AGO-trial. Breast Cancer Res Treat 100:S20
16. Venturini M, Del Mastro L, Aitini E et al (2005) Dose-dense adjuvant chemotherapy in early breast cancer patients: results from a randomized trial. J Natl Cancer Inst 97:1724–1733
17. Goldie J, Coldman A (1979) A mathematic model for relating the drug sensitivity of tumors to their spontaneous mutation rate. Cancer Treat Rep 63:1727–1733
18. Bonadonna G, Valagussa P, Moliterni A, Zambetti M, Brambilla C (1995) Adjuvant cyclophosphamide, methotrexate, and fluorouracil in node-positive breast cancer. The results of 20 years of follow-up. N Engl J Med 332:901–906
19. Chirivella I, Bermejo B, Insa A et al (2009) Optimal delivery of anthracycline-based chemotherapy in the adjuvant setting improves outcome of breast cancer patients. Breast Cancer Res Treat 114:479–484
20. Budman D, Berry D, Cirrincione C et al (1998) Dose and dose intensity as determinants of outcome in the adjuvant treatment of breast cancer. The Cancer and Leukemia Group B. J Natl Cancer Inst 90:1205–1211
21. Bonneterre J, Rocha H, Kerbrat P et al (2005) Epirubicin increases long-term survival in adjuvant chemotherapy of patients with poor-prognosis, node-positive, early breast cancer: 10-year follow-up results of the French Adjuvant Study Group 05 Randomized Trial. J Clin Oncol 23:2686–2693
22. Untch M, Moebus V, Kuhn W et al (2009) Intensive dose-dense compared with conventionally scheduled preoperative chemotherapy for high-risk primary breast cancer. J Clin Oncol 27:2938–2945
23. Moebus V, Jackisch C, Lueck H-J et al (2010) Intense dose-dense sequential chemotherapy with epirubicin, paclitaxel, and cyclophosphamide compared with conventionally scheduled chemotherapy in high-risk primary breast cancer: mature results of an AGO phase III study. J Clin Oncol 28:2874–2880
24. Moore M, Welte K, Gabrilove J, Souza L (1987) Biological activities of recombinant human granulocyte colony stimulating factor (rhG-CSF) and tumor necrosis factor: in vivo and in vitro analysis. Haematol Blood Transfus 31:210–220
25. Tamura M, Hattori K, Nomura H et al (1987) Induction of neutrophilic granulocytosis in mice by administration of purified human native granulocyte colony-stimulating factor (G-CSF). Biochem Biophys Res Commun 142:454–460
26. Welte K, Bonilla M, Gabrilove J et al (1987) Recombinant human granulocyte-colony stimulating factor: in vitro and in vivo effects on myelopoiesis. Blood Cells 13:17–30
27. Bronchud M, Scarffe J, Thatcher N et al (1987) Phase I/II study of recombinant human granulocyte colony-stimulating factor in patients receiving intensive chemotherapy for small cell lung cancer. Br J Cancer 56:809–813
28. Gabrilove J, Jakubowski A, Fain K et al (1988) Phase I study of granulocyte colony-stimulating factor in patients with transitional cell carcinoma of the urothelium. J Clin Invest 82:1454–1461
29. Morstyn G, Souza LM, Keech J et al (1988) Effect of granulocyte colony stimulating factor on neutropenia induced by cytotoxic chemotherapy. Lancet 331:667–672
30. Dührsen U, Villeval J, Boyd J, Kannourakis G, Morstyn G, Metcalf D (1988) Effects of recombinant human granulocyte colony-stimulating factor on hematopoietic progenitor cells in cancer patients. Blood 72:2074–2081
31. Gabrilove JL, Jakubowski A, Scher H et al (1988) Effect of granulocyte colony-stimulating factor on neutropenia and associated morbidity due to chemotherapy for transitional-cell carcinoma of the urothelium. N Engl J Med 318:1414–1422
32. Bronchud M, Potter M, Morgenstern G et al (1988) In vitro and in vivo analysis of the effects of recombinant human granulocyte colony-stimulating factor in patients. Br J Cancer 58:64–69

33. Crawford J, Ozer H, Stoller R et al (1991) Reduction by granulocyte colony-stimulating factor of fever and neutropenia induced by chemotherapy in patients with small-cell lung cancer. N Engl J Med 325:164–170
34. Trillet-Lenoir V, Green J, Manegold C et al (1993) Recombinant granulocyte colony stimulating factor reduces the infectious complications of cytotoxic chemotherapy. Eur J Cancer 29A:319–324
35. Kinstler O, Molineux G, Treuheit M, Ladd D, Gegg C (2002) Mono-N-terminal poly(ethylene glycol)-protein conjugates. Adv Drug Deliv Rev 54:477–485
36. Tamada T, Honjo E, Maeda Y et al (2006) Homodimeric cross-over structure of the human granulocyte colony-stimulating factor (GCSF) receptor signaling complex. Proc Natl Acad Sci U S A 103:3135–3140
37. Yang B-B, Lum PK, Hayashi MM, Roskos LK (2004) Polyethylene glycol modification of filgrastim results in decreased renal clearance of the protein in rats. J Pharm Sci 93:1367–1373
38. Maack T, Johnson V, Kau S, Figueriedo J, Sigulem D (1979) Renal filtration, transport, and metabolism of low-molecular-weight proteins: a review. Kidney Int 16:251–270
39. Molineux G, Kinstler O, Briddell B et al (1999) A new form of filgrastim with sustained duration in vivo and enhanced ability to mobilize PBPC in both mice and humans. Exp Hematol 27:1724–1734
40. Johnston E, Crawford J, Blackwell S et al (2000) Randomized, dose-escalation study of SD/01 compared with daily filgrastim in patients receiving chemotherapy. J Clin Oncol 18:2522–2528
41. Holmes FA, OÂ'Shaughnessy JA, Vukelja S et al (2002) Blinded, randomized, multicenter study to evaluate single administration pegfilgrastim once per cycle versus daily filgrastim as an adjunct to chemotherapy in patients with high-risk stage II or stage III/IV breast cancer. J Clin Oncol 20:727–731
42. Green MD, Koelbl H, Baselga J et al (2003) A randomized double-blind multicenter phase III study of fixed-dose single-administration pegfilgrastim versus daily filgrastim in patients receiving myelosuppressive chemotherapy. Ann Oncol 14:29–35
43. Misset JL, Dieras V, Gruia G et al (1999) Dose-finding study of docetaxel and doxorubicin in first-line treatment of patients with metastatic breast cancer. Ann Oncol 10:553–560
44. Crawford J, Wolff D, Culakova E et al (2005) First-cycle risk of severe and febrile neutropenia in cancer patients receiving systemic chemotherapy: results from a prospective nationwide study. J Support Oncol 3:52–53
45. Pettengell R, Schwenkglenks M, Leonard R et al (2008) Neutropenia occurrence and predictors of reduced chemotherapy delivery: results from the INC-EU prospective observational European neutropenia study. Support Care Cancer 16:1299–1309
46. Lyman G (2005) Guidelines of the National Comprehensive Cancer Network on the use of myeloid growth factors with cancer chemotherapy: a review of the evidence. J Natl Compr Canc Netw 3:557–571
47. Kuderer NM, Dale DC, Crawford J, Lyman GH (2007) Impact of primary prophylaxis with granulocyte colony-stimulating factor on febrile neutropenia and mortality in adult cancer patients receiving chemotherapy: a systematic review. J Clin Oncol 25:3158–3167
48. Sung L, Nathan P, Alibhai S, Tomlinson G, Bevene J (2007) Meta-analysis: effect of prophylactic hematopoietic colony-stimulating factors on mortality and outcomes of infection. Ann Intern Med 147:400–411
49. Bohlius J, Herbst C, Reiser M, Schwarzer G, Engert A (2008) Granulopoiesis-stimulating factors to prevent adverse effects in the treatment of malignant lymphoma. Cochrane Database Syst Rev 4:CD003189
50. Pettengell R, Gurney H, Radford J et al (1992) Granulocyte colony-stimulating factor to prevent dose-limiting neutropenia in non-Hodgkin's lymphoma: a randomized controlled trial. Blood 80:1430–1436

51. Blayney DW, McGuire BW, Cruickshank SE, Johnson DH (2005) Increasing chemotherapy dose density and intensity: phase I trials in non-small cell lung cancer and non-Hodgkin's lymphoma. Oncologist 10:138–149
52. Burstein HJ, Parker LM, Keshaviah A et al (2005) Efficacy of pegfilgrastim and darbepoetin alfa as hematopoietic support for dose-dense every-2-week adjuvant breast cancer chemotherapy. J Clin Oncol 23:8340–8347
53. Siena S, Piccart MJ, Holmes FA, Glaspy J, Hackett J, Renwick JJ (2003) A combined analysis of two pivotal randomized trials of a single dose of pegfilgrastim per chemotherapy cycle and daily filgrastim in patients with stage II-IV breast cancer. Oncol Rep 10:715–724
54. Pinto L, Liu Z, Doan Q, Bernal M, Dubois R, Lyman G (2007) Comparison of pegfilgrastim with filgrastim on febrile neutropenia, grade IV neutropenia and bone pain: a meta-analysis of randomized controlled trials. Curr Med Res Opin 23:2283–2295
55. Morrison V, Wong M, Hershman D, Campos L, Ding B, Malin J (2007) Observational study of the prevalence of febrile neutropenia in patients who received filgrastim or pegfilgrastim associated with 3–4 week chemotherapy regimens in community oncology practices. J Manag Care Pharm 13:337–348
56. Weycker D, Hackett J, Edelsberg JS, Oster G, Glass AG (2006) Are shorter courses of filgrastim prophylaxis associated with increased risk of hospitalization? Ann Pharmacother 40:402–407
57. Bronchud M, Howell A, Crowther D, Hopwood P, Souza L, Dexter T (1989) The use of granulocyte colony-stimulating factor to increase the intensity of treatment with doxorubicin in patients with advanced breast and ovarian cancer. Br J Cancer 60:121–125
58. Morstyn G, Campbell L, Lieschke G et al (1989) Treatment of chemotherapy-induced neutropenia by subcutaneously administered granulocyte colony-stimulating factor with optimization of dose and duration of therapy. J Clin Oncol 7:1554–1562
59. Neidhart J, Mangalik A, Kohler W et al (1989) Granulocyte colony-stimulating factor stimulates recovery of granulocytes in patients receiving dose-intensive chemotherapy without bone marrow transplantation. J Clin Oncol 7:1685–1692
60. Masuda N, Fukuoka M, Furuse K (1992) CODE chemotherapy with or without recombinant human granulocyte colony-stimulating factor in extensive-stage small cell lung cancer. Oncol 49:19–24

Use of rHuG-CSF for the Treatment of Myeloid Leukemia and in Targeting Leukemia Stem Cells

Fumihiko Ishikawa

1 Introduction

The utility of recombinant human granulocyte colony-stimulating factor (rHuG-CSF) in leukemia treatment has been well established, in supporting granulopoiesis after chemotherapy and in mobilizing hematopoietic stem/progenitor cells (HSPC) for both allogeneic and autologous transplantation. Successes in these two objectives have led to significantly improved care of patients with leukemia. This chapter discusses another potential application of this agent as a part of targeted therapy against leukemia stem cells (LSC). Within the functional hierarchy of leukemia cells, LSC are the cells capable of initiating disease. Reports have shown that LSC are cell-cycle quiescent and chemotherapy resistant, suggesting their crucial role in acute myeloid leukemia (AML) relapse. The use of rHuG-CSF may induce quiescent AML stem cells to cycle, thereby rendering them susceptible to chemotherapy. Such application of rHuG-CSF may become an important part of curative therapeutic strategies for AML.

2 rHuG-CSF for Therapy-Related Neutropenia

As with the treatment for other hematologic malignancies, induction regimens for AML result in severe suppression of normal hematopoiesis. With the exception of promyelocytic leukemia (AML M3), idarubicin/daunorubicin and cytosine arabinocide (Ara-C) are frequently chosen for inducing patients into remission at initial diagnosis. After successful induction, AML patients experience further

F. Ishikawa (✉)
Research Unit for Human Disease Models, RIKEN Research Center for Allergy and Immunology, Box 230-0045, 1-7-22 Suehiro-cho, Tsurumi-ku, Yokohama City, Kanagawa 230-0045, Japan
e-mail: f_ishika@rcai.riken.jp

G. Molineux et al. (eds.), *Twenty Years of G-CSF*, Milestones in Drug Therapy,
DOI 10.1007/978-3-0348-0218-5_9,

myelosuppression through postremission consolidation and/or maintenance therapy. At some point during the course of AML treatment, almost all patients experience moderate (absolute neutrophil count [ANC] $< 1.0 \times 10^9$/L) to severe (ANC $< 0.5 \times 10^9$/L) peripheral blood neutropenia.

Neutrophils are responsible for an important part of innate immunity against bacterial and fungal infection. When neutrophil counts are $< 0.5 \times 10^9$/L, the risks for febrile neutropenia as well as bacterial and fungal infection are greatly increased [1, 2]. Until neutrophil recovery occurs at approximately 3 weeks after treatment, the patients are at an increased risk of febrile neutropenia and opportunistic infections by a broad species of bacteria including *Pseudomonas aeruginosa* and other Gram-negative bacilli. Compared with bacteremia, fungal infection after chemotherapy is less frequently observed in these patients, but *Aspergillus* infection is sometimes fatal in patients with severe neutropenia.

rHu-GCSF was identified as a myeloid growth factor that directs the development of neutrophils and other myeloid subsets [3–5] In patients with AML who are undergoing initial or repeat induction chemotherapy, several clinical studies have shown that the administration of rHuG-CSF produces modest decreases in the duration of neutropenia associated with variable and modest decreases in the duration of hospitalization and the incidence of severe infection [6–11]. These findings have led to recommendations by the American Society of Clinical Oncology (ASCO) that the use of rHuG-CSF is reasonable during induction, although no advantage in remission rate, remission duration, or survival is reported [12] (see chapter "Practice Guidelines for the Use of rHuG-CSF in an Oncology Setting" by Saraf and Ozer for more information). On the contrary, the effect of rHuG-CSF in reducing the duration of neutropenia after consolidation therapy appears more important. In two large randomized controlled trials, the use of rHuG-CSF was associated with marked reduction in the duration of severe neutropenia and a decreased rate of severe infections [8, 9].

3 rHuG-CSF in Mobilization of Normal Hematopoietic Stem Cells

rHuG-CSF plays another crucial role in AML therapy as a mobilizer of hematopoietic stem cells (HSC) for collection before allogeneic and autologous HSC transplantation. rHuG-CSF treatment results in mobilization of bone marrow HSC into peripheral blood through reduced expression of matrix metalloproteinase (MMP)-9, stromal cell-derived factor (SDF)-1, and other molecules in stromal cells constituting the bone marrow HSC niche [13–15]. Treatment of mice with rHuG-CSF significantly reduces stromal cell-derived factor (SDF)-1 leading to an inhibition of HSC binding to their niche [14]. In addition to its direct action on HSPC, in vivo experiments revealed the effect of rHuG-CSF on bone metabolism resulting in dynamic changes in adhesion/migration of HSC. By studying the UDP-galactose ceramide galactosyltransferase-deficient ($Cgt^{-/-}$) mice, it was determined that such mobilization of HSC and hematopoietic progenitor cells (HPC) is critically regulated by sympathetic nervous system [15].

4 rHuG-CSF as a Regulator of Cell Cycle in Normal Hematopoietic Stem Cells

In the steady-state bone marrow, normal HSC residing in their niche remain cell cycle quiescent and metabolically inactive [16, 17]. The relatively dormant state of HSC is considered important for protection against various stimuli such as toxic agents or infection. During states of hematopoietic stress, such as infection, inflammation, or bleeding, HSC enter the cell cycle and initiate the cascade of proliferation and differentiation to provide mature hematopoietic and immune effector cells. G-CSF has been found to play a role in the stress response by normal HSC. rHuG-CSF, alone or in combination with other cytokines or with chemotherapeutic agents, has been shown to increase cell-cycle entry by murine and human HSC and leads to the proliferation of myeloid progenitors, both in vitro and in vivo [18–23]. Coadministration of rHuG-CSF and recombinant human interleukin-3 (rHuIL-3) in patients with lymphoid malignancies increased progenitors in the S-phase [18]. These findings suggest that G-CSF plays a role in the modulation of murine and human HSC cell cycle.

5 rHuG-CSF and Chemotherapy Responsiveness of AML Cells

The development of chemotherapy regimens containing anthracyclin derivatives (e.g., daunorubicin, idarubicin) and Ara-C has resulted in successful induction of traditionally defined morphologic complete remission in many cases of AML with favorable and standard risk [24–27]. However, 47–91% of patients who achieve complete remission eventually experience relapse, resulting in $<$ 50% 5-year survival [25, 26]. Minimal residual disease (MRD), referring to the presence of treatment-resistant leukemia cells that escape identification by the traditional definition for complete remission, may cause relapse after complete remission [28]. To prevent disease relapse, identifying and understanding the nature of MRD is necessary. By definition, MRD must consist of rare leukemia cells that escape killing by chemotherapeutic agents used to induce remission. One possible mechanism by which leukemia cells evade chemotherapy may be through metabolic and proliferative inactivity, since many anticancer agents target actively proliferating cells. This possibility has led investigators to consider agents that stimulate leukemia cells to become metabolically active and/or to proliferate as sensitizers for conventional chemotherapy. The observation that AML blasts express G-CSF receptor (G-CSFR) made rHuG-CSF a promising candidate molecule [29].

Several in vitro studies have been performed to determine the effects of hematopoietic growth factors, including rHuG-CSF, on proliferation, metabolism, and chemotherapy responsiveness of AML cells. rHuG-CSF and recombinant human granulocyte-macrophage colony-stimulating factor (rHuGM-CSF), alone or in combination, enhanced in vitro colony formation by primary human AML cells [30]. The addition of hematopoietic growth factors, including rHuG-CSF, was found to stimulate human primary AML cells to proliferate in vitro as evidenced by increased thymidine uptake [31]. Augmentation of chemotherapy responsiveness through priming by hematopoietic growth factors has been observed with cytokines including rHuG-CSF, rHuGM-CSF, rHuIL-3, and recombinant human stem cell factor (rHuSCF) [32–40]. Enhanced in vitro sensitivity of human AML cells has been reported by rHuG-CSF, rHuIL-3, and rHuGM-CSF correlating with Ara-C incorporation into DNA, at initial diagnosis and, to a lesser extent, at relapse [41]. In the case of rHuGM-CSF, increased ^{3}H-Ara-C incorporation into the DNA of primary human AML cells was associated with increased DNA polymerase activity and to a lesser extent, thymidine kinase and deoxycytidine kinase activity, suggesting that increased cellular metabolism of Ara-C may be one mechanism for this effect [38].

Given the promising in vitro findings, clinical investigations using these agents in uncontrolled settings were performed. In one such study, rHuGM-CSF priming induced AML cells to enter the cell cycle and the degree of cytoreduction correlated with the degree of S-phase entry by AML cells in elderly patients with de novo AML and myelodysplastic syndromes (MDS)/AML [42].

Based on such findings suggesting enhanced chemotherapy responsiveness of leukemia cells through metabolic activation and cell-cycle entry, controlled clinical trials were conducted to elucidate whether cytokine-combined chemotherapy improves clinical outcomes such as relapse and survival rates in patients with AML. A multicenter randomized controlled study by the Eastern Cooperative Oncology Group (ECOG) compared rHuGM-CSF with placebo in association with induction therapy containing daunorubicin, idarubicin, or mitoxantrone with a standard dose of Ara-C in 245 older adults with AML, and found no differences in disease-free and overall survivals or in toxicity [43]. The Acute Leukemia French Association Group multicenter randomized controlled trial examined the effect of GM-CSF given during induction and consolidation/post-remission chemotherapy in 258 young adults with newly diagnosed AML and found a significant improvement in event-free survival but not in overall survival [44].

As for the use of rHuG-CSF in combination with chemotherapy for AML, several randomized controlled studies have been published [45–50] (Table 1). In a large multicenter trial, rHuG-CSF-combined chemotherapy reduced the rate of relapse in standard-risk patients but did not affect overall survival in favorable risk, standard risk, or unfavorable risk groups [48]. However, others reported no significant improvement in patient outcomes in various chemotherapy regimens in conjunction with rHuG-CSF. While the interpretation of these studies as a whole is

Table 1 Randomized controlled studies examining chemotherapy regimens containing rHuG-CSF in acute myeloid leukemia are summarized

References	*n*	Diagnosis	Median age [range] (years)	G-CSF dose	G-CSF schedule	Chemotherapy regimen	Duration of follow-up	Complete remission	Relapse	Disease-free survival	Overall survival	Days to PMN recovery
Lowenberg et al. [48]	640	Newly diagnosed	G 44.0/No G 44.9 [18–60]	150 μg/m^2	Day −1 to PMN recovery during induction	Induction: Cycle 1 – IDA/AraC; Cycle 2 – Amsacrin/AraC Consolidation: MIT/ETP or auto-SCT or allo-SCT	4 years	79% vs. 83% (p = NS) Standard risk subgroup: 87% vs. 86% (p = NS)	46% vs. 54% (p = 0.04) Standard risk subgroup: 44% vs. 52% (p = 0.02)	42% vs. 33% (p = 0.02) Standard risk subgroup: 45% vs. 33% (p = 0.006)	40% vs. 35% (p = NS) Standard risk subgroup: 45% vs. 35% (p = 0.02)	Cycle 1 G 26 vs. No G 26 (p = NS) Cycle 2 G 24 vs. No G 23 (p = NS)
Ohno et al. [50]	58	Relapsed/refractory	G 43/No G 47 [16–66]	200 μg/m^2	Day −2 to PMN recovery in each cycle	Induction and consolidation: BHAC/MIT/ ETP	30 months	50% vs. 37% (p = NS)	NR	p = NS	p = NS	G 25 vs. No G 32 (p = 0.0001)
Estey et al. [47]	215	Newly diagnosed and refractory	65 [NR]	200 μg/m^2	Day −1, day +1 or day +2 to PMN recovery in each cycle	Induction: FAI +/− ATRA Consolidation: alternated FAI +/− ATRA	~2 years	Multivariate G > no G (p = 0.018)	NR	p = NS	~15% (p = NS)	G 24 vs. No G 29 ($p < 0.001$)
Buchner et al. [46]	895	Newly diagnosed	NR [16–83]	150 μg/m^2	Day −2 to PMN recovery in each cycle	Induction: TAD + HAM or HAM × 2 Consolidation: TAD × 1	22 months	p = NS	NR	p = NS	p = NS	NR
Amadori et al. [45]	722	Newly diagnosed	68 [61–80]	150 μg/m^2	Day +1 to +7 or day +1 to PMN recovery during induction	Induction: MICE Consolidation: mini-ICE	3 years	58% vs. 49% (p = 0.009)	~73% (p = NS)	~16% (p = NS)	~13% (p = NS)	NR
Milligan et al. [49]	356	Relapsed/refractory	NR [15- >71]	5 μg/kg	Day +1 to PMN recovery in each cycle	Induction and consolidation: ADE or FLA +/− ATRA	4 years	58% vs. 61% (p = NS)	58% vs. 66% (p = NS)	31% vs. 28% (p = NS)	22% vs. 22% (p = NS)	Course 1 G 20 vs. No G 24 (p = 0.01) Course 2 G 21 vs. No G 22 (p = 0.04)

ADE Ara-C/daunorubicin/ETP; *auto-SCT* autologous stem cell transplantation; *allo-SCT* allogeneic stem cell transplantation; *ATRA* all-trans retinoic acid; *BHAC* behenoyl cytosine arabinoside; *ETP* etoposide; *FAI* fludarabine/AraC/IDA; *FLA* fludarabine/AraC; *HAM* high-dose AraC/MIT; *ICE* IDA/AraC/ETP; *IDA* idarubicin, AraC, cytarabine; *MICE* MIT/AraC/ETP; *MIT* mitoxantorone; *NR* not reported; *NS* not significant; *PMN* polymorphonuclear cells; *TAD* 6-thioguanine/daunorubicin/MIT made difficult by the heterogeneity of patient population, chemotherapy regimens used, and the variability in study designs, the equivocal and seemingly conflicting findings of these clinical trials have led clinicians caring for AML patients to view cytokine-combined chemotherapy as possibly promising but controversial.

6 rHuG-CSF and AML Stem Cells

The existence of functional hierarchy within human AML has become more and more accepted [51], which has led to the concept of LSC that initiate AML and produce nonstem progeny that make up the bulk of total body leukemia burden in patients. These rare and specialized leukemia cells, not the more abundant nonstem AML blasts, may be the functional component of MRD, responsible for disease relapse. To date, clinical trials have not directly examined the effect of therapeutic strategies on LSC and rigorous examination of in vivo biology of primary human AML stem cells awaited the development of robust animal models in the form of xenotransplantation.

An in vivo model was developed to evaluate human stem cell properties, the model was called "SCID-repopulating assay" and functions by injecting human hematopoietic stem cells or leukemic cells into CB17.SCID or NOD/SCID mice [52–54]. The SCID mice have severe defects in acquired immunity, and allow the engraftment of both normal and diseased human hematopoietic cells to a certain extent. Primary human AML $CD34^{+}CD38^{-}$ cells preferentially engraft and initiate leukemia in immune-compromised recipients [55]. This pioneering work established the presence of functional heterogeneity among AML cells. Since that time, the creation of more-sensitive SCID repopulating assays that allow more robust recapitulation of human AML in mice has extended the capacity of the SCID-repopulating system to understand the mechanisms underlying AML relapse in addition to leukemia initiation [56, 57].

One such xenotransplantation system with improved sensitivity allowing the evaluation of "stemness" of human cells is the newborn NOD/SCID/IL2rgKO (NSG) transplantation system. The NSG mouse model, with dysfunctional innate immune subsets through blocked γc signaling, supports more efficient long-term engraftment of human normal and leukemic cells compared with NOD/SCID or NOD/SCID/b2mKO mice [56, 58]. In this model, $CD34^{+}CD38^{-}$ AML cells not only engraft the recipient NSG mice efficiently, but also generate $CD34^{+}CD38^{+}$ and $CD34^{-}$ leukemic cells in the recipient bone marrow [56]. Moreover, $CD34^{+}CD38^{-}$ AML cells harvested from the bone marrow of these recipient mice initiate AML in secondary and tertiary recipients demonstrating self-renewal capacity of these cells. Importantly, the robust recapitulation of patient AML status in mice has allowed us to create an in vivo therapeutic model by treating AML-engrafted mice by chemotherapy. In this model, $CD34^{+}CD38^{-}$ human primary AML cells exhibited significant chemotherapy resistance compared with $CD34^{+}CD38^{+}$ and $CD34^{-}$ AML cells.

Histologic examination demonstrated that chemotherapy-resistant AML stem cells preferentially reside in the endosteal region of the bone marrow.

To clarify the mechanism responsible for chemotherapy resistance of AML stem cells, we examined the expression of molecules associated with apoptosis pathways, those responsible for drug efflux and also cell-cycle status of these cells. This investigation revealed the presence of a significantly greater proportion of $CD34^{+}CD38^{-}$ AML cells in G0 phase compared with $CD34^{+}CD38^{+}$ AML cells [59]. Confocal imaging of AML-engrafted recipient bone marrow demonstrated the differential cell-cycle status of human AML cells according to the location within the bone marrow. Leukemia cells in the center of the bone marrow were actively cycling ($Ki67^{+}$), whereas leukemia cells in the endosteal region were found to be cell cycle quiescent ($Ki67^{-}$) (Fig. 1a). Mouse studies have established the importance of both the endosteal region containing the osteoblastic niche and the perivascular region as niches for normal hematopoietic stem cells. Our findings suggest that cell-cycle quiescent human primary AML cells may be more frequently located in the osteoblastic stem cell niche.

These findings indicate that cell-cycle quiescence may be a mechanism for chemotherapy resistance of adult human AML stem cells residing in the endosteal region of the bone marrow, offering a possible explanation for AML relapse occurring despite successful cytoreduction through chemotherapy that target actively cycling cells. Therefore, we set out to examine whether it is possible to modulate the cell-cycle status of human primary AML stem cells at their niche and whether such modification leads to enhanced chemotherapy responsiveness.

To do so, we treated AML-engrafted mice by subcutaneous injection of 300 mg/kg rHuG-CSF daily for 5 days. When cell-cycle status of AML stem cells was examined by flow cytometry and confocal imaging on the day after the final injection, we found that the frequency of LSC in G0 phase decreased significantly and that a significant proportion of LSC in the endosteal region entered cell cycle (Fig. 1b, c). While these effects were statistically significant in seven cases of AML examined, the degree of cell-cycle induction was variable among the cases. Next, to elucidate whether successful cell-cycle entry increases the susceptibility of human AML stem cells to chemotherapeutic agents in vivo, we compared the effect of chemotherapy (Ara-C) alone and chemotherapy after cell-cycle induction by rHuG-CSF. Again, both flow cytometry and histopathologic examination demonstrated that cell-cycle induction by rHuG-CSF enhances the elimination of LSC within the endosteal niche by conventional chemotherapy.

An in vivo therapeutic model for primary human AML allowed us to examine the effect of rHuG-CSF on LSC cell cycle and on chemotherapy responsiveness. Our findings suggest that rHuG-CSF induces cell-cycle entry of quiescent LSC at the bone marrow niche, leading to increased susceptibility to chemotherapy. At the same time, some critical issues remain regarding the use of rHuG-CSF as a modulator of chemotherapy response in AML LSC.

First, rHuG-CSF has been found to modify cell-cycle status of normal HSC in vitro, potentially resulting in reduced capacity for normal hematopoiesis [22]. While this is a legitimate concern, clinical experience using rHuG-CSF as a part of

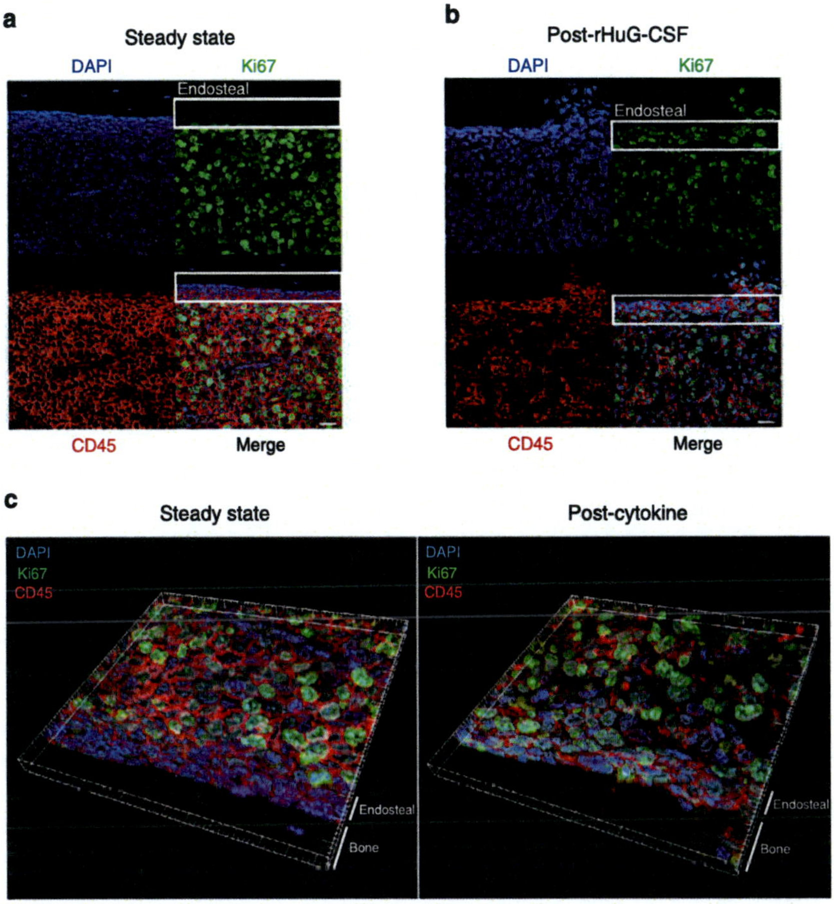

Fig. 1 rHuG-CSF facilitates cell cycle entry of quiescent human AML cells at the bone marrow endosteum. **(a)** Immunofluorescence labeling for human CD45 (*red*) and Ki67 (*green*) demonstrates cell-cycle quiescence of AML stem cells at the endosteal region of the bone marrow. **(b)** After rHuG-CSF treatment for 5 days, considerable numbers of AML cells showed increasing amounts of Ki67 expression suggesting cell-cycle entry. **(c)** Three-dimensional images demonstrated distinct cell-cycle status of human AML stem cells at the niche. AML stem cells are largely cell-cycle quiescent at steady state, while AML stem cells have entered cell cycle after rHuG-CSF treatment

the conditioning regimen for HSC transplantation in AML patients offer some reassurance [60, 61]. Second, while cell-cycle induction by rHuG-CSF appears to significantly improve chemotherapy responsiveness of LSC, it does not guarantee a complete elimination of all LSC present in the body. In our study, the frequency of remaining LSC diminished by approximately 100-fold on the average. However, remaining LSC, albeit at significantly reduced numbers, retained the capacity to initiate AML in serial transplantation. Third, while the efficacy of HSC

mobilization by rHuG-CSF is well accepted both in healthy HSC donors and in AML patients undergoing autologous HSC transplantation, its capacity for LSC cell-cycle induction in AML patients has not been directly examined. Such investigation may be useful to obtain preliminary findings to design definitive clinical trials examining the efficacy of rHuG-CSF combined chemotherapy for the elimination of AML LSC.

In addition to rHuG-CSF, the effects of other cytokines on AML stem cells have been investigated. For example, interferon (IFN)α is a promising candidate molecule potentially modulating cell cycle and other properties of AML stem cells. Studies have demonstrated that IFNα breaks the dormancy of mouse HSC, suggesting the use of IFNα for the treatment of chronic myeloid leukemia (CML) and other hematologic malignancies in combination with chemotherapeutic agents [62].

Disruption of stem–niche interaction is another approach that has been actively investigated. AMD3100, an inhibitor of CXCR4-SDF1 binding, has been reported to effect cell-cycle induction, increased transcription of cell-cycle regulators and subsequent alteration in chemotherapy sensitivity of primary human AML cells and mouse leukemic cells [63]. An approach using rHuG-CSF, other cytokines, and small molecules in combination may be required to achieve optimal elimination of LSC in AML patients.

7 Future Prospective of rHuG-CSF in Treatment of Myeloid Leukemia

Beginning with its early characterization as a white blood cell growth factor, rHuG-CSF has become established as an important part of the arsenal in the treatment of both hematologic and nonhematologic malignancies. Recent observations suggest that it may also be effective in modulating the cell-cycle status of primary human AML stem cells in vivo. One possible therapeutic approach may be to include rHuG-CSF in the setting of postremission therapy, where LSC remaining at low or undetectable levels may lead to disease relapse (Fig. 2). In this setting, targeted elimination of residual LSC may be a key to preventing relapse and improving patient outcomes. The degree of LSC response to rHuG-CSF among patients with AML is expected to be variable. Therefore, it will be critical to clarify how best to use rHuG-CSF in combination with chemotherapy and/or with other cytokines and small molecules. Determination of the patient population and the clinical setting where chemotherapy-sensitization with rHuG-CSF is most effective will be necessary to achieve improved long-term outcomes in these patients.

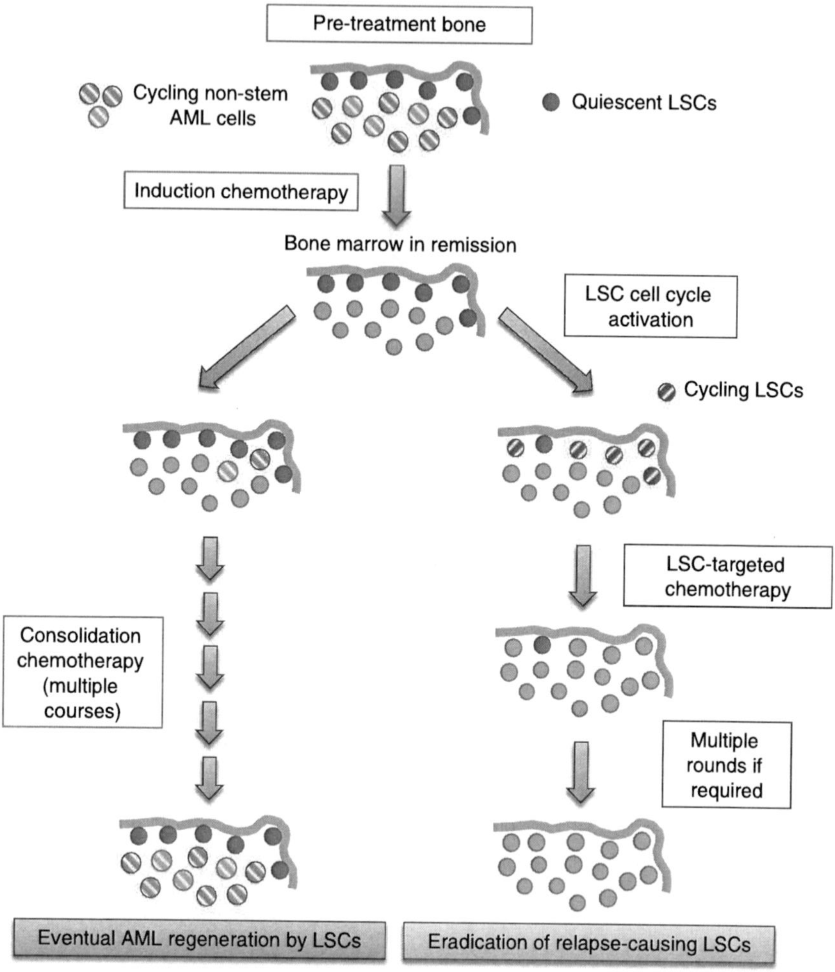

Fig. 2 Postremission anti-LSC therapy to prevent AML relapse. Before induction chemotherapy, most of the bone marrow consists of nonstem AML cells in AML patients. When complete remission is achieved by induction chemotherapy, cell-cycle active nonstem AML cells are eliminated but cell-cycle quiescent LSC survive. If subsequent rounds of chemotherapy do not eliminate LSC, AML relapse occurs through regeneration of nonstem AML cells. Postremission therapy that includes LSC cell-cycle activation may help eliminate LSC and prevent AML relapse

Acknowledgments I thank Y. Saito for critical review and comments on this manuscript; S. Takagi and Y. Najima for preparation of Table 1; and N. Suzuki, A. Sone, and M. Tomizawa for technical assistance with primary data presented in this manuscript. Figures 1 and 2 are reproduced from [59] (Saito et al. Nat Biotechnol 2010).

References

1. Bodey GP, Buckley M, Sathe YS, Freireich EJ (1966) Quantitative relationships between circulating leukocytes and infection in patients with acute leukemia. Ann Intern Med 64:328–340
2. Bodey GP, Rodriguez V, Chang HY, Narboni G (1978) Fever and infection in leukemic patients: a study of 494 consecutive patients. Cancer 41:1610–1622
3. Nagata S, Tsuchiya M, Asano S et al (1986) Molecular cloning and expression of cDNA for human granulocyte colony-stimulating factor. Nature 319:415–418
4. Souza LM, Boone TC, Gabrilove J et al (1986) Recombinant human granulocyte colony-stimulating factor: effects on normal and leukemic myeloid cells. Science 232:61–65
5. Welte K, Bonilla MA, Gillio AP et al (1987) Recombinant human granulocyte colony-stimulating factor. Effects on hematopoiesis in normal and cyclophosphamide-treated primates. J Exp Med 165:941–948
6. Rowe JM (1998) Treatment of acute myeloid leukemia with cytokines: effect on duration of neutropenia and response to infections. Clin Infect Dis 26:1290–1294
7. Godwin JE, Kopeky KJ, Head DR et al (1998) A double-blind placebo-controlled trial of granulocyte colony-stimulating factor in elderly patients with previously untreated acute myeloid leukemia: a Southwest oncology group study. Blood 91:3607–3615
8. Harousseau JL, Witz B, Lioure B et al (2000) Granulocyte colony-stimulating factor after intensive consolidation chemotherapy in acute myeloid leukemia: results of a randomized trial of the Groupe Ouest-Est Leucemies Aigues Myeloblastiques. J Clin Oncol 18:780–787
9. Heil G, Hoelzer D, Sanz MA et al (1997) A randomized, double-blind, placebo-controlled, phase III study of filgrastim in remission induction and consolidation therapy for adults with de novo acute myeloid leukemia. Blood 90:4710–4718
10. Ottmann OG, Bug G, Krauter J (2007) Current status of growth factors in the treatment of acute myeloid and lymphoblastic leukemia. Semin Hematol 44:183–192
11. Usuki K, Urabe A, Masaoka T et al (2002) Efficacy of granulocyte colony-stimulating factor in the treatment of acute myelogenous leukaemia: a multicentre randomized study. Br J Haematol 116:103–112
12. Smith TJ, Khatcheressian J, Lyman GH et al (2006) 2006 update of recommendations for the use of white blood cell growth factors: an evidence-based clinical practice guideline. J Clin Oncol 24:3187–3205
13. Heissig B, Hattori K, Dias S et al (2002) Recruitment of stem and progenitor cells from the bone marrow niche requires MMP-9 mediated release of kit-ligand. Cell 109:625–637
14. Kollet O, Dar A, Shivtiel S et al (2006) Osteoclasts degrade endosteal components and promote mobilization of hematopoietic progenitor cells. Nat Med 12:657–664
15. Katayama Y, Battista M, Kao WM et al (2006) Signals from the sympathetic nervous system regulate hematopoietic stem cell egress from bone marrow. Cell 124:407–421
16. Arai F, Hirao A, Ohmura M et al (2004) Tie2/angiopoietin-1 signaling regulates hematopoietic stem cell quiescence in the bone marrow niche. Cell 118:149–161
17. Fleming HE, Janzen V, Lo Celso C et al (2008) Wnt signaling in the niche enforces hematopoietic stem cell quiescence and is necessary to preserve self-renewal in vivo. Cell Stem Cell 2:274–283
18. Lemoli RM et al (1995) Proliferative response of human marrow myeloid progenitor cells to in vivo treatment with granulocyte colony-stimulating factor alone and in combination with interleukin-3 after autologous bone marrow transplantation. Exp Hematol 23:1520–1526
19. Lemoli RM, Fortuna A, Fogli M et al (1997) Cycling status of CD34+ cells mobilized into peripheral blood of healthy donors by recombinant human granulocyte colony-stimulating factor. Blood 89:1189–1196
20. Morrison SJ, Wright DE, Weissman IL (1997) Cyclophosphamide/granulocyte colony-stimulating factor induces hematopoietic stem cells to proliferate prior to mobilization. Proc Natl Acad Sci U S A 94:1908–1913

21. Rosti V, Malabarba L, Ramajoli I et al (2000) Cord blood-derived hematopoietic progenitor cells: in vitro response to hematopoietic growth factors and their recruitment into the S-phase of the cell cycle. Haematologica 85:18–25
22. van Os R, Robinson S, Sheridan T, Mauch PM (2000) Granulocyte-colony stimulating factor impedes recovery from damage caused by cytotoxic agents through increased differentiation at the expense of self-renewal. Stem Cells 18:120–127
23. Yang FC, Watanabe S, Tsuji K et al (1998) Human granulocyte colony-stimulating factor (G-CSF) stimulates the in vitro and in vivo development but not commitment of primitive multipotential progenitors from transgenic mice expressing the human G-CSF receptor. Blood 92:4632–4640
24. Byrd JC, Mrozek K, Dodge RK et al (2002) Pretreatment cytogenetic abnormalities are predictive of induction success, cumulative incidence of relapse, and overall survival in adult patients with de novo acute myeloid leukemia: results from Cancer and Leukemia Group B (CALGB 8461). Blood 100:4325–4336
25. Grimwade D, Walker H, Harrison G et al (2001) The predictive value of hierarchical cytogenetic classification in older adults with acute myeloid leukemia (AML): analysis of 1065 patients entered into the United Kingdom Medical Research Council AML11 trial. Blood 98:1312–1320
26. Grimwade D, Walker H, Oliver F et al (1998) The importance of diagnostic cytogenetics on outcome in AML: analysis of 1,612 patients entered into the MRC AML 10 trial. Blood 92:2322–2333
27. Slovak ML, Kopecky KJ, Cassileth PA et al (2000) Karyotypic analysis predicts outcome of preremission and postremission therapy in adult acute myeloid leukemia: a Southwest Oncology Group/Eastern Cooperative Oncology Group Study. Blood 96:4075–4083
28. Hokland P, Ommen HB (2011) Towards individualized follow-up in adult acute myeloid leukemia in remission. Blood 117:2577–2584
29. Beekman R, Touw IP (2010) G-CSF and its receptor in myeloid malignancy. Blood 115:5131–5136
30. Vellenga E, Young DC, Wagner K, Wiper D, Ostapovicz D, Griffin JD (1987) The effects of GM-CSF and G-CSF in promoting growth of clonogenic cells in acute myeloblastic leukemia. Blood 69:1771–1776
31. Delwel R, Salem M, Pellens C et al (1988) Growth regulation of human acute myeloid leukemia: effects of five recombinant hematopoietic factors in a serum-free culture system. Blood 72:1944–1949
32. Bhalla K, Birkhofer M, Arlin Z, Grant S, Lutzky J, Graham G (1988) Effect of recombinant GM-CSF on the metabolism of cytosine arabinoside in normal and leukemic human bone marrow cells. Leukemia 2:810–813
33. Bhalla K, Holladay C, Arlin Z, Grant S, Ibrado AM, Jasiok M (1991) Treatment with interleukin-3 plus granulocyte-macrophage colony-stimulating factors improves the selectivity of Ara-C in vitro against acute myeloid leukemia blasts. Blood 78:2674–2679
34. Cannistra SA, Groshek P, Griffin JD (1989) Granulocyte-macrophage colony-stimulating factor enhances the cytotoxic effects of cytosine arabinoside in acute myeloblastic leukemia and in the myeloid blast crisis phase of chronic myeloid leukemia. Leukemia 3:328–334
35. Inatomi Y, Toyama K, Clark SC, Shimizu K, Miyauchi J (1994) Combinations of stem cell factor with other hematopoietic growth factors enhance growth and sensitivity to cytosine arabinoside of blast progenitors in acute myelogenous leukemia. Cancer 54:455–462
36. Lista P, Porcu P, Avanzi GC, Pegoraro L (1988) Interleukin 3 enhances the cytotoxic activity of 1-beta-D-arabinofuranosylcytosine (ara-C) on acute myeloblastic leukaemia (AML) cells. Br J Haematol 70:121–123
37. Miyauchi J, Kelleher CA, Wang C, Minkin S, McCulloch EA (1989) Growth factors influence the sensitivity of leukemic stem cells to cytosine arabinoside in culture. Blood 73:1272–1278

38. Reuter C, Auf der Landwehr U, Schleyer E et al (1994) Modulation of intracellular metabolism of cytosine arabinoside in acute myeloid leukemia by granulocyte-macrophage colony-stimulating factor. Leukemia 8:217–225
39. Tanaka M (1993) Recombinant GM-CSF modulates the metabolism of cytosine arabinoside in leukemic cells in bone marrow. Leuk Res 17:585–592
40. te Boekhorst PA, Lowenberg B, Vlastuin M, Sonneveld P (1993) Enhanced chemosensitivity of clonogenic blasts from patients with acute myeloid leukemia by G-CSF, IL-3 or GM-CSF stimulation. Leukemia 7:1191–1198
41. te Boekhorst PA, Lowenberg B, Sonneveld P (1994) Hematopoietic growth factor stimulation and cytarabine cytotoxicity in vitro: effects in untreated and relapsed or primary refractory acute myeloid leukemia cells. Leukemia 8:1480–1486
42. Frenette PS, Desforges JF, Schenkein DP, Rabson A, Slapack CA, Miller KB (1995) Granulocyte-macrophage colony stimulating factor (GM-CSF) priming in the treatment of elderly patients with acute myelogenous leukemia. Am J Hematol 49:48–55
43. Rowe JM, Neuberg D, Friedenberg W et al (2004) A phase 3 study of three induction regimens and of priming with GM-CSF in older adults with acute myeloid leukemia: a trial by the Eastern Cooperative Oncology Group. Blood 103:479–485
44. Thomas X, Raffoux E, Botton S et al (2007) Effect of priming with granulocyte-macrophage colony-stimulating factor in younger adults with newly diagnosed acute myeloid leukemia: a trial by the Acute Leukemia French Association (ALFA) Group. Leukemia 21:453–461
45. Amadori S, Suciu S, Jehn U et al (2005) Use of glycosylated recombinant human G-CSF (lenograstim) during and/or after induction chemotherapy in patients 61 years of age and older with acute myeloid leukemia: final results of AML-13, a randomized phase-3 study. Blood 106:27–34
46. Buchner T, Berdel WE, Hiddemann W (2004) Priming with granulocyte colony-stimulating factor – relation to high-dose cytarabine in acute myeloid leukemia. N Engl J Med 350:2215–2216
47. Estey EH, Thall PF, Pierce S et al (1999) Randomized phase II study of fludarabine + cytosine arabinoside + idarubicin +/– all-trans retinoic acid +/– granulocyte colony-stimulating factor in poor prognosis newly diagnosed acute myeloid leukemia and myelodysplastic syndrome. Blood 93:2478–2484
48. Lowenberg B, Van Putten W, Theobald M et al (2003) Effect of priming with granulocyte colony-stimulating factor on the outcome of chemotherapy for acute myeloid leukemia. N Engl J Med 349:743–752
49. Milligan DW, Wheatley K, Littlewood T, Craig JI, Burnett AK (2006) Fludarabine and cytosine are less effective than standard ADE chemotherapy in high-risk acute myeloid leukemia, and addition of G-CSF and ATRA are not beneficial: results of the MRC AML-HR randomized trial. Blood 107:4614–4622
50. Ohno R, Naoe T, Kanamaru A et al (1994) A double-blind controlled study of granulocyte colony-stimulating factor started two days before induction chemotherapy in refractory acute myeloid leukemia. Kohseisho Leukemia Study Group. Blood 83:2086–2092
51. Dick JE (2008) Stem cell concepts renew cancer research. Blood 112:4793–4807
52. McCune JM, Namikawa J, Kaneshima H, Shultz LD, Lieberman M, Weissman IL (1988) The SCID-hu mouse: murine model for the analysis of human hematolymphoid differentiation and function. Science 241:1632–1639
53. Mosier DE, Gulizia RJ, Baird SM, Wilson DB (1988) Transfer of a functional human immune system to mice with severe combined immunodeficiency. Nature 335:256–259
54. Shultz LD, Schweiter PA, Christianson SW et al (1995) Multiple defects in innate and adaptive immunologic function in NOD/LtSz-scid mice. J Immunol 154:180–191
55. Lapidot T, Sirad C, Vormoor J et al (1994) A cell initiating human acute myeloid leukaemia after transplantation into SCID mice. Nature 367:645–648
56. Ishikawa F, Yohsida S, Saito Y et al (2007) Chemotherapy-resistant human AML stem cells home to and engraft within the bone-marrow endosteal region. Nat Biotechnol 25:1315–1321

57. Ninomiya M, Abe A, Katsumi A et al (2007) Homing, proliferation and survival sites of human leukemia cells in vivo in immunodeficient mice. Leukemia 21:136–142
58. Ishikawa F, Yasukawa M, Lyons B et al (2005) Development of functional human blood and immune systems in NOD/SCID/IL2 receptor gamma chain(null) mice. Blood 106:1565–1573
59. Saito Y, Uchida N, Tanaka S et al (2010) Induction of cell cycle entry eliminates human leukemia stem cells in a mouse model of AML. Nat Biotechnol 28:275–280
60. Mori T, Aisa Y, Watanabe R et al (2008) Long-term follow-up of allogeneic hematopoietic stem cell transplantation for de novo acute myelogenous leukemia with a conditioning regimen of total body irradiation and granulocyte colony-stimulating factor-combined high-dose cytarabine. Biol Blood Marrow Transplant 14:651–657
61. Takahashi S, Iseki T, Ooi J et al (2004) Single-institute comparative analysis of unrelated bone marrow transplantation and cord blood transplantation for adult patients with hematologic malignancies. Blood 104:3813–3820
62. Essers MA, Offner S, Blanco-Bose WE et al (2009) IFNalpha activates dormant haematopoietic stem cells in vivo. Nature 458:904–908
63. Zeng Z, Shi YX, Samudio U et al (2009) Targeting the leukemia microenvironment by CXCR4 inhibition overcomes resistance to kinase inhibitors and chemotherapy in AML. Blood 113:6215–6224

Use of rHuG-CSF in Myelodysplastic Syndromes

Mojtaba Akhtari and Lori Maness

1 Introduction

The myelodysplastic syndromes (MDS) represent a series of clonal hematologic neoplasm characterized by morphologic dysplasia, aberrant hematopoiesis, and peripheral blood refractory cytopenias. They are accompanied by increased risk of symptomatic anemia, infectious complications, and bleeding diathesis, as well as having a propensity for progressing to acute myeloid leukemia (AML), particularly in patients who have the higher-grade MDS. These syndromes show notable biologic and clinical heterogeneity, and their complex pathobiology is not clearly understood. Ineffective hematopoiesis; however, is a well-recognized feature, wherein normal blood cell maturation, differentiation, function, and survival are impaired. These abnormalities contribute to the development of peripheral blood pancytopenia, and most patients succumb to complications of bone-marrow failure. The common presenting symptoms include fatigue, dyspnea, bleeding, and infection [1–3]. The epidemiology of MDS has recently become clearer. MDS is primarily a disease of older adults (median age: 69 years), and the average annual age-adjusted incidence rate of MDS for 2001 through 2003 was 3.3 per 100,000; and the annual incidence among individuals older than age 70 years exceeds between 22 and 45 per 100,000 persons. Overall, MDS affects approximately 1 in 500 persons aged over 60 years, making it the most common hematologic malignancy in this age group [4–6]. The common causes of death in a cohort of 216 patients

M. Akhtari (✉) • L. Maness
Division of Hematology and Oncology, Department of Internal Medicine, University of Nebraska Medical Center, 987680 Nebraska Medical Center, Omaha, NE 681980-7680, USA
e-mail: mojtaba.akhtari@unmc.edu

G. Molineux et al. (eds.), *Twenty Years of G-CSF*, Milestones in Drug Therapy,
DOI 10.1007/978-3-0348-0218-5_10, © Springer Basel AG 2012

with MDS included bone-marrow failure (infection/hemorrhage) (88%) and AML transformation (28%) [7].

2 Diagnosis

The diagnosis and classification of MDS starts with a comprehensive history and physical examination; complete blood count with leukocyte differential, reticulocyte count, bone marrow aspiration and biopsy with iron stain and cytogenetic studies; and assay of serum ferritin, vitamin B12, folate, and erythropoietin (EPO) values. MDS is probably a common cause of mild-to-moderate chronic anemia in elderly patients, and it is mislabeled as "anemia of chronic disease" or "anemia of renal insufficiency." MDS is sometimes overdiagnosed, since morphologic dysplasia is not specific for MDS, and there are reactive causes of dysplasia including megaloboastic anemias, exposure to toxins such as arsenic and alcohol, and after recent cytotoxic and growth factor therapy, recent intercurrent illness, human immunodeficiency virus (HIV), and copper deficiency [4]. Two classification systems are used: French-American-British (FAB) [8] and World Health Organisation (WHO) [9], and of the several prognostic scoring systems, the most commonly used is the International Prognostic Scoring System (IPSS) [10].

3 Management

Management decisions can be challenging since patients with MDS are elderly individuals with comorbidities, so they should be categorized based on three clinical features: age, performance status, and IPSS-defined risk category [11]. Anemia, neutropenia, and thrombocytopenia, particularly refractory cytopenias, are major causes of morbidity in these patients; thus therapeutic goals in patients with MDS should include control of symptoms due to cytopenias.

4 Supportive Care

Supportive care is the mainstay of management for patients with MDS, particularly for patients with low-risk MDS (defined as MDS with $< 5\%$ bone marrow blasts or an IPSS score of ≤ 1.0), and patients with poor prognosis whose age or performance status excludes them from receiving more high-intensity treatment such as intensive combination chemotherapy and hematopoietic stem cell transplantation. The major goals are to reduce morbidity and mortality, improve quality of life, and minimize treatment-related toxicities. Symptomatic anemia and associated fatigue are usually a major problem in MDS, which cause significant morbidity and most of the patients would need red blood cell transfusion during the course of their disease. It has been observed that recombinant human erythropoietin (rHuEPO) can correct anemia in MDS patients [12]. Neutropenia is another common cytopenia in MDS, and infection is a serious complication in this patient population. No evidence, however, supports the routine use of prophylactic antibacterial or antifungal drugs in the treatment of neutropenic patients with MDS. Most patients with MDS and neutropenia respond to recombinant human granulocyte colony-stimulating factor (rHuG-CSF) [13–16]. Bleeding is common particularly when there is severe thrombocytopenia. Platelets develop intrinsic functional defect in MDS, which contributes to the bleeding tendency in these patients. Patients with MDS who are thrombocytopenic and who are not symptomatic, do not need platelet transfusion; however, they should receive platelet transfusion if they are bleeding or if they are undergoing interventional procedures.

5 Recombinant Human Granulocyte Colony-Stimulating Factor

G-CSF is the growth factor that is necessary for proliferation and differentiation of myeloid precursor cells into neutrophils; it also intensifies multiple neutrophil functions [17]. rHuG-CSF is frequently used in the clinical management of patients with MDS to improve their anemia and neutropenia; and it is considered a beneficial therapeutic intervention, which has a prominent place in the supportive care plan of these patients. rHuG-CSF does not affect the thrombocytopenia or the propensity to progress to AML in MDS patients. It has not been shown that rHuG-CSF alone can improve survival in patients with MDS [14–16, 18–20]. rHuG-CSF can be used for the management of neutropenia after chemotherapy, in association with a hypomethylating agents or lenalidomide, or in case of severe infection. Most patients with MDS who are treated with rHuG-CSF show increments in their peripheral blood neutrophil counts, which is indicative of presence of some degrees of bone-marrow reserve in those patients. No evidence from randomized studies shows an advantage of using maintenance rHuG-CSF in MDS patients. In vitro evidence suggests that rHuG-CSF is able to repair functional abnormalities of neutrophils in patients with MDS [21]. A preliminary report on five MDS patients treated with intravenous rHuG-CSF (50–1,600 $\mu g/m^2$) showed improvement of neutropenia [14]. Eighteen patients were enrolled in a phase 1/2 clinical study of subcutaneous injection of rHuG-CSF (0.1–0.3 μg/kg/day), and 16 of them showed an increase in their neutrophil counts from 5- to 40-fold [19]. Subsequently, 11 patients from the cohort participated in a long-term maintenance program with subcutaneous injection of rHuG-CSF, and ten patients achieved improved neutrophil counts for up to 16 months. This study showed that patients with neutrophil counts maintained at $>1.5 \times 10^9/L$ had fewer bacterial infections than patients with lower neutrophil counts [20]. Severely neutropenic patients may benefit from prophylactic therapy with low-dose rHuG-CSF [13]. A preliminary phase 3 multiinstitutional randomized trial report of 102 high-risk patients with MDS (RAEB [refractory anemia with excess blasts] or RAEB-t [refractory anemia with excess blasts in transformation]) showed no increased risk of AML-evolution in the treatment group of 50 patients [22]. In vitro evidence suggests that rHuG-CSF is associated with a growth advantage of an existing subclone of cells with monosomy 7 over diploid cells [23].

6 Recombinant Human Granulocyte Colony-Stimulating Factor and Erythropoietin

Erythropoietin (EPO) is the primary stimulus of normal erythropoiesis [24]. Serum EPO concentrations usually show an inverse relationship with the degree of anemia in patients with MDS, with the highest concentrations being found in patients with erythroid hypoplasia [25]. The hallmark of MDS is increased apoptosis; and there is in vitro evidence that EPO functions as a survival factor with antiapoptotic properties [26]. EPO induces antiapoptotic protein Bcl-X_L [27]; it also activates the antiapoptotic PI3-kinase [28]. Similarly, G-CSF has antiapoptotic effects both in vitro and in vivo [29–31]. In vitro evidence suggests that G-CSF has antiapoptotic function through inhibiting Fas-triggered caspase activity in bone marrow cells isolated from RARS (refractory anemia with ringed sideroblasts) patients; it also promotes erythroid colony growth and differentiation of stem cells from RARS patients [32]. G-CSF changes the survival capacity of the mobilized $CD34^+$ cells, and peripheral blood stem cell mobilization with rHuG-CSF can significantly reduce the number of apoptotic $CD34^+$ cells in comparison with apoptotic $CD34^+$ cells from unstimulated mobilization [33]. Additionally, it has been shown by in vitro experiments that G-CSF blocks spontaneous cytochrome *c* release and mitochondria-dependent apoptosis in hematopoietic progenitor cells from RARS patients [34].

rHuEPO has been used extensively to treat anemia in patients with MDS [35–37]. The efficacy of rHuEPO alone is relatively low, and overall erythroid response rates from 7.5 to 36% have been reported [12, 38]. In vitro evidence exists for synergistic effects of the combination of rHuG-CSF and rHuEPO on erythropoiesis [39]. A clinical response to the combination of rHuG-CSF and rHuEPO has been demonstrated. Morphologically, bone-marrow biopsies of MDS patients show reduced number of apoptotic precursors compared with the pretreatment samples [40]. Hematopoietic growth factors such as EPO and G-CSF not only block apoptosis of erythroid precursors but also promote growth of cytogenetically normal progenitors in MDS patients [41].

Clinical responses have been investigated with the combination of rHuEPO and rHuG-CSF in patients with MDS (Table 1). Two phase 2 clinical trials of rHuG-CSF and rHuEPO in MDS patients (mainly RA, RARS, and RAEB) demonstrated

Table 1 Clinical studies of rHuG-CSF and rHuEPO in MDS

References	No. of patients	Response rate (%)
[42]	24	42
[43]	21	38
[13]	55	48
[44]	10	Nil
[45]	98	36
[46]	56	38
[47]	32	50
[48]	33	61
[49]	281	45
[50]	53	42
[51]	60	42
[52]	129	39
[55]	403	50
[56]	121	39
[57]	24	47
[58]	12	31

erythroid response rates of 42% (10 of 24 patients) and 38% (8 of 21 patients), respectively [42, 43]. These studies showed strong erythroid responses in terms of improved hemoglobin values and reduced red blood cell transfusion requirements, and the response rates were considerably better than with rHuEPO alone. Subsequently, another clinical trial with 55 patients showed a 48% erythroid response rate (21 of 44 evaluable patients), and 81% of responders maintained their response during an 8-week maintenance period. This study revealed an interesting finding that approximately 50% of the patients with MDS who respond to the rHuG-CSF and rHuEPO combination lost their response with rHuG-CSF withdrawal, and some of those patients regained a response when rHuG-CSF was restarted [13]. A small phase II clinical study of rHuG-CSF and rHuEPO combination in MDS patients has also been reported from Japan, which did not show promising results; 10 patients: received the combination for 10 weeks, and only one patient had a delayed erythroid response, although 80% (eight of ten patients) had a neutrophil response [44]. An American–Scandinavian study of 98 MDS patients treated with rHuG-CSF and rHuEPO showed a similar response rate of 36%. This study revealed that patients with serum EPO concentrations <500 U/L had a favorable response to rHuG-CSF and rHuEPO if the red blood cell transfusion need was <2 units per month [45]. A subsequent phase 2 randomized clinical trial of 56 MDS patients (RA, RARS, RAEB) treated with rHuG-CSF and rHuEPO showed an overall erythroid response of 38%. The response rates for patients with RA, RARS, and RAEB were 20%, 46%, and 37%, respectively. In this study, patients were randomly assigned to two treatment groups: group A was primed with rHuG-CSF for 4 weeks followed by the rHuG-CSF and rHEPO for 12 weeks; and group B started with rHEPO for 8 weeks followed by the combination for 10 weeks. The response rates were identical in the two treatment groups. This study clearly confirmed the in vivo synergy between rHuG-CSF and rHuEPO, and this synergistic effect was more pronounced in RARS patients [46]. A Spanish nonrandomized clinical trial of 32 patients with MDS (RA and RARS) treated with rHuG-CSF and rHuEPO showed an erythroid response rate of 50%, and a multivariate analysis confirmed the predictive value of the American–Scandinavian scoring system [47]. A German nonrandomized clinical trial of 33 MDS patients (RA, RARS, and RAEB) treated with rHuG-CSF and rHuEPO demonstrated an erythroid response rate of 61% after 12 weeks, which increased to 80% after 36 weeks [48]. The predictive value of low serum EPO concentrations (<150 U/L) was observed in a Greek phase 2 clinical study in 281 MDS patients (RA, RARS, and RAEB), and an overall erythroid response rate of 45.1% was reported [49]. The Scandinavian MDS Group has published the results of a prospective study of 53 MDS patients treated with rHuG-CSF and rHuEPO, which showed an overall erythroid response rate of 42%, and it validated the American–Scandinavian predictive model and scoring system. It demonstrated response rates of 61% in the good predictive group and 14% in the intermediate group. This study also showed that responding patients had a significantly better quality of life [50]. However, a French randomized controlled clinical trial did not show any significant difference in quality of life between the treatment group and supportive care group; the study demonstrated an erythroid response rate of 42% [51]. The Scandinavian group published the results of 129 MDS patients treated with rHuG-CSF and rHuEPO that were followed for up to 42 months, and it showed an erythroid response rate of 39% and median response duration of 23 months. They did not find any difference in survival between treated and untreated patients [52]. An Italian randomized prospective study compared rHuEPO with the combination of rHuG-CSF and rHuEPO in 30 low-risk MDS patients; it showed an erythroid response in 6 of 15 (40%) patients in the rHuEPO group and in 11 of 15 (73.3%) patients in the rHuG-CSF and rHuEPO combination group. In 4 of 9 (44.4%) patients who did not have a response to rHuEPO, adding rHuG-CSF induced an erythroid response at 16 weeks [53]. A pooled analysis of retrospective data from 162 published articles consisting of 2,592 MDS patients with RA and RARS suggested that growth factors may improve survival in MDS [54]. A French–Belgian retrospective study examined erythroid response rate and overall survival in 403 patients with MDS who were treated with rHuEPO with or without rHuG-CSF; and 62% and 50% response rates were seen in the rHuEPO-alone group and rHuG-CSF and rHuEPO combination group, respectively. This study reported a better overall survival in the rHuEPO-alone arm; but results were not adjusted for all currently employed prognostic factors and red cell transfusion requirement in the multivariate analysis, and selection bias is another potential confounding factor [55]. A Scandinavian retrospective study compared 121 MDS patients treated with rHuG-CSF and rHuEPO with 237 untreated patients, and an erythroid response rate of 39% and median response duration of 23 months were reported. This study demonstrated an encouraging survival benefit in a multivariate analysis, but its results are potentially confounded by patient selection bias. No increased rate of AML was observed [56]. An American phase 2 intrapatient dose-escalation clinical trial of rHuEPO with or without rHuG-CSF in 24 MDS patients showed that addition of rHuG-CSF resulted in an erythroid response of 47% in patients who did not have a good response to rHuEPO. A weekly weight-based rHuEPO regimen was used in the study [57]. A phase 3 prospective randomized clinical trial of 73 MDS patients treated with rHuEPO with or without rHuG-CSF plus supportive care versus supportive care alone demonstrated an erythroid response of 31% in 12 of 39 patients who received the rHuG-CSF and rHuEPO combination [58]. This study did not reveal any difference in overall survival between the rHuEPO and supportive care cohorts with a median follow-up of 5.8 years; it did not also show an increased incidence of transformation to AML. However, a survival benefit was observed in the erythroid responders versus nonresponders. A meta-analysis of 15 clinical trials in MDS patients treated with rHuEPO with or without rHuG-CSF or rHu granulocyte-macrophage colony-stimulating factor (GM-CSF) suggested that higher doses of rHuEPO (60–80,000 U/week) may be more effective than standard dose rHuEPO (30–40,000 U/week) [59].

7 Predictors of Erythroid Response to rHuG-CSF and rHuEPO Combination

The major predictors of a positive erythroid response to growth factor treatment include pretreatment lower serum EPO concentrations (<500 U/L) and lower red blood cell transfusion requirements (<2 units/month). A multivariate analysis of an American–Scandinavian clinical study demonstrated that two pretreatment clinical variables, baseline serum EPO values, and initial low red cell transfusion need, had prognostic values to predict response to the combination of rHuG-CSF and rHuEPO. By using pretreatment serum EPO values as a ternary variable (<100, 100–500, or >500 U/L), and red blood cell transfusion requirement as a binary variable (<2 or ≥2 units per month), a predictive model, and scoring system for erythroid response was developed. Patients were separated into three predicted erythroid response-rate groups of high (74%), intermediate (23%), and poor (7%) (Table 2). This study revealed that patients with serum EPO concentrations <500 U/L had a favorable response to the combination of rHuG-CSF and rHuEPO if the transfusion need was <2 units per month [45]. This scoring system was validated in another prospective study, in which patients with 1 or both of these positive predictors showed erythroid response rates of 14% versus 61%, respectively [50]. The predictive model and scoring system were derived for patients treated with a therapeutic trial of 12-week duration, and 39% of patients with a high predictive score did not achieve an erythroid response [60]. A Dutch study demonstrated that flow cytometric analysis of myeloblasts in the bone marrow samples can be used in predicting response to growth factor treatment in patients with MDS. In a cohort of 46 patients (low- and intermediate-risk), the predictive model and scoring system were associated with an erythroid response to the rHuG-CSF and rHuEPO combination; however, aberrant phenotype of myeloblasts was

Table 2 Predictive model and scoring system for erythroid response to the combination of rHuG-CSF and rHuEPO in patients with MDS

	Score					
Variable	−3	−2	−1	0	+1	+2
Serum EPO (U/L)	>500				100–500	<100
RBC transfusions (units/month)	≥				<2	

	Response group	
Predictive score	Type	% Responders (no. of patients)
>+1	High	74 (22/29)
±1	Intermediate	23 (7/31)
<−1	Low	7 (3/34)

RBC red blood cell; *rHuEPO* recombinant human erythropoietin; *rHuG-CSF* recombinant human granulocyte colony-stimulating factor. From Hellström-Lindberg et al. [45] with permission from John Wiley and Sons

highly associated with treatment failure among patients with the greatest response probability according to the predictive model and scoring system [61].

8 Conclusion

Our increased understanding of the biology, behavior, and pertinent signaling pathways of MDS has led to the identification of specific subtypes with distinct clinical behavior and different therapeutic requirements. Much progress has been made in the treatment of MDS; however, there is great room for improvement in treatments used in the management of patients with MDS. rHuG-CSF is an effective therapeutic modality in MDS patients, and it should be considered for the management of refractory symptomatic cytopenias. The combination of rHuG-CSF and rHuEPO can improve survival in appropriate MDS patients such as low-risk MDS and RARS patients. Recombinant growth factors have created a significant opportunity for improvement in the care of patients with MDS.

References

1. Albitar M, Manshouri T, Shen Y et al (2002) Myelodysplastic syndrome is not merely "preleukemia". Blood 100:791–798
2. Nimer SD (2008) Myelodysplastic syndrome. Blood 111:4841–4851
3. Tefferi A, Vardiman JW (2009) Myelodyspalstic syndromes. N Engl J Med 361:1872–1885
4. Bowen D, Gulligan D, Jowitt S, Kelsey S, Mufti G, Oscie D, Parker J (2003) Guidelines for the diagnosis and therapy of adult myelodysplastic syndromes. Br J Haematol 120:187–200
5. Bennett JM, Catovsky MT, Flandrin DAG, Galton HR, Sultan C (1982) Proposals for the classification of the myelodysplastic syndromes. Br J Haematol 51:189–199

6. Vardiman JW, Thiele J, Arber DA et al (2009) The 2008 revision of the World Health Organization classification of myeloid neoplasms and acute leukemia: rationale and important changes. Blood 114:937–951
7. Greenberg P, Cox C, LeBeau MM et al (1997) International scoring system for evaluating prognosis in myelodysplastic syndromes. Blood 89:2079–2088
8. Rollison DE, Howlader N, Smith MT et al (2008) Epidemiology of myelodysplastic syndromes and chronic myeloproliferative disorders in the United States, 2001–2004, using data from the NAACCR and SEER programs. Blood 112:45–52
9. Hoffman WK, Ottman OG, Ganser F, Hoeltzer D (1996) Myelodysplatic syndromes: clinical features. Semin Hematol 33:177–185
10. Sanz GF, Sanz MA, Vallespi T et al (1989) Two regression models and scoring system for predicting survival and planning treatment in myelodysplastic syndromes: a multivariate analysis of prognostic factors in 370 patients. Blood 74:395–408
11. Myelodysplastic syndromes v.2.2010. NCCN Clinical Practice Guidelines in Oncology™. Available at: http://www.nccn.org
12. Geissler RG, Schulte P, Ganse A (1997) Treatment with growth factors in myelodysplastic syndromes. Pathol Biol (Paris) 45:656–667
13. Negrin RS, Stein R, Doherty K et al (1996) Maintenance treatment of the anemia of myelodysplastic syndromes with recombinant human granulocyte colony-stimulating factor and erythropoietin: evidence for in vivo synergy. Blood 87:4076–4081
14. Kobayashi Y, Okabe T, Ozawa K et al (1989) Treatment of myelodysplastic syndromes with recombinant human granulocyte colony-stimulating factor: a preliminary report. Am J Med 86:178–182
15. Yoshida Y, Hirashima K, Asano S, Takaku F (1991) A phase II trial of recombinant human granulocyte colony-stimulating factor in the myelodysplastic syndromes. Br J Haematol 78:378–384
16. Kaczmarski RS, Mufti GJ (1993) Low-dose filgrastim therapy for chronic neutropenia. N Engl J Med 329:1280–1281
17. Panopoulos AD, Watowich SS (2008) Granulocyte colony-stimulating factor: molecular mechanisms of action during steady state and emergency hematopoiesis. Cytokine 42:277–288
18. Ohyashiki K, Ohyashik JH, Toyama K, Takaku F (1989) Hematologic and cytogenetic findings in myelodysplastic syndromes treated with recombinant granulocyte colony-stimulating factor. Jpn J Cancer Res 80:848–854
19. Negrin RS, Haeber DH, Naler A et al (1989) Treatment of myelodysplastic syndromes with recombinant granulocyte colony-stimulating factor. Ann Intern Med 110:966–984
20. Negrin RS, Haeber DH, Nehler A et al (1990) Maintenance treatment of patients with myelodysplastic syndromes using recombinant human granulocyte colony-stimulating factor. Blood 76:36–43
21. You A, Kitagawa S, Okabe T et al (1987) Recombinant human granulocyte colony-stimulating factor repairs the abnormalities of neutrophils in patients with myelodysplastic syndromes and chronic myeloid leukemia. Blood 70:404–411
22. Greenberg P, Taylor K, Larson R et al (1993) Phase III randomized multicenter trial of G-CSF vs observation for myelodysplastic syndromes (MDS). Blood 82(Suppl 1):196a; abstract
23. Sloand EM, Yong AS, Ramkissoon S et al (2006) Granulocyte colony-stimulating factor preferentially stimulates proliferation of monosomy 7 cells bearing the isoform IV receptor. Proc Natl Acad Sci U S A 103:14483–14488
24. Erslev AJ (1991) Erythropoietin. N Engl J Med 324:1339–1344
25. Jacobs A, Janowska-Wieczorek A, Caro J, Bowen DT, Lewis T (1989) Circulating erythropoietin in patients with myelodysplastic syndrome. Br J Haematol 73:36–39
26. Silva M, Grillot D, Benito A, Richard C, Nunez G, Fernandez-Luna JL (1996) Erythropoietin can promote erythroid progenitor survival by repressing apoptosis through Bcl-XL and Bcl-2. Blood 88:1576–1582

27. Gregory T, Yu C, Ma A, Orkin SH, Blobel GA, Weiss MJ (1999) GATA-1 and erythropoietin cooperate to promote erythroid cell survival by regulating bcl-xL expression. Blood 94:87–96
28. Uddin S, Kottegoda S, Stigger D, Platanias LC, Wickrema A (2000) Activation of the Akt/FKHRL1 pathway mediates the antiapoptotic effects of erythropoietin in primary human erythroid progenitors. Biochem Biophys Res Commun 275:16–19
29. Watson RW, O'Neill A, Brannigen AE et al (1999) Regulation of Fas antibody induced neutrophil apoptosis is both caspase and mitochondrial dependent. FEBS Lett 453:67–71
30. Hassan Z, Fadeel B, Zhivotovsky B, Hellström-Lindberg E (1999) Two pathways of apoptosis induced with all-trans retinoic acid and etoposide in the myeloid cell line P39. Exp Hematol 27:1322–1329
31. Colotta F, Re F, Polentarutti N, Sozzani S, Mantovani A (1992) Modulation of granulocyte survival and programmed cell death by cytokines and bacterial products. Blood 80:2012–2020
32. Schmidt-Mende J, Tehranchi R, Forsblom AM et al (2001) Granulocyte colony-stimulating factor inhibits Fas-triggered apoptosis in bone marrow cells isolated from patients with refractory anemia with ringed sideroblasts. Leukemia 15:742–751
33. Philpott NJ, Prue RL, Marsh JC, Gordon-Smith EC, Gibson FM (1997) CSF-mobilized CD34 peripheral blood stem cells are significantly less apoptotic than unstimulated peripheral blood CD34 cells: role of G-CSF as survival factor. Br J Haematol 97:146–152
34. Tehranchi R, Fadeel B, Forsblom AM et al (2003) Granulocyte colony-stimulating factor inhibits spontaneous cytochrome c release and mitochondria-dependent apoptosis of myelodysplastic syndrome hematopoietic progenitors. Blood 101:1080–1110
35. Bessho M, Jinnai I, Matsuda A, Saito M, Hirashima K (1990) Improvement of anaemia by recombinant erythropoietin in patients with myelodysplastic syndromes and aplastic anaemia. Int J Cell Cloning 8:445–458
36. Hellström E, Birgegård G, Lockner D, Wide L, Helmers C, Öst Å (1990) Treatment of myelodysplastic syndromes with recombinant human erythropoietin. Eur J Haematol 47:355–360
37. Rose EH, Abels RI, Nelson RA, McCullough DM, Lessin L (1995) The use of r-HUEpo in the treatment of anaemia related to myelodysplasia (MDS). Br J Haematol 89:831–837
38. Hellström Lindberg E (1995) Efficacy of erythropoietin in the myelodysplastic syndromes. An analysis of 205 patients in 17 studies. Br J Haematol 89:67–71
39. Greenberg PL, Negrin RS, Ginzton N (1992) In vitro-in vivo correlations of erythroid responses to G-CSF plus erythropoietin in myelodysplastic syndromes. Exp Hematol 20:733; abstract
40. Hellstrom-Lindberg E, Kanter-Lewensohn L, Ost A (1997) Morphological changes and apoptosis in bone marrow from patients with myelodysplastic syndromes treated with granulocyte-CSF and erythropoietin. Leuk Res 21:415–425
41. Tehranchi R, Fadeel B, Schmidt-Mende J et al (2005) Antiapoptotic role of growth factors in the myelodysplastic syndromes: concordance between in vitro and in vivo observations. Clin Cancer Res 11:6291–6299
42. Negrin RS, Stein R, Doherty K et al (1993) Treatment of the anemia of myelodysplastic syndromes using human granulocyte colony-stimulating factor in combination with erythropoietin. Blood 82:737–743
43. Hellström-Lindberg E, Birgegård G, Carlsson M et al (1993) A combination of granulocyte-colony-stimulating factor and erythropoietin may synergistically improve the anaemia in patients with myelodysplastic syndromes. Leuk Lymphoma 11:221–228
44. Imamura M, Kobayashi M, Kobayashi S et al (1994) Failure of combination therapy with recombinant granulocyte colony-stimulating factor and erythropoietin in myelodysplastic syndromes. Ann Hematol 68:163–166
45. Hellström-Lindberg E, Negrin R, Stein R et al (1997) Erythroid response to treatment with G-CSF plus erythropoietin for the anaemia of patients with myelodysplastic syndromes: proposal for a predictive model. Br J Haematol 99:344–351

46. Hellström-Lindberg E, Ahlgren T, Beguin Y et al (1988) Treatment of anemia in myelodysplastic syndromes with granulocyte colony-stimulating factor plus erythropoietin: results from a randomized phase II study and long-term follow-up of 71 patients. Blood 92:68–75
47. Remacha AF, Arrizabalaga B, Villegas A et al (1999) Erythropoietin plus granulocyte colony-stimulating factor in the treatment of myelodysplastic syndromes. Identification of a subgroup of responders. The Spanish Erythropathology Group. Haematologica 84:1058–1064
48. Matovani L, Lentini G, Hentschel B et al (2000) Treatment of anaemia in myelodysplastic syndromes with prolonged administration of recombinant human granulocyte colony-stimulating factor and erythropoietin. Br J Haematol 109:367–375
49. Terpos E, Mougiou A, Kouraklis A et al (2002) Prolonged administration of erythropoietin increases erythroid response rate in myelodysplastic syndromes: a phase II trial in 281 patients. Br J Haematol 118:174–180
50. Hellström-Lindberg E, Gulbrandsen N, Lindberg G et al (2003) A validated decision model for treating the anaemia of myelodysplastic syndromes with erythropoietin + granulocyte colony-stimulating factor: significant effects on quality of life. Br J Haematol 120:1037–1046
51. Casadevall N, Durieux P, Dubois S et al (2004) Health, economic, and quality-of-life effects of erythropoietin and granulocyte colony-stimulating factor for the treatment of myelodysplastic syndromes: a randomized, controlled trial. Blood 91:179–188
52. Jädersten M, Montgomery SM, Dybedal I, Porwit-MacDonald A, Hellström-Lindberg E (2005) Long-term outcome of treatment of anemia in MDS with erythropoietin and G-CSF. Blood 106:803–811
53. Balleari E, Rossi E, Clavio M et al (2005) Erythropoietin plus granulocyte colony-stimulating factor is better than erythropoietin alone to treat anemia in low-risk myelodysplastic syndromes: results from a randomized single-centre study. Ann Hematol 85:174–180
54. Golshayan AR, Jin T, Maciejewski J et al (2007) Efficacy of growth factors compared to other therapies for low-risk myelodysplastic syndromes. Br J Haematol 137:125–132
55. Park S, Grabar S, Kelaidi C et al (2008) Predictive factors of response and survival in myelodysplastic syndrome treated with erythropoietin and G-CSF: the GFM experience. Blood 111:574–582
56. Jädersten M, Malcovati L, Dybedal I et al (2008) Erythropoietin and granulocyte-colony stimulating factor treatment associated with improved survival in myelodysplastic syndrome. J Clin Oncol 26:3607–3617
57. Gotlib J, Lavori P, Quesada S, Stein RS, Shahnia S, Greenberg PL (2009) A Phase II intra-patient dose-escalation trial of weight-based darbepoetin alfa with or without granulocyte colony-stimulating factor in myelodysplastic syndromes. Am J Hematol 84:15–20
58. Greenberg PL, Sun Z, Miller KB et al (2009) Treatment of myelodysplastic syndrome patients with erythropoietin with or without granulocyte colony-stimulating factor: results of a prospective randomized phase 3 trial by the Eastern Cooperative Oncology Group (E1996). Blood 114:2393–2400
59. Mundle S, Lefebvre P, Vekeman F, Duh MS, Rastogi R, Moyo V (2009) An assessment of erythroid response to epoetin alpha as a single agent versus in combination with granulocyte- or granulocyte-macrophage-colony-stimulating factor in myelodysplastic syndromes using a meta-analysis approach. Cancer 115:706–715
60. Bowen DT (2006) Hematopoietic growth factors. In: Deeg HJ, Bowen DT, Gore SD, Haferlach T, Le Beau MM, Niemeyer C (eds) Hematologic malignancies: myelodysplastic syndromes. Springer, New York, pp 99–109
61. Westers TM, Alhan C, Chamuleau ME et al (2010) Aberrant immunophenotype of blasts in myelodysplastic syndromes is a clinically relevant biomarker in predicting response to growth factor treatment. Blood 115:1779–1784
62. Gotlib J, Greenberg PL (2005) Supportive care in myelodysplastic syndromes: hemopoietic cytokine and iron chelation therapy. In: Greenberg PL (ed) Myelodysplastic syndromes. Cambridge University Press, Cambridge, pp 209–242

The Role of Hematopoietic Growth Factors in Aplastic Anemia: An Evidence-Based Perspective

Pia Raanani, Ronit Gurion, Anat Gafter-Gvili, Isaac Ben-Bassat, and Ofer Shpilberg

1 The Use of Evidence-Based Medicine in Hematology

1.1 The Role of Evidence-Based Medicine in Hematology

Evidence-based medicine, which tries to directly base clinical practice and policy decision on supporting evidence, is becoming more and more popular in the establishment of practice guidelines, and is now part of the process of decision making in medicine [1]. Evidence-based medicine aims to apply the best available evidence gained from the scientific research to medical decision making. Evidence-based medicine uses, in a balanced manner, the most updated and best evidence for treating the patient, and represents the transition from obtaining clinical decisions based on previous clinical experience and physicians' "gut feeling" to decision making based on relevant clinical trials with precise research objectives, ranked by their relevance and validity according to established criteria formed specifically for the use of evidence-based medicine.

Several approaches compare therapy outcomes including: randomized controlled trials, observational studies, i.e., nonrandomized trials, and expert opinion. Randomized controlled trials are considered the "gold standard" because randomization ensures comparability of subjects in all cohorts, and data are collected prospectively. Unfortunately, well-designed randomized controlled trials are not common in hematologic disorders, due to rarity of these disorders (e.g., aplastic anemia, acute leukemia) and their dire prognosis. Furthermore, individual studies rarely provide definitive answers concerning the use of a certain intervention or treatment. The strongest evidence for therapeutic interventions is provided by

P. Raanani (✉) • R. Gurion • A. Gafter-Gvili • I. Ben-Bassat • O. Shpilberg
Hemato-oncology Hospitalization Unit, Institute of Hematology, Davidoff Cancer Center, Beilinson Hospital, Rabin Medical Center, Petah-Tikva 49100, Israel
e-mail: praanani@012.net.il

G. Molineux et al. (eds.), *Twenty Years of G-CSF*, Milestones in Drug Therapy,
DOI 10.1007/978-3-0348-0218-5_11, © Springer Basel AG 2012

systematic reviews of well-designed randomized clinical trials. A systematic review uses a predefined, explicit methodology to help identify which forms of healthcare are useful or harmful by bringing together results from similar randomized controlled trials. Systematic reviews assemble results in a way that reduces the likelihood that chance observations unjustifiably affect clinical practice. Meta-analyses combine results of several individual studies that address a set of related research hypotheses to produce a significant effect size. Meta-analyses can increase power and precision of estimates of treatment effects [2, 3]. Their main use in hematologic disorders is to increase power when individual studies are too small and cannot detect an effect size. Observational studies represent another method, also used in the context of hematologic disorders and used to compare treatment outcomes. Their main advantage is the large number of participants included [2]. Due to rarity of patients, observational studies can be a good solution for decision making in hematology. Observational studies and randomized clinical trials were found to be concordant in over 90% of cases. Another way to compare therapy outcomes is by using expert opinion. This approach is regarded by some as the highest level of evidence in therapy decision making, while others, like the Grading of Recommendations, Assessment, Development and Evaluations (GRADE) system (http://clinicalevidence.bmj.com/ceweb/about/about-grade.jsp) rank it very low. Expert opinion, including guidelines and recommendations, is often used in hematologic disorders. An update of the 2003 guidelines for the diagnosis and treatment of aplastic anemia has been published [4]. The guidelines issued many recommendations, most of them based on expert opinion.

In conclusion, evidence-based medicine has an important role in choosing the right treatment approach, yet in hematologic disorders, trade-off and compromises have to be done due to the rarity of patients and often their poor prognosis. Aplastic anemia is a classic example for this concern.

1.2 Limitations of the Use of Evidence-Based Medicine in Hematology

Despite the increasing role of evidence-based medicine in the process of decision making in medicine, we should be aware of the limitations of the various approaches used, especially in the field of blood disorders.

The main limitation of randomized controlled trials concerning blood disorders is the small number of participants in each treatment group, because results of these trials are most convincing when they have large numbers of subjects allowing for the detection of significant differences between treatment groups. Another limitation of randomized controlled trials in hematologic disorders is that due to the rarity of the disorders and the small number of patients, it is necessary to conduct multicenter trials with a small number of participants in each center. The rarity of blood disorders, like aplastic anemia, contributes to the slow rate of accrual to trials and was the reason for early closure of some of the studies in aplastic anemia, before the end of enrolment of the needed number of patients. A large European Group for Blood and Marrow Transplantation (EBMT) trial addressing the issue of hematopoietic growth factors in aplastic anemia was closed prematurely due to slow recruitment and its results are pending [5]. Furthermore, trials may take many years to conduct and additionally, follow-up should sometimes be long so that the evaluated interventions or treatments are no more relevant when they are finally published. Another limitation stems from the fact that randomized clinical trials report on a cohort of subjects and thus, it is not appropriate to extrapolate their conclusions to the individual patient under question [6, 7]. Randomized clinical trials are also prone to methodological biases, since they rely heavily on adequate randomization methods such as randomization generation and allocation concealment [2, 8]. Systematic reviews and meta-analyses have their own limitations, which include pooling of biases of the included studies, biases introduced by the process of selecting trials for inclusion, heterogeneity between the trials included, and publication bias with trials reporting on favorable outcomes more likely to be published than those with unfavorable outcomes. Individual patient data meta-analyses might help to standardize outcomes and analyses across trials and might also overcome the problem of inconsistent reporting of outcomes across trials. The main limitation of these meta-analyses, however, is problems in accessing the "raw" data from the original clinical trials [9]. Observational studies have their own limitations, which include mainly heterogeneity of variables of patients, disorders, diagnostic criteria, and treatments. Furthermore, like randomized clinical trials, observational studies are often behind the times in a field because of the need to accumulate sufficient numbers of subjects with adequate follow-up. The main drawback of observational studies is that unrecognized confounding factors may distort results [2].

Recommendations for rare disorders, such as severe aplastic anemia, are often based on expert opinion rather than on multiple, well-designed, randomized clinical trials, as well as systematic reviews and meta-analyses of such trials [4]. The conduct of randomized clinical trials should be encouraged, even in such rare diseases as aplastic anemia, to establish future guidelines based on the highest level of evidence possible.

The limitations of evidence-based medicine reports should be taken into consideration when interpreting results. Several methods have been developed and published, aiming to improve the quality of reporting of evidence-based medicine-based publications. The CONSORT (CONsolidated Standards Of Reporting Trials) guidelines published in 1996 and revised in 2001 were developed to make the reporting of randomized clinical trials more transparent [10, 11]. These guidelines were followed in 1999 by the QUOROM (QUality Of Reporting Of Meta-analyses) and by the STARD (STAndards for Reporting of Diagnostic accuracy) and the STROB (STrengthening the Reporting of OBservational studies in Epidemiology) in 2003 and 2007, respectively [12]. The PRISMA (Preferred Reporting Items for Systematic Reviews and Meta-Analyses) Statement, published in 2009 is an update and expansion of the QUOROM Statement [13].

In conclusion, various methods of evidence-based medicine are used; each one has its own limitations. These limitations are more prominent in publications evaluating interventions in hematologic disorders such as aplastic anemia due to the relative paucity of patients. Measures to improve assessment of these limitations are being taken and their results should be taken into consideration by physicians treating patients with hematologic disorders.

2 Aplastic Anemia: Diagnosis and Management

Aplastic anemia is characterized by pancytopenia and hypoplastic bone marrow [14]. It is a rare condition with an annual incidence rate of approximately two patients per million population in the general population in North America and Europe [15]. Its incidence is two to three times higher in East Asia than in the Western population [16]. The disease is most commonly diagnosed between the ages of 15–25 years, presenting with another peak around age 60–65 years [17]. The diagnosis of acquired aplastic anemia requires the exclusion of other conditions associated with pancytopenia, among which are congenital bone marrow failure as Fanconi anemia and myelodysplastic syndromes (MDS). The usual presentation of patients with aplastic anemia is weakness and fatigue as a result of the anemia and mucosal and skin bleeding resulting from the thrombocytopenia. Infection is less common at presentation. In 70–80% of patients, aplastic anemia is categorized as idiopathic because the primary etiology is unknown [17].

Aplastic anemia is divided into two major categories: severe aplastic anemia (SAA) and nonsevere aplastic anemia (NSAA). SAA is defined by the presence of a hypocellular bone marrow and any two of the following criteria: reticulocyte count $< 20 \times 10^9$/L and/or neutrophil count $<0.5 \times 0^9$/L and/or platelet count $< 20 \times 10^9$/L [18]. Very SAA (VSAA) is a subcategory of SAA defined when the neutrophil count is $<0.2 \times 10^9$/L [19]. Patients not fulfilling the criteria for SAA or VSAA are considered as having NSAA.

Allogeneic hematopoietic cell transplantation (alloHCT) and immunosuppressive therapy (IST) are the main therapeutic modalities currently used to treat patients with aplastic anemia, with success rates ranging from 60 to 80% [20].

AlloHCT is the only curative treatment for SAA. AlloHCT from an HLA-identical sibling donor is the initial treatment recommended by current guidelines for newly diagnosed patients who have SAA or VSAA, are younger than 40 years old, and have an HLA-compatible sibling donor. Similarly, HLA-matched sibling donor transplant should be the first choice in children with NSAA who need treatment [4]. This recommendation is based on observational studies and retrospective cohort studies only, which showed impressive results of about 80–90% overall survival at 10 years with this approach as first-line treatment in SAA [21–23]. The data support matched-sibling donor transplantation only up to the age of 40 years. It seems likely that with the improvement in supportive care and reduced intensity conditioning regimens, patients aged between 40 and 55 years may also benefit from transplantation as first-line therapy [24].

Although alloHCT is regarded as first-line therapy, due to age restriction of the recipients and unavailability of matched HLA donors, alloHCT is not always a feasible option. IST is indicated for patients who are not eligible for sibling-donor alloHCT, namely, patients with NSAA who are transfusion dependent; patients with NSAA who although not transfusion-dependent, may have significant neutropenia and are at risk for infection; patients with SAA or VSAA who are older than 40 years; and younger patients with SAA or VSAA who lack an HLA-compatible sibling donor [4].

Concerning the optimal IST regimen, the current guidelines base their recommendations in favor of a combination of antithymocyte globulin and cyclosporine A over antithymocyte globulin alone, only on improved hematologic response rate, ranging between 60 and 80% [21, 25–27], while individual randomized trials could not show a survival benefit for IST combination therapy over antithymocyte globulin alone. A systematic review and meta-analysis including four trials and randomizing 313 patients compared the combination of antithymocyte globulin and cyclosporine A to antithymocyte globulin alone [28]. Results of this meta-analysis showed that for patients with SAA, the combination of antithymocyte globulin and cyclosporine A is superior to antithymocyte globulin alone, not only in terms of response rates but also in terms of all-cause mortality at 3 months, 1, and 5 years [(RR 0.50, 95% CI 0.29–0.85), (RR 0.54, 95% CI 0.30–0.99), and (RR 0.58, 95% CI 0.36–0.93), respectively] [28]. These findings, based on the highest level of evidence, further support the current recommendations of using combined IST rather than antithymocyte globulin only, for patients with SAA. As for patients with NSAA, the meta-analysis could not demonstrate superiority of combination therapy with antithymocyte globulin and cyclosporine A over antithymocyte globulin alone in terms of all-cause mortality, hematologic response, and relapse rate [28]. These results might imply that NSAA patients may not benefit from a combination therapy compared with antithymocyte globulin alone. While the meta-analysis compared the combination of antithymocyte globulin and cyclosporine A to antithymocyte globulin alone, only one randomized clinical trial compared the combination of antithymocyte globulin and cyclosporine A and cyclosporine A alone. This study included patients with transfusion-dependent NSAA and showed a significantly higher overall response rate at 6 months (74%) in the combination group compared with the cyclosporine A-alone group (46%) [29]. Thus, while many authorities use a watch-and-wait supportive approach for transfusion-independent patients with NSAA, they consider antithymocyte globulin and cyclosporine A as first-line therapy also for transfusion-dependent NSAA patients.

In conclusion, AlloHCT is the only curative treatment for patients with SAA, thus it should be regarded as first-line therapy. For patients ineligible for transplant, IST, and specifically the combination of antithymocyte globulin and cyclosporine A, is recommended based on better survival rates and response rates.

3 Supportive Care Apart from Hematopoietic Growth Factors

Supportive care is an integral and crucial part of the treatment of patients with aplastic anemia. It involves transfusional support and infection prevention. Current guidelines recommend administering prophylactic platelet transfusions when the platelet count is $<10 \times 10^9$/L or $<20 \times 10^9$/L in the presence of fever, rather than giving platelets only in response to bleeding manifestations. Red cell transfusions should be given to maintain safe hemoglobin concentration depending on additional comorbidities, i.e., usually >8 g/dL [4].

Current guidelines recommend the administration of irradiated blood products to SAA patients undergoing alloHCT as well as those receiving IST, although there is no evidence to support this practice [4]. A published survey performed in 2008 including 12 centers of EBMT and 2 United States centers demonstrated that the administration of irradiated blood products, to prevent transfusion-associated graft-versus-host disease (TA-GVHD), was actually the common practice in 12 of the 14 centers [30]. Based on the results of this survey, the EBMT aplastic anemia working party proposed that patients with aplastic anemia should receive irradiated blood products during and after antithymocyte globulin treatment. This policy should probably be continued for at least as long as patients are receiving IST, such as cyclosporine A [30].

Patients with aplastic anemia who are severely neutropenic (neutrophil count $<0.5 \times 10^9$/L) should ideally be placed in isolation when in hospital and should receive prophylactic antibiotics and antifungals [4]. Antiviral prophylaxis with acyclovir is recommended for all transplanted patients and is commonly administered during IST with antithymocyte globulin and for the first 3–4 weeks afterward [4]. These recommendations for infection prophylaxis are based on several meta-analyses and data from randomized clinical trials that relate to immunocompromised hematologic patients in general and not specifically to those with aplastic anemia [31–33].

The use of antibacterial prophylaxis was evaluated in a systematic review and meta-analysis of 95 randomized controlled trials comparing antibiotic prophylaxis with placebo or no intervention or another antibiotic, in afebrile neutropenic patients. Most of the trials included patients with hematologic malignancies, mainly acute leukemia, but several trials of patients with aplastic anemia were included. Death from all causes was reduced by 33% (95% CI, 0.55–0.81) in neutropenic patients who received any antibiotic prophylaxis and by 48% (95% CI, 0.35–0.77) in patients who received quinolones for prophylaxis, compared with placebo or no intervention [31].

In another systematic review and meta-analysis of 64 randomized clinical trials including patients with hematologic malignancies and patients undergoing alloHCT for various indications, systemic antifungals were compared with placebo, no intervention, or other antifungal agents for prophylaxis in neutropenic cancer patients after chemotherapy or alloHCT [32]. Also in this meta-analysis, several trials included patients with aplastic anemia. Antifungal prophylaxis significantly decreased all-cause mortality significantly at the end of follow-up (RR, 0.84; 95% CI, 0.74–0.95).

The present guidelines recommend hospitalization in isolation for patients with SAA and VSAA [4]. These recommendations are based on a systematic review of prospective comparative studies that assessed infection control measures for high-risk cancer patients undergoing chemotherapy and for alloHSCT recipients [32]. Patients with aplastic anemia were included. Overall, protective isolation with control of air quality in most studies, combined with additional measures such as barrier isolation or endogenous flora suppression, significantly reduced the risk of death at 30 days RR 0.60 (95% CI 0.50–0.72;29 studies) [32].

In conclusion, supportive care involving transfusional support and infection prevention is an important part of the treatment of patients with aplastic anemia. Prophylactic platelet transfusions when the platelet count is $<10 \times 10^9$/L and prophylactic red cell transfusions when the hemoglobin concentration is <8 g/dL are recommended. In addition, patients with SAA who have a neutrophil count $<0.5 \times 10^9$/L should be hospitalized in isolation and should receive both prophylactic antibiotics and antifungals.

4 Evidence-Based Use of Hematopoietic Growth Factors

Recombinant hematopoietic growth factors include both myeloid colony-stimulating factors, i.e., granulocyte colony-stimulating factor (rHuG-CSF), and granulocyte-macrophage colony-stimulating factor (rHuGM-CSF), as well as erythropoiesis-stimulating agents (ESA), i.e., erythropoietin (rHuEPO). The rationale for using hematopoietic growth factors in aplastic anemia is based on their ability to regulate proliferation, differentiation, and function of neutrophils and to ameliorate neutropenia and its complications. Moreover, their use may improve response to IST, as they may act in concert with endogenous hematopoietic growth factors to stimulate hematopoietic stem cells [33–37]. The addition of ESA, especially rHuEPO, to hematopoietic growth factors is aimed at encouraging hemoglobin production and synergizing the stimulation of other lineage precursors [38].

Several prospective randomized clinical trials evaluated the role of hematopoietic growth factors in aplastic anemia. Their main target was to examine whether the addition of hematopoietic growth factors to IST increases response rate, decreases infection rate, and prolongs survival. These trials included patients with SAA and NSAA who received IST and were randomly assigned to receive hematopoietic growth factors or control. Table 1 summarizes the randomized clinical trials published on the subject.

The first randomized clinical trial was published in 1991 [39] and included 27 patients with SAA. All patients received antithymocyte globulin as IST and were randomly assigned to receive rHuGM-CSF or placebo. The addition of hematopoietic growth factors significantly increased the neutrophil count and was accompanied with fewer febrile days ($p < 0.05$). However, the response and the

Table 1 Randomized clinical trials evaluating the role of hematopoietic growth factors in aplastic anemia patients treated with IST

References	IST	HGF (type, dose, schedule)	No. of patients	Age, years median (range)	Diagnosis (no. of pts)	Baseline ANC ($\times 10^9$/L)	Long-term all-cause mortality (no. of pts)	Overall response (no. of pts)	Clonal evolution (no. of pts)
[39]	IV ATG (source varied) + corticosteroids	Placebo	14	42	NA	0.54	4 (1 y)	2 (3 mo)	NA
		IV continuous rHuGM-CSF 300 μg/SC; rHuGM-CSF 150 μg × 2/d for 28 d	13	32	NA	0.69	2 (1 y)	7 (12 mo)	NA
[40]	IV equine	ATG + corticosteroids		11	32 (21–67)	NA	NA	5 (1 y)	4 (12 mo)
		0							
	SC rHuGM-CSF 300 μg + IV EPO 6000 units for 3 mo	11	34	(23–63)	NA	NA	0 (1 y)	10 (12 mo)	0
	CsA		8	26 (9–45)	NA	0.41	5 (1 y)	1 (12 mo)	0
SC		rHuGM-CSF 300 μg + IV EPO 6000 units for 3 mo	8	28 (12–42)	NA	0.40	4 (1 y)	3 (12 mo)	0
[41]	IV horse ATG + CsA + corticosteroids		31	9 (1–15)	NSAA = 13 SAA = 18 VSAA = 0	0.46	0 (1 y)	27 (12 mo)	NA
		SC/IV G-CSF 400 μg/m^2 for 90 d	33	8 (2–16)	NSAA = 15 SAA = 18 VSAA = 0	0.48	2 (1 y)	21 (12 mo)	NA
[42]	IV ATG (source varied) + CsA + corticosteroids		49	22 (1–82)	NSAA = 0 SAA = 30 VSAA = 19	0.2	13 (5 y)	31 (5 y)	1
		SC lenograstim 5 μg/kg/d for 98 d	53	26 (2–71)	NSAA = 0 SAA = 27 VSAA = 26	0.2	13 (5 y)	31 (5 y)	4
[43]	IV horse ATG + CsA + corticosteroids		47	35 (8–71)	NSAA = 0 SAA = 33 VSAA = 14	Neutrophils 0.39	9 (5 y)	37 (6 mo)	1
		SC rHuGM-CSF 5 μg/kg/d + SC rHuEPO 100 units/kg/d) 3 d/wk for first mo, 2 d/wk for second mo, and 1 d/wk for third mo	30	36 (5–68)	NSAA = 0 SAA = 19 VSAA = 11	Neutrophils 0.43	6 (5 y)	22 (6 mo)	3

[44]	IV horse ATG + CsA + corticosteroids		47	53 (19–74)	NSAA = 0 SAA = 36 VSAA = 11	Neutrophils 0.30	6 (4 y)	31 (12 mo)	2
		IV filgrastim 400 μg/d or lenograstim 50 μg/d, every other day for 23 d and then once or twice a week until d84	48	54 (19–75)	NSAA = 0 SAA = 29 VSAA = 19	Neutrophils 0.32	4 (4 y)	37 (12 mo)	3
[5]	IV horse ATG + CsA		97	46 (2–81)	NSAA = 0 SAA = 0 VSAA = 12	NA	23 (6 y)	60 (12 mo)	1
		SC rHuG-CSF 150 mg/m^2/d from d8 until d120	95		NSAA = 0 SAA = 3 VSAA = 69	NA	21 (6 y)	69 (12 mo)	1

ANC absolute neutrophil count; *ATG* anti-thymocyte globulin; *CsA* cyclosporine A; *d* day; *IST* immunosuppressive therapy; *rHuEPO* recombinant human erythropoietin; *IV* intravenous; *mo* month; *NA* not available; *no.* number; *NSAA* nonsevere aplastic anemia; *pts* patients; *rHuG-CSF* recombinant human granulocyte colony-stimulating factor; *rHuGM-CSF* recombinant human granulocyte-macrophage colony-stimulating factor; *SAA* severe aplastic anemia; *SC* subcutaneous; *VSAA* very severe aplastic anemia; *wk* week; *y* year survival rates were comparable in both groups. Furthermore, adverse events, none of them of grade 3 or 4, were more common in the group of patients who received treatment with hematopoietic growth factors, including low-grade fever, arthralgias, myalgias, and mostly bone pain. Clonal evolution to MDS, acute myeloid leukemia (AML), or paroxysmal nocturnal hemoglobinuria, a much-feared complication, was not reported [39].

The next study was published in 1998 on 38 patients with SAA randomly assigned to four groups: antithymocyte globulin only, cyclosporine A only, antithymocyte globulin with hematopoietic growth factors, and cyclosporine A with hematopoietic growth factors [40]. The hematopoietic growth factors administered included rHuGM-CSF and rHuEPO. At 1 year, the complete remission rate was in favor of the group treated with hematopoietic growth factor, i.e., antithymocyte globulin or cyclosporine A with the addition of hematopoietic growth factors, compared with the IST-only treatment group (74% vs. 26%, respectively, $p < 0.05$). These results were more pronounced for patients receiving antithymocyte globulin compared with those receiving cyclosporine A (90.9% vs. 36.4%, $p < 0.05$ compared with 37.5% vs. 12.5%, respectively). Overall survival at 1 year was better in the antithymocyte globulin and hematopoietic growth factor groups compared with the antithymocyte globulin-alone group (100% vs. 54% respectively, $p < 0.05$). Conversely, as regards to the cyclosporine A groups, although there was a trend in favor of the group which was treated with hematopoietic growth factors, it was not statistically significant (50% vs. 37.5%, $p > 0.05$). Unfortunately, long-term overall survival results were not reported in this study. The growth factors were well tolerated, except for fever which resolved spontaneously. Clonal evolution was not reported in this trial as well [40]. Since both randomized controlled trials were small, they did not have enough power to detect a statistically significant difference between the two groups.

The third study was published in 2000 [41]. This study was larger and included 119 patients with aplastic anemia, who had received antithymocyte globulin, cyclosporine A, and danazole as IST. All 50 patients with VSAA received rHuG-CSF without randomization [41]. Sixty-nine patients with NSAA and SAA were randomly assigned to receive or not to receive rHuG-CSF. Despite a more rapid increase in neutrophil count in the group receiving hematopoietic growth factors during the first 3 months, there was no statistically significant difference in the other major outcomes between the two groups, including response rate at 3 and at 12 months, infection rate, the number of febrile days, and overall survival (91% $\pm$ 5% vs. 93% $\pm$ 6%). There were no adverse events attributed to hematopoietic growth factors. With regard to clonal evolution, there was no report of paroxysmal nocturnal hemoglobinuria, but three patients were diagnosed with MDS ($n = 2$ in the hematopoietic growth factor group; $n = 1$ in the control group) [41]. The major limitation of this study is that more than one-third of the participants were diagnosed with NSAA, which could introduce a bias to the results.

The fourth study on the subject was published in 2002 and included 102 patients with SAA [42]. All patients received antithymocyte globulin and cyclosporine A and were randomly assigned to receive or not to receive rHuG-CSF (lenograstim) subcutaneously. Although the addition of hematopoietic growth factor significantly increased the neutrophil count (83% vs. 44.9%; $p < 0.001$), there was no statistically significant difference between the two groups with regard to infectious episodes during the study period. Furthermore, there was no difference in overall survival at 5 years (75.5% in the growth factor group vs. 73.5% in the control). During a follow-up period of 5 years, five patients developed clonal evolution: three patients treated with growth factors developed paroxysmal nocturnal hemoglobinuria and two patients developed MDS, one in each group. These differences were not statistically significant [42].

Four years later, in 2006, the fifth randomized study on the issue was published and included 142 patients randomly assigned to 1 of 4 groups (I to IV) according to the IST administered and the addition of hematopoietic growth factors (rHuEPO and rHuGM-CSF) [43]. Patients in groups I and II did not receive hematopoietic growth factors, while patients in groups III and IV did. Patients randomly assigned to treatment in group I received horse antithymocyte globulin alone; patients in group II received horse antithymocyte globulin plus cyclosporine A; patients in group III received horse antithymocyte globulin, cyclosporine A, and hematopoietic growth factors; and patients in group IV received the same regimen as patients in group III but with rabbit instead of horse antithymocyte globulin. The overall response rate (e.g., complete and partial responses) was in favor of groups II and III compared with groups I and IV. When group II was compared with group III, the addition of growth factors to horse antithymocyte globulin and cyclosporine A did not alter the response rate (78.7% vs. 73.3%), early infection-related mortality (6% vs. 13%), and overall survival (81% vs. 80%). Four patients with paroxysmal nocturnal hemoglobinuria were identified during the follow-up period, with three patients in group III (patients who received hematopoietic growth factors), and one patient in group II (patients who did not receive hematopoietic growth factors). Clonal evolution to MDS or AML did not occur during the follow-up period [43].

The sixth study was published in 2007 and included 101 patients with SAA [44]. All patients received antithymocyte globulin, cyclosporine A, and prednisolone, and were randomly assigned to 1 of 2 groups: rHuG-CSF or no additional treatment. Although there was a statistically significant difference in favor of the rHuG-CSF treatment with regard to response rate at 6 months (77% vs. 57%, $p = 0.03$), at 1 year there was no statistically difference between the two groups (76% vs. 79%, $p = 0.46$). Furthermore, there was no statistically significant difference in the rate of documented infections, overall survival at 4 years (94% vs. 88%, $p = 0.44$) and in the risk for developing clonal evolution to MDS or AML ($p = 0.63$) between both groups. However, the addition of rHuG-CSF decreased the relapse rate compared with the control group ($p = 0.01$) [44].

Results of the largest conducted clinical trial on the issue have been published by the aplastic anemia working party of the EBMT [5]. The EBMT trial included 205 patients with SAA and VSAA. All patients received IST with antithymocyte globulin and cyclosporine A and were randomly assigned to receive rHuG-CSF or no additional treatment. The addition of rHuG-CSF to IST significantly increased the neutrophil count, decreased infection rate (24% in the rHuG-CSF group vs. 36% in the control; $p = 0.006$) and hospitalization days ($p = 0.03$), especially in the VSAA patients. It did not alter other important outcomes, however, such as response rate (73% in the rHuG-CSF group vs. 62% in the control; $p = 0.64$), relapse rate, and overall survival (76% in the rHuG-CSF group vs. 77% in the control group; $p = 0.64$). During a follow-up period of 6 years, two cases of MDS or AML were detected – one in each group [5].

Results of this large study confirm those of a systematic review and meta-analysis published in 2009 that included the previous five randomized controlled trials and 414 patients [45]. The IST regimen for most trials in the meta-analysis consisted of antithymocyte globulin, cyclosporine A, and corticosteroids. The hematopoietic growth factors used in three trials was rHuG-SCF, in 1 trial rHuGM-CSF; and in 2 trials rHuGM-CSF and rHuEPO. The addition of hematopoietic growth factors to IST, compared with IST alone yielded no difference in all-cause mortality at 100 days [RR 1.33 (95% CI 0.56–3.18)]; at 1 year [RR 0.90 (95% CI 0.50–1.63)]; and at 5 years [RR 0.89 (95% CI 0.55–1.46)] (Fig. 1). There was no difference in overall hematologic response between the two groups at 3 months [RR 1.13 (95% CI 0.88–1.45)] and at 12 months [RR 1.21 (95% CI 0.78–1.86)]. There was no increase in the incidence of clonal evolution in the hematopoietic growth factor group [RR 1.59 (95% CI 0.39–6.51)]. Furthermore, there was no difference in the number of clinically documented infections between both groups [RR 1.10 (95% CI 0.90–1.33)]. Thus, according to this meta-analysis, considered to be the highest level of evidence in the field, hematopoietic growth factors should not be given routinely to patients with aplastic anemia receiving IST because main outcomes such as overall survival, response rate, or infectious parameters are not affected by their administration.

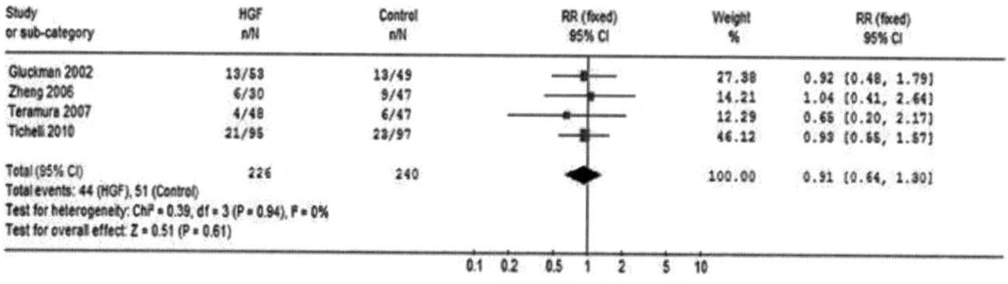

Fig. 1 All-cause mortality for patients with severe aplastic anemia receiving immunosuppressive therapy with or without Very bad quality. Actually, cannot be read rHuG-CSF [45]

The updated guidelines for the diagnosis and treatment of aplastic anemia published in 2009 issued many recommendations, most of them based on expert opinion [4]. According to these guidelines, hematopoietic growth factors should not be used on their own in an attempt to "treat" the aplastic anemia and their routine long-term use is not recommended outside the setting of clinical trial. Regarding the use of hematopoietic growth factors for the prevention and treatment of systemic infections, while the guidelines suggest that a short course may be considered for severe systemic infection not responding to intravenous antibiotics and antifungal drugs (grade C recommendation, level IV evidence as outlined in appendix 3 of the Procedure for Guidelines Commissioned by the BCSH [http://www.bcshguidelines.com/process1.asp#App3]), the systematic review published in the same year could not show their usefulness for the prevention of clinically documented infections [RR 1.10 (95% CI 0.90–1.33)] or severe infections [RR 0.88 (95% CI 0.58–1.34)] (Grade A recommendation, level Ia evidence) [45].

5 Summary and Future Perspectives

We base our recommendations for treatment and interventions on evidence-based medicine data as opposed to intuitive and personal beliefs based on personal experience. This decision may be difficult in the case of rare disorders where there are not enough randomized clinical trials to provide us with these data. With this respect, hematologic disorders in general and aplastic anemia in particular are notorious and guidelines are based mainly on expert opinion. Treating physicians are advised to follow recommendations based on the highest level of evidence possible, namely those obtained from multiple, well-designed, randomized clinical trials, as well as systematic reviews and meta-analyses of such trials (grade A recommendation, level Ia and Ib evidence). Data obtained from publications based on evidence-based medicine in aplastic anemia support the notion that hematopoietic growth factors should not be recommended routinely as an adjunct to IST for these patients, but their administration to the specific patient can be considered on an individual basis.

The conduction of further randomized controlled trials and meta-analyses based on individual patient data should be encouraged in aplastic anemia despite the paucity of patients to establish future guidelines in general and the role of hematopoietic growth factors in particular.

References

1. Woolf SH (2000) Evidence-based medicine and practice guidelines: an overview. Cancer Control 7:362–367
2. Gale RP, Lazarus HM (2011) How helpful are meta-analyses in determining the best therapy of blood diseases? Acta Haematologica 125:91–101

3. Harbour R, Miller J (2001) A new system for grading recommendations in evidence based guidelines. Br Med J 323:334–336
4. Marsh JC, Ball SE, Cavenagh J et al (2009) Guidelines for the diagnosis and management of aplastic anaemia. Br J Haematol 147:43–70
5. Tichelli A, Schrezenmeier H, Socie G et al (2009) Use of G-CSF in patients with severe aplastic anemia treated with ATG and cyclosporine increases neutrophils and decreases infection rates and hospitalization days but does not improve long-term outcome: results of a prospective, randomized clinical trial of the EBMT. Blood 114:205
6. Rothwell PM (1995) Can overall results of clinical trials be applied to all patients? Lancet 345:1616–1619
7. Mant D (1999) Can randomised trials inform clinical decisions about individual patients? Lancet 353:743–746
8. Kunz R, Oxman AD (1998) The unpredictability paradox: review of empirical comparisons of randomised and non-randomised clinical trials. Br Med J 317:1185–1190
9. Bohlius J, Schwarzer G (2011) Twist and shout: one decade of meta-analyses on erythropoiesis-stimulating agents in cancer patients. Acta Haematologica 125:55–67
10. Rennie D (1996) How to report randomized controlled trials. The CONSORT statement. JAMA 276:649
11. Begg C, Cho M, Eastwood S et al (1996) Improving the quality of reporting of randomized controlled trials. The CONSORT statement. JAMA 276:637–639
12. Moher D, Cook DJ, Eastwood S, Olkin I, Rennie D, Stroup DF (1999) Improving the quality of reports of meta-analyses of randomised controlled trials: the QUOROM statement. Lancet 354:1896–1900
13. Moher D, Liberati A, Tetzlaff J, Altman DG, and the PRISMA Group (2009) Preferred reporting items for systematic reviews and meta-analyses: the PRISMA statement. Ann Intern Med 151:264–269
14. Brodsky RA (1998) Biology and management of acquired severe aplastic anemia. Curr Opin Oncol 10:95–99
15. Montane E, Ibanez L, Vidal X et al (2008) Epidemiology of aplastic anemia: a prospective multicenter study. Haematologica 93:518–523
16. Issaragrisil S, Kaufman DW, Anderson T et al (2006) The epidemiology of aplastic anemia in Thailand. Blood 107:1299–1307
17. Young NS, Calado RT, Scheinberg P (2006) Current concepts in the pathophysiology and treatment of aplastic anemia. Blood 108:2509–2519
18. Camitta BM, Thomas ED, Nathan DG et al (1979) A prospective study of androgens and bone marrow transplantation for treatment of severe aplastic anemia. Blood 53:504–514
19. Bacigalupo A, Hows J, Gluckman E et al (1998) Bone marrow transplantation (BMT) versus immunosuppression for the treatment of severe aplastic anaemia (SAA): a report of the EBMT SAA working party. Br J Haematol 70:177–182
20. Ades L, Mary JY, Robin M et al (2004) Long-term outcome after bone marrow transplantation for severe aplastic anemia. Blood 103:2490–2497
21. Storb R, Leisenring W, Anasetti C et al (1997) Long-term follow-up of allogeneic marrow transplants in patients with aplastic anemia conditioned by cyclophosphamide combined with antithymocyte globulin. Blood 89:3890–3891
22. Anasetti C, Doney KC, Storb R et al (1986) Marrow transplantation for severe aplastic anemia. Long-term outcome in fifty "untransfused" patients. Ann Intern Med 104:461–466
23. Horowitz MM (2000) Current status of allogeneic bone marrow transplantation in acquired aplastic anemia. Semin Hematol 37:30–42
24. Armand P, Antin JH (2007) Allogeneic stem cell transplantation for aplastic anemia. Biol Blood Marrow Transplant 13:505–516
25. Locasciulli A, Oneto R, Bacigalupo A et al (2007) Outcome of patients with acquired aplastic anemia given first line bone marrow transplantation or immunosuppressive treatment in the last decade: a report from the European Group for Blood and Marrow Transplantation (EBMT). Haematologica 92:11–18

26. Fuhrer M, Rampf U, Baumann I et al (2005) Immunosuppressive therapy for aplastic anemia in children: a more severe disease predicts better survival. Blood 106:2102–2104
27. Bacigalupo A, Bruno B, Saracco P et al (2000) Antilymphocyte globulin, cyclosporine, prednisolone, and granulocyte colony-stimulating factor for severe aplastic anemia: an update of the GITMO/EBMT study on 100 patients. European Group for Blood and Marrow Transplantation (EBMT) Working Party on Severe Aplastic Anemia and the Gruppo Italiano Trapianti di Midolio Osseo (GITMO). Blood 95:1931–1934
28. Gafter-Gvili A, Ram R, Gurion R et al (2008) ATG plus cyclosporine reduces all-cause mortality in patients with severe aplastic anemia – systematic review and meta-analysis. Acta Haematol 120:237–243
29. Marsh J, Schrezenmeier H, Marin P et al (1999) Prospective randomized multicenter study comparing cyclosporin alone versus the combination of antithymocyte globulin and cyclosporin for treatment of patients with nonsevere aplastic anemia: a report from the European Blood and Marrow Transplant (EBMT) Severe Aplastic Anaemia Working Party. Blood 93:2191–2195
30. Marsh J, Socie G, Tichelli A et al (2010) Should irradiated blood products be given routinely to all patients with aplastic anaemia undergoing immunosuppressive therapy with antithymocyte globulin (ATG)? A survey from the European Group for Blood and Marrow Transplantation Severe Aplastic Anaemia Working Party. Br J Haematol 150:377–379
31. Gafter-Gvili A, Paul M, Fraser A, Leibovici L (2007) Antibiotic prophylaxis in neutropenic patients. Isr Med Assoc J 9:460–462
32. Robenshtok E, Gafter-Gvili A, Goldberg E et al (2007) Antifungal prophylaxis in cancer patients after chemotherapy or hematopoietic stem-cell transplantation: systematic review and meta-analysis. J Clin Oncol 25:5471–5489
33. Vadhan-Raj S, Broxmeyer HE, Hittelman WN (1992) Use of granulocyte-macrophage colony-stimulating factor in hematopoietic disorders: biology and nature of response. Semin Hematol 29:4–13
34. Hord JD, Gay JC, Whitlock JA et al (1995) Long-term granulocyte-macrophage colony-stimulating factor and immunosuppression in the treatment of acquired severe aplastic anemia. J Pediatr Hematol Oncol 7:140–144
35. Locasciulli A, Arcese W, Locatelli F, Di Bona E, Bacigalupo A (2001) Treatment of aplastic anaemia with granulocyte-colony stimulating factor and risk of malignancy. Italian Aplastic Anaemia Study Group. Lancet 357:43–44
36. Symeonidis A, Kouraklis-Symeonidis A, Seimeni U et al (2002) Ticlopidine-induced aplastic anemia: two new case reports, review, and meta-analysis of 55 additional cases. Am J Hematol 71:24–32
37. Marsh JC, Ball SE, Darbyshire P et al (2003) Guidelines for the diagnosis and management of acquired aplastic anaemia. Br J Haematol 123:782–801
38. Negrin RS, Stein R, Doherty K et al (1996) Maintenance treatment of the anemia of myelodysplastic syndromes with recombinant human granulocyte colony-stimulating factor and erythropoietin: evidence for in vivo synergy. Blood 87:4076–4081
39. Gordon-Smith EC, Yandle A, Milne A et al (1991) Randomised placebo controlled study of RH-GM-CSF following ALG in the treatment of aplastic anaemia. Bone Marrow Transplant 7 (Suppl 2):78–80
40. Shao Z, Chu Y, Zhang Y, Chen G, Zheng Y (1998) Treatment of severe aplastic anemia with an immunosuppressive agent plus recombinant human granulocyte-macrophage colony-stimulating factor and erythropoietin. Am J Hematol 59:185–191
41. Kojima S, Hibi S, Kosaka Y et al (2000) Immunosuppressive therapy using antithymocyte globulin, cyclosporine, and danazol with or without human granulocyte colony-stimulating factor in children with acquired aplastic anemia. Blood 96:2049–2054
42. Gluckman E, Rokicka-Milewska R, Hann I et al (2002) Results and follow-up of a phase III randomized study of recombinant human-granulocyte stimulating factor as support for immunosuppressive therapy in patients with severe aplastic anaemia. Br J Haematol 119:1075–1082

43. Zheng Y, Liu Y, Chu Y (2006) Immunosuppressive therapy for acquired severe aplastic anemia (SAA): a prospective comparison of four different regimens. Exp Hematol 34:826–831
44. Teramura M, Kimura A, Iwase S et al (2007) Treatment of severe aplastic anemia with antithymocyte globulin and cyclosporin A with or without G-CSF in adults: a multicenter randomized study in Japan. Blood 10:1756–1761
45. Gurion R, Gafter-Gvili A, Paul M et al (2009) Hematopoietic growth factors in aplastic anemia patients treated with immunosuppressive therapy-systematic review and meta-analysis. Haematologica 94:712–719

Using rHuG-CSF in Multiple Myeloma: Consolidated Data, Evolutions, and New Concepts

Pellegrino Musto

1 General Aspects

The major risk of infection in cancer patients is correlated with duration and severity of chemotherapy-induced neutropenia, which depend on several factors, including the underlying disease, patient's related characteristics, the intensity of chemotherapy, bone marrow reserve, the use of hematopoietic growth factors, and the stem cell dose in case of transplantation [1, 2].

Granulocyte colony-stimulating factor (G-CSF) is the primary regulator of granulopoiesis. At least three forms of commercially available recombinant human (rHu)G-CSF are largely employed in the clinical practice: glycosylated rHuG-CSF (lenograstim), another nonglycosylated form (filgrastim), and a long-acting variant (pegfilgrastim), characterized by the attachment of a polyethylene glycol moiety that reduces renal excretion and masks proteolytic cleavage sites, resulting in increased G-CSF serum concentrations for up to 14 days after a single injection [3, 4]. As in other cancers, all these rHuG-CSF have shown efficacy in reducing the incidence and the duration of severe neutropenia, as well as the frequency of neutropenic fever after chemotherapy or myelosuppressive therapy in multiple myeloma. In these situations, rHuG-CSF, by accelerating granulocyte reconstitution, may enable a significant reduction in the incidence, duration, and severity of infections [3, 4].

According to recently published National Comprehensive Cancer Network (NCCN) and European Organization for the Research on the Treatment of Cancer (EORTC) guidelines [1, 2], several patient-related factors can facilitate the occurrence of various degrees of chemotherapy-induced neutropenia and, in particular, of

P. Musto (✉)
Department of Onco-Hematology, Unit of Hematology and Stem Cell Transplantation, IRCCS, Centro di Riferimento Oncologico della Basilicata (CROB), Via Padre Pio 1, Rionero in Vulture (PZ) 85028, Italy
e-mail: p.musto@crob.it; p.musto@tin.it

G. Molineux et al. (eds.), *Twenty Years of G-CSF*, Milestones in Drug Therapy,
DOI 10.1007/978-3-0348-0218-5_12, © Springer Basel AG 2012

febrile neutropenia also in patients with multiple myeloma, thus providing a rationale for using rHuG-CSF in this specific setting. Age, for example, may be relevant, as the median age of these patients is approximately 65–70 years. Furthermore, bone marrow infiltration by neoplastic plasma cells is generally extensive in these patients, while the frequent long history of the disease allows multiple lines and cycles of treatment, including repeated chemotherapy and radiotherapy (frequently on sites actively producing hematopoietic marrow). Multiple myeloma frequently shows different degrees of renal failure, low albumin concentrations, and poor performance status, often due to bone pain and movement impairment. These conditions may significantly contribute to increase the risk of developing febrile neutropenia. Finally, immunoparesis is a peculiar immunologic defect of multiple myeloma that may further facilitate infections.

Despite these observations, however, the literature regarding the specific issue of preventing or reducing infections by rHuG-CSF in patients with multiple myeloma is scarce, and this is probably due to the fact that, in the past, treatments for multiple myeloma (above all the combination of melphalan and prednisone) were not generally associated with a significant (i.e., >10–20%) risk of febrile neutropenia, thus not justifying the use of rHuG-CSF. Other more recent and myelotoxic drug combinations, such as VAD, DCEP, or DT-PACE, may require rHuG-CSF to prevent or reduce febrile neutropenia or to maintain dose intensity of cycles. These combinations, however, are rarely used to day.

2 How to Use rHuG-CSF in the Era of Novel Agents

During the last years, the wide diffusion of autologous stem cell transplantation (AuSCT) as a standard treatment for eligible patients with multiple myeloma and, more recently, the availability of the so-called "novel agents" (thalidomide, bortezomib, and lenalidomide) have completely changed the therapeutic scenario of multiple myeloma [5–7], also raising the question of whether and how to prevent and manage a possible higher risk of febrile neutropenia induced by these newer approaches. Of note, neither NCCN nor EORTC guidelines examine specifically the role of rHuG-CSF in patients with multiple myeloma who receive novel agents, so that current indications are indirectly derived from clinical studies and from daily practice [8] (see chapter "Use of rHuG-CSF in New Chemotherapy Strategies" by Renwick for further information on some new agents).

Severe neutropenia and febrile neutropenia are rarely observed if thalidomide is used as single agent; infections may occur when thalidomide is combined with dexamethasone, but, in this case, the main cause is not neutropenia, which instead may be seen more frequently using other combinations (i.e., thalidomide with alkilating agents or anthracyclines) [9]. Myelotoxicity due to bortezomib is more frequently represented by thrombocytopenia, rather than neutropenia. However, as for thalidomide, combinations with other drugs, in particular melphalan, cyclophosphamide, and anthracyclines, may significantly increase the risk of neutropenia [8].

In all these situations the use of rHuG-CSF should be managed, according to current guidelines, on a single-case basis, also using drug dose adjustments and antibiotics if necessary [1, 2]. In particular, also in multiple myeloma, secondary prophylaxis should be strongly recommended in patients who experience a previous episode of infection, severe neutropenia, or febrile neutropenia.

Neutropenia is a common adverse effect during treatments with lenalidomide at currently used doses (25 mg/day) [10]. The incidence of grade 3/4 neutropenia in relapsed patients treated with the combination of lenalidomide and dexamethasone ranges from 30 to 46%, but decreases to 12–20% in newly diagnosed patients. It is, however, higher in patients previously undergoing an AuSCT or with an impaired renal function, or when lenalidomide is combined with other myelotoxic agents. On the other hand, myelotoxicity of the lenalidomide and dexamethasone combination is generally not cumulative, as the rates decrease over the 21 days of treatment and return to normal in the 7-day rest between cycles. In the event of neutropenia, if the absolute neutrophil count (ANC) is $<1.0 \times 10^9$/L, the use of rHuG-CSF is recommended to prevent febrile neutropenia and to avoid dosage reductions, which could reduce the efficacy of the treatment. Along with rHuG-CSF, drug interruption and following appropriate dose adjustments (5–15 mg/day) are required if ANC becomes $<0.5 \times 10^9$/L. In general, prophylactic use of rHuG-CSF is recommended in cases where severe myelotoxicity is anticipated, such as elderly patients who are highly treated or concomitantly on melphalan, cyclophosphamide, or anthracyclines. Likewise, a history of febrile neutropenia or the presence of other patient-related risk factors should be always taken into account in this setting.

3 rHuG-CSF and AuSCT

At present, AuSCT remains a therapeutic procedure of choice in younger patients with multiple myeloma. It includes both mobilization processes and conditioning regimens with high-dose chemotherapy (generally high-dose cyclophosphamide and melphalan, respectively) that ensure the development in all treated cases of grade 4 neutropenia. Thus, AuSCT is counted among those treatments in which the risk of febrile neutropenia constantly requires the prophylactic use of rHuG-CSF to induce acceleration of ANC recovery, to shorten the duration of hospitalization, and to reduce infections and intravenous antibacterial therapy after high-dose chemotherapy.

rHuG-CSF is a fundamental component of the peripheral blood stem cell (PBSC)-mobilizing procedure. Sometimes used as a single agent, rHuG-CSF is more frequently employed in combination with other chemotherapy-mobilizing drugs, in particular high-dose cyclophosphamide in the multiple myeloma setting.

3.1 rHuG-CSF for Mobilizing PBSC

Overall, by analyzing different neoplastic disorders and a great variety of regimens, all forms of rHuG-CSF have been found to be useful in mobilizing CD34$^+$ PBSC in clinical trials, generally showing comparable activity; however, in the specific setting of multiple myeloma, some possible peculiarities have emerged.

Ria et al. [11] compared the efficiency of glycosylated (lenograstim) and nonglycosylated (filgrastim) rHuG-CSF at the daily dose of 10 μg/kg after high-dose cyclophosphamide (3–4 g/m^2) in mobilizing hematopoietic PBSC in 44 patients with multiple myeloma. In this study, a significantly higher collection of CD34$^+$ PBSC was obtained from patients receiving lenograstim compared with patients receiving filgrastim (14.21 ± 2.8 vs. $10.82 \pm 2.33 \times 10^6$/kg, $p < 0.01$). The percentage of patients who reached the minimum collection target $>3 \times 10^6$/kg CD34$^+$ PBSC after two aphereses was 73% in lenograstim-treated patients vs. 46% in filgrastim-treated patients ($p < 0.001$). No significant differences between the two regimens were observed with regard to toxicity and days to bone marrow recovery.

It should be understood that glycosylation significantly impacts G-CSF's pharmacokinetic parameters through different pathways: a non-saturable mechanism in spleen and kidney, and a saturable mechanism in neutrophils [12, 13]. Studies have shown that rHuG-CSF is degraded by serum enzymes, particularly elastases, and that glycosylation reduces this elastase-dependent inactivation, prolonging rHuG-CSF activity and making it more effective [14]. It has been suggested a role for matrix metalloprotease (MMP)-9, as its serum concentrations increase on days 4 and 5 after rHuG-CSF administration, thus facilitating PBSC mobilization by degrading SDF-1, upregulating the CXCR4 expression of the CD34$^+$ cells, and increasing their migration ability [15, 16]. Of interest, a reduced lenograstim dose of 7.5 μg/kg was as effective as a filgrastim dose of 10 μg/kg for PBSC mobilization in a prospective, randomized study including patients with multiple myeloma [17], while lenograstim 10 μg/kg mobilized multiple myeloma patients more efficiently than 5 μg/kg, without influencing transplants procedures and engraftment time in another retrospective study [18]. Other trials, however, including variable proportions of myeloma patients, failed to show any significant difference in mobilizing PBSC comparing lenograstim versus filgrastim [19, 20].

Single doses of pegfilgrastim have been investigated for mobilizing autologous PBSC in patients with multiple myeloma [21–25]. Generally, the results have shown similar efficacy compared to unconjugated rHuG-CSF in terms of blood CD34$^+$ cell count, stem cell yields, as well as engraftment of after reinfusion and side effects. However, using the same mobilization chemotherapy in two consecutive protocols and comparing either twice-a-day filgrastim versus two doses of pegfilgrastim, Tricot et al. showed some statistically significant advantages for pegfilgrastim, such as higher percentage of patients collecting at least 15×10^6/kg PBSC, higher median number of PBSC/kg collected, faster post-transplantation neutrophil and platelet recovery [26].

Two dose levels of pegfilgrastim were examined for PBSC mobilization in 30 patients with multiple myeloma [27]. Four days after cytotoxic therapy with cyclophosphamide (4 g/m^2), a single dose of either 6 mg pegfilgrastim ($n = 15$) or 12 mg pegfilgrastim ($n = 15$) or daily doses of 8 μg/kg unconjugated rHuG-CSF ($n = 15$) were administered. Pegfilgrastim was equally effective at 6 and 12 mg with regard to mobilization and yield of CD34$^+$ cells. Pegfilgrastim in either dose was associated with a more rapid leukocyte recovery ($p = 0.03$) and an earlier performance of the first apheresis procedure ($p < 0.05$) in comparison to unconjugated rHuG-CSF. No difference regarding CD34$^+$ cell maximum and yield could be observed.

3.2 The Role of rHuG-CSF After Autologous SCT

Several studies have compared the use of multiple, daily injections of filgrastim with a single dose (6 mg) of pegfilgrastim in patients with multiple myeloma after AuSCT. In a review of six papers (five case–control, one randomized study) including 153 patients, most of whom had multiple myeloma, no clear differences emerged, although some favorable trends were seen in terms of incidence and duration of febrile neutropenia, documented infections, marrow recovery supportive cares and costs when pegfilgrastim was employed [28].

In a study that evaluated 72 patients with multiple myeloma who underwent 92 AuSCT, the median duration of grade 4 neutropenia and hospitalization were slightly, but significantly shorter with pegfilgrastim compared with filgrastim (5 vs. 6 days, $p = 0.0079$; and 14.5 vs. 15.5 days, $p = 0.024$) [29]. Pegfilgrastim-treated patients also had fewer red blood cell transfusions ($p = 0.00065$) and reduced cost of the treatment ($p = 0.031$).

In a study of 164 consecutive patients, 75 with multiple myeloma, patients who received pegfilgrastim had faster engraftment (9.6 days compared with 10.9 days for filgrastim, $p < 0.0001$), a lower incidence of febrile neutropenia (59% compared with 78%, $p = 0.015$), as well as shorter hospital stay, fewer days of treatment with intravenous antibiotics (6.3 days compared with 9.6 days, $p = 0.006$), and fewer radiographic tests, which translated to an estimated total cost savings of more than US$8000 per patient [30]. Overall, no differences in toxicity were reported with these two agents.

In a randomized trial of 80 patients including 12 patients with multiple myeloma and aiming to demonstrate the non-inferiority of pegfilgrastim compared with filgrastim after AuSCT, similar results were obtained in terms of mean duration of neutropenia, time to reach ANC >0.5 or 1×10^9/L, incidence of fever (62% vs. 56%), documented infections (31% vs. 25%), and duration of antibiotic therapy (5.7 and 4.0 days, respectively) [31].

Another phase 3, randomized, double-blinded, placebo-controlled trial compared efficacy, costs, and safety of single-dose pegfilgrastim (6 mg) versus daily filgrastim (5 μg/kg/day) after AuSCT [32]. Seventy-eight patients (23 with multiple myeloma) matched for age, sex, underlying disease, stage, and CD34/kg transplant dose were enrolled. Both filgrastim and pegfilgrastim were given on day +1 after transplant; filgrastim was continued to an ANC of at least 5×10^9/L for 3 days. The median time to neutrophil engraftment was the same in both groups (12 days). No differences in platelet engraftment (11 vs. 13 days), number of platelet transfusions (5 vs. 4), percent with positive cultures for bacterial pathogens (23% vs. 15%), days of fever (1 vs. 2), deaths before engraftment (1 vs. 1), or duration of hospital stay (19 vs. 19 days) were seen between the pegfilgrastim and filgrastim groups, respectively. A per-patient savings of US$961 for the pegfilgrastim group was realized ($p < 0.001$).

Another group investigated 20 patients with multiple myeloma and 20 with lymphoma, who received pegfilgrastim 6 mg on day +1 after AuSCT [33]. Forty patients treated with daily filgrastim starting at median day +7 (range: 5–7), matched by age, sex, diagnosis, high-dose chemotherapy schedule, $CD34^+$ cell-dose, and previous therapy lines were used for comparison. Median time to neutrophil engraftment was 9.5 versus 11 days for pegfilgrastim and filgrastim, respectively ($p < 0.0001$). Likewise, duration of neutropenia, intravenous antibiotic use, and hospitalization favored pegfilgrastim, while platelet engraftment, transfusion requirement, and fever duration were equivalent in both groups. No grade ≥ 3 toxicities were observed. Interestingly, different levels of advantages were seen in patients with multiple myeloma receiving pegfilgrastim, who experienced improved neutrophil recovery kinetics without translating, however, to outcome improvements.

rHuG-CSF and, in particular pegfilgrastim, have been demonstrated to be of help in performing AuSCT on an outpatient basis, resulting in no different outcome in terms of safety and efficacy. In this setting, 38 patients with multiple myeloma (48 autografts) received pegfilgrastim, given at a single dose of 6 mg at day +5 from stem cell infusion, while 81 (113 autografts) received rHuG-CSF from day +2 up to stable neutrophil recovery (median: 8 days) [34]. No statistically significant difference was noted in terms of hospital readmission: in the pegfilgrastim group, readmission was needed in 6 of 48 autografts (12%) compared with 30 of 113 (26%) in the rHuG-CSF subgroup, $p = 0.06$. The median time of hospital stay for readmitted patients was identical for the two subgroups (9 days).

Glycosylation modifies the chemical properties of G-CSF that translate to a higher plasma half-life, confer pharmacokinetic advantages, and induce a higher affinity for specific receptors, causing an increment in bioavailability and molecular activity [11, 35]. Furthermore, neutrophils exposed in vitro to nonglycosylated rHuG-CSF present reduced motility, morphologic abnormalities, increased spontaneous actin polymerization, and RhoA activation, as well as a more immature phenotype and a slight reduction in the release of reactive oxygen species compared with those exposed to glycosylated rHuG-CSF [11, 35]. As a consequence, neutrophils mobilized by lenograstim (glycosylated rHuG-CSF) maintain all their functions in vitro, displaying a higher expression of the maturity markers involved in recognition, adhesion, phagocytosis, and interaction with immunoglobulins while filgrastim (nonglycosylated)-exposed neutrophils present functional defects due to higher adhesiveness, cytoskeletal alterations, and a more immature phenotype. All these features contribute to the impairment of both chemotaxis and the capability to respond correctly to further stimulation. Consistent with these findings, neutrophils exposed to nonglycosylated rHuG-CSF may be less effective in preventing febrile neutropenia in patients with chemotherapy-induced neutropenia compared with those exposed to glycosylated rHuG-CSF.

Aiming to verify in vivo these concepts, Orciuolo et al. performed a prospective, multicenter, randomized clinical trial to test occurrence of febrile episodes in 176 patients with multiple myeloma who were randomly assigned to receive lenograstim or filgrastim after high-dose cyclophosphamide in the context of PBSC mobilization and harvest phases [36]. The incidence of febrile episodes was 9.1% with filgrastim and 1.1% with lenograstim, indirectly confirming a possible functional block on neutrophil activity as a consequence of filgrastim exposure. The patients in the lenograstim group also presented a significantly higher absolute $CD34^+$ cell number compared with the filgrastim patients, but no differences were detected for global collection efficacy. In this study, filgrastim and lenograstim had similar safety profiles.

3.3 G-CSF Response as Predictive Indicator of Outcome

Straka et al. prospectively investigated the potential of rHuG-CSF responsiveness as a predictor of various parameters related to chemotherapy-induced neutropenia [37]. A total of 168 patients, 104 of whom had multiple myeloma, received a single dose of subcutaneous rHuG-CSF (lenograstim, 263 μg) after high-dose chemotherapy. Highly variable leukocyte peaks were measured and grouped as low (100–10,100/μL), medium (>10,100–18,300/μL), and high (>18,300–44,800/μL). rHuG-CSF responsiveness (low vs. medium vs. high) was inversely correlated with febrile neutropenia (77% vs. 60% vs. 48%; $p = 0.0037$), rate of infection, including fever of unknown origin (91% vs. 67% vs. 54%; $p < 0.0001$), days with intravenous antibiotics (9 vs. 6 vs. 5; $p < 0.0001$), and antifungal therapy ($p = 0.042$). In multivariate analysis, rHuG-CSF responsiveness remained the only factor significantly associated with infection ($p = 0.016$), while baseline leukocyte count, previous radiotherapy or chemotherapy received, conditioning regimens and number of $CD34^+$ cells infused did not. In addition, rHuG-CSF responsiveness was inversely correlated with grade 3/4 oral mucositis (67% vs. 33% vs. 23%; $p < 0.0001$). The authors concluded that rHuG-CSF responsiveness could represent an indirect signature of the myeloid marrow reserve predicting defense against neutropenic infection after intensive chemotherapy.

4 Summary of Evidences and Future Directions: Biosimilars, Plerixafor, and Specific Properties of rHuG-CSF-Mobilized Hematopoietic Precursors

Some cautions are necessary in interpreting biologic and clinical data from the described studies as most of them were case–control, retrospective, and not randomized trials. In addition, some trials also included patients with hematologic malignancies other than multiple myeloma, mainly lymphomas. Furthermore, the modalities of administration of rHuG-CSF, in particular in the transplant setting, were not homogeneous.

Notwithstanding, evidence suggests that all available forms of rHuG-CSF are safe and effective and, therefore, the use of any of these agents to prevent febrile neutropenia and febrile neutropenia-related complications, where indicated, is recommended in the setting of multiple myeloma, as in other malignancies [38, 39]. Choice of formulation remains a matter for individual clinical judgment. Some biologic and clinical differences are possible, but still not firmly demonstrated.

There is also a substantial equivalence of different rHuG-CSF in efficacy after high-dose chemotherapy, probably with some not clinically relevant advantages for pegfilgrastim. Costs may depend upon local prices and schedules used; pegfilgrastim, however, has shown to be potentially cost-effective in some specific settings.

To date, rHuG-CSF remains the most favored cytokine administered for PBSC mobilization in multiple myeloma because of its great efficacy and lack of serious toxicity. Again, some advantages have been reported for pegfilgrastim or lenograstim with respect to filgrastim, but these advantages do not have yet robust confirmation.

Biosimilars are officially approved versions of innovator biotechnology products made after patent and exclusivity expiration [40]. Biosimilar products containing filgrastim have been recently available also for use in patients with multiple myeloma. In a study, a biosimilar was used to treat 414 multiple myeloma or lymphoma patients subjected to AuSCT for mobilization of progenitors (5 days at 300 μg/day) and for the recovery of neutropenia after transplantation (100 μg/day, starting on day +5) [41]. A mean dose of 9.47×10^6 $CD34^+$ cells/kg was infused; time required to ANC of 0.1, 0.5, and 1.0×10^9/L was 5, 6 and 7 days of treatment, respectively.

Direct comparisons of the different rHuG-CSF and filgrastim biosimilars are not currently available. Although some data show the physicochemical and biologic comparability of biosimilar rHuG-CSF and their reference product, the amount of clinical information about these drugs is scanty, at least for use in multiple myeloma. In this setting, the Italian Societies of Hematology, Experimental Hematology and Marrow Transplantation [42] feel that the decision on appropriate clinical use of these drugs, including the multiple myeloma setting, should be supported by a larger evidence on safety and efficacy, in particular for less common and delicate indications, such as PBSC mobilization [43].

Plerixafor is a small bicyclam molecule that reversibly and selectively antagonizes the CXCR4 chemokine receptor and blocks binding of its cognate ligand, stromal cell-derived factor-1-α (SDF-1-α or CXCL12), resulting in mobilization of CD34$^+$ cells to the peripheral blood [44, 45]. The chemokine receptor CXCR4 and ligand SDF-1 are integrally involved in homing and mobilization of hematopoietic progenitor cells. Disruption of the CXCR4/SDF-1 axis by the CXCR4 antagonist, plerixafor, has been demonstrated in phase 2 and phase 3 trials, including double-blind, placebo-controlled studies [46, 47], to improve mobilization when used in conjunction with rHuG-CSF. This approach is safe, with few adverse events and produces significantly greater numbers of CD34$^+$ cells compared with rHuG-CSF alone, with a reduced number of aphereses. Plerixafor has been made available to patients with multiple myeloma who are poor mobilizers through compassionate use programs, allowing a 70–85% successful collection rate [48–51]

Plerixafor in combination with chemotherapy for PBSC mobilization has not been extensively studied, although preliminary studies in non-Hodgkin's lymphoma and multiple myeloma suggest that plerixafor may be added safely to chemotherapy with rHuG-CSF-based mobilization regimens and may accelerate the rate of increase in CD34$^+$ cells [50, 52, 53].

After the introduction of plerixafor, stem cell mobilization issues for transplantation in patients with multiple myeloma have been reviewed by the International Myeloma Working Group [54]. It is now recommended, in particular, to perform studies to look at optimizing collection strategies after exposure to novel therapies (particularly lenalidomide-based combinations) with plerixafor and rHuG-CSF or plerixafor plus chemotherapy. In fact, the ability to mobilize greater numbers of CD34$^+$ cells may provide more opportunities to deliver optimal cell doses at transplant with faster engraftment and, potentially, better long-term outcomes. Increased CD34$^+$ cell yield, in addition, may allow cells to be stored for tandem or salvage transplantation, avoiding the need to attempt mobilization at a time when mobilization could be challenging for the patient [55].

No evidence has been reported of tumor cell mobilization or increases in the relative number of peripheral blood multiple myeloma tumor cells after rHuG-CSF plus plerixafor treatment [56, 57]. On the other hand, an additional application of plerixafor to a standard rHuG-CSF mobilization regimen leads to not only more CD34$^+$ cells, but also to a significant increase in the proportion of more primitive stem cell subsets with CD34$^+$, Lin$^-$, CD38$^-$, aldehyde dehydrogenase (high) phenotype PBSC, with increased proliferative activity and marked repopulation capacity after clinical transplantation [58, 59].

Finally, the CD34$^+$ cells mobilized by pegylated rHuG-CSF show higher expression levels of genes indicative of early hematopoiesis, including HOXA9, MEIS1, and GATA3, with lower expression of genes characteristic of erythroid and later stages of myeloid differentiation and a lower functional BFU-E/CFU-GM ratio [60]. Consistently, greater numbers of hematopoietic stem cells and common myeloid progenitors and fewer megakaryocyte–erthrocyte progenitors were found in the pegfilgrastim-mobilized $CD34^+$ cells, suggesting that stimulation with pegylated or unconjugated rHuG-CSF results in different expression of key regulatory genes and different functional properties of mobilized hematopoietic stem cells as well as their progeny. This finding might be relevant for future specific applications of these cells in blood stem cell transplantation.

References

1. NCCN Clinical Practice Guidelines in Oncology (2010) Myeloid Growth Factors. V.1.2011. http://www.nccn.org
2. Aapro MS, Bohlius J, Cameron DA, European Organisation for Research and Treatment of Cancer et al (2011) 2010 update of EORTC guidelines for the use of granulocyte-colony stimulating factor to reduce the incidence of chemotherapy-induced febrile neutropenia in adult patients with lymphoproliferative disorders and solid tumours. Eur J Cancer 47:8–32
3. Renwick W, Pettengell R, Green M (2009) Use of filgrastim and pegfilgrastim to support delivery of chemotherapy: twenty years of clinical experience. BioDrugs 23:175–186
4. Keating GM (2011) Lenograstim: a review of its use in chemotherapy-induced neutropenia, for acceleration of neutrophil recovery following haematopoietic stem cell transplantation and in peripheral blood stem cell mobilization. Drugs 71:679–707
5. Morabito F, Gentile M, Mazzone C et al (2010) Therapeutic approaches for newly diagnosed multiple myeloma patients in the era of novel drugs. Eur J Haematol 85:181–191
6. Musto P, D'Auria F, Pietrantuono G et al (2009) First-line treatment of multiple myeloma in elderly patients: the GIMEMA multiple myeloma working party perspective. Curr Drug Targets 10:906–922
7. Offidani M, Corvatta L, Morabito F et al (2011) How to treat patients with relapsed/refractory multiple myeloma: evidence-based information and opinions. Expert Opin Investig Drugs 20(6):779–793
8. Palumbo A, Mateos MV, Bringhen S, San Miguel JF (2011) Practical management of adverse events in multiple myeloma: can therapy be attenuated in older patients? Blood Rev 25:181–191
9. Offidani M, Corvatta L, Polloni C et al (2011) Infectious complications in patients with multiple myeloma treated with new drug combinations containing thalidomide. Leuk Lymphoma 52:776–785
10. Persona EP, Mesa MG, Sánchez PJ, Rodríguez AP (2011) Lenalidomide treatment for patients with multiple myeloma: diagnosis and management of most frequent adverse events. Adv Ther 28(Suppl 1):11–16
11. Ria R, Gasparre T, Mangialardi G et al (2010) Comparison between filgrastim and lenograstim plus chemotherapy for mobilization of PBPCs. Bone Marrow Transplant 45:277–281
12. Kuwabara T, Kobayashi S, Sugiyama Y (1996) Pharmacokinetics and pharmacodynamics of a recombinant human granulocyte colony-stimulating factor. Drug Metab Rev 28:625–658
13. Hayashi N, Kinoshita H, Yukawa E, Higuchi S (1999) Pharmacokinetic and pharmacodynamic analysis of subcutaneous recombinant human granulocyte colony stimulating factor (lenograstim) administration. J Clin Pharmacol 39:583–592
14. Carter CR, Whitmore KM, Thorpe R (2004) The significance of carbohydrates on G-CSF: differential sensitivity of G-CSFs to human neutrophil elastase degradation. J Leukoc Biol 75:515–522

15. Saito T, Usui N, Asai O et al (2007) Elevated serum levels of human matrix metalloproteinase-9 (MMP-9) during the induction of peripheral blood stem cell mobilization by granulocyte colony-stimulating factor (G-CSF). J Infect Chemother 13:426–428
16. Carion A, Benboubker L, Hérault O et al (2003) Stromal-derived factor 1 and matrix metalloproteinase 9 levels in bone marrow and peripheral blood of patients mobilized by granulocyte colony-stimulating factor and chemotherapy. Relationship with mobilizing capacity of haematopoietic progenitor cells. Br J Haematol 122:918–926
17. Ataergin S, Arpaci F, Turan M et al (2008) Reduced dose of lenograstim is as efficacious as standard dose of filgrastim for peripheral blood stem cell mobilization and transplantation: a randomized study in patients undergoing autologous peripheral stem cell transplantation. Am J Hematol 83:644–648
18. Romeo A, Chierichini A, Spagnoli A et al (2010) Standard- versus high-dose lenograstim in adults with hematologic malignancies for peripheral blood progenitor cell mobilization. Transfusion 50:2432–2446
19. Lefrère F, Bernard M, Audat F et al (1999) Comparison of lenograstim vs filgrastim administration following chemotherapy for peripheral blood stem cell (PBSC) collection: a retrospective study of 126 patients. Leuk Lymphoma 35:501–505
20. Kopf B, De Giorgi U, Vertogen B et al (2006) A randomized study comparing filgrastim versus lenograstim versus molgramostim plus chemotherapy for peripheral blood progenitor cell mobilization. Bone Marrow Transplant 38:407–412
21. Keating GM (2011) Plerixafor: a review of its use in stem-cell mobilization in patients with lymphoma or multiple myeloma. Drugs 71:1623–1647
22. Jantunen E (2011) Novel strategies for blood stem cell mobilization: special focus on plerixafor. Expert Opin Biol Ther 11:1241–1248
23. Putkonen M, Rauhala A, Pelliniemi TT, Remes K (2009) Single-dose pegfilgrastim is comparable to daily filgrastim in mobilizing peripheral blood stem cells: a case-matched study in patients with lymphoproliferative malignancies. Ann Hematol 88:673–680
24. Hosing C, Qazilbash MH, Kebriaei P et al (2006) Fixed-dose single agent pegfilgrastim for peripheral blood progenitor cell mobilisation in patients with multiple myeloma. Br J Haematol 133:533–537
25. Steidl U, Fenk R, Bruns I et al (2005) Successful transplantation of peripheral blood stem cells mobilized by chemotherapy and a single dose of pegylated G-CSF in patients with multiple myeloma. Bone Marrow Transplant 35:33–36
26. Tricot G, Barlogie B, Zangari M et al (2008) Mobilization of peripheral blood stem cells in myeloma with either pegfilgrastim or filgrastim following chemotherapy. Haematologica 93:1739–1742
27. Bruns I, Steidl U, Kronenwett R et al (2006) A single dose of 6 or 12 mg of pegfilgrastim for peripheral blood progenitor cell mobilization results in similar yields of CD34+ progenitors in patients with multiple myeloma. Transfusion 46:180–185
28. Musto P, Scalzulli PR, Terruzzi E et al (2007) Peg-filgrastim versus filgrastim after autologous stem cell transplantation: case-control study in patients with multiple myeloma and review of the literature. Leuk Res 31:1487–1493
29. Samaras P, Blickenstorfer M, Siciliano RD et al (2011) Pegfilgrastim reduces the length of hospitalization and the time to engraftment in multiple myeloma patients treated with melphalan 200 and auto-SCT compared with filgrastim. Ann Hematol 90:89–94
30. Mathew S, Adel N, Rice RD et al (2010) Retrospective comparison of the effects of filgrastim and pegfilgrastim on the pace of engraftment in auto-SCT patients. Bone Marrow Transplant 45:1522–1527
31. Castagna L, Bramanti S, Levis A et al (2010) Pegfilgrastim versus filgrastim after high-dose chemotherapy and autologous peripheral blood stem cell support. Ann Oncol 21:1482–1485
32. Gerds A, Fox-Geiman M, Dawravoo K et al (2010) Randomized phase III trial of pegfilgrastim versus filgrastim after autologus peripheral blood stem cell transplantation. Biol Blood Marrow Transplant 16:678–685

33. Wannesson L, Luthi F, Zucca E et al (2011) Pegfilgrastim to accelerate neutrophil engraftment following peripheral blood stem cell transplant and reduce the duration of neutropenia, hospitalization, and use of intravenous antibiotics: a phase II study in multiple myeloma and lymphoma and comparison with filgrastim-treated matched controls. Leuk Lymphoma 52:436–443
34. Ferrara F, Izzo T, Criscuolo C et al (2010) Comparison of fixed dose pegfilgrastim and daily filgrastim after autologous stem cell transplantation in patients with multiple myeloma autografted on an outpatient basis. Hematol Oncol 29:1–5
35. Ribeiro D, Veldwijk MR, Benner A et al (2007) Differences in functional activity and antigen expression of granulocytes primed in vivo with filgrastim, lenograstim, or pegfilgrastim. Transfusion 47:969–980
36. Orciuolo E, Buda G, Marturano E et al (2011) Lenograstim reduces the incidence of febrile episodes, when compared with filgrastim, in multiple myeloma patients undergoing stem cell mobilization. Leuk Res 35:899–903
37. Straka C, Sandherr M, Salwender H et al (2011) Testing G-CSF responsiveness predicts the individual susceptibility to infection and consecutive treatment in recipients of high-dose chemotherapy. Blood 17:2121–2128
38. Gunzer K, Clarisse B, Lheureux S, Delcambre C, Joly F (2010) Contribution of glycosylated recombinant human granulocyte colony-stimulating factor (lenograstim) use in current cancer treatment: review of clinical data. Expert Opin Biol Ther 10:615–630
39. Aapro M, Crawford J, Kamioner D (2010) Prophylaxis of chemotherapy-induced febrile neutropenia with granulocyte colony-stimulating factors: where are we now? Support Care Cancer 18:529–541
40. Sörgel F, Lerch H, Lauber T (2010) Physicochemical and biologic comparability of a biosimilar granulocyte colony-stimulating factor with its reference product. BioDrugs 24:347–357
41. Ferro HH, Juni M, Bello R, Vidal A, Diez RA, Pavlovsky S (2009) Utilization study of filgrastim (Neutromax) during autologous haematopoietic precursor transplantation for myeloma and lymphoma patients. Transfus Apher Sci 41:87–93
42. Barosi G, Bosi A, Abbracchio MP, et al (2011) Key concepts and critical issues on biosimilars in hematology. A position paper of the Italian Society of hematology, Italian Society of Experimental Hematology, and Italia group for Bone Marrow transplantation. Haematologica 96:937–942
43. Lefrère F, Brignier AC, Elie C et al (2011) First experience of autologous peripheral blood stem cell mobilization with biosimilar granulocyte colony-stimulating factor. Adv Ther 28:304–310
44. Mohty M, Duarte RF, Croockewit S, Hübel K, Kvalheim G, Russell N (2011) The role of plerixafor in optimizing peripheral blood stem cell mobilization for autologous stem cell transplantation. Leukemia 25:1–6
45. Steinberg M, Silva M (2010) Plerixafor: a chemokine receptor-4 antagonist for mobilization of hematopoietic stem cells for transplantation after high-dose chemotherapy for non-Hodgkin's lymphoma or multiple myeloma. Clin Ther 32:821–843
46. DiPersio JF, Stadtmauer EA, Nademanee A et al (2009) Plerixafor and G-CSF versus placebo and G-CSF to mobilize hematopoietic stem cells for autologous stem cell transplantation in patients with multiple myeloma. Blood 113:5720–5726
47. D'Addio A, Curti A, Worel N et al (2011) The addition of plerixafor is safe and allows adequate PBSC collection in multiple myeloma and lymphoma patients poor mobilizers after chemotherapy and G-CSF. Bone Marrow Transplant 46:356–363
48. Duarte RF, Shaw BE, Marín P et al (2011) Plerixafor plus granulocyte CSF can mobilize hematopoietic stem cells from multiple myeloma and lymphoma patients failing previous mobilization attempts: EU compassionate use data. Bone Marrow Transplant 46:52–58
49. Calandra G, McCarty J, McGuirk J et al (2008) AMD3100 plus G-CSF can successfully mobilize CD34+ cells from non-Hodgkin's lymphoma, Hodgkin's disease and multiple

myeloma patients previously failing mobilization with chemotherapy and/or cytokine treatment: compassionate use data. Bone Marrow Transplant 41:331–338
50. Worel N, Rosskopf K, Neumeister P et al (2011) Plerixafor and granulocyte-colony-stimulating factor (G-CSF) in patients with lymphoma and multiple myeloma previously failing mobilization with G-CSF with or without chemotherapy for autologous hematopoietic stem cell mobilization: the Austrian experience on a named patient program. Transfusion 51:968–975
51. Basak GW, Jaksic O, Koristek Z, et al (2011) Haematopoietic stem cell mobilization with plerixafor and G-CSF in patients with multiple myeloma transplanted with autologous stem cells. Eur J Haematol 86:488–495
52. Dugan MJ, Maziarz RT, Bensinger WI et al (2010) Safety and preliminary efficacy of plerixafor (Mozobil) in combination with chemotherapy and G-CSF: an open-label, multicenter, exploratory trial in patients with multiple myeloma and non-Hodgkin's lymphoma undergoing stem cell mobilization. Bone Marrow Transplant 45:39–47
53. Attolico I, Pavone V, Ostuni A, et al (2011) Plerixafor Added to Chemotherapy Plus G-CSF Is Safe and Allows Adequate PBSC Collection in Predicted Poor Mobilizer Patients with Multiple Myeloma or Lymphoma. Biol Blood Marrow Transplant. Jul 24. [Epub ahead of print] PMID: 21791194
54. Giralt S, Stadtmauer EA, Harousseau JL et al (2009) IMWG International Myeloma Working Group (IMWG) consensus statement and guidelines regarding the current status of stem cell collection and high-dose therapy for multiple myeloma and the role of plerixafor (AMD 3100). Leukemia 23:1904–1912
55. Basak GW, Jaksic O, Koristek Z et al (2011) Haematopoietic stem cell mobilization with plerixafor and G-CSF in patients with multiple myeloma transplanted with autologous stem cells. Eur J Haematol 86(6):488–495
56. Fruehauf S, Ehninger G, Hübel K et al (2010) Mobilization of peripheral blood stem cells for autologous transplant in non-Hodgkin's lymphoma and multiple myeloma patients by plerixafor and G-CSF and detection of tumor cell mobilization by PCR in multiple myeloma patients. Bone Marrow Transplant 45:269–275
57. Tricot G, Cottler-Fox MH, Calandra G (2010) Safety and efficacy assessment of plerixafor in patients with multiple myeloma proven or predicted to be poor mobilizers, including assessment of tumor cell mobilization. Bone Marrow Transplant 45:63–68
58. Fruehauf S, Veldwijk MR, Seeger T et al (2009) A combination of granulocyte-colony-stimulating factor (G-CSF) and plerixafor AMD3100 (plerixafor) mobilizes more primitive peripheral blood progenitor cells than G-CSF alone: results of a European phase II study. Cytotherapy 11:992–1001
59. Taubert I, Saffrich R, Zepeda-Moreno A et al (2011) Characterization of hematopoietic stem cell subsets from patients with multiple myeloma after mobilization with plerixafor. Cytotherapy 13:459–466
60. Bruns I, Steidl U, Fischer JC et al (2008) Pegylated granulocyte colony-stimulating factor mobilizes CD34+ cells with different stem and progenitor subsets and distinct functional properties in comparison with unconjugated granulocyte colony-stimulating factor. Haematologica 93:347–355

Use of rHuG-CSF in New Chemotherapy Strategies

William Renwick

1 Introduction

The discovery of recombinant human granulocyte colony-stimulating factor (rHuG-CSF) and the development of filgrastim as a clinical therapeutic tool allowed physicians to broaden their scope with respect to chemotherapeutic options and goals. From the initial registration trials, filgrastim demonstrated the ability to limit the duration and severity of neutropenia and to reduce the frequency of febrile neutropenia [1, 2]. Filgrastim allowed the maintenance of dose density, enabling the completion of the full doses of chemotherapy on schedule [3]. Furthermore, it enabled the development of more aggressive regimens with increased dose densities and dose intensities.

The advent of filgrastim allowed investigators to test the efficacy of dose-dense chemotherapy, enabling neutrophil recovery within 2 weeks before the next cycle of chemotherapy is administered. The first large randomized controlled trial of dose-dense chemotherapy was Cancer and Leukemia Group B Trial (CALGB) 9741, which enrolled women with node-positive breast cancer [4]. The dose-dense regimens significantly prolonged disease-free survival (relative risk [RR], 0.74; $p = 0.01$), the primary endpoint, and overall survival (RR, 0.69; $p = 0.013$) compared with the conventional regimens.

The last 2 decades have seen the advent of a number of new classes of chemotherapeutic agents. As each novel agent reaches the clinic from the bench-top, clinicians are excited and enthralled at the possibilities that they bring with them. Whether it is the tyrosine kinase inhibitors (TKI), heralded by imatinib [5] that transformed a lethal disease into a manageable chronic illness, or the ever-expanding list of monoclonal antibodies, these new chemotherapeutic agents

W. Renwick (✉)
Department of Haematology and Medical Oncology, Western Hospital, 2nd Floor South, Main Building, Gordon Street, Footscray 3011, Melbourne, VIC, Australia
e-mail: William.Renwick@wh.org.au

G. Molineux et al. (eds.), *Twenty Years of G-CSF*, Milestones in Drug Therapy,
DOI 10.1007/978-3-0348-0218-5_13, © Springer Basel AG 2012

bring opportunity and hope that the fight against cancer will have a new weapon. As they complete early trials to show safety and clinical efficacy, clinicians and scientists are thinking ahead to the possible combinations of therapy, which ideally will bring additive efficacy. With combinations, there is always the possibility of additive toxicities. Hematologic toxicities, including febrile neutropenia, are watched for, and rHuG-CSF has been used and will continue to be used when required, as these new combinations are developed and tested.

This chapter reviews the novel chemotherapeutic strategies currently been tested, or recently been accepted into practice, and the role that rHuG-CSF has in enabling or enhancing the fight against cancer. It is divided into new regimens of well-established drugs, and the new agents and their combinations.

2 New Chemotherapy Regimens Using rHuG-CSF

2.1 Solid Tumor Regimens

A phase 3 trial compared gemcitabine (G) in combination with carboplatin (C) or paclitaxel (P) versus paclitaxel plus carboplatin in patients with advanced or metastatic nonsmall-cell lung cancer (NSCLC) [6]. These agents are all myelosuppressive, and neutropenia grade 3/4 was 39% in the GC group, 20% in the GP group, and 34.7% in the PC group. Incidence of febrile neutropenia was quite low with grade 3/4 being 2.8%, 3.9%, and 3.0%, respectively, for GC, GP, and PC groups. rHuG-CSF was allowed if there was persistent neutropenia despite dose reductions in the previous course, in accordance with American Society of Clinical Oncology (ASCO) guidelines [7]. Combination chemotherapy is often used in treating breast cancer, and the combination of doxorubicin and cyclophosphamide with docetaxel in sequential and concurrent regimens was tested as adjuvant chemotherapy in patients with operable, node-positive, early-stage breast cancer [8]. The patients were randomly assigned to receive 4 cycles of doxorubicin (60 mg/m^2) and cyclophosphamide (600 mg/m^2) every 3 weeks, followed by 4 cycles of docetaxel (100 mg/m^2) every 3 weeks (sequential ACT), 4 cycles of doxorubicin (60 mg/m^2) and docetaxel (60 mg/m^2) every 3 weeks, or 4 cycles of doxorubicin (60 mg/m^2), cyclophosphamide (600 mg/m^2), and docetaxel (60 mg/m^2) (concurrent ACT). After five early deaths were reported in the concurrent ACT group, the doses were modified in this group to doxorubicin (50 mg/m^2), cyclophosphamide (500 mg/m^2), and docetaxel (75 mg/m^2); and the group containing doxorubicin and docetaxel only to 50 mg/m^2 and 75 mg/m^2 as well. The protocol was amended to include primary prophylaxis with either rHuG-CSF or rHu granulocyte–macrophage colony-stimulating factor (GM-CSF). These changes enabled completion of the study at the new doses without further major hematologic toxicity. The trial showed improved overall survival in the sequential ACT group compared with that in doxorubicin and docetaxel (Hazard Ratio [HR] for death, 0.83; $p = 0.03$), but a nonsignificant reduction compared with the concurrent ACT group (HR for death, 0.86; $p = 0.09$).

Drullinsky et al. reported a dose-dense regimen of cyclophosphamide, methotrexate, and 5-fluorouracil (5-FU) in patients with early-stage breast cancer receiving rHuG-CSF support [9]. A total of 38 patients with median age 52 years (range: 38–78 years) were treated with cyclophosphamide (600 mg/m^2), methotrexate (40 mg/m^2), and fluorouracil (600 mg/m^2) with pegfilgrastim 6 mg on day 2. Of the 38 patients, 29 completed 8 cycles at the prescribed 14-day interval, showing its tolerability and feasibility.

Dose-dense neoadjuvant gemcitabine, epirubicin, and albumin-bound paclitaxel with pegfilgrastim was studied in patients with locally advanced breast cancer [10]. In this phase 2 multicenter trial, 123 patients were enrolled and treated with neoadjuvant gemcitabine (2,000 mg/m^2), epirubicin (50 mg/m^2), and albumin-bound paclitaxel (175 mg/m^2) at 14-day intervals. Pegfilgrastim 6 mg was administered on day 2 after each dose of chemotherapy. The patients then underwent mastectomy or breast conservation surgery, and postoperatively received four doses of gemcitabine (2,000 mg/m^2) and albumin-bound paclitaxel (220 mg/m^2) with pegfilgrastim support. In all, 116 patients (95%) completed neoadjuvant chemotherapy and subsequent surgical resection. Only 11% had grade 3/4 neutropenia, with 1 episode of febrile neutropenia. The 3-year progression-free survival rate was 48% and overall survival rate was 86% in this well-tolerated regimen.

A dose-escalation trial of biweekly docetaxel and gemcitabine using filgrastim or pegfilgrastin investigated increasing dose intensity of these agents in patients with advanced solid tumors [11]. Patients received gemcitabine (3,000 mg/m^2) and increasing doses of docetaxel starting at 55 mg/m^2, in 10 mg/m^2 increments, every 14 days with filgrastim or pegfilgrastim. The maximum tolerated dose was 75 mg/m^2 with the dose-limiting toxicity of fatigue/dyspnea and diarrhea. Filgrastim was given to the initial 25 patients with an average of seven doses per cycle, and the last ten patients received pegfilgrastim with no differences with respect to toxicities or effectiveness between the two groups. This combination showed signs of clinical activity and was well tolerated.

In a phase 1/2 pilot study of capecitabine with or without vinorelbine after sequential dose-dense epirubicin and paclitaxel, the initial chemotherapy was delivered using filgrastim support [12]. Patients with high-risk early breast cancer ($\geq$4 positive nodes) were given epirubicin (150 mg/m^2) every 2 weeks for 3 cycles, followed by paclitaxel (225 mg/m^2) every 2 weeks for 3 cycles. Filgrastim was given at 5 μg/kg on days 3–10. The patients were then treated with oral capecitabine (1,000 mg/m^2 or 1,250 mg/m^2 twice daily) daily from day 1 to 14, either alone, or in combination with vinorelbine (25 mg/m^2) on days 1 and 8 in an incremental fashion with dose level 1 being capecitabine 1,000 mg/m^2 twice daily, and dose level 2, adding in vinorelbine. At dose level 2, 5 of the 10 patients experienced dose-limiting toxicities and evaluation of the vinorelbine was abandoned. Febrile neutropenia occurred in 12% of patients. At 35 months median follow-up, the estimated relapse-free and overall survival rates were 82% and 91%, respectively.

A randomized pilot phase 2 study studied doxorubicin and cyclophosphamide (AC) or epirubicin and cyclophosphamide (EC) given 2 weekly with pegfilgrastim, or 3 weekly for patients with early breast cancer [13]. Patients were randomly assigned to 1 of 4 groups: (1) standard AC – 4 cycles of doxorubicin (60 mg/m^2) and cyclophosphamide (600 mg/m^2) on day 1, every 21 days; (2) accelerated AC – same dose but every 2 weeks with pegfilgrastim on day 2; (3) standard EC – epirubicin (90 mg/m^2) and cyclophosphamide (600 mg/m^2) on day 1, every 21 days; and (4) accelerated EC – same dose but every 2 weeks with pegfilgrastim on day 2. Chemotherapy was delayed by weekly intervals (maximum 3 weeks for 3-weekly and 2 weeks for 2-weekly regimens) until the neutrophil count was $\geq 1 \times 10^9$/L and platelet count was $\geq 100 \times 10^9$/L. If febrile neutropenia occurred, for standard groups, the dose of anthracycline and cyclophosphamide was reduced by 20% and rHuG-CSF was administered as per ASCO guidelines [14]. In the accelerated-dose groups, 20% dose reductions were made. Significantly worse day-1, grade 3/4 neutropenia was seen in the standard group (6 of 61, 10%) compared to that in the accelerated group (0 of 65) ($p = 0.01$). No significant difference was noted in episodes of neutropenic sepsis. The authors concluded that AC and EC given 2 weekly with filgrastim support are as well-tolerated as the 3-weekly regimens. No survival advantage was noted, unlike that in the similar trial by Citron et al. in 2003 [4].

3 Hematologic Regimens

The largest dose-dense trial to date was a phase 3 trial that compared standard CHOP-21 (cyclophosphamide, doxorubicin, vincristine, and prednisone every 21 days) with CHOP-21 plus etoposide (CHOEP-21), CHOP-14 plus filgrastim, and CHOEP-14 plus filgrastim in patients with non-Hodgkin's lymphoma (NHL). Two distinct study populations were enrolled: 710 good-prognosis patients aged 18–60 years (designated the NHL-B1 trial) [15] and 689 patients aged 61–75 years (the NHL-B2 trial) [16]. The efficacy data were obtained at median follow-up of 58 months. In the good-prognosis patients, the dose-dense regimens did not improve the primary endpoint, event-free survival, compared with the standard regimens [15]. However, the dose-dense regimens significantly improved overall survival (RR, 0.70; $p = 0.044$). In the older patients, when the three intensified treatment groups were compared with CHOP-21, only CHOP-14 plus filgrastim significantly improved 5-year event-free-survival (RR, 0.66; $p = 0.003$) and overall survival (RR, 0.58; $p < 0.001$) [16]. CHOP-14 plus filgrastim is now considered an appropriate treatment for patients aged 18–75 years who have aggressive NHL.

The RICOVER-60 trial confirmed the value of CHOP-14 plus rHuG-CSF and showed the benefit of adding rituximab for elderly patients with aggressive NHL [17]. This trial compared 6 and 8 cycles of CHOP-14 with or without rituximab (all with rHuG-CSF support) in more than 1,200 older patients (aged 61–80 years) with CD20$^+$ diffuse large B-cell lymphoma. In a multivariate analysis that compared 6 cycles of

CHOP-14 with the other three groups, CHOP-14 plus rituximab for 6 cycles was the only regimen that significantly improved event-free survival (the primary endpoint), progression-free survival, and overall survival. It did not compare R-CHOP-14 with R-CHOP-21 with or without filgrastim, and to this date, no one has published data on CHOP-like regimens on a 14-day schedule without growth factor support.

Hodgkin lymphoma has excellent cure rates with modern treatment, and even in advanced cases, very good progression-free and overall survival rates are achieved. These require dose-intense chemotherapy as evidenced by the need for rHuG-CSF support with the BEACOPP-14 [18] and BEACOPP-escalated regimens [19].

The Hyper-CVAD [20], ICE [21], and ESHAP [22] regimens are all very commonly used in aggressive and relapsed lymphomas. They all mandate rHuG-CSF support, although there are no randomized studies proving this requirement as it would be unethical to attempt such a trial.

Newer regimens have been developed that use dose intensities that would be unable to be sustained without rHuG-CSF support. Combinations of vinorelbine, gemcitabine, and ifosfamide have been tried with adjuvant rHuG-CSF. A multicenter phase 2 study used a risk-adjusted salvage chemotherapy for relapsed and refractory lymphoma [23]. Patients were divided into three risk groups: Group 1 comprised good-risk patients in first relapse, after a durable first complete remission (>12 months for follicular NHL, or beyond 6 months for all other subtypes of NHL and Hodgkin lymphoma). Group 2 comprised patients with poor-risk disease, who were either primary refractory or had relapsed within 12 months (for follicular NHL), or 6 months (for all other subtypes of NHL or Hodgkin lymphoma) of initial therapy. Group 3 comprised patients who had relapsed at least 6 months after autologous peripheral blood stem cell (PBSC) transplantation. Groups 1 and 3 received vinorelbine (25 mg/m^2) on days 1 and 8, gemcitabine (1,000 mg/m^2) on days 1 and day 8, and dexamethasone (16 mg/m^2) on days 1 and 8 with pegfilgrastim on day 9 (VGF). Group 2 received the same regimen but with the addition of ifosfamide (3,000 mg/m^2) on day 9 (FGIV). Each cycle was 21 days and given as an outpatient, and 4 cycles were planned. After 2 cycles, restaging was performed and if a satisfactory result (>50% reduction in tumor bulk at all sites and negative functional imaging) was achieved, then the next 2 cycles were given. If less than a satisfactory result was achieved, treatment was escalated (VGF to FGIV or FGIV to IVAC). The patients were able to exit the study after cycle 2 to undergo PBSC transplantation at the discretion of their attending physician.

VGF was well tolerated, with few unplanned hospital admissions and grade 3/4 neutropenia occurring in 30% of patients and febrile neutropenic episodes in 6% of patients. The FGIV regimen was more toxic, with grade 3/4 neutropenia occurring in 60% of patients and febrile neutropenic episodes in 19% of patients. Median progression-free survival was 21 months, 5 months, and 4 months for groups 1, 2, and 3, respectively. A total of 30 patients from groups 1 and 2 underwent autologous PBSC transplantation either during their treatment or subsequently. Stem cell mobilization was not impeded. Estimated 4-year overall survival was 49%, 28%, and 31%, respectively, for groups 1, 2, and 3. The authors concluded that VGF and FGIV are effective and well-tolerated salvage approaches that can be delivered in an ambulatory care setting. The aspect of therapy escalation on the basis of interim analysis was suggested to be further studied, and new positron emission tomography (PET)-directed trials are currently ongoing.

Santoro et al. also investigated these agents in patients with relapsed and refractory Hodgkin lymphoma [24]. They treated 91 patients with ifosfamide (2,000 mg/m^2) on days 1–4, gemcitabine (800 mg/m^2) on days 1 and 4, vinorelbine (20 mg/m^2) on day 1, and prednisolone 100 mg on days 1–4 (IGEV). Each cycle was 21 days, and four courses were planned. rHuG-CSF was administered from days 7 to 12, or up to apheresis in the course of mobilization. Neutropenia grade 3 was noted in 22.7% and grade 4 in 5.7% of cycles. Documented infection was observed in 3.5%. Only 1 patient did not mobilize stem cells. Complete response rate was 53.8% and partial response was 27.5%. Of 74 patients in complete remission or partial remission, 64 continued to single or tandem high-dose chemotherapy with PBSC support. Five of the remaining patients had previously relapsed from prior transplants and were allocated to receive nonmyeloablative allogeneic transplants. The 3-year freedom-from-progression and overall survival rates were 53.0% and 70.0%, respectively. Thus, IGEV was found to be effective as a salvage-induction regimen with high response rates, favorable toxicity, and good mobilizing potential.

Dose-adjusted EPOCH is another newer regimen used in treatment of aggressive lymphomas [25]. It uses filgrastim to support the dose-intensity regimen. The same group (Dunleavy et al.) has reported on a modification of EPOCH by addition of dose-dense rituximab in human immunodeficiency virus (HIV)-associated diffuse large B-cell lymphoma [26]. This phase 2 study incorporated interim PET imaging in the decision process for the number of cycles given.

Patients were given the treatment as a 96-h continuous infusion of etoposide (50 mg/m^2/day), doxorubicin (10 mg/m^2/day), and vincristine (0.4 mg/m^2/day with no cap) and oral prednisone (60 mg/m^2/day) with cyclophosphamide (750 mg/m^2) on day 5. Rituximab (375 mg/m^2) was administered on days 1 and 5, after the infusion and before the cyclophosphamide. All patients received filgrastim 300 μg daily from day 6 until neutrophils were $>5.0 \times 10^9$/L. Cycles were repeated every 21 days. Patients withheld their antiretroviral therapy from the beginning of the treatment until completion. If after two cycles, the computed tomography (CT)/PET was negative, only 1 further cycle was given. If the CT/PET was still positive, a further 2–3 cycles were given with a CT/PET at the start of each cycle. The authors make the point that evaluating PET scans in HIV-positive patients has its own issues because of HIV-associated reactive changes, and special assessments were made. Neutropenia $<0.5 \times 10^9$/L occurred in 46% of cycles and febrile neutropenia occurred on 31% of cycles. At a median 5-year follow-up, progression-free and overall survival rates were 84% and 68%, respectively. Most patients were able to cease treatment after 3 cycles. Interestingly, there was a marked difference in the outcomes between germinal center B-cell-derived tumors and nongerminal center B-cell-derived tumors, with the latter faring far worse.

Burkitt lymphoma and Burkitt-like NHL are the most aggressive lymphomas. The standard-of-care treatment has been the combination chemotherapy CODOX-M/IVAC [27] since the mid-to-late 1990s. This regimen is very myelosuppressive, and rHuG-CSF support is required.

4 New Agents

4.1 *Cabazitaxel*

Cabazitaxel is a novel tubulin-binding taxane with antitumor activity in docetaxel-resistant cancers [28]. In a phase 2 trial, it was noted that neutropenia was a significant side effect [29]. Cabazitaxel was evaluated in patients with metastatic breast cancer who were resistant to taxane therapy. Grades 3 and 4 neutropenia were 73%.

In a phase 3 randomized trial in metastatic castration-resistant prostate cancer, patients received cabazitaxel 25 mg/m^2 every 3 weeks plus prednisolone, or mitoxantrone 12 mg/m^2 every 3 weeks plus prednisolone for 10 cycles [28]. Neutropenia grade 3/4 occurred more often in the cabazitaxel group (82% vs. 58%), as did febrile neutropenia (8% vs. 1%). Seven deaths from neutropenia/sepsis occurred in this group compared with 1 death in the mitoxantrone group. The monitoring committee advised strict protocol adherence with regard to dose modification and treatment of neutropenia as per ASCO guidelines [6]. Subsequent trials have recommended primary prophylaxis with rHuG-CSF in patients with high-risk clinical features (aged >65 years, poor performance status, previous episodes of febrile neutropenia, etc.).

4.2 *Ixabepilone*

Ixabepilone is a novel semisynthetic epothilone B analog that induces cell death by apoptosis by microtubule stabilization, similar to taxanes. It can overcome common mechanisms of resistance that limit many chemotherapeutic agents such as anthracyclines, taxanes, and capecitabine [30]. Thomas et al. examined ixabepilone plus capecitabine in anthracycline- and taxane-resistant metastatic breast cancer [31]. Patients received ixabepilone (40 mg/m^2) on day 1 of a 21-day cycle, plus capecitabine (2,000 mg/m^2) orally on days 1–14, or capecitabine alone (2,500 mg/m^2). Grade 3/4 neutropenia was markedly increased in the combination group (68% vs. 11%), with febrile neutropenia also increased (5% vs. 2%). Growth factor support with filgrastim was administered to 20% of the patients in the combination group and to 3% in the capecitabine-alone group.

4.3 Bendamustine

Bendamustine is a novel alkylator used in chronic lymphocytic leukemia (CLL), and more recently in low-grade lymphomas as a single agent, and also in combination with rituximab [32]. In a multicenter study of single-agent bendamustine in patients with rituximab-refractory, indolent B-cell NHL, bendamustine was given at a dose of 120 mg/m^2 on days 1 and 2, every 21 days, for 6–8 cycles. The overall response rate was 75%, with 14% complete response, 3% unconfirmed response, and 58% partial response. Grade 3/4 neutropenia was noted in 61% patients with 6% febrile neutropenia. rHuG-CSF was not allowed as primary prophylaxis during cycle 1, but was allowed in patients who had grade 4 neutropenia for >1 week, failure of white cells to recover to <grade 1 by the next scheduled dose, or febrile neutropenia in a previous cycle. A total of 38% patients required filgrastim or pegfilgrastim.

An interim report of bendamustine–rituximab combination compared to cyclophosphamide, doxorubicin, vincristine, and prednisolone with rituximab (R-CHOP) as first-line treatment in patients with advanced follicular, low-grade, and mantle cell NHL appears to show superior efficacy with reduced toxicities [33]. It, however, showed a much-reduced incidence of grade 3/4 neutropenia compared with R-CHOP (10.7% vs. 46.5%). Growth factor support was not generally recommended, and it was used more commonly in the R-CHOP group (20% of all cycles vs. 4.0%). This agent appears to not have overt myelotoxicity.

5 Novel Classes of Chemotherapeutic Agents

5.1 Tyrosine Kinase Inhibitors

In 1988, Yaish et al. provided the first description of tyrphostins – compounds inhibiting tyrosine phosphorylation [34]. Later generically described as tyrosine kinase inhibitors (TKI), these compounds represent a novel pathway to disrupt the growth of malignant cells. The first paper described the inhibition of the catalytic activity of the epidermal growth factor receptor (EGFR). The number of TKI is ever expanding; a small number are reviewed here.

5.2 Imatinib, Dasatinib, and Nilotinib

The first of this new class of agents to make it from the bench-top to the clinic was imatinib, directed against the constitutively activated bcr-abl tyrosine kinase [5, 35]. In this trial, patients who were refractory to previous interferon (IFN) were treated with oral imatinib 400 mg/day. Complete hematologic responses occurred in 95% of patients; 60% had a major cytogenetic response and 41% had a complete cytogenetic response. Grade 3 neutropenia occurred in 27% and grade 4 neutropenia in 8.1% of the patients. Febrile neutropenia was <1%. These results completely changed the landscape in medicine, with a new targeted approach based on scientific principles derived from knowledge of the biology of the disease process, leading to a well-tolerated, effective treatment.

Further development has led to second-generation bcr-abl TKI, dasatinib and nilotinib. Imatinib was also shown to be efficacious against gastrointestinal stromal tumors (GIST), hypereosinophilic syndrome, systemic mastocytosis, and dermatofibrosarcoma protuberans.

The use of dasatinib in imatinib-resistant or -intolerant patients was reported [36]. Neutropenia was very common (76%), but was easily reversible on dose interruption or reduction. Similar results were reported with nilotinib [37]. Neutropenia grade 3/4 was noted in 9% at 400-mg twice-a-day dose and 22% at 600-mg twice-a-day dose.

Combinations of imatinib with chemotherapy have been reported in the setting of blastic transformation and in Philadelphia chromosome-positive acute lymphocytic leukemia (ALL), without any major increase in toxicity [38]. Studies are underway assessing the second-generation agents in combination with chemotherapy [39, 40].

5.3 Lapatinib

Lapatinib is an oral dual inhibitor of EGFR and HER2 tyrosine kinases. It is approved by the US Food and Drug Administration (FDA) for use in combination with capecitabine for the treatment of HER2-positive metastatic breast cancer [41]. Neutropenia was not a noted side effect in monotherapy [42]. Geyer et al. reported its use in combination with capecitabine [43]. Patients with HER2-positive, locally advanced or metastatic breast cancer that had progressed after treatment containing an anthracycline, a taxane, and trastuzumab were randomly assigned to single-agent capecitabine at 2,500 mg/m^2 on days 1–14 of a 21-day cycle, or capecitabine at 2,000 mg/m^2 with lapatinib 1,250 mg/day continuously. Time to progression was significantly reduced in the combination therapy group, with median time to progression being 8.4 months, compared with 4.4 months. Neutropenia and infections were not noted to be different in the two groups, neither were they significant.

Neutropenia was not noted to be a problem when lapatinib was assessed in combination with letrozole in patients with advanced cancer [44]. Lapatinib was studied in combination with trastuzumab in a phase 1 dose-escalation study without any significant neutropenia [45]. In combination with docetaxel, however, neutropenia was a major problem. A dose-escalation trial with this combination had to be amended to include pegfilgrastim to alleviate the high incidence of neutropenia [41]. The starting dose of lapatinib was 1,250 mg/day with docetaxel at 60 mg/m^2 with the plan to increase in cohorts of 3 to a maximum of 1,500 mg/day lapatinib and 100 mg/m^2 docetaxel. All three patients in the first cohort had grade 4 neutropenia and the next cohort had a reduced dose of docetaxel of 50 mg/m^2. One patient in this cohort had grade 4 neutropenia, and thus an amendment was made to include pegfilgrastim 6 mg 24 h after the docetaxel infusion. The first cohort had a 1,000-mg/day dose of lapatanib with 60 mg/m^2, and no neutropenia was observed. The next cohort had a docetaxel dose of 75 mg/m^2, and no problems were noted. The doses were able to be increased to an optimal treatment regimen of lapatinib 1,250 mg/day, with docetaxel 75 mg/m^2, with rash being the dose-limiting toxicity. The effect of lapatinib on the relationship between neutropenia and docetaxel concentration suggests that lapatinib increased sensitivity to this toxicity.

This finding is supported by a phase 3 trial comparing paclitaxel with and without lapatinib [46]. Patients were assigned to paclitaxel 175 mg/m^2 on day 1 every 3 weeks with either placebo daily or lapatinib 1,500 mg/day. The patients received therapy (paclitaxel up to 6 cycles) until disease progression. Febrile neutropenia occurred at a rate of 12% for paclitaxel alone, compared with 18% in the combination. Growth factors were not used in this trial.

5.4 *Erlotinib*

Erlotinib is an EGFR inhibitor used mainly in the treatment of NSCLC and pancreatic cancer. As monotherapy, it shows no major myelotoxicity when used in relapsed patients with NSCLC [47]. Herbst et al. reported on its combination with carboplatin and paclitaxel in a randomized trial in patients with advanced NSCLC without any significant difference in myelotoxicity between the two groups [48]. Erlotinib has been combined with gemcitabine and capecitabine doublet chemotherapy, as well as with bevacizumab in a dose-finding study [49]. The neutropenia levels were high, but tolerable (grade 3/4 neutropenia rate of 45%). Growth factors were not required as part of the protocol. Further evaluation is continuing in a phase 3 trial.

5.5 *Gefitinib*

Gefitinib is the first selective inhibitor of EGFR's tyrosine kinase domain. It has shown excellent results in NSCLC setting, including superior progression-free survival when compared to carboplatin with paclitaxel in first-line treatment. The HR was 0.48 (CI, 0.36–0.64, $p < 0.0001$) in patients with EGFR mutation-positive tumors and significantly worse in patients with EGFR mutation-negative tumors (HR, 2.85; CI, 2.05–3.98; $p < 0.0001$) [50]. Neutropenia was 3.7% in the gefitinib group compared to 67.1% in the chemotherapy group. This report was the first example of a targeted agent being superior to doublet chemotherapy in solid tumors. In combination with chemotherapy, gefitinib has not shown increased myelotoxicity, suggesting that growth factor support will be dependent on the combination chemotherapy [51].

5.6 Sunitinib

Sunitinib is a multitargeted receptor TKI that inhibits signaling of c-KIT, platelet-derived growth factor receptor (PDGFR), all three vascular growth factor receptors (VGFR), FLT3, and the ret proto-oncogene (*RET*). It has been used primarily in GIST cancers and renal cell carcinoma. It showed good activity in patients with imatinib-resistant GIST in a phase 3, placebo-controlled trial. Neutropenia was low, with grade 3/4 incidences of 10% vs. 4% [52].

A dose-escalation study was completed with sunitinib in combination with docetaxel in patients with advanced solid tumors [53]. Neutropenia was significant with grade 3/4 neutropenia occurring in 60% in both schedules, and grade 3/4 febrile neutropenia occurring in 30% in one group and 10% in the other. Growth factor support was used in 48% of the patients on the maximum tolerated dose cohort, either pegfilgrastim or filgrastim, particularly in cycles 2, 3, or 4.

5.7 Monoclonal Antibodies

Monoclonal antibodies have become integral in the fight against cancer. As antibodies are a key component of the adaptive immune response, in the role of both recognition of foreign antigens and the stimulation of an immune response to them, the advent of monoclonal antibodies has allowed raising antibodies against specific antigens presented on the surfaces of tumors [54].

The first reports in the literature of their use were in the early 1980s by Nadler et al. [55] and by Ritz et al. [56]. Ritz et al. reported that there was "Rapid and specific clearance of leukemia cells from the peripheral blood. Intravenous antibody also rapidly binds to bone marrow lymphoblasts, and in one instance, resulted in partial regression of tumor cell infiltrates in lymph nodes and skin." They noted, however, that clinically significant responses in general had not been achieved. The results, however, were proof of principle, and much work went into further development.

The first FDA-approved monoclonal antibody was the transplant rejection drug OKT3 (muromonab) directed against CD3 in 1986 [57], and 1994 saw the approval of abciximab, the potent GPIIb/IIIa inhibitor for use in acute coronary syndrome. These approvals were followed by the advent of the first monoclonal antibody approved for use in cancer, rituximab, directed against the pan-B-cell antigen, CD20 [58]. Trastuzumab was approved shortly after for treatment of breast cancer [59]. The growth of monoclonal antibodies became exponential in the 2000s. In 2007, eight of the best-selling biotechnology drugs in the USA were therapeutic monoclonal antibodies [60]. As each of the monoclonal antibodies was developed, they were tested as single agents before being added to chemotherapy regimens. While very few had myelotoxic effects individually (not surprising given their targeted nature), some trials suggested increased hematologic toxicity, some thrombocytopenia, and some neutropenia when given in combinations.

5.8 *Rituximab*

Rituximab binds to the cluster of differentiation 20 (CD20), which is widely expressed on B cells, from early pre-B cells to cells later in differentiation. The exact mechanism of action of rituximab remains unclear, but it induces apoptosis of $CD20^+$ cells by a combination of antibody-dependent cellular cytotoxicity (ADCC) and complement-dependent cytotoxicity (CDC) [61]. Rituximab has shown great efficacy in treating B-cell malignancies and is being actively investigated in any diseases caused by the production of antibodies. It can be used as monotherapy, but mostly it is used in combination with other well-established regimens. It seems to add little toxicity to the chemotherapy regimen with which it is combined, but it has been noted to cause late-onset neutropenia of unknown mechanism, usually occurring >2 months after last treatment [62]. This neutropenia tends to be self-limiting and not dangerous. The need for filgrastim in treatment regimens containing rituximab is generally based on the underlying regimen.

A recent publication by the German Chronic Lymphocytic Leukaemia Study Group examined the use of the fludarabine and cyclophosphamide, with or without rituximab, in a randomized trial [63]. Patients were randomly assigned to receive six courses of fludarabine (25 mg/m^2/day) and cyclophosphamide (250 mg/m^2/day) for the first 3 days of each 28-day cycle, with or without rituximab (375 mg/m^2 on day 0 of the first course and 500 mg/m^2 on day 1 of the second to sixth cycle). Prophylaxis with antiviral drugs or rHuG-CSF was not recommended. A total of 817 patients were treated, and at 3 years, 65% of patients in the chemoimmunotherapy group were free from progression, compared with 45% in the chemotherapy group (HR, 0.56; $p < 0.0001$). It was the first time that overall survival had been shown to be improved in a CLL trial, with rates of 87% versus 83% (HR, 0.67; CI 0.48–0.92; $p = 0.01$). The chemoimmunotherapy group was noted to have an increase in the incidence of grade 3/4 neutropenia (34% vs. 21%; $p < 0.0001$) and leukocytopenia (24% vs. 12%; $p < 0.0001$). Serious infections were not increased. Patients aged 65 years or older had more hematologic toxicity in the chemoimmunotherapy group, but with no increase in the total rate of infections. rHuG-CSF was administered in 86 treatment cycles, for a median of 7 days in the chemotherapy group and 6 days in the chemoimmunotherapy group, but far more frequently in the chemoimmunotherapy group (75 vs. 11). In 40 treatment cycles, rHuG-CSF was given as prophylaxis without any sign of neutropenia; in 46 cycles, it was administered to treat an adverse event (neutropenia or leukocytopenia).

The second-generation anti-CD20 antibodies ofatumumab and GA101 are currently undergoing clinical trials, and ofatumumab is licensed for CLL refractory to fludarabine and alemtuzumab. Neutropenia does not appear to be a major issue in monotherapy with ofatumumab [64].

5.9 Radioimmunotherapy (RIT)

The use of antibodies labeled with a radionuclide to deliver cytotoxic radiation to a target cell offers substantial advantages. The ability of the antibody to bind specifically to a tumor-associated antigen increases the dose delivered to the tumor cells while decreasing the dose to normal tissues. Ibritumomab tiuxetan and tositumomab I^{131} are the only FDA-approved radioimmunotherapy (RIT) agents. Ibritumomab is an anti-CD20 monoclonal antibody that is attached to a chelator tiuxetan to which a radioactive isotope (yttrium-90 or indium-111) is added [65]. The radioactive isotope releases beta radiation, which kills the attached cell and some nearby cells. Rituximab is given before each dose to pre-deplete B lymphocytes. It has shown improved response rates compared with rituximab and is approved treatment in relapsed or refractory, low-grade, or follicular B-cell NHL, and as consolidation therapy in patients with follicular NHL who achieve a partial or complete response to initial therapy [65].

Tositumomab is another anti-CD20 monoclonal antibody that is attached to the radionuclide iodine-131. Iodine-131 emits both beta and gamma radiation. Tositumomab has been studied in relapsed or chemotherapy/rituximab refractory follicular lymphoma [65]. Both agents have been shown to cause marked cytopenias, particularly neutropenia. Ibritumomab has been shown to cause a median nadir of 0.6×10^9/L; incidence per patient of neutrophil count $<1.0 \times 10^9$/L, 74%; incidence per patient of neutrophil count $<0.5 \times 10^9$/L, 35%; and median duration of neutrophil count $<1.0 \times 10^9$/L, 29 days. The median time to the neutrophil nadir was 62 days. Infections had an incidence of 29% in the first 3 months; 3% were serious and 2% life-threatening [66]. The recommendation is to avoid filgrastim for 2 weeks before and after treatment because of the theoretical potential for increased sensitivity of the dividing myeloid cells to radiation. Filgrastim has been used in treatment of infections while neutropenic after treatment with these agents.

5.10 Trastuzumab

Trastuzumab is directed against the HER2/neu receptor, important in cell growth, and is particularly effective in cancers that overexpress HER2. Between 20 and 30% of early stage breast cancers overexpress HER2 [67]. Trastuzumab was approved by the FDA in 1998 and has been used in varying methods and combinations. As monotherapy, neutropenia is not a recognized side effect. In metastatic disease, trastuzumab in combination with chemotherapy is superior to trastuzumab alone [68], making further investigations into possible combinations important. Trastuzumab was assessed in combination with docetaxel. Patients received 6 cycles of docetaxel 100 mg/m^2 with or without trastuzumab 4 mg/kg loading dose, followed by 2 mg/kg weekly until disease progression [69]. There was a higher incidence of grade 3/4 leucopenia and neutropenia (20% vs. 15%, and 32% vs. 22%, respectively) in the combination group, as was the incidence of febrile neutropenia (23% vs. 17%), suggestive of synergistic effect of the agents on marrow toxicity. Only four patients in the combination group and three in the docetaxel-only group received rHuG-CSF. Neutropenia is likely to be predominantly dependent on the myelotoxicity of the chemotherapy with which trastuzumab is combined, as evidenced by the trial investigating trastuzumab with docetaxel and vinorelbine in HER2-positive stage IV breast cancer [70]. Filgrastim was required from days 2 to 21 of each 21-day cycle. Grade 4 neutropenia was the most common grade 4 event.

5.11 *Cetuximab*

Cetuximab is a chimeric monoclonal antibody directed against EGFR. It was approved by the FDA in 2004 and is used in the treatment of patients with colorectal cancer and head-and-neck cancer. Cetuximab has been combined with chemotherapy in phase 3 trials and does not seem to enhance myelotoxicity. In a phase 3 trial of cetuximab plus irinotecan after fluorouracil and oxaliplatin failure, patients received irinotecan (350 mg/m^2) every 3 weeks alone or with cetuximab (400 mg/m^2 day 1, followed by 250 mg/m^2 weekly) [71]. No difference was seen in the rates of neutropenia. FOLFIRI (irinotecan, fluorouracil, and leucovorin) has been used with or without cetuximab in the initial treatment of patients with metastatic colorectal cancer [72]. The incidence of neutropenia in this study was no different in the two groups (28.2% vs. 24.6%). These trials suggest that cetuximab does not add to myelotoxicity, at least in the regimens tested.

5.12 *Bevacizumab*

Bevacizumab is a humanized monoclonal antibody directed against vascular endothelial growth factor A (VEGF-A). VEGF-A stimulates angiogenesis, particularly in cancer and in retinal proliferation in the diabetic eye. Bevacizumab was approved by the FDA in 2004 for use in combination with standard chemotherapy for the treatment of patients with metastatic colorectal cancer and NSCLC. Bevacizumab does not show myelotoxicity as monotherapy, but neutropenia can be a concern in combination therapy. A number of clinical trials have shown increased incidence of neutropenia and febrile neutropenia with the addition of bevacizumab. Miles et al. reporting on the AVADO trial in locally recurrent or metastatic breast cancer noted increased incidence of febrile neutropenia (grade $\geq$3) [73]. The patients were randomly assigned to docetaxel (100 mg/m^2) alone or with bevacizumab (either 7.5 mg/kg or 15 mg/kg). The incidence of febrile neutropenia in the docetaxel-alone group was 12% compared with 15.2% and 16.6% in the other groups. Cohen et al. reported increased neutropenia and febrile neutropenia in first-line treatment of advanced/metastatic recurrent NSCLC [74]. Patients received carboplatin (AUC 6) and paclitaxel (200 mg/m^2), with or without bevacizumab (15 mg/kg). Grade 4 neutropenia occurred in 17% compared with 26% for the combination group, and febrile neutropenia incidence was 2% vs. 5%. In the RIBBON-1 trial, Robert et al. reported on the use of bevacizumab in combination with 1 of 3 chemotherapy regimens: capecitabine, a taxane, or an anthracycline in first-line treatment of HER2-negative locally recurrent or metastatic breast cancer [75]. The febrile neutropenia rates for capecitabine with or without bevacizumab were both 0%. For the taxane group, the group containing taxane plus bevacizumab, it was 3.8% vs. 5.0% for the taxane-only group, suggesting no additional risk with the addition of the bevacizumab. However, the anthracycline-based therapy plus bevacizumab had a febrile neutropenia rate of 7.9% vs. 2% for the chemotherapy alone, suggesting some synergistic effect with regard to myelotoxicity.

Hecht et al. reported in 2010 on the use of adjuvant pegfilgrastim to reduce the incidence of neutropenia and febrile neutropenia in patients with colorectal cancer [76]. They noted that adding irinotecan and/or oxaliplatin to every-2-week 5-FU/leucovorin (LV) prolongs survival in patients with colorectal cancer, but increases neutropenia. Patients were randomly assigned to pegfilgrastim 6 mg or placebo per cycle on day 4, in conjunction with either every-2-week FOLFOX (5 FU/LV/oxaliplatin), FOLFIRI (5-FU/LV/irinotecan), or FOIL (5-FU/LV/oxaliplatin/irinotecan). The grade 3/4 neutropenia odds ratio was 0.19 (CI, 0.10–0.37); grade 3/4 febrile neutropenia was 2% vs. 8% ($p = 0.04$). In long-term follow-up, both groups had similar progression-free and overall survivals. On the basis of this, the Pegfilgrastim Anti-VEGF Evaluation Study (PAVES) trial has been started [77]. This study is a phase 3, randomized, double-blind, placebo-controlled multicenter study evaluating the efficacy of pegfilgrastim to reduce the incidence of febrile neutropenia in patients with newly diagnosed, locally advanced, or metastatic colorectal cancer who are receiving first-line treatment with bevacizumab and either 5-fluorouracil, oxaliplatin, leucovorin (FOLFOX); or 5-fluoruracil, irinotecan, leucovorin (FOLFIRI). It will also investigate the effect on overall survival, progression-free survival, and overall response rate in each group.

5.13 *Thalidomide and Lenalidomide*

The treatment of myeloma has been revolutionized in the last 5–10 years. Previous treatment consisted of an alkylator with a steroid (e.g., melphalan and prednisolone) for elderly patients, and induction chemotherapy with autologous PBSC transplantation for those young/fit enough to tolerate the procedure. The choice of induction regimen did not seem to matter as the major benefit appeared to come from the transplant. Newer agents have now appeared (thalidomide, bortezomib, and lenalidomide) and shown improvements in patient outcomes. Trials to establish the best combinations and order in which to treat patients are ongoing. However, at this stage, myeloma is still considered an incurable disease. No randomized trials have examined the use of rHuG-CSF as primary or secondary prophylaxis in treatment of myeloma, except in the post-autograft setting. Information regarding the risk of febrile neutropenia needs to be derived from the original trials incorporating these new agents. The exact mechanism by which thalidomide and lenalidomide achieve their antitumor activity remains unclear. They are antiangiogenic, but most of their effect seems to be immunomodulatory [78].

5.14 Thalidomide, Melphalan, and Prednisolone

Two randomized trials comparing melphalan and prednisolone (MP) with melphalan, prednisolone, and thalidomide (MPT) [79, 80] did not use rHuG-CSF as part of the protocol. Neutropenia rates were no different in the Palumbo trial [79] (16% MPT vs. 17% MP) but with an increased infection rate (10% MPT vs. 2% MP). No information was provided about febrile neutropenia. In the Facon trial [80], neutropenia rates were higher (48% MPT vs. 26% MP) but infection rates similar, albeit higher than that in the Palumbo trial (13% MPT vs. 9% MP). No information was given regarding episodes of febrile neutropenia.

5.15 Lenalidomide

Lenalidomide's major toxicities are hematopoiesis and, in particular, neutropenia. In the registration trials for lenalidomide, lenalidomide plus dexamethasone (Len/Dex) was compared with dexamethaone only. In the trial by Weber et al. [81], in the Len-Dex group, there was 35% grade 3 neutropenia and 6.2% grade 4 neutropenia, but febrile neutropenia was only 2.8% grade 3 and 0.6% grade 4. Grade 3/4 infections were noted in 21.5% of patients. rHuG-CSF was given as first response to neutropenia if no other adverse events were noted, and dose reduction as a second-line response. A total of 60 patients (33.9%) were given rHuG-CSF; 28 of these 60 patients (46.7%) were given rHuG-CSF as first response. Of these 28 patients, 12 (42.9%) were able to maintain a 25-mg dose. In the simultaneous trial for Europe/Australasia [82], the incidence of grade 3 neutropenia was 25% and grade 4 neutropenia of 4.5%, but febrile neutropenia was only 2.8% for grade 3 and 0.6% for grade 4. Grade 3/4 infections were noted in 11.3% of the patients. rHuG-CSF was given to 38 (21.6%) patients; 23 (60.5%) of these patients received rHuG-CSF as first response to grade 3 or 4 neutropenia. Twelve of the 23 patients (52.2%) were able to maintain a 25-mg dose.

Mateos et al. reported, anecdotally, three cases using rHuG-CSF for secondary prophylaxis to enable continued dose-intense lenalidomide therapy [83]. The patients were able to continue their prescribed treatment, but the authors noted that this continuation is not evidence that the patients' outcomes were improved. Lenalidomide has been used in relapsed/refractory CLL with good responses and a reasonable toxicity profile [84]. Neutropenia was a frequent occurrence with 41% grade 3/4 events. The incidence of infections was low with pneumonia being the most common, complicating 3% of the courses. rHuG-CSF use was described as being according to common practice guidelines without any specifics given. Lenalidomide has been used as monotherapy in relapsed or refractory, aggressive B-cell lymphomas with good success [85, 86]. Wiernik et al. reported response rates of 35%, with 12% complete response/unconfirmed complete response. The incidence of neutropenia was noted to be 24.5% for grade 3 and 8.2% for grade 4. In an update from the same group in a larger international trial, Witzig et al. reported similar results with overall response rates of 35%, with complete responses of 13%. Grade 4 neutropenia was 17%. At this time, rHuG-CSF will continue to be used ad hoc with these agents.

5.16 Proteasome Inhibitors

Bortezomib is a first-in-class proteasome inhibitor licensed for use in patients with relapsed myeloma and mantle cell lymphoma, and is currently being investigated in other diseases [87]. The ubiquitin–proteasome signaling pathway mediates the degradation of proteins involved in the regulation of cell growth. This component is essential for cellular proliferation and survival. The proteasome activates nuclear factor-κB signaling, resulting in tumor growth and metastasis. Bortezomib is a small molecule – synthetic, boronic acid dipeptide – which blocks this pathway. The major side effects noted with this agent are peripheral neuropathy and myelosuppression, particularly thrombocytopenia [88]. In a phase 2 study in relapsed, refractory myeloma, grade 3/4 neutropenia occurred in 14% patients with <1% febrile neutropenia, and growth factor support was not required [89].

Bortezomib plus melphalan and prednisolone (VMP) was compared to melphalan and prednisolone only (MP) [90]. Neutropenia grade 3/4 was seen in 30% and 10%, respectively, in VMP compared to 23% and 15%, respectively, in the MP group. Febrile neutropenia rates were not reported. rHuG-CSF was not used in the protocol, only dose reduction of melphalan or bortezomib. Bortezomib has been used in combination with chemotherapy in patients with lymphoma without significant additional myelotoxicity, but in these are regimens that are highly myelotoxic and requiring growth factor support [91].

5.17 *mTOR Inhibitors*

Another novel class of agents in cancer therapy is the mammalian target of rapamycin (mTOR) inhibitors. They have long been used as immunosuppressants in organ transplantation. The mTOR pathway is aberrantly activated in many malignancies. Everolimus and temsirolimus are mTOR inhibitors that have been investigated in a number of malignancies. Neutropenia has been noted, but thrombocytopenia has been the most significant side effect [87, 92].

5.18 *PARP Inhibitors*

Poly (adenosine diphosphate-ribose) polymerase-1 (PARP-1) is an important regulator of the DNA base excision-repair pathway. Several forms of cancer are more dependent on PARP than regular cells, making it an attractive target for chemotherapeutic cancer therapy. These include tumors with BRCA1 or BRCA2 mutations, often triple-negative breast cancers [93].

Iniparib is the first PARP inhibitor to be tested in phase 3 trials, in both breast and squamous cell lung cancer. In the phase 2 trial in the setting of triple-negative metastatic breast cancer, it was added to gemcitabine 1,000 mg/m^2 and carboplatin at a dose equivalent to AUC of 2 on days 1 and 8 every 21 days. Iniparib was added at a dose of 4.0 mg/m^2 (and later amended to 5.6 mg/m^2 after more phase 1 data) on days 1, 4, 8, and 11 on a 21-day cycle [94]. Good improvement was noted in progression-free survival and overall survival data, with no apparent difference in toxicity. Thus, to date, neutropenia does not appear to be a major toxicity for this class of drugs.

References

1. Crawford J, Ozer H, Stoller R et al (1991) Reduction by granulocyte colony-stimulating factor of fever and neutropenia induced by chemotherapy in patients with small-cell lung cancer. N Engl J Med 325:164–70
2. Trillet-Lenoir V, Green J, Manegold C et al (1993) Recombinant granulocyte colony stimulating factor reduces the infectious complications of cytotoxic chemotherapy. Eur J Cancer 29A:319–24
3. Renwick W, Pettengall R, Green M (2009) The use of filgrastim for supporting chemotherapy delivery: two decades of experience. Biodrugs 23:175–186
4. Citron M, Berry D, Cirrincione C et al (2003) Randomized trial of dose-dense versus conventially scheduled and sequential versus concurrent combination chemotherapy as post-operative adjuvant treatment of node-positive primary breast cancer: first report of Intergroup Trial C9741/Cancer and Leukemia Group B Trial 9741. J Clin Oncol 21:1431–1439
5. Druker B, Tamara S, Buchdunger E et al (1996) Effects of a selective inhibitor of the Abl tyrosine kinase on the growth of Bcr-Abl positive cells. Nat Med 2:561–566

6. Treat J, Gonin R, Socinski M et al (2010) A randomised, phase III multicenter trial of gemcitabine in combination with carboplatin or paclitaxel versus paclitaxel plus carboplatin in patients with advanced non-small-cell lung cancer. Ann Oncol 21:540–547
7. Smith T, Khatcheressian J, Lyman G et al (2006) 2006 update of recommendations for the use of white blood cell growth factors: an evidenced-based clinical practice guideline. J Clin Oncol 24:3187–3205
8. Swain S, Jeong J, Geyer C et al (2010) Longer therapy, iatrogenic amenorrhea, and survival in early breast cancer. N Engl J Med 362:2053–2065
9. Drullinsky P, Sugarman S, Fornier M et al (2010) Dose dense cyclophosphamide, methotrexate, flurouracil is feasible at 14-day intervals: a pilot study of every-14-day dosing as adjuvant therapy for breast cancer. Clin Breast Cancer 10:440–444
10. Yardley D, Zubkus J, Inhorn R et al (2010) A phase II trial of dose-dense neoadjuvant gemcitabine, epirubicin, and albumin-bound paclitaxel with pegfilgrastim in the treatment of patients with locally advanced breast cancer. Clin Breast Cancer 10:367–372
11. Dragnev K, Hardin S, Pipas J et al (2010) A dose escalation trial of biweekly docetaxel and gemcitabine with filgrastim or pegfilgrastim for the treatment of patients with advanced solid tumors. Chemotherapy 56:135–141
12. Muller V, Thomssen C, Schmidt M et al (2010) Final results of a phase I/II pilot study of capecitabine with or without vinorelbine after sequential dose-dense epirubicin and paclitaxel in high-risk early breast cancer. BMC Cancer 10:430–437
13. Jones R, Walsh G, Ashley S et al (2009) A randomised pilot phase II study of doxorubicin and cyclophosphamide (AC) or epirubicin and cyclophosphamide (EC) given 2 weekly with pegfilgrastim (accelerated) vs 3 weekly (standard) for women with early breast cancer. Br J Cancer 100:305–310
14. Ozer H, Armitage J, Bennett C et al (2000) American Society of Clinical Oncology 2000 update of recommendations for the use if hematopoietic colony-stimulating factors: evidence-based, clinical practice guidelines. J Clin Oncol 18:3558–3585
15. Pfreundschuh M, Trumper L, Kloess M et al (2004) Two-weekly or 3-weekly CHOP chemotherapy with or without etoposide for the treatment of young patients with good-prognosis (normal LDH) aggressive lymphomas: results of the NHL-B1 trial of the DSHNHL. Blood 104:626–633
16. Pfreundschuh M, Trumper L, Kloess M et al (2004) Two-weekly or 3-weekly CHOP chemotherapy with or without etoposide for the treatment of elderly patients with aggressive lymphomas: results of the NHL-B2 trial of the DSHNHL. Blood 104:634–641
17. Pfreundschuh M, Schubert J, Ziepert M et al (2008) Six versus eight cycles of bi-weekly CHOP-14 with or without rituximab in elderly patients with aggressive CD20+ B-cell lymphomas: a randomised controlled trial (RICOVER-60). Lancet Oncol 9:105–16
18. Sieber M, Bredenfeld H, Josting A et al (2003) 14-day variant of the bleomycin, etoposide, doxorubicin, cyclophosphamide, vincristine, procarbazine, and prednisone regimen in advanced-staged Hodgkin's lymphoma: results of a pilot study of the German Hodgkin's Lymphoma Study Group. J Clin Oncol 21:1734–1739
19. Diehl V, Franklin J, Pfreundschuh M et al (2003) Standard and increased-dose BEACOPP chemotherapy compared with COPP-ABVD for advanced Hodgkin's disease. N Engl J Med 348:2386–2395
20. Khouri I, Romaguera J, Kantarjian H (1998) Hyper-CVAD and high-dose methotrexate/cytarabine followed by stem-cell transplantation: an active regimen for aggressive mantle-cell lymphoma. J Clin Oncol 16:3803–3809
21. Moskowitz C, Bertino J, Glassman J (1999) Ifosfamide, carboplatin, and etoposide: a highly effective cytoreduction and peripheral-blood progenitor-cell mobilization regimen for transplant-eligible patients with non-Hodgkin's lymphoma. J Clin Oncol 17:3776–3785
22. Velasquez W, McLaughlin P, Tucker S et al (1994) ESHAP – an effective chemotherapy regimen in refractory and relapsing lymphoma: a 4-year follow-up study. J Clin Oncol 12:1169–1176

23. Pasricha S, Grigg A, Catalano J et al (2008) A multicenter phase 2 study of risk-adjusted salvage chemotherapy incorporating vinorelbine and gemcitabine for relapsed and refractory lymphoma. Cancer 113:3192–3198
24. Santoro A, Magagnoli M, Spina M et al (2007) Ifosfamide, gemcitabine, and vinorelbine: a new induction regimen for refractory and relapsed Hodgkin's lymphoma. Haematologica 92:35–41
25. Wilson W, Grossbard M, Pittaluga S et al (2002) Dose-adjusted EPOCH chemotherapy for untreated large B-cell lymphomas: a pharmacodynamic approach with high efficacy. Blood 99:2685–2693
26. Dunleavy K, Healey Bird B, Pittaluga S et al (2007) Efficacy and toxicity of dose-adjusted EPOCH-Rituximab in adults with newly diagnosed Burkitt lymphoma. J Clin Oncol 25(18 Suppl):8035
27. Mead G, Sydes M, Walewski J et al (2002) An international evaluation of CODOX-M and CODOX-M alternating with IVAC in adult Burkitt's lymphoma: results of United Kingdom Lymphoma Group LY06 study. Ann Oncol 13:1264–1274
28. De Bono J, Oudard S, Ozguroglu M et al (2010) Prednisone plus cabazitaxel or mitoxantrone for metastatic castration-resistant prostate cancer progressing after docetaxel treatment: a randomised open-label trial. Lancet 376:1147–1154
29. Pivot X, Koralewski P, Hidalgo J et al (2008) A multicenter phase II study of XRP6258 administered as a 1-hr IV infusion every 3 weeks in taxane-resistant metastatic breast cancer patients. Ann Oncol 19:1547–1552
30. Egerton N (2010) Optimizing ixabepilne treatment schedules in patients with advance or metastatic cancer. Cancer Chemother Pharmacol 66:1005–1012
31. Thomas E, Gomez H, Li R et al (2007) Ixabepilone plus capecitabine for metastatic breast cancer progressing after anthracycline and taxane treatment. J Clin Oncol 25:5210–5217
32. Kahl B, Bartlett N, Leonard J et al (2010) Bendamustine is effective therapy in patients with rituximab-refractory, indolent B-cell non-Hodgkin lymphoma. Cancer 116:106–114
33. Rummel M, Niederle N, Maschmeyer G, et al (2009) Bendamustine plus rituximab is superior in respect of progression free survival and CR rate when compared to CHOP plus Rituximab as first-line treatment of patients with advanced follicular, indolent and mantle cell lymphomas: final results of a randomised phase III study of the StiL (Study Group Indolent Lymphomas, Germany). Blood (ASH Annual Meeting Abstracts) Nov 2009; 114:405
34. Yaish P, Gazit A, Gilon C et al (1988) Blocking of EGF-dependent cell proliferation by EGF receptor kinase inhibitors. Science 242:933–935
35. Kantarjian H, Sawyers C, Hochhaus A et al (2002) Hematologic and cytogenetic responses to imatinib mesylate in chronic myelogenous leukemia. N Engl J Med 346:645–652
36. Guilhot F, Apperley J, Kim D et al (2007) Dastatinib induces significant hematologic and cytogenetic responses in patients with imatinib-resistant or -intolerant chronic myeloid leukemia in chronic phase. Blood 109:4143–4150
37. Kantarjian H, Giles F, Wunderle L et al (2006) Nilotinib in imatinib-resistant CML and Philadelphia chromosome-positive ALL. N Engl J Med 354:2542–2551
38. de Labarthe A, Rousselot P, Huguet-Rigal F et al (2007) Imatinib combined with induction or consolidation chemotherapy in patients with de novo Philadelphia chromosome-positive acute lymphoblastic leukemia: results of the GRAAPH-2003 study. Blood 109(4):1408–1413
39. First-line Dasatinib plus Conventional Chemotherapy in adults with newly diagnosed Ph-positive ALL (2009). http://clinicaltrials.gov/show/NCT01004497. Last accessed 18th September 2011
40. Nilotinib and combination chemotherapy in treating patients with newly diagnosed acute lymphoblastic leukemia (2009). http://clinicaltrials.gov/show/NCT00844298. Last accessed 18th September 2011
41. LoRusso P, Jones S, Koch K et al (2008) Phase I and pharmacokinetic study of lapatinib and docetaxel in patients with advanced cancer. J Clin Oncol 26:3051–3056
42. Paul B, Trovato J, Thompson J (2008) Lapatinib: a dual tyrosine kinase inhibitor for metastatic breast cancer. Am J Health Syst Pharm 65(18):1703–10

43. Geyer C, Forster J, Lindquist D et al (2006) Lapatinib plus capecitabine for HER2-positive advanced breast cancer. N Engl J Med 355:2733–2743
44. Chu Q, Cianfrocca M, Goldstein L et al (2008) A phase I and pharmacokinetic study of lapatinib in combination with letrozole in patients with advanced cancer. Clin Cancer Res 14:4484–4490
45. Stornido A, Pegram M, Overmoyer B et al (2008) Phase I dose escalation and pharmacokinetic study of lapatinib in combination with trastuzumab in patients with advanced ErbB2-positive breast cancer. J Clin Oncol 26:3317–3323
46. Di Leo A, Gomex H, Aziz Z et al (2008) Phase III, double-blind, randomized study comparing lapatinib plus paclitaxel with placebo plus paclitaxel as first-line treatment for metastatic breast cancer. J Clin Oncol 26:5544–5552
47. Shepherd F, Pereira J, Ciuleanu T et al (2005) Erlotinib in previously treated non-small-cell lung cancer. N Engl J Med 353:123–132
48. Herbst R, Prager D, Hermann R et al (2005) TRIBUTE: a phase III trial of erlotinib hydrochloride (OSI-774) combined with carboplatin and paclitaxel chemotherapy in advanced non-small-cell lung cancer. J Clin Oncol 23:5892–5899
49. Starling N, Watkins D, Cunningham D et al (2009) Dose finding and early efficacy study of gemcitabine plus capecitabine in combination with bevacizumab plus erlotinib in advanced pancreatic cancer. J Clin Oncol 27:5499–5405
50. Mok T, Wu Y, Thongprasert S et al (2009) Gefitinib or carboplatin-paclitaxel in pulmonary adenocarcinoma. N Engl J Med 361:947–957
51. Herbst R, Giaccone G, Schiller J et al (2004) Gefitinib in combination with paclitaxel and carboplatin in advanced non-small-cell lung cancer: a phase III trial. INTACT 2. J Clin Oncol 22:785–794
52. Demetri G, Oosterom A, Garrett C et al (2006) Efficacy and safety of sunitinib in patients with advanced gastrointestinal stromal tumour after failure of imatinib: a randomised controlled trial. Lancet 368:1329–1338
53. Robert F, Sandler A, Schiller J et al (2010) Sunitinib in combination with docetaxel in patients with advance solid tumors: a phase I dose-escalation study. Cancer Chemother Pharmacol 66:669–680
54. Janeway CA Jr et al (2005) Immunobiology, 6th edn. Garland Science, New York
55. Nadler L, Stashenko P, Hardy R et al (1980) Serotherapy of a patient with a monoclonal antibody directed against a human lymphoma-associated antigen. Cancer Res 40:31473154
56. Ritz J, Schlossman S (1982) Utilization of monoclonal antibodies in the treatment of leukaemia and lymphoma. Blood 59:1–11
57. Hooks M, Wade C, Millikan W (1991) Murumonab CD-3: a review of its pharmacology, pharmacokinetics, and clinical use in transplantation. Pharmacotherapy 11:26–37
58. Maloney D, Grillo-Lopez A, White C et al (1997) IDEC-C2B8 (Rituximab) anti-CD20 monoclonal antibody therapy in patients with relapsed low-grade non-Hodgkin's Lymphoma. Blood 90:2188–2195
59. Hudis C (2007) Trastuzumab – mechanism of action and use in clinical practice. N Engl J Med 357:39–51
60. Scolnik P (2009) mAbs: a business perspective. MAbs 1:179–184
61. Shaw T, Quan J, Totoritis M (2003) B cell therapy for rheumatoid arthritis: the rituximab (anti-CD20) experience. Ann Rheum Dis 62(Suppl 2):55–59
62. Voog E, Morschhauser F, Solal-Celigny P (2003) Neutropenia in patients treated with rituximab. N Engl J Med 348:2691–2694
63. Hallek M, Fischer K, Fingerle-Rowson G et al (2010) Addition of rituximab to fludarabine and cyclophosphamide in patients with chronic lymphocytic leukaemia: a randomised, open-label, phase 3 trial. Lancet 376:1164–1174
64. Coiffier B, Lepetre S, Pedersen L et al (2008) Safety and efficacy of ofatumumab, a fully human monoclonal anti-CD20 antibody, in patients with relapsed or refractory chronic lymphocytic leukaemia: a phase 1–2 study. Blood 111:1094–1100

65. Morschhauser F, Dreyling M, Rohatiner A et al (2009) Rationale for consolidation to improve progression-free survival in patients with non-Hodgkin's lymphoma: a review of the evidence. Oncologist 14(Suppl 2):17–29
66. Ibritumomab tiuxetan (Zevalin®) Product information. Spectrum Pharmaceuticals
67. Piccart-Gebhart M, Procter M, Leyland-Jones B et al (2005) Trastuzumab after adjuvant chemotherapy in HER2-positive breast cancer. N Engl J Med 353:1659–1672
68. Vogel C, Tan-Chui E (2005) Trastuzumab plus chemotherapy: convincing survival benefit or not? J Clin Oncol 23:4247–4250
69. Marty M, Cognetti D, Maraninchi D et al (2005) Efficacy and safety of trastuzumab combined with docetaxel in patient with human epidermal growth factor receptor 2-positive metastatic breast cancer given as first-line treatment: results of a randomized phase II trial by the M7701 Study Group. J Clin Oncol 23:4265–4274
70. Kash J, Barlow W, Albain K, et al (2008) Phase II Southwest Oncology Group study of docetaxel and vinorelbine plus filgrastim with weekly trastuzumab for HER2-positive, stage IV breast cancer. J Clin Oncol 26(suppl) ASCO Meeting Abstracts Aug 18 2008:1033
71. Sobrero A, Maurel J, Fehrenbacher L et al (2008) EPIC: phase III trial of cetuximab plus irinotecan after fluoropyrimidine and oxaliplatin failure in patients with metastatic colorectal cancer. J Clin Oncol 6:2311–2319
72. Cutsern E, Kohne C, Hitre E et al (2009) Cetuximab and chemotherapy as initial treatment for metastatic colorectal cancer. N Engl J Med 360:1408–1417
73. Miles D, Chan A, Dirix L et al (2010) Phase III Study of bevacizumab plus docetaxel compared with placebo plus docetaxel for the first-line treatment of human epidermal growth factor receptor 2-negative metastatic breast cancer. J Clin Oncol 28:3239–3947
74. Cohen M, Gootenberg J, Keegan P et al (2007) FDA drug approval summary: bevacizumab (Avastin®) plus carboplatin and paclitaxel as first-line treatment of advanced/metastatic recurrent nonsquamous non-small cell lung cancer. Oncologist 12:713–718
75. Robert N, Dieras V, Glaspy J et al (2011) RIBBON-1: randomized, double-blind, placebo-controlled, phase III trial of chemotherapy with or without bevacizumab for first-line treatment of human epidermal growth factor receptor 2-negative, locally recurrent or metastatic breast cancer. J Clin Oncol 29:1252–1260
76. Hecht J, Pillai M, Gollard R et al (2010) A randomised, placebo-controlled phase II study evaluating the reduction of neutropenia and febrile neutropenia in patients with colorectal cancer receiving pegfilgrastim with every-2-week chemotherapy. Clin Colorectal Cancer 2:95–101
77. PAVES: pegfilgrastim anti-VEGF evaluation study. http://clinicaltrials.gov/ct2/show/NCT00911170 2009 Last accessed 18th September 2011
78. Anderson K (2005) Lenalidomide and thalidomide: mechanisms of action-similarities and differences. Semin Hematol 42(Suppl 4):S3–8
79. Palumbo A, Bringhen S, Caravita T et al (2006) Oral melphalan and prednisone chemotherapy plus thalidomide compared with melphalan and prednisone alone in elderly patients with myeloma: randomised controlled trial. Lancet 367:825–831
80. Facon T, Mary J, Hulin C et al (2007) Melphalan and prednisone plus thalidomide versus melphalan and prednisone alone or reduced-intensity autologous stem cell transplantation in elderly patients with multiple myeloma (IFM 99–06): a randomised trial. Lancet 370:1209–1218
81. Weber D, Chen C, Niesvizky R et al (2007) Lenalidomide plus dexamethasone for relapsed multiple myeloma in North America. N Engl J Med 357:2133–2142
82. Dimopoulos M, Spencer A, Attal M et al (2007) Lenalidomide plus dexamethasone for relapsed or refractory multiple myeloma. N Engl J Med 357:2123–2132
83. Mateos M, Garcia-Sanz R, Colado E et al (2008) Should prophylactic granulocyte-colony stimulating factor be used in multiple myeloma patients developing neutropenia under lenalidomide-based therapy? Br J Haematol 140:324–26

84. Ferrajoli A, Lee B, Schlette E et al (2008) Lenalidomide induces complete and partial remissions in patients with relapsed and refractory chronic lymphocytic leukemia. Blood 111:5291–97
85. Wiernik P, Lossos I, Tuscano J et al (2008) Lenalidomide monotherapy in relapsed or refractory aggressive non-Hodgkin's lymphoma. J Clin Oncol 26:4952–4957
86. Witzig T, Vose J, Zinani L et al (2011) An international phase II trial of single-agent lenalidomide for relapsed or refractory aggressive B-cell non-Hodgkin's lymphoma. Ann Oncol 22(7):1622–1627
87. Johnston P, Yuan R, Cavalli F et al (2010) Targeted therapy in lymphoma. J Hematol Oncol 3:45–54
88. San Miguel J, Blade J, Boccadoro M et al (2006) A practical update on the use of bortezomib in the management of multiple myeloma. Oncologist 11:51–61
89. Richardson P, Barlogie B, Berensen J et al (2003) A phase 2 study of bortezomib in relapsed, refractory myeloma. N Engl J Med 348:2609–2617
90. San Miguel J, Schlag R, Khuageva N et al (2008) Bortezomib plus melphalan and prednisone for initial treatment of multiple myeloma. N Engl J Med 359:906–917
91. Weigert O, Weidmann E, Mueck R et al (2009) A novel regimen combining high dose cytarabine and bortezomib has activity in multiply relapsed and refractory mantle cell lymphoma – long-term results of a multicenter observation study. Leuk Lymphoma 50:716–22
92. Andre F, Campone M, O'Regan R et al (2010) Phase I study of Everolimus plus weekly paclitaxel and trastuzumab in patients with metastatic breast cancer pretreated with trastuzumab. J Clin Oncol 28:2110–2115
93. Farmer H, McCabe N, Lord CJ et al (2005) Targeting the DNA repair defect in BRCA mutant cells as a therapeutic strategy. Nature 434:917–921
94. O'Shaughnessy J, Osborne C, Pippen J et al (2011) Iniparib plus chemotherapy in metastatic triple-negative breast cancer. N Engl J Med 364:205–214

rHuG-CSF in Peripheral Blood Progenitor Cell Transplantation

Ashanka Beligaswatte, Ian Lewis, and Luen Bik To

1 Introduction: An Historical Note

The initial indications for the therapeutic application for recombinant human G-CSF (rHuG-CSF) were the treatment of chronic neutropenias, and neutropenia after chemotherapy. Its efficacy in peripheral blood stem cell (PBSC) and peripheral blood progenitor cell (PBPC) mobilization was quite unexpected. The Haematopoiesis Research team headed by Professor Don Metcalf at the Walter and Eliza Hall Institute of Medical Research, Melbourne, Australia had a proactive role, in association with Amgen in Thousand Oaks, California, in the early clinical development of rHuG-CSF. It was studies in Metcalf's laboratory that first raised the possibility that rHuG-CSF may lead to mobilization of stem and progenitor cells into the blood. At that time, George Morstyn, William Sheridan, and others had just completed a study showing that rHuG-CSF reduced the duration and severity of postautologous transplantation neutropenia, but not thrombocytopenia. Six hundred kilometers west of Melbourne, Christopher Juttner and L Bik To at the Institute of Medical and Veterinary Science/Royal Adelaide Hospital had demonstrated mobilization by myelosuppressive chemotherapy, the feasibility of transplantation using mobilized blood cells [1–3], and a cell dose effect on hematopoietic recovery [4].

The transplant community was initially skeptical about whether mobilized blood cells have a role in transplantation in spite of the initial report by Richman et al. [5]. This skepticism was influenced by the poor engraftment seen in murine and canine studies, and from two case reports in humans (reviewed in [6]). Two definitive reports in 1992 overcame this skepticism. The Melbourne and Adelaide groups

A. Beligaswatte
Royal Adelaide Hospital and Institute of Medical and Veterinary Science, Adelaide, Australia

I. Lewis • L.B. To (✉)
Royal Adelaide Hospital, Institute of Medical and Veterinary Sciences, and University of Adelaide, Frome Rd, Adelaide, SA 5000, Australia
e-mail: Bik.To@health.sa.gov.au

G. Molineux et al. (eds.), *Twenty Years of G-CSF*, Milestones in Drug Therapy,
DOI 10.1007/978-3-0348-0218-5_14, © Springer Basel AG 2012

joined forces to conduct the first rHuG-CSF mobilization and autologous transplantation study that showed a reduction in posttransplant thrombocytopenia [7]. Separately, the Adelaide team reported on the rapid hematopoietic recovery and reduced resource utilization and hospitalization associated with mobilized peripheral blood cell transplantation (PBCT) [8]. Subsequent reports from the Melbourne Group were among the first cluster of studies using rHuG-CSF-mobilized blood cells for allogeneic transplantation [9]. The two groups collaborated again in defining the role of recombinant human stem cell factor (rHuSCF) in mobilization [10–13].

2 The Mechanisms of Stem Cell Mobilization

G-CSF is a cytokine that is classically associated with myeloid differentiation. Its role, if any, in the physiological trafficking of hematopoietic stem cells is unclear. At therapeutic doses, however, it enforces stem cell release from the marrow for clinically useful periods of time. Unraveling the mechanisms by which G-CSF mobilizes hematopoietic stem and progenitor cells is an area of active research. Initially, it was believed that G-CSF directly expanded the pool of hematopoietic stem cells [14]; however, it is now clear that this is not true, since chimeric mice carrying both G-CSF receptor (G-CSFR)-negative and -positive progenitor cells are able to mobilize both cell types in response to rHuG-CSF. Moreover, human $CD34^+$ cells expressing G-CSFR produce only myeloid colonies in clonal cultures whereas $CD34^+$ $G\text{-}CSFR^-$ cells give rise to multilineage colonies [15]. Research into the pathways of rHuG-CSF-induced mobilization has, therefore, also explored how rHuG-CSF alters the microenvironment of the stem cell niche, leading to loss of adhesion and transendothelial migration of progenitor cells.

2.1 Summary of Potential Mechanisms of Stem Cell Mobilization by rHuG-CSF

Table 1 summarizes the potential mechanism of stem cell mobilization by rHuG-CSF. Initial investigations of the mechanism of mobilization focused on the role of integrins in the retention of progenitor cells in the marrow. The most significant of these was the $\alpha_4\beta_1$ integrin, very late antigen-4 (VLA-4), which is involved in constitutive adhesion of $CD34^+$ cells to the extracellular matrix through vascular cell adhesion molecule-1 (VCAM-1) and fibronectin. This interaction is augmented by β_2 and $\alpha_5\beta_1$ integrins [16], and negatively modulated by platelet/endothelial cell adhesion molecule-1 (PECAM-1) [17]. Mobilized $CD34^+$ cells also show reduced expression of VLA-4, ICAM-1, LFA-1, and LFA-3, and significantly higher levels of L-selectin compared with bone marrow cells [18]. Although G-CSF is able to disrupt integrin-mediated retention of marrow hematopoietic progenitors, evidence

Table 1 Summary of potential mechanisms of stem cell mobilization by rHuG-CSF

1. Increased myeloid differentiation and proliferation within the marrow
• Proteolytic marrow microenvironment due to increases in neutrophil elastase, cathepsins, matrix metalloproteinases, complement cascade
• Marrow hypoxia, leading to stabilization of HIF-1α, and increased secretion of VEGF
2. Modulation of CXCR4/SDF-1 interaction via
• CD26
• Gfi-1
• Reduced SDF-1 production by osteoblasts
• Cleavage of CXCR4
• Sympathetic innervation
3. Disruption of integrin VLA-4/VCAM-1 binding
• Proteolytic marrow environment
• Via SCF/c-kit
4. Activation of macrophages leading to inhibition of osteoblasts involved in stem cell retention
5. Reduced expression of stem cell maintenance genes by mesenchymal stromal cells
6. Migration
• Movement of progenitor cells closer to vessels
• Increased transendothelial migration facilitated by VEGF
• Activation of stem cell motility by c-kit
• Activation of complement, neutrophil egress, and release of S1P

from several lines of work indicates that this is not a direct effect. Mobilization resulted in reduced amounts of VCAM-1 due to cleavage of VCAM-1 by neutrophil elastase and cathepsin G [19]. Moreover, mobilization using anti-VLA-4 and anti-VCAM-1 antibodies was shown to be possible in mice deficient in G-CSFR. However, functional *c-kit* was required [20], confirming other work suggesting an interaction between integrins and the *c-kit*/SCF pathway [13, 21].

Stromal-cell derived factor-1 (SDF-1), also known as CXCL12, is a chemokine secreted by reticulin cells and osteoblasts. It is a ligand for the 7-transmembrane G-protein-linked receptor CXCR4, and was the first chemoattractant reported for hematopoietic progenitor cells [22]. SDF-1-mediated signaling in $CD34^+$ cells occurs partly through atypical PKC-zeta which appears to be involved in proliferation and matrix metalloproteinase (MMP)-9 secretion [23]. Furthermore, suppressor of cytokine signaling (SOCS) family proteins are upregulated by SDF-1 through *JAK-STAT* activation. These proteins attenuate the mobilization effect of cytokines and contribute toward retention of progenitor cells. In mice whose marrows were reconstituted with a tetracycline-regulated, SOCS-expressing lentiviral vector, doxycycline treatment resulted in extensive mobilization [24].

Stem cell egress can be induced by altering the SDF-1 gradient. Increasing the amount of plasma SDF-1 in mice using an adenoviral vector resulted in stem cell mobilization [25]. It is likely that during mobilization, SDF-1 levels in the marrow are less than those in the peripheral blood. G-CSF decreases marrow SDF-1 mRNA synthesis in mice, with marrow SDF-1 levels correlating inversely with stem cell mobilization. Cell sorting experiments have shown that most marrow SDF-1 is synthesized by osteoblasts – cells that are not known to express G-CSFR.

Hence, this effect of G-CSF is possibly mediated indirectly [26]. The dipeptidyl peptidase CD26, which is able to cleave SDF-1, may play an important role here, since knockout mice deficient in CD26 were unable to mobilize in response to rHuG-CSF [27]; however, the CXCR4 antagonist plerixafor has been shown to mobilize progenitor cells in $CD26^{-/-}$ knockout mice, suggesting that CD26 is upstream of the CXCR4/SDF-1 axis but downstream of G-CSF in the events that lead to mobilization [28]. The CXCR4/SDF-1 axis is not the sole mediator of stem cell retention within the marrow as $CXCR4^{-/-}$ chimeras can mobilize in response to VLA-4 antagonists [29].

Apart from reducing amounts of SDF-1, rHuG-CSF also causes a decrease in marrow CXCR4 by several mechanisms, including signaling through transcriptional repressors, and indirectly by promoting a highly proteolytic microenvironment within the marrow. Expression of growth factor independence-1 (Gfi-1), a transcriptional repressor, is increased by rHuG-CSF, and in turn leads to a reduction in CXCR4 synthesis [30]. Knockout mice conditionally deficient for the related transcription factor Gfi-1b have an expanded pool of progenitor cells in their marrow that retain their ability to self-renew and initiate multilineage differentiation [31].

Generating a proteolytic microenvironment appears to be a common denominator in several pathways of stem cell mobilization. rHuG-CSF increases amounts of proteases, including neutrophil elastase and cathepsins, which are released by the accumulation of granulocyte precursors in the marrow. Elastase inhibition prevented mobilization in one study [32]. Mobilization has been shown to correlate with cleavage of the *N*-terminus of CXCR4 [33]. Additionally, downregulation of the serine protease inhibitors serpina1 and serpina3 has been reported after rHuG-CSF or cyclophosphamide [34]. MMP also contribute toward establishing this proteolytic environment. Peripheral blood $CD34^+$ cells were found to express MMP-2 and MMP-9 irrespective of mobilization. In contrast, bone marrow $CD34^+$ cells did not constitutively express these, but could be induced to do so by various mobilization regimens [35]. Membrane type-1 MMP (MT-1 MMP) was increased, and its inhibitor RECK was reduced in circulating $CD34^+$ cells compared with marrow cells. This difference was upregulated by rHuG-CSF, which also promoted the incorporation of MT-1 MMP into membrane lipid rafts, resulting in activation of pro-MMP-2 and facilitation of mobilization [36, 37]. However, many of these proteases do not appear to be essential to mobilization. rHuG-CSF was able to mobilize mice deficient in MMP-9, neutrophil elastase, cathepsin G, or dipeptidyl peptidase I, pointing to the presence of both protease-dependent and -independent pathways of mobilization [38].

Considerable attention has been paid to the role of bone in stem cell retention. Stimulating osteoclasts with different stressors mobilizes progenitor cells. RANK ligand resulted in reduction of SDF-1, SCF, and osteopontin along the endosteum, leading to CXCR4- and MMP-9-dependent mobilization. Calcitonin, which inhibits osteoclast activity, also reduced mobilization [39]. In humans, rHuG-CSF treatment led to increased evidence of bone turnover (reduced osteocalcin and increased urinary deoxypyridinoline) that was proportional to the degree of mobilization.

Treatment with pamidronate abolished the increase in deoxypyridinoline without affecting stem cell release [40], suggesting that rHuG-CSF-mediated mobilization depends on osteoclast activity, but not on bone resorption. rHuG-CSF may cause these effects through its action on osteoblasts. To eliminate the potential for confounding by other cell lines, mouse osteoblasts were isolated using a novel flow-cytometric approach. rHuG-CSF stimulation resulted in phenotypic and functional changes that directly influenced stem cell proliferation and maintenance of reconstitution potential. These effects were found to depend on the ataxia telangiectasia mutated (*ATM*) gene [41].

Research reveals the intimate link between progenitor cells, osteoblasts, mesenchymal stromal cells, and the adrenergic system. Mesenchymal stromal cells expressing nestin contain the marrow's colony-forming unit (CFU)-fibroblast activity. These cells have been shown to be organized around progenitor cells and adrenergic nerve fibers, and express stem cell maintenance genes. Mobilization or adrenergic stimulation results in downregulation of these genes. Parathormone increased nestin$^+$ cells in the marrow, and promoted their osteoblastic differentiation. In vivo depletion of nestin$^+$ cells mobilized progenitor cells [42]. The vitamin D receptor may be involved, since deficient mice show a severe impairment of rHuG-CSF-mediated osteoblast suppression and stem cell mobilization [43].

G-CSF leads to spatial changes in the stem cell niche. Progenitor cell location within the marrow depends on the degree of maturation, with more mature cells positioned closer to vessels. During rHuG-CSF-induced mobilization, progenitors migrate closer to blood flow [44]. Vascular endothelial growth factor (VEGF) is a mediator of vascular permeability and progenitor mobilization. Unstimulated CD34$^+$ cells are weak VEGF producers, but could be induced to increase production after incubation with cytokines such as rHuG-CSF and rHuSCF. VEGF, in turn, stimulated release of G-CSF from endothelial cells and increased SDF-1-mediated transendothelial migration, thus acting as a paracrine loop [45]. Confocal laser scanning microscopy has shown that the endosteal–bone marrow interface in mice is hypoxic, with constitutive expression of hypoxia inducible factor-1α (HIF-1α). Mobilization with rHuG-CSF or cyclophosphamide caused these hypoxic areas to spread centrally, associated with an increase in HIF-1α and VEGF-A. The accumulation of myeloid precursors in the marrow during mobilization may lead to hypoxia, stabilization of HIF-1α transcription, and increased VEGF-A expression, facilitating transendothelial migration of stem cells [46].

Among all the postulated mechanisms of mobilization, one possible link emerges that provides a unifying hypothesis of how rHuG-CSF mobilizes stem cells. A study showed that rHuG-CSF depletes endosteal osteoblasts and the expression of factors that support stem cell retention. The authors noted a concomitant decrease in a population of trophic endosteal macrophages. Subsequent in vivo depletion of this specific population resulted in stem cell egress, strongly suggesting that macrophages play an important role in stem cell retention and rHuG-CSF-induced mobilization [47]. This concept is supported by another study demonstrating downregulation of stem cell retention genes, reduction of SDF-1 levels, and stem cell mobilization in mice that were depleted of CD169$^+$

macrophages [48]. Stem cell mobilization is possible in transgenic mice in which expression of G-CSFR is restricted to cells of the monocytic lineage [49]. The use of rHuG-CSF in this model led to osteoblast suppression and reduction in SDF-1 levels similar to that seen when rHuG-CSF is administered to wild-type animals. These studies indicate that G-CSF signaling occurs through receptors on monocytes and macrophages, leading to inhibition of osteoblasts involved in stem cell retention.

Other small studies add further details regarding the mechanism of stem cell mobilization. For example, the complement cascade interacts with the SDF-1/CXCR4 axis. C5a is a strong granulocyte chemoattractant, causing a wave of neutrophils to exit the marrow and release agents that prime hematopoietic stem cells to respond to an increasing SDF-1 gradient [50]. Activation of the complement cascade may result in increased amounts of S1P released by the interaction of the membrane attack complex with erythrocytes. Red cells are the major reservoir of S1P which, at physiologically relevant concentrations, is a much more potent progenitor cell chemoattractant than SDF-1 [51]. Norepinephrine mediates the interaction between the sympathetic nervous system and the CXCR4/SDF-1 axis, which causes SDF-1 and CXCR4 downregulation, and establishes a circadian rhythm of stem cell mobilization [52, 53]. rHuG-CSF-induced mobilization was inhibited by cannabinoid antagonists and impaired in cannabinoid receptor knock-out mice, indicating that endocannabinoids may play a significant role [54]. Forward genetic mapping studies in mice have shown that the epidermal growth factor receptor (*EGFR*) gene was in a region modifying rHuG-CSF-mediated mobilization. *EGFR* deletion or pharmacological inhibition using erlotinib resulted in better mobilization [55]. Additional pathways/molecules that may be involved in rHuG-CSF-mediated stem cell mobilization include the hepatocyte growth factor (HGF)/c-MET axis [56, 57]; alterations in cholesterol [58] and levels of the adipokines visfatin and resistin [59]; src family kinases [60]; inactivation of the retinoblastoma protein [61]; CD99 [62]; the 67 kDa laminin receptor [63]; and the Rho GTPases Rac1 and Rac2 [64, 65]. These pathways have not been clinically exploited, and therefore are not described in greater detail. Currently, however, the macrophage–osteoblast pathway appears most likely to be the central mechanism of rHuG-CSF-induced stem cell mobilization. A review of the mechanisms of mobilization has recently been published [174].

3 rHuG-CSF Use in Stem Cell Mobilization

Protocols for the use of rHuG-CSF in mobilization depend on the underlying disease, whether the aim is allogeneic or autologous transplantation, and the success of previous mobilization efforts. They also show some variation among centers. Most healthy donors are mobilized with rHuG-CSF 10 μg/kg/day for 4 days, with apheresis beginning on day 5. The same dose of rHuG-CSF is used to attempt steady-state mobilization for autologous transplantation. In cases where mobilization follows a cycle of chemotherapy, rHuG-CSF 5 μg/kg/day is often used, with apheresis commencing once a predefined threshold of $CD34^+$ cells is mobilized [175, 176].

Studies on allogeneic donors have shown that higher doses of rHuG-CSF result in greater progenitor cell yields. Among 75 donors given rHuG-CSF at either 10 μg/kg/day × 5 days or 12 μg/kg/12 h × 4 days, yields were better with the higher dose (3.7×10^6/kg/apheresis vs. 2.0×10^6/kg/apheresis) [66]. Administering rHuG-CSF in divided doses may contribute to increased mobilization. Fifty donors were given rHuG-CSF either as 10 μg/kg/day or 5 μg/kg/12 h. The target of collecting $>3.0 \times 10^6$/kg $CD34^+$ cells in one apheresis was achieved in 68% vs. 96% in these groups [67]. A small crossover study was performed to investigate the mechanism of better mobilization with split doses. Four volunteers received 5 μg/kg/12 h and 10 μg/kg/day, for 4 days each, with an intervening washout period of 3 months. The twice-daily schedule resulted in a higher day-5 $CD34^+$ cell count (94.5 vs. 47/μL, $p = 0.05$). Minimum serum G-CSF amounts were higher with the split-dose regimen, raising the possibility that better mobilization is related to a more continuous exposure to rHuG-CSF [68]; however, dividing the rHuG-CSF dose has not always resulted in better mobilization. No difference in $CD34^+$ counts was seen when 81 normal donors were given 3 days of rHuG-CSF at either 12 μg/kg/day or 6 μg/kg/12 h [69]. Currently most centers administer 10 μg/kg rHuG-CSF as a once-daily dose as this is most effective for most patients, and also reduces the need for multiple injections.

Combining rHuG-CSF with chemotherapy allows the twin aims of mobilization and antitumor activity to be attempted simultaneously, and has been shown to enhance progenitor cell yield. High-dose cyclophosphamide (7 g/m^2) with rHuG-CSF 10 μg/kg was compared to rHuG-CSF alone in a cohort of patients with lymphoma. The $CD34^+$ yields were 6.41 vs. 2.89×10^6/kg ($p = 0.009$), respectively. No difference was seen in $CD34^+/CD38^-$, $CD34^+/Thy1^+$, $CD34^+/HLADR^-$ counts, or in eventual engraftment [70]. Higher $CD34^+$ cell yields were reported with rHuG-CSF plus cyclophosphamide compared with rHuG-CSF alone in a retrospective study of 706 myeloma patients. These patients engrafted slower, unless the stem cells were infused >30 days after the first apheresis, an effect attributed to reversible marrow damage caused by cyclophosphamide [71]. One reason for better $CD34^+$ yields when rHuG-CSF is combined with chemotherapy may be the higher collection efficiency when white cell counts are lower. In a review of 415 aphereses in 201 patients undergoing mobilization with either rHuG-CSF alone or in combination with chemotherapy, collection efficiency was 53% when the white cell count was $<20 \times 10^9$/L compared to 40% when it was higher; 61% of aphereses after chemotherapy plus rHuG-CSF were performed when the white cell count was $<20 \times 10^9$/L compared to only 21% after rHuG-CSF alone [72]. Increased mobilization with chemotherapy is achieved at the price of a longer mobilization procedure, the need for red cell and platelet transfusions, and frequently, the requirement for hospitalization and antibiotics for febrile neutropenia. These findings have led to a critical reappraisal of the need for chemotherapy when mobilizing patients with myeloma. In a review of 201 patients treated in a single center, the use of cyclophosphamide with rHuG-CSF did not increase complete remission rates or time to progression, despite reducing the number of aphereses compared with growth factors alone [73].

Despite several small studies suggesting a benefit of higher rHuG-CSF doses in combination with chemotherapy, a prospective study of 131 cancer patients randomly assigned to 5 vs. 10 μg/kg/day rHuG-CSF after standard chemotherapy reported no statistically significant difference in $CD34^+$ cell collection [74]. A double-blind, dose-finding study with a two-stage Bayesian design showed increasing estimated probabilities of a successful collection with increasing doses: 84%, 87.7%, 91%, 93.9%, and 96.4%, with doses of 50, 75, 100, 125, and 150 $\mu g/m^2$/day, respectively [75], suggesting that most patients will mobilize with smaller doses, and rHuG-CSF 5 μg/kg/day is the regimen used in most units at present.

The pegylated form of rHuG-CSF (pegfilgrastim) has been evaluated in stem cell mobilization. When compared with rHuG-CSF, $CD34^+$ cells mobilized by pegfilgrastim had higher expression of genes indicative of early hematopoiesis, including HOXA9, MEIS1, and GATA3, and lower expression of genes characteristic of erythroid and later stages of myeloid differentiation. Greater numbers of immature cells and common myeloid progenitors, and fewer megakaryocyte–erythrocyte precursors were reported [76]. The feasibility of using pegfilgrastim was demonstrated in 25 healthy donors who were given 12 mg pegfilgrastim. Only one person required additional rHuG-CSF, and 80% achieved the target collection in a single apheresis. The kinetics of mobilization was similar to rHuG-CSF with maximum mobilization of $CD34^+$ cells occurring on day 5 [77]. A phase 1/2 study of steady-state mobilization in normal donors reported suboptimal mobilization with a dose of 6 mg, but good effects when the dose was increased to 12 mg [78]. Such dose dependence was not seen when pegfilgrastim was used with cyclophosphamide [79].

The efficacy of different formulations and dosing regimens rHuG-CSF has been investigated in an attempt to optimize stem cell mobilization. Lenograstim is a glycosylated form of rHuG-CSF produced in Chinese hamster ovaries, while filgrastim is a nonglycosylated version synthesized in *E. coli*. Several prospective studies have shown improved mobilization with lenograstim. A crossover study randomly assigned 32 healthy volunteers to 10 μg/kg/day of either formulation with a minimum washout period of 4 weeks. Peak $CD34^+$ cells and the number of CFU-GM (granulocyte–macrophage progenitor cells) at day 6 were higher with lenograstim (104 vs. 38/μL and 14.6 vs. 10.2/μL, respectively, $p < 0.0001$ for both comparisons) [80]. A prospective study involving 501 donors randomly assigned to 10 μg/kg/day of either lenograstim or filgrastim confirmed mobilization of 11.5% more $CD34^+$ cells with lenograstim. Post hoc analysis indicated that this result was due to better mobilization with lenograstim in men [81]. In contrast to healthy donors, $CD34^+$ cell yield appears similar in patients receiving either lenograstim or filgrastim after chemotherapy [82]. It is probable that the formulation of rHuG-CSF makes little difference in the context of enhanced mobilization with myelosuppressive chemotherapy combined with cytokines.

4 Harvesting Mobilized Stem Cells

Proper timing of the collection is an important aspect of optimizing the progenitor cell yield while minimizing the number of apheresis sessions. In patients given cyclophosphamide with or without recombinant human granulocyte-macrophage colony-stimulating factor (rHuGM-CSF), long-term culture-initiating cells (LTC-IC) increased sixfold, and colony-forming cells (CFC) increased 26-fold above normal, at the time of white cell recovery [83]. An analysis of 163 apheresis collections from 26 lymphoma patients mobilized with cyclophosphamide plus rHuG-CSF indicated that peak harvests tended to occur when the white cell count was $>10 \times 10^9$/L. CD34$^+$ cell counts and CFU-GM generally peaked within 24 h of each other in most patients [84]. In patients with acute myeloid leukemia (AML) undergoing chemotherapy, the more primitive CFU-Mix was not found to be superior to CFU-GM in predicting hematopoietic reconstitution [85]. Another study confirmed that CFU-A (multipotent), CFU-GM, and BFU-E (erthyroid blast-forming units) numbers peaked simultaneously in lymphoma patients undergoing chemotherapy-based mobilization [86]. These studies suggest that different functional classes of progenitors are ontogenically close to each other, and that they all contribute to marrow regeneration. Similar kinetics were seen in healthy donors mobilized with rHuG-CSF alone. Administration of rHuG-CSF daily results in peak GM-CFC and CD34$^+$ cell counts on day 5. The CD34$^+$ cell peak was broad with elevated numbers seen during days 4–6 [9].

The peripheral blood CD34$^+$ cell count, but not the white cell count, correlates with the apheresis yield [87]. If the circulating CD34$^+$ cell count is $\geq 0.2 \times 10^9$/L, 94% of collections performed the next day would be expected to yield $\geq 2.0 \times 10^6$ CD34$^+$ cells/kg [88]. Mobilization measured by CD34$^+$ counts has been shown to occur 11 days after chemotherapy in 97% of patients. The median peak CD34$^+$ counts occurred on days 14 and 15 irrespective of the underlying malignancy and type of chemotherapy, although heavily pretreated patients mobilized later. The authors recommended measuring CD34$^+$ counts from day 11 onwards in patients undergoing mobilization with a combination of chemotherapy and rHuG-CSF [89]. Using CD34$^+$ counts to guide apheresis has been shown to reduce costs [90]. Based on such studies, most centers begin monitoring peripheral blood CD34$^+$ cell counts when white cell count recovery is seen, and initiate apheresis when a predetermined threshold is exceeded.

Large volume apheresis (defined as processing ≥ 3 times the blood volume or ≥ 15 L) results in higher progenitor cell yields and may reduce the number of sessions in some cases. Most studies have not shown a significant decline in progenitor levels or collection efficiency with time [91].

Contamination of the apheresis product with tumor cells has been reported in a number of malignancies, including myeloma, leukemia, lymphoma, and neuroblastoma. Neoplastic cells were detected in the blood of 27 myeloma patients before mobilization with cyclophosphamide and rHuG-CSF. After mobilization, these cells increased in number with kinetics similar to that of CD34$^+$ cells [92].

In chronic myeloid leukemia, up to 87% of progenitor cell colonies from the CD34$^+$ CD38$^-$ fraction in mobilized peripheral blood were *BCR-ABL* positive [93]. The role of such contaminating cells in posttransplant relapse is uncertain, although they do represent a potential source.

Diverse attempts have been made at selection of CD34$^+$ cells to reduce tumor contamination. Despite CD34$^+$ cell selection using an avidin–biotin immunoadsorption technique, abnormal IgH was undetectable by polymerase chain reaction (PCR) in only 3 of 15 informative cases with non-Hodgkin's lymphoma [94]. A median 5-log purging efficiency was reported in a study of leukemia patients undergoing double B-cell depletion of the harvested progenitor cells using immunomagnetic CD34$^+$ cell selection followed by a negative step with anti-CD19/20/23/37-labeled immunomagnetic beads. All transplanted patients achieved a rapid and durable engraftment, with 17 of 20 maintaining complete remission at 20 months [95]. In a group of myeloma patients, immunomagnetic bead separation of CD34$^+$ cells resulted in undetectable levels of IgH rearrangement by PCR in 13 of 24 evaluable patients, without affecting engraftment [96]. The feasibility of sequential in vivo followed by in vitro purging of tumor cells was demonstrated in 34 myeloma patients who underwent 2 sequential cycles of high-dose chemotherapy, stem cell mobilization, CD34$^+$ cell selection, and transplantation. The second cycle resulted in a fivefold reduction of tumor cells compared with the first mobilization [97].

Such selection procedures are not without risk. Ten breast cancer patients transplanted with purified CD34$^+$ cells had significantly prolonged times to marrow recovery compared with patients who received unmanipulated products [98]. Enrichment of CD34$^+$Thy-1$^+$Lin$^-$ cells by high-speed fluorescence-activated cell sorting (FACS) resulted in no contamination detected by PCR in three of nine samples from myeloma patients. However, neutrophil and platelet engraftment was significantly delayed (16 and 39 days, respectively) [99]. Fourteen of 20 lymphoma patients transplanted with CD34$^+$Thy-1$^+$ cells purified by immunomagnetic separation followed by high-speed flow cytometric sorting had infectious complications despite timely engraftment [100]. Moreover, the benefit of such selection procedures is not clearly established. An adequate yield was obtained by rHuG-CSF plus cyclophosphamide mobilization, followed by CD34$^+$ selection, in 20 of 36 myeloma patients. They were compared with the remaining 16 patients who received unselected cells. Tumor contamination, as assessed by IgH fingerprinting, was present in 7 of 20 vs. 5 of 9 patients in these groups, respectively. No difference was seen in event-free survival [101]. For patients with B-cell lymphomas, chemoimmunotherapy with rituximab can be considered an important alternative, without such drawbacks, to ex vivo selection. Fifteen patients with mantle cell or follicular lymphoma with marrow involvement and detectable molecular rearrangements were mobilized after chemoimmunotherapy. PCR of apheresis collections was negative in 93% of patients compared with 40% of control patients treated with chemotherapy alone [102].

5 Factors Affecting Progenitor Cell Mobilization and Yield

Not all persons administered rHuG-CSF with or without chemotherapy will mobilize adequate numbers of progenitor cells to allow transplantation. In a review of 1,834 patients treated at a single institution, failure rates using either rHuG-CSF alone, or in combination with chemotherapy, were 18.8% and 18.6%, respectively [103]. Identifying patients at risk of poor mobilization would allow optimization of management strategies, and has been the focus of considerable research. Younger age and male sex have been positively correlated with stem cell harvest in multivariate analysis [104]. In a series of 137 harvests from 68 healthy donors, mobilization was suboptimal in 57% of cases when either white cell count was $<25 \times 10^9$/L or platelets were $<100 \times 10^9$/L before apheresis [105].

Damage to stem cells has been repeatedly shown to impact negatively on progenitor cell mobilization. Among 243 cancer patients, radiotherapy, marrow involvement, and the number of cycles of chemotherapy impaired mobilization [106]. In a study involving 57 myeloma patients, mobilization $\geq$12 months from diagnosis, $\geq$6 cycles of alkylating agents, and marrow plasmacytosis >20% were adverse factors, with only 38% of patients with all three achieving adequate stem cell harvests [107]. Among 307 patients mobilized with rHuG-CSF and cyclophosphamide, diagnosis, previous chemotherapy exposure, treatment with mitoxantrone, and the preapheresis platelet count were independently associated with inadequate collection [108]. Stem cell toxic treatments that had been used in lymphoma patients mobilizing poorly included nitrogen mustard, procarbazine, melphalan, carmustine, and >7.5 g cytarabine [109]. Other agents shown to impair mobilization include fludarabine [110] and platinum compounds [111]. Use of rituximab does not appear to impair mobilization [112]. In a review of 302 myeloma patients, use of lenalidomide was associated with an odds ratio of 5.9 of failure to mobilize, although most patients could be successfully remobilized [113]. In contrast, the impact of thalidomide or bortezomib appears to be small. These issues have been discussed in detail by an expert panel of the International Myeloma Working Group [114].

Steady-state peripheral blood $CD34^+$ counts correlate with apheresis yields, although the correlation coefficient was 0.52 in a series of 100 cancer patients receiving chemotherapy and rHuG-CSF [115], indicating that a substantial degree of the variation in yields was not accounted for by steady-state $CD34^+$ cell counts. The response to a test dose of 12 μg/kg of rHuG-CSF was assessed as a predictor of mobilization in 22 patients, who then received chemotherapy and G-CSF. A peak CFU-GM $\geq$ 250/mL, and a $CD34^+$ cell increment $\geq$2.5/μL, were associated with subsequent good mobilization [116]. Changes in cytokine levels may be indicative of poor mobilization, with a small study showing that premobilization tumor necrosis factor (TNFα) <3.6 pg/mL increased risk (hazard ratio = 19.9, $p = 0.0002$) [117]. Increased serum ferritin may be an adverse risk factor [118]. Given the modest ability of such assessments to capture the wide variability in responses to rHuG-CSF, they are not part of routine clinical practice.

The impact of genetic factors on stem cell mobilization has been explored. In murine studies, a quantitative trait locus has been found on chromosome 1 with highly significant linkage for splenomegaly after rHuG-CSF administration. Mice lacking C57BL/6 alleles in this region had diminished stem cell mobilization with rHuG-CSF [119]. In humans, a source of genetic variability has been identified in the G $\rightarrow$ A transition at position 801 in the 3′-untranslated region of the SDF-1 gene (SDF1-3′A allele). Carriers of the SDF1-3′A allele, in particular SDF1-3’AA homozygotes, had higher mobilization compared with SDF1-3′GG homozygotes ($p < 0.005$ for both comparisons) [120]. In 63 cancer patients undergoing mobilization, 67% of good (≥50 $CD34^+$ cells/μL) mobilizers, but only 36% of intermediate/poor mobilizers were SDF1-3’A allele carriers [121].

6 Management of Poor Mobilizers

"Poor mobilizers" and "failed mobilization" are terms that have been variously defined in the literature, based either on preapheresis $CD34^+$ cell counts or on target $CD34^+$ cell yields. A preapheresis peripheral blood $CD34^+$ cell count <20/μL has been used to indicate inadequate mobilization, although a level of 10/μL may be sufficient to yield 2×10^6 $CD34^+$ cells/kg. The target apheresis yield would also depend on the circumstances, such as a plan to collect sufficient stem cells for two autologous rescue procedures in patients with myeloma. At present, almost all remobilization strategies include the use of rHuG-CSF [174, 175].

One approach has been to reattempt mobilization with a higher rHuG-CSF dose. A 60% success rate was reported in patients in whom rHuGC-SF failed to mobilize enough cells, and were given an increased dose of rHuG-CSF as an immediate salvage remobilization [122]. Another study reported that of 138 patients who failed to be mobilized with chemotherapy and rHuG-CSF (5 μg/kg/day), 107 (77%) had at least partially successful collections with higher dose (10 μg/kg/day) rHuG-CSF alone [123]. The use of rHuG-CSF-stimulated bone marrow harvests is another option. As would be expected based on our current understanding of the mechanism of mobilization, rHuG-CSF does not increase the number of stem cells in the marrow, but rather induces an expansion of myeloid precursors [124]. Bone marrow priming with rHuG-CSF 15–16 μg/kg/day × 3 days was used on 86 heavily pretreated patients who had previously failed to be mobilized. The median number of $CD34^+$ cells harvested was 0.83×10^6/kg. All patients proceeded to transplant. Engraftment appeared to be prolonged but transplant-related mortality was acceptable at 4.6% [125]. Changing chemotherapy regimens has been explored. The use of an intermediate dose (375 mg/m^2) of etoposide with rHuG-CSF allowed successful collection in all 152 myeloma patients given this regimen at a single institution. Significantly, this regimen was able to overcome the effects of age and prior chemotherapy exposure on progenitor cell mobilization [126].

Considerable work has been done in relation to the use of rHuSCF. Clinical studies have demonstrated a dose–response relationship and a possible benefit with rHuSCF pretreatment. In a trial involving 62 patients with breast cancer randomly assigned to rHuG-CSF and rHuSCF or rHuG-CSF alone, rHuSCF was found to sustain the peripheral blood progenitor cell count (up to 100-fold in some cases) after cessation of rHuG-CSF, leading to increased $CD34^+$ collections. In contrast, there was a rapid decrease in $CD34^+$ cells in the group receiving rHuG-CSF alone. In a subsequent cohort, pretreatment with rHuSCF for 3 days resulted in an earlier wave of progenitor cells [10]. In a different study, breast cancer patients underwent three cycles of high-dose chemotherapy supported by autologous transplants. They were assigned to receive rHuG-CSF alone or in combination with rHuSCF in varying doses (including rHuSCF pretreatment). The use of rHuSCF at $\geq$10 μg/kg/day led to better mobilization, an effect that was enhanced by rHuSCF pretreatment [11]. In both the previous studies, phenotyping of mobilized $CD34^+$ cells with CD38, Thy-1, and MDR-1 revealed no difference in maturity compared with cell mobilized with rHuG-CSF alone [12].

The utility of rHuSCF has been shown in patients who fail to be mobilized. Forty-four such patients treated with rHuG-CSF alone or rHuG-CSF and chemotherapy were remobilized with the same regimen plus rHuSCF. Target collections were achieved in 54% and 45% of these groups [13]. These findings have been confirmed by other studies [127–129]. The combination of rHuG-CSF and rHuSCF enhanced mobilization even in patients with lymphoma previously exposed to fludarabine [130]. Mast cell degranulation reactions are the most significant side effect, with severe but nonfatal reactions reported in about 10% of patients in one study [129].

The roles of rHuGM-CSF and recombinant human interleukin-3 (rHuIL-3) have been evaluated. However, these are not used since no advantage has been demonstrated over rHuG-CSF alone in terms of efficacy, and they are associated with more toxicity [131]. Other agents with potential utility in poor mobilizers include erythropoietin [132], thrombopoietin [133], growth hormone [134], parathormone [135], and the retinoic acid receptor alpha agonist VTP195183 [136]. The place of these agents in clinical practice remains to be clarified.

The development of the partial CXCR4 agonist plerixafor has been the most significant advance in the management of poor mobilizers. On its own, plerixafor appears to mobilize different subsets of cells compared with rHuG-CSF. In a rhesus macaque model, plerixafor-mobilized collections had more B, T and mast cell precursors compared with rHuG-CSF-mobilized collections [137]. Compared with rHuG-CSF, more plerixafor-mobilized $CD34^+$ cells were in G1, and had higher expression of CXCR4 and VLA-4 [138]. At present, most patients are mobilized with a combination of both these agents. In these situations, plerixafor may further enhance mobilization of primitive progenitor cells such as the $CD34^+$ $CD38^-$ and aldehyde dehydrogenase (ALDH) (bright) $CD34^+$ $CD38^-$ subsets [139].

The efficacy of plerixafor in combination with rHuG-CSF has been established in several phase 3 multicenter, placebo-controlled randomized studies. In a trial involving 302 myeloma patients randomized to rHuG-CSF with either plerixafor or placebo, 71.6% in the plerixafor group vs. 34.4% in the placebo group reached the primary endpoint of collecting $\geq 6 \times 10^6$ $CD34^+$ cells/kg within two aphereses [140].

In a similarly designed trial involving 298 lymphoma patients, 59% of the plerixafor group vs. 20% of the placebo group achieved the primary endpoint of $\geq 5 \times 10^6$ CD34$^+$ cells/kg within four aphereses [141]. In this latter study, 62 poor mobilizers (placebo, $n = 52$; plerixafor, $n = 10$), defined as either $<0.8 \times 10^6$ CD34$^+$ cells/kg in two collections or $<2.0 \times 10^6$ CD34$^+$ cells/kg in four collections, were enrolled in an open-label rescue protocol. Four of ten patients from the plerixafor group and 33 of 52 patients from the placebo group mobilized $>2.0 \times 10^6$ CD34$^+$ cells/kg. In those proceeding to transplant, engraftment was similar to those who had not failed initial mobilization [142]. In a study involving 60 myeloma patients, plerixafor appeared to overcome the adverse impact of previous therapy with lenalidomide [143]. Analyses of plerixafor use in compassionate use programs on poor mobilizers, or predicted poor mobilizers, have shown adequate collections in about 67% of patients [144].

7 Engraftment of Mobilized Progenitor Cells

The transfused CD34$^+$ cell count has consistently been shown to be the most significant predictor of engraftment. Murine studies have proposed that 1.0–2.0 $\times 10^6$ CD34$^+$ cells/kg is a minimum threshold for short-term engraftment, while at least 0.5–2.0 $\times 10^6$ CD34$^+$ cells/kg would be needed for long-term reconstitution. The general consensus is that 2×10^6 CD34$^+$ cells/kg would reliably lead to rapid and sustained engraftment [6]. On the other hand, it has been suggested, based on a Cox regression analysis of 692 patients mobilized with rHuG-CSF with or without cyclophosphamide, that a CD34$^+$ cell dose $\geq 5.0 \times 10^6$/kg would be optimal [145]. CD34$^+$ cell doses within these ranges have been shown to lead to similar short-term engraftment. In a post hoc analysis of 438 patients from two multicenter trials of rHuG-CSF with or without plerixafor, the effect of the transfused CD34$^+$ cell dose on engraftment was analyzed irrespective of the mobilization regimen. Short-term neutrophil and platelet recovery were similar in each of the groups receiving 2–4, 4–6, and $\geq 6 \times 10^6$ CD34$^+$ cells/kg. However, a significant linear relationship between CD34$^+$ cells infused and the platelet count at 6 and 12 months was reported in patients with both myeloma and lymphoma [146].

The short- and medium-term engraftment potential of mobilized blood cells seems to be higher than bone marrow. Mobilized blood contained higher frequencies of CFC (1.6×), week-5 cobblestone area-forming cells (CAFC) (8.4×), and week 8 CAFC (10.3×) [147]. Another study reported that mobilized peripheral blood contained similar numbers of CD34$^+$ cells, but 2.5× more early myeloid progenitors (CD34$^+$CD33$^+$CD15$^-$), compared with bone marrow [148].

The effect of CD34$^+$ cell subsets on engraftment was assessed in a cohort of 27 patients undergoing autologous transplantation. The CD34$^+$L-selectin$^+$ cell count showed a better, statistically significant, correlation with time to platelet recovery than the total CD34$^+$ cell count ($r = -0.86$ vs. -0.55) [149]. L-selectin-expressing subsets were evaluated in 86 patients undergoing autologous PBPC transplants.

$CD34^+$L-selectin$^+$ cell counts were found to be only a minor improvement on total $CD34^+$ cell counts in predicting short-term engraftment. In this study, the transplanted $CD34^+$Thy-1$^+$ cell dose was the only subset correlating with long-term engraftment [150]. In a large cohort of 410 patients undergoing autografting, $CD34^+CD33^-$ cell dose was an independent predictor of engraftment kinetics in addition to total $CD34^+$ cells transfused. It was the only predictor of red cell and platelet transfusions [151]. The expression of CD133 has shown conflicting correlation with engraftment kinetics. CD26 expression correlated with engraftment but not with the $CD34^+$ cell count, suggesting that it may be an independent predictor of engraftment, but awaits formal evaluation [152]. In a small study, a $CD34^+CXCR4^+$ cell count $>2.5 \times 10^6$/kg was associated with a significantly shorter time to platelet engraftment (median 10 vs. 14.5 days, $p = 0.02$) [153]. Overall, no $CD34^+$ cell subset has been proven to be significantly better than the total $CD34^+$ cell count in clinical practice.

Ex vivo expansion of harvested progenitor cells represents an attempt to shorten the period of cytopenias until engraftment occurs [154]. Among the cytokines used in the culture medium, G-CSF, SCF, and IL-3 were found to be the most important [155]. The feasibility of this approach was shown in a report of 19 breast cancer patients whose myeloid and erythroid progenitors expanded a median 46-fold at 21 days [156]. Efficient megakaryocyte expansion could be achieved with cytokine combinations containing thrombopoietin [157]. Clinical benefit was demonstrated in a study involving 34 myeloma patients who were transplanted with expanded (rHuG-CSF, rHuSCF, thrombopoietin) $CD34^+$ cells, with or without additional unmanipulated cells. Posttransplant neutropenia was abrogated in all but one patient [158]. At present, cytokines are not routinely used for ex vivo expansion of progenitor cells, mainly due to loss of primitive cells during progenitor cell differentiation. In contrast, recent work in the cord blood setting has shown that Notch ligands can expand hematopoietic stem cells that retain longer-term differentiation capability [159].

8 rHuG-CSF Mobilization and the Immune System

The risk of acute graft-versus-host disease (aGVHD) with PBSC transplants is approximately 20%, which is similar to bone marrow transplants, despite the infusion of up to 10 times more T lymphocytes with mobilized blood. In contrast, at a median follow up of 3 years, the risk of chronic GVHD (cGVHD) was shown to be increased from 55 to 73% in a multicenter study of 350 patients [160]. The risks of both aGVHD [161] and cGVHD [162] are greater in patients transfused a higher $CD34^+$ cell dose, which is likely to be a consequence of changes in immune effector cells paralleling the intensity of the mobilization process, rather than a direct effect mediated by $CD34^+$ cells. These changes can impact transplant outcomes. In a cohort of 78 patients, relapse-free survival at 59 months was lower in the group receiving $\geq 8.3 \times 10^6$ $CD34^+$ cells/kg, primarily because of the mortality associated with cGVHD. These findings indicate the need to balance the risk of poor engraftment with that of extensive cGVHD [162].

Compared with 28 steady-state donors, rHuG-CSF-mobilized cells from 104 healthy donors showed reduced natural killer (NK)- and lymphokine-activated killer cell-mediated cytotoxicity, associated with a diminished B- and T-cell mitogen response [163]. Moreover, compared with unprimed bone marrow, the rHuG-CSF-mobilized $CD34^+$ cell fraction contained six times fewer NK cell progenitors when cultured in limiting-dilution assays [164]. In another study on six healthy donors, rHuG-CSF therapy resulted in fewer $CD25^+$, $CD56^+$, and $CD57^+$ cells. In comparison to steady-state cells, freshly isolated rHuG-CSF-mobilized mononuclear cells demonstrated similar allogeneic responses to irradiated HLA-incompatible mononuclear cells, but no NK activity against K562 targets. These results suggest that rHuG-CSF-mobilized mononuclear cells may prevent normal Th1 alloresponses, maintain alloreactivity to HLA-mismatched antigens, and have impaired NK activity [165]. These findings were corroborated clinically in a cohort of 27 patients undergoing HLA-matched sibling stem cell transplant (SCT), where the $CD56^+CD16^+$ cell dose correlated inversely with the risk of aGVHD, and a lower $CD56^+CD16^+$ to $CD34^+$ ratio was associated with increased cGVHD risk [166].

Microfluorometry studies on peripheral blood from healthy donors have shown an increase in type-2 dendritic cells (DC2) that induce Th2 differentiation of naive T cells. DC1 cell counts were not affected [167]. This finding was confirmed in another study on healthy donors that reported that rHuG-CSF decreased IFN-γ production, caused a polarization from a Th1 to a Th2 phenotype, and increased IL-4 and IL-17 secretion by helper T cells [168]. Th2 polarization has less propensity to cause aGVHD. Such polarization has been referred to as a Th17 phenotype, and was found to be dependent on G-CSF-induced IL-21 signaling. Donor $CD8^+$ cells were the main source of IL-17 and controlled the skin infiltration by macrophages that eventually led to sclerodermatous changes in cGVHD [169]. SOCS-3, a key regulator of intracellular signaling by G-CSF, has been shown to dampen immune responses leading to GVHD [170], suggesting another mechanism to explain the reduced risk of aGVHD after rHuG-CSF-mobilized SCT.

Despite a reduction in GVHD risk, immune cells in rHuG-CSF-mobilized collections retain the ability to mediate an immunologic clearance of tumor cells in the recipient. This phenomenon of graft-versus-tumor effect forms a cornerstone of the success of allogeneic SCT in treating several malignancies. A metaanalysis of several studies has shown that graft-versus-leukemia effects are enhanced by the use of peripheral blood compared with bone marrow transplantation [171]. A discussion of the complex mechanisms by which G-CSF achieves this partial separation of GVHD from graft-versus-tumor effect is beyond the scope of this chapter, but is reviewed by Morris et al. [172].

Improved survival has been reported in patients achieving higher lymphocyte counts after SCT. For example, in a study of 122 myeloma patients, a higher day-30 lymphocyte count translated to better overall and progression-free survival, and correlated with the lymphocyte count in the apheresis product. More lymphocytes were present in harvests mobilized with rHuG-CSF alone compared with rHuG-CSF with chemotherapy [173], illustrating once again that optimizing outcomes after PBSC transplantation requires a more considered approach than simply attempting to increase the $CD34^+$ cell dose. The growing appreciation of the complex and multifaceted effects of rHuG-CSF on the SCT process should allow this goal to be realized.

9 Concluding Remarks

Most hematopoietic SCT performed today use peripheral blood. rHuG-CSF remains the cornerstone of protocols developed to mobilize progenitor cells. The mechanisms by which mobilization occurs is being actively researched, but significant progress has been made, leading to the clinical development of agents that can enhance mobilization in cases where rHuG-CSF with or without chemotherapy is inadequate. The $CD34^+$ cell dose has emerged as the most widely used predictor of engraftment, and hence determines the adequacy of attempts at mobilization. Clinical and genetic factors that affect rHuG-CSF-based mobilization have been defined. Mobilization has wide ranging effects on the content and function of immune cells in the stem cell collection. An improved understanding of these changes could be expected to lead to strategies to mitigate the risk of GVHD and optimize transplant outcomes.

References

1. To LB, Haylock DN, Kimber RJ, Juttner CA (1984) High levels of circulating haemopoietic stem cells in very early remission from acute non-lymphoblastic leukaemia and their collection and cryopreservation. Br J Haematol 58:399–410
2. Juttner CA, To LB, Haylock DN, Branford A, Kimber RJ (1995) Circulating autologous stem cells collected in very early remission from acute non-lymphoblastic leukaemia produce prompt but incomplete haemopoietic reconstitution after high dose melphalan or supralethal chemoradiotherapy. Br J Haematol 61:739–745
3. To LB, Dyson PG, Branford AL et al (1987) Peripheral blood stem cells collected in very early remission produce rapid and sustained autologous haemopoietic reconstitution in acute non-lymphoblastic leukaemia. Bone Marrow Transplant 2:103–108
4. To LB, Dyson PG, Juttner CA (1986) Cell-dose effect in circulating stem-cell autografting. Lancet 2:404–405
5. Richman CM, Weiner RS, Yankee RA (1976) Increase in circulating stem cells following chemotherapy in man. Blood 47:1031–1039
6. To LB, Haylock DN, Simmons PJ, Juttner CA (1997) The biology and clinical uses of blood stem cells. Blood 89:2233–2258
7. Sheridan WP, Begley CG, Juttner CA et al (1992) Effect of peripheral-blood progenitor cells mobilised by filgrastim (G-CSF) on platelet recovery after high-dose chemotherapy. Lancet 339:640–644

8. To LB, Roberts MM, Haylock DN et al (1992) Comparison of haematological recovery times and supportive care requirements of autologous recovery phase peripheral blood stem cell transplants, autologous bone marrow transplants and allogeneic bone marrow transplants. Bone Marrow Transplant 9:277–284
9. Grigg AP, Roberts AW, Raunow H et al (1995) Optimizing dose and scheduling of filgrastim (granulocyte colony-stimulating factor) for mobilization and collection of peripheral blood progenitor cells in normal volunteers. Blood 86:4437–4445
10. Begley CG, Basser R, Mansfield R et al (1997) Enhanced levels and enhanced clonogenic capacity of blood progenitor cells following administration of stem cell factor plus granulocyte colony-stimulating factor to humans. Blood 90:3378–3389
11. Basser RL, To LB, Begley CG et al (1998) Rapid hematopoietic recovery after multicycle high-dose chemotherapy: enhancement of filgrastim-induced progenitor-cell mobilization by recombinant human stem-cell factor. J Clin Oncol 16:1899–1908
12. Roberts MM, Swart BW, Simmons PJ, Basser RL, Begley CG, To LB (1999) Prolonged release and c-kit expression of haemopoietic precursor cells mobilized by stem cell factor and granulocyte colony stimulating factor. Br J Haematol 104:778–784
13. To LB, Bashford J, Durrant S et al (2003) Successful mobilization of peripheral blood stem cells after addition of ancestim (stem cell factor) in patients who had failed a prior mobilization with filgrastim (granulocyte colony-stimulating factor) alone or with chemotherapy plus filgrastim. Bone Marrow Transplant 31:371–378
14. Morrison SJ, Wright DE, Weissman IL (1997) Cyclophosphamide/granulocyte colony-stimulating factor induces hematopoietic stem cells to proliferate prior to mobilization. Proc Nat Acad Sci U S A 94:1908–1913
15. Ebihara Y, Xu MJ, Manabe A et al (2000) Exclusive expression of G-CSF receptor on myeloid progenitors in bone marrow CD34+ cells. Br J Haematol 109:153–161
16. Papayannopoulou T, Priestly GV, Nakamoto B, Zafiropoulos V, Scott LM, Harlan JM (2001) Synergistic mobilization of hemopoietic progenitor cells using concurrent beta1 and beta2 integrin blockade or beta2-deficient mice. Blood 97:1282–1288
17. Leavesley DI, Oliver JM, Swart BW, Berndt MC, Haylock DN, Simmons PJ (1994) Signals from platelet/endothelial cell adhesion molecule enhance the adhesive activity of the very late antigen-4 integrin of human CD34+ hemopoietic progenitor cells. J Immunol 153:4673–4683
18. Watanabe T, Dave B, Heimann DG, Lethaby E, Kessinger A, Talmadge JE (1997) GM-CSF-mobilized peripheral blood CD34+ cells differ from steady-state bone marrow CD34+ cells in adhesion molecule expression. Bone Marrow Transplant 19:1175–1181
19. Lévesque JP, Takamatsu Y, Nilsson SK, Haylock DN, Simmons PJ (2001) Vascular cell adhesion molecule-1 (CD106) is cleaved by neutrophil proteases in the bone marrow following hematopoietic progenitor cell mobilization by granulocyte colony-stimulating factor. Blood 98:1289–1297
20. Papayannopoulou T, Priestly GV, Nakamoto B (1998) Anti-VLA4/VCAM-1-induced mobilization requires cooperative signaling through the kit/mkit ligand pathway. Blood 91:2231–2239
21. To LB, Haylock DN, Dowse T et al (1994) A comparative study of the phenotype and proliferative capacity of peripheral blood (PB) CD34+ cells mobilized by four different protocols and those of steady-phase PB and bone marrow CD34+ cells. Blood 84:2930–2939
22. Aiuti A, Webb IJ, Bleul C, Springer T, Gutierrez-Ramos JC (1997) The chemokine SDF-1 is a chemoattractant for human CD34+ hematopoietic progenitor cells and provides a new mechanism to explain the mobilization of CD34+ progenitors to peripheral blood. J Exp Med 185:111–120
23. Petit I, Goichberg P, Spiegel A et al (2005) Atypical PKC-zeta regulates SDF-1-mediated migration and development of human CD34+ progenitor cells. J Clin Invest 115:68–76
24. Pello OM, Moreno-Ortiz Mdel C, Rodríguez-Frade JM et al (2006) SOCS up-regulation mobilizes autologous stem cells through CXCR4 blockade. Blood 108:3928–3937

25. Hattori K, Heissig B, Tashiro K et al (2001) Plasma elevation of stromal cell-derived factor-1 induces mobilization of mature and immature hematopoietic progenitor and stem cells. Blood 97:3354–3360
26. Semerad CL, Christopher MJ, Liu F et al (2005) G-CSF potently inhibits osteoblast activity and CXCL12 mRNA expression in the bone marrow. Blood 106:3020–3027
27. Christopherson KW, Cooper S, Hangoc G, Broxmeyer HE (2003) CD26 is essential for normal G-CSF-induced progenitor cell mobilization as determined by CD26−/− mice. Exp Hematol 31:1126–1134
28. Paganessi LA, Walker AL, Tan LL et al (2011) Effective mobilization of hematopoietic progenitor cells in G-CSF mobilization defective CD26(−/−) mice through AMD3100-induced disruption of the CXCL12-CXCR4 axis. Exp Hematol 39:384–390
29. Christopher MJ, Liu F, Hilton MJ, Long F, Link DC (2009) Suppression of CXCL12 production by bone marrow osteoblasts is a common and critical pathway for cytokine-induced mobilization. Blood 114:1331–1339
30. De La Luz Sierra M, Gasperini P, McCormick PJ, Zhu J, Tosato G (2007) Transcription factor Gfi-1 induced by G-CSF is a negative regulator of CXCR4 in myeloid cells. Blood 110:2276–2285
31. Khandanpour C, Sharif-Askari E, Vassen L et al (2010) Evidence that growth factor independence 1b regulates dormancy and peripheral blood mobilization of haematopoietic stem cells. Blood 116:5149–5161
32. Petit I, Szyper-Kravitz M, Nagler A et al (2002) G-CSF induces stem cell mobilization by decreasing bone marrow SDF-1 and up-regulating CXCR4. Nat Immunol 3:687–694
33. Lévesque JP, Hendy J, Takamatsu Y, Simmons PJ, Bendall LJ (2003) Disruption of the CXCR4/CXCL12 chemotactic interaction during hematopoietic stem cell mobilization induced by GCSF or cyclophosphamide. J Clin Invest 111:187–196
34. Winkler IG, Hendy J, Coughlin P, Horvath A, Lévesque JP (2005) Serine protease inhibitors serpina1 and serpina3 are down-regulated in bone marrow during hematopoietic progenitor mobilization. J Exp Med 201:1077–1088
35. Janowska-Wieczorek A, Marquez LA, Nabholtz JM et al (1999) Growth factors and cytokines upregulate gelatinase expression in bone marrow CD34(+) cells and their transmigration through reconstituted basement membrane. Blood 93:3379–3390
36. Vagima Y, Avigdor A, Goichberg P et al (2009) MT1-MMP and RECK are involved in human CD34+ progenitor cell retention, egress, and mobilization. J Clin Invest 119:492–503
37. Shirvaikar N, Marquez-Curtis LA, Shaw AR, Turner AR, Janowska-Wieczorek A (2011) MT1-MMP association with membrane lipid rafts facilitates G-CSF-induced hematopoietic stem/progenitor cell mobilization. Exp Hematol 38:823–835
38. Lévesque JP, Liu F, Simmons PJ et al (2004) Characterization of hematopoietic progenitor mobilization in protease-deficient mice. Blood 104:65–72
39. Kollet O, Dar A, Shivtiel S et al (2006) Osteoclasts degrade endosteal components and promote mobilization of hematopoietic progenitor cells. Nat Med 12:657–664
40. Takamatsu Y, Simmons PJ, Moore RJ, Morris HA, To LB, Lévesque JP (1998) Osteoclast-mediated bone resorption is stimulated during short-term administration of granulocyte colony-stimulating factor but is not responsible for hematopoietic progenitor cell mobilization. Blood 92:3465–3473
41. Mayack SR, Wagers AJ (2008) Osteolineage niche cells initiate hematopoietic stem cell mobilization. Blood 11:519–531
42. Méndez-Ferrer S, Michurina TV, Ferraro F et al (2010) Mesenchymal and haematopoietic stem cells form a unique bone marrow niche. Nature 466:829–834
43. Kawamori Y, Katayama Y, Asada N et al (2010) Role for vitamin D receptor in neuronal control of hematopoietic stem cell niche. Blood 116:5528–5535
44. Winkler IG, Barbier V, Wadley R, Zannettino A, Williams S, Lévesque JP (2010) Positioning of bone marrow hematopoietic and stromal cells relative to blood flow in vivo: serially

reconstituting hematopoietic stem cells reside in distinct non-perfused niches. Blood 116:375–385
45. Bautz F, Rafii S, Kanz L, Möhle R (2000) Expression and secretion of vascular endothelial growth factor-A by cytokine-stimulated hematopoietic progenitor cells. Possible role in the hematopoietic microenvironment. Exp Hematol 28:700–706
46. Lévesque JP, Winkler IG, Hendy J et al (2007) Hematopoietic progenitor cell mobilization results in hypoxia with increased hypoxia-inducible transcription factor-1 alpha and vascular endothelial growth factor A in bone marrow. Stem Cells 25:1954–1965
47. Winkler IG, Sims NA, Pettit AR et al (2010) Bone marrow macrophages maintain hematopoietic stem cell (HSC) niches and their depletion mobilizes HSC. Blood 116:4815–4828
48. Chow A, Lucas D, Hidalgo A et al (2011) Bone marrow CD169+ macrophages promote the retention of hematopoietic stem and progenitor cells in the mesenchymal stem cell niche. J Exp Med 208:261–271
49. Christopher MJ, Rao M, Liu F, Woloszynek JR, Link DC (2011) Expression of the G-CSF receptor in monocytic cells is sufficient to mediate hematopoietic progenitor mobilization by G-CSF in mice. J Exp Med 208:251–260
50. Jalili A, Shirvaikar N, Marquez-Curtis L et al (2010) Fifth complement cascade protein (C5) cleavage fragments disrupt the SDF-1/CXCR4 axis: further evidence that innate immunity orchestrates the mobilization of hematopoietic stem/progenitor cells. Exp Hematol 38:321–332
51. Ratajczak MZ, Lee H, Wysoczynski M et al (2010) Novel insight into stem cell mobilization-plasma sphingosine-1-phosphate is a major chemoattractant that directs the egress of hematopoietic stem progenitor cells from the bone marrow and its level in peripheral blood increases during mobilization. Leukemia 24:976–985
52. Katayama Y, Battista M, Kao WM et al (2006) Signals from the sympathetic nervous system regulate hematopoietic stem cell egress from bone marrow. Cell 124:407–421
53. Méndez-Ferrer S, Battista M, Frenette PS (2010) Cooperation of beta(2)- and beta(3)-adrenergic receptors in hematopoietic progenitor cell mobilization. Ann N Y Acad Sci 1192:139–144
54. Jiang S, Alberich-Jorda M, Zagozdzon R et al (2011) Cannabinoid receptor 2 and its agonists mediate hematopoiesis and hamatopoietic stem and progenitor cell mobilization. Blood 117:827–838
55. Ryan MA, Nattamai KJ, Xing E et al (2010) Pharmacological inhibition of EGFR signaling enhances G-CSF-induced hematopoietic stem cell mobilization. Nat Med 10:1141–1146
56. Jalili A, Shirvaikar N, Marquez-Curtis LA, Turner AR, Janowska-Wieczorek A (2010) The HGF/c-Met axis synergizes with G-CSF in the mobilization of hematopoietic stem/progenitor cells. Stem Cells Dev 8:1143–1151
57. Tesio M, Golan K, Corso S et al (2011) Enhanced c-Met activity promotes G-CSF-induced mobilization of hematopoietic progenitor cells via ROS signaling. Blood 117:419–428
58. Gomes AL, Carvalho T, Serpa J, Torre C, Dias S (2010) Hypercholesterolemia promotes bone marrow cell mobilization by perturbing the SDF-1:CXCR4 axis. Blood 115:3886–3894
59. Tanaka Y, Yujiri T, Tanaka M, Mitani N, Tanimura A, Tanizawa Y (2009) Alteration of adipokines during peripheral blood stem cell mobilization induced by granulocyte colony-stimulating factor. J Clin Apher 24:205–208
60. Borneo J, Munugalavadla V, Sims EC et al (2007) Src family kinase-mediated negative regulation of hematopoietic stem cell mobilization involves both intrinsic and microenvironmental factors. Exp Hematol 35:1026–1037
61. Walkley CR, Shea JM, Sims NA, Purton LE, Orkin SH (2007) Rb regulates interactions between hematopoietic stem cells and their bone marrow microenvironment. Cell 129:1081–1095
62. Imbert AM, Belaaloui G, Bardin F, Tonnelle C, Lopez M, Chabannon C (2006) CD99 expressed on human mobilized peripheral blood CD34+ cells is involved in transendothelial migration. Blood 108:2578–2586

63. Selleri C, Ragno P, Ricci P et al (2006) The metastasis-associated 67-kDa laminin receptor is involved in G-CSF-induced hematopoietic stem cell mobilization. Blood 108:2476–2484
64. Gu Y, Filippi MD, Cancelas JA et al (2003) Hematopoietic cell regulation by Rac1 and Rac2 guanosine triphosphatases. Science 302:445–449
65. Cancelas JA, Lee AW, Prabhakar R, Stringer KF, Zheng Y, Williams DA (2005) Rac GTPases differentially integrate signals regulating hematopoietic stem cell localization. Nat Med 11:886–891
66. Engelhardt M, Bertz H, Afting M, Waller CF, Finke J (1999) High-versus standard-dose filgrastim (rhG-CSF) for mobilization of peripheral-blood progenitor cells from allogeneic donors and CD34(+) immunoselection. J Clin Oncol 17:2160–2172
67. Kröger N, Renges H, Krüger W et al (2000) A randomized comparison of once versus twice daily recombinant human granulocyte colony-stimulating factor (filgrastim) for stem cell mobilization in healthy donors for allogeneic transplantation. Br J Haematol 111:761–765
68. Kröger N, Sonnenberg S, Cortes-Dericks L, Freiberger P, Mollnau H, Zander AR (2004) Kinetics of G-CSF and CD34+ cell mobilization after once or twice daily stimulation with rHu granulocyte-stimulating factor (lenograstim) in healthy volunteers: an intraindividual crossover study. Transfusion 44:104–110
69. Anderlini P, Donato M, Lauppe MJ et al (2000) A comparative study of once-daily versus twice-daily filgrastim administration for the mobilization and collection of CD34+ peripheral blood progenitor cells in normal donors. Br J Haematol 109:770–772
70. Dazzi C, Cariello A, Rosti G et al (2000) Is there any difference in PBPC mobilization between cyclophosphamide plus G-CSF and G-CSF alone in patients with non-Hodgkin's Lymphoma? Leuk Lymphoma 39:301–310
71. Gertz MA, Kumar SK, Lacy MQ et al (2009) Comparison of high-dose CY and growth factor with growth factor alone for mobilization of stem cells for transplantation in patients with multiple myeloma. Bone Marrow Transplant 43:619–625
72. Gidron A, Verma A, Doyle M et al (2003) Can the stem cell mobilization technique influence CD34+ cell collection efficiency of leukapheresis procedures in patients with hematologic malignancies? Bone Marrow Transplant 35:243–246
73. Dingli D, Nowakowski GS, Dispenzieri A et al (2006) Cyclophosphamide mobilization does not improve outcome in patients receiving stem cell transplantation for multiple myeloma. Clin Lymphoma Myeloma 6:384–388
74. André M, Baudoux E, Bron D et al (2003) Phase III randomized study comparing 5 or 10 microg per kg per day of filgrastim for mobilization of peripheral blood progenitor cells with chemotherapy, followed by intensification and autologous transplantation in patients with nonmyeloid malignancies. Transfusion 43:50–57
75. Lefrère F, Zohar S, Bresson JL et al (2006) A double-blind low dose-finding phase II study of granulocyte colony-stimulating factor combined with chemotherapy for stem cell mobilization in patients with non-Hodgkin's lymphoma. Haematologica 91:550–553
76. Bruns I, Steidl U, Fischer JC et al (2008) Pegylated granulocyte colony-stimulating factor mobilizes CD34+ cells with different stem and progenitor subsets and distinct functional properties in comparison with unconjugated granulocyte colony-stimulating factor. Haematologica 93:347–355
77. Kroschinsky F, Hölig K, Poppe-Thiede K et al (2005) Single-dose pegfilgrastim for the mobilization of allogeneic CD34+ peripheral blood progenitor cells in healthy family and unrelated donors. Haematologica 90:1665–1671
78. Hill GR, Morris ES, Fuery M et al (2006) Allogeneic stem cell transplantation with peripheral blood stem cells mobilized by pegylated G-CSF. Biol Blood Marrow Transplant 12:603–607
79. Bruns I, Steidl U, Kronenwett R et al (2006) A single dose of 6 or 12 mg of pegfilgrastim for peripheral blood progenitor cell mobilization results in similar yields of CD34+ progenitors in patients with multiple myeloma. Transfusion 46:180–185

80. Höglund M, Smedmyr B, Bengtsson M et al (1997) Mobilization of CD34+ cells by glycosylated and nonglycosylated G-CSF in healthy volunteers – a comparative study. Eur J Haematol 59:177–183
81. Fischer JC, Frick M, Wassmuth R, Platz A, Punzel M, Wernet P (2005) Superior mobilisation of haematopoietic progenitor cells with glycosylated G-CSF in male but not female unrelated stem cell donors. Br J Haematol 130:740–746
82. Lefrère F, Bernard M, Audat F et al (1999) Comparison of lenograstim vs filgrastim administration following chemotherapy for peripheral blood stem cell (PBSC) collection: a retrospective study of 126 patients. Leuk Lymphoma 35:501–505
83. Sutherland HJ, Eaves CJ, Lansdorp PM, Phillips GL, Hogge DE (1994) Kinetics of committed and primitive blood progenitor mobilization after chemotherapy and growth factor treatment and their use in autotransplants. Blood 83:3808–3814
84. Benjamin RJ, Linsley L, Axelrod JD et al (1995) The collection and evaluation of peripheral blood progenitor cells sufficient for repetitive cycles of high-dose chemotherapy support. Transfusion 35:837–844
85. To LB, Dyson PG, Branford AL, Haylock DN, Kimber RJ, Juttner CA (1987) CFU-mix are no better than CFU-GM in predicting hemopoietic reconstitutive capacity of peripheral blood stem cells collected in the very early remission phase of acute nonlymphoblastic leukemia. Exp Hematol 15:351–354
86. Hepburn MD, Nagesh K, Heppleston AD, Cachia PG, Pippard MJ (2001) Timing of the appearance of multipotential and committed haemopoietic progenitors in peripheral blood after mobilization in patients with lymphoma. Clin Lab Haematol 23:119–124
87. Yu J, Leisenring W, Bensinger WI, Holmberg LA, Rowley SD (1999) The predictive value of white cell or CD34+ cell count in the peripheral blood for timing apheresis and maximizing yield. Transfusion 39:442–450
88. Armitage S, Hargreaves R, Samson D, Brennan M, Kanfer E, Navarrete C (1997) CD34 counts to predict the adequate collection of peripheral blood progenitor cells. Bone Marrow Transplant 20:587–591
89. Seggewiss R, Buss EC, Herrmann D, Goldschmidt H, Ho AD, Fruehauf S (2003) Kinetics of peripheral blood stem cell mobilization following G-CSF-supported chemotherapy. Stem Cells 21:568–574
90. Gutensohn K, Magens MM, Kuehnl P, Zeller W (2010) Increasing the economic efficacy of peripheral blood progenitor cell collections by monitoring peripheral blood CD34+ concentrations. Transfusion 50:656–662
91. Desikan KR, Jagannath S, Siegel D et al (1998) Collection of more hematopoietic progenitor cells with large volume leukapheresis in patients with multiple myeloma. Leuk Lymphoma 28:501–508
92. Lemoli RM, Fortuna A, Motta MR et al (1996) Concomitant mobilization of plasma cells and hematopoietic progenitors into peripheral blood of multiple myeloma patients: positive selection and transplantation of enriched CD34+ cells to remove circulating tumor cells. Blood 87:1625–1634
93. Lewis ID, Haylock DN, Moore S, To LB, Hughes TP (1997) Peripheral blood is a source of BCR-ABL-negative pre-progenitors in early chronic phase chronic myeloid leukemia. Leukemia 11:581–587
94. Lopez M, Lemoine FM, Firat H et al (1997) Bone marrow versus peripheral blood progenitor cells CD34 selection in patients with non-Hodgkin's lymphomas: different levels of tumor cell reduction. Implications for autografting. Blood 90:2830–2838
95. Dreger P, Viehmann K, von Neuhoff N et al (2000) A prospective study of positive/negative ex vivo B-cell depletion in patients with chronic lymphocytic leukemia. Exp Hematol 28:1187–1196
96. Abonour R, Scott KM, Kunkel LA et al (1998) Autologous transplantation of mobilized peripheral blood CD34+ cells selected by immunomagnetic procedures in patients with multiple myeloma. Bone Marrow Transplant 22:957–963

97. Dyson PG, Horvath N, Joshua D et al (2000) CD34+ selection of autologous peripheral blood stem cells for transplantation following sequential cycles of high-dose therapy and mobilization in multiple myeloma. Bone Marrow Transplant 25:1175–1184
98. Prince HM, Wall D, Rischin D et al (2002) CliniMACS CD34-selected cells to support multiple cycles of high-dose therapy. Cytotherapy 4:147–155
99. Tricot G, Gazitt Y, Leemhuis T et al (1998) Collection, tumor contamination, and engraftment kinetics of highly purified hematopoietic progenitor cells to support high dose therapy in multiple myeloma. Blood 91:4489–4495
100. Vose JM, Bierman PJ, Lynch JC et al (2001) Transplantation of highly purified CD34 + Thy-1+ hematopoietic stem cells in patients with recurrent indolent non-Hodgkin's lymphoma. Biol Blood Marrow Transplant 7:680–687
101. Gupta D, Bybee A, Cooke F et al (1999) CD34+-selected peripheral blood progenitor cell transplantation in patients with multiple myeloma: tumour cell contamination and outcome. Br J Haematol 104:166–177
102. Magni M, Di Nicola M, Devizzi L et al (2000) Successful in vivo purging of CD34-containing peripheral blood harvests in mantle cell and indolent lymphoma: evidence for a role of both chemotherapy and rituximab infusion. Blood 96:864–869
103. Pusic I, Jiang SY, Landua S et al (2008) Impact of mobilization and remobilization strategies on achieving sufficient stem cell yields for autologous transplantation. Biol Blood Marrow Transplant 14:1045–1056
104. Vasu S, Leitman SF, Tisdale JF et al (2008) Donor demographic and laboratory predictors of allogeneic peripheral blood stem cell mobilization in an ethnically diverse population. Blood 112:2092–2100
105. Tomblyn M, Gordon LI, Singhal S et al (2005) Use of total leukocyte and platelet counts to guide stem cell apheresis in healthy allogeneic donors treated with G-CSF. Bone Marrow Transplant 36:663–666
106. Bensinger W, Appelbaum F, Rowley S et al (1995) Factors that influence collection and engraftment of autologous peripheral-blood stem cells. J Clin Oncol 13:2547–2555
107. Perea G, Sureda A, Martino R et al (2001) Predictive factors for a successful mobilization of peripheral blood CD34+ cells in multiple myeloma. Ann Hematol 80:592–597
108. Mendrone A Jr, Arrais CA, Saboya R, Chamone Dde A, Dulley FL (2008) Factors affecting hematopoietic progenitor cell mobilization: an analysis of 307 patients. Transfus Apher Sci 39:187–192
109. Moskowitz CH, Glassman JR, Wuest D et al (1998) Factors affecting mobilization of peripheral blood progenitor cells in patients with lymphoma. Clin Cancer Res 4:311–316
110. Ketterer N, Salles G, Moullet I et al (1998) Factos associated with successful mobilization of peripheral blood progenitor cells in 200 patients with lymphoid malignancies. Br J Haematol 103:235–242
111. Ford CD, Green W, Warenski S, Petersen FB (2004) Effect of prior chemotherapy on hematopoietic stem cell mobilization. Bone Marrow Transplant 33:901–905
112. Hosing C, Saliba RM, Körbling M et al (2006) High-dose rituximab does not negatively affect peripheral blood stem cell mobilization kinetics in patients with intermediate-grade non-Hodgkin's lymphoma. Leuk Lymphoma 47:1290–1294
113. Popat U, Saliba R, Thandi R et al (2009) Impairment of filgrastim-induced stem cell mobilization after prior lenalidomide in patients with multiple myeloma. Biol Blood Marrow Transplant 15:718–723
114. Giralt S, Stadtmauer EA, Harousseau JL et al (2009) International myeloma working group (IMWG) consensus statement and guidelines regarding the current status of stem cell collection and high-dose therapy for multiple myeloma and the role of plerixafor (AMD 3100). Leukemia 23:1904–1912
115. Fruehauf S, Schmitt K, Veldwijk MR et al (1999) Peripheral blood progenitor cell (PBPC) counts during steady-state haemopoiesis enable the estimation of the yield of mobilized

PBPC after granulocyte colony-stimulating factor supported cytotoxic chemotherapy: an update on 100 patients. Br J Haematol 105:786–794
116. Mijovic A, Pagliuca A, Mufti GJ (1999) The "G-CSF test": the response to a single dose of granulocyte colony-stimulating factor predicts mobilization of hemopoietic progenitors in patients with hematologic malignancies. Exp Hematol 27:1204–1209
117. Lysák D, Hrabětová M, Vrzalová J et al (2011) Changes of cytokine levels during granulocyte-colony-stimulating factor stem cell mobilization in healthy donors: association with mobilization efficiency and potential predictive significance. Transfusion 51:319–327
118. Ozkurt ZN, Yegin ZA, Suyanı E et al (2010) Factors affecting stem cell mobilization for autologous hematopoietic stem cell transplantation. J Clin Apher 25(5):280–286
119. Roberts AW, Hasegawa M, Metcalf D, Foote SJ (2000) Identification of a genetic locus modulating splenomegaly induced by granulocyte colony-stimulating factor in mice. Leukemia 14:657–661
120. Bogunia-Kubik K, Gieryng A, Dlubek D, Lange A (2009) The CXCL12-3′A allele is associated with a higher mobilization yield of CD34 progenitors to the peripheral blood of healthy donors for allogeneic transplantation. Bone Marrow Transplant 44:273–278
121. Benboubker L, Watier H, Carion A et al (2001) Association between the SDF1-3′A allele and high levels of CD34(+) progenitor cells mobilized into peripheral blood in humans. Br J Haematol 113:247–250
122. Lie AK, Hui CH, Rawling T et al (1998) Granulocyte colony-stimulating factor (G-CSF) dose-dependent efficacy in peripheral blood stem cell mobilization in patients who had failed initial mobilization with chemotherapy and G-CSF. Bone Marrow Transplant 22:853–857
123. Lefrère F, Lévy V, Makke J, Audat F, Cavazzana-Calvo M, Micléa JM (2004) Successful peripheral blood stem cell harvesting with granulocyte colony-stimulating factor alone after previous mobilization failure. Haematologica 89:1532–1534
124. Johnsen HE, Hansen PB, Plesner T et al (1992) Increased yield of myeloid progenitor cells in bone marrow harvested for autologous transplantation by pretreatment with recombinant human granulocyte-colony stimulating factor. Bone Marrow Transplant 10:229–234
125. Lemoli RM, de Vivo A, Damiani D et al (2003) Autologous transplantation of granulocyte colony-stimulating factor-primed bone marrow is effective in supporting myeloablative chemotherapy in patients with hematologic malignancies and poor peripheral blood stem cell mobilization. Blood 102:1595–1600
126. Wood WA, Whitley J, Moore D et al (2011) Chemomobilization with etoposide is highly effective in patients with multiple myeloma and overcomes the effects of age and prior therapy. Biol Blood Marrow Transplant 17:141–146
127. Dawson MA, Schwarer AP, Muirhead JL, Bailey MJ, Bollard GM, Spencer A (2005) Successful mobilization of peripheral blood stem cells using recombinant human stem cell factor in heavily pretreated patients who have failed a previous attempt with a granulocyte colony-stimulating factor-based regimen. Bone Marrow Transplant 36:389–396
128. Shpall EJ, Wheeler CA, Turner SA et al (1999) A randomized phase 3 study of peripheral blood progenitor cell mobilization with stem cell factor and filgrastim in high-risk breast cancer patients. Blood 93:2491–2501
129. Stiff P, Gingrich R, Luger S et al (2000) A randomized phase 2 study of PBPC mobilization by stem cell factor and filgrastim in heavily pretreated patients with Hodgkin's disease or non-Hodgkin's lymphoma. Bone Marrow Transplant 26:471–481
130. Herbert KE, Morgan S, Prince HM et al (2009) Stem cell factor and high-dose twice daily filgrastim is an effective strategy for peripheral blood stem cell mobilization in patients with indolent lymphoproliferative disorders previously treated with fludarabine: results of a phase II study. Leukemia 23:305–312
131. Spitzer G, Adkins D, Mathews M et al (1997) Randomized comparison of G-CSF+ GM-CSF vs G-CSF alone for mobilization of peripheral blood stem cells: effects on hematopoietic recovery after high-dose chemotherapy. Bone Marrow Transplant 20:921–930

132. Hart C, Grassinger J, Andreesen R, Hennemann B (2009) EPO in combination with G-CSF improves mobilization effectiveness after chemotherapy with ifosfamide, epirubicin and etoposide and reduces costs during mobilization and transplantation of autologous hematopoietic progenitor cells. Bone Marrow Transplant 43:197–206
133. Linker C, Anderlini P, Herzig R et al (2003) Recombinant human thrombopoietin augments mobilization of peripheral blood progenitor cells for autologous transplantation. Biol Blood Marrow Transplant 9:405–413
134. Carlo-Stella C, Di Nicola M, Milani R et al (2004) Use of recombinant human growth hormone (rhGH) plus recombinant human granulocyte colony-stimulating factor (rhG-CSF) for the mobilization and collection of CD34+ cells in poor mobilizers. Blood 103:3287–3295
135. Ballen KK, Shpall EJ, Avigan D et al (2007) Phase I trial of parathyroid hormone to facilitate stem cell mobilization. Biol Blood Marrow Transplant 13:838–843
136. Herbert KE, Walkley CR, Winkler IG et al (2007) Granulocyte colony-stimulating factor and an RARalpha specific agonist, VTP195183, synergize to enhance the mobilization of hematopoietic progenitor cells. Transplantation 83:375–384
137. Donahue RE, Jin P, Bonifacino AC et al (2009) Plerixafor (AMD3100) and granulocyte colony-stimulating factor (G-CSF) mobilize different CD34+ cell populations based on global gene and microRNA expression signatures. Blood 114:2530–2541
138. Larochelle A, Krouse A, Metzger M et al (2006) AMD3100 mobilizes hematopoietic stem cells with long-term repopulating capacity in nonhuman primates. Blood 107:3772–3778
139. Taubert I, Saffrich R, Zepeda-Moreno A et al (2011) Characterization of hematopoietic stem cell subsets from patients with multiple myeloma after mobilization with plerixafor. Cytotherapy 13:459–466
140. DiPersio JF, Stadtmauer EA, Nademanee A et al (2009) Plerixafor and G-CSF versus placebo and G-CSF to mobilize hematopoietic stem cells for autologous stem cell transplantation in patients with multiple myeloma. Blood 113:5720–5726
141. DiPersio JF, Micallef IN, Stiff PJ et al (2009) Phase III prospective randomized double-blind placebo-controlled trial of plerixafor plus granulocyte colony-stimulating factor compared with placebo plus granulocyte colony-stimulating factor for autologous stem-cell mobilization and transplantation for patients with non-Hodgkin's lymphoma. J Clin Oncol 27:4767–4773
142. DiPersio JF, Micallef IN, Stiff PJ et al (2009) Successful stem cell remobilization using plerixafor (mozobil) plus granulocyte colony stimulating factor in patients with non-hodgkin lymphoma: results from the plerixafor NHL phase 3 study rescue protocol. Biol Blood Marrow Transplant 15:1578–1586
143. Micallef IN, Ho AD, Klein LM, Marulkar S, Gandhi PJ, McSweeney PA (2011) Plerixafor (Mozobil) for stem cell mobilization in patients with multiple myeloma previously treated with lenalidomide. Bone Marrow Transplant 46:350–355
144. Basak GW, Knopinska-Posluszny W et al (2010) Hematopoietic stem cell mobilization with the reversible CXCR4 receptor inhibitor plerixafor (AMD3100) – Polish compassionate use experience. Ann Hematol 90:557–568
145. Weaver CH, Hazelton B, Birch R et al (1995) An analysis of engraftment kinetics as a function of the CD34 content of peripheral blood progenitor cell collections in 692 patients after the administration of myeloablative chemotherapy. Blood 86:3961–3969
146. Stiff PJ, Micallef I, Nademanee AP et al (2011) Transplanted CD34(+) cell dose is associated with long-term platelet count recovery following autologous peripheral blood stem cell transplant in patients with non-Hodgkin's lymphoma or multiple myeloma. Biol Blood Marrow Transplant 17(8):1146–1153
147. Theilgaard-Mönch K, Raaschou-Jensen K, Andersen H et al (1999) Single leukapheresis products collected from healthy donors after the administration of granulocyte colony-stimulating factor contain ten-fold higher numbers of long-term reconstituting hematopoietic

progenitor cells than conventional bone marrow allograft. Bone Marrow Transplant 23:243–249
148. Theilgaard-Mönch K, Raaschou-Jensen K, Schjødt K et al (2003) Pluripotent and myeloid-committed CD34+ subsets in hematopoietic stem cell allografts. Bone Marrow Transplant 32:1125–1133
149. Dercksen MW, Gerritsen WR, Rodenhuis S et al (1995) Expression of adhesion molecules on CD34+ cells: CD34+ L-selectin+ cells predict a rapid platelet recovery after peripheral blood stem cell transplantation. Blood 85:3313–3319
150. Pratt G, Rawstron AC, English AE et al (2001) Analysis of CD34+ cell subsets in stem cell harvests can more reliably predict rapidity and durability of engraftment than total CD34+ cell dose, but steady state levels do not correlate with bone marrow reserve. Br J Haematol 114:937–943
151. Pecora AL, Preti RA, Gleim GW et al (1998) CD34+ CD33- cells influence days to engraftment and transfusion requirements in autologous blood stem-cell recipients. J Clin Oncol 16:2093–2104
152. Prabhash K, Khattry N, Bakshi A et al (2010) CD26 expression in donor stem cell harvest and its correlation with engraftment in human haematopoietic stem cell transplantation: potential predictor of early engraftment. Ann Oncol 21:582–588
153. Spencer A, Jackson J, Baulch-Brown C (2001) Enumeration of bone marrow 'homing' haemopoietic stem cells from G-CSF-mobilised normal donors and influence on engraftment following allogeneic transplantation. Bone Marrow Transplant 28:1019–1022
154. Haylock DN, To LB, Dowse TL, Juttner CA, Simmons PJ (1992) Ex vivo expansion and maturation of peripheral blood CD34+ cells into the myeloid lineage. Blood 80:1405–1412
155. Makino S, Haylock DN, Dowse T et al (1997) Ex vivo culture of peripheral blood CD34+ cells: effects of hematopoietic growth factors on production of neutrophilic precursors. J Hematother 6:475–489
156. Shapiro F, Yao TJ, Raptis G, Reich L, Norton L, Moore MA (1994) Optimization of conditions for ex vivo expansion of CD34+ cells from patients with stage IV breast cancer. Blood 84:3567–3574
157. Lazzari L, Henschler R, Lecchi L, Rebulla P, Mertelsmann R, Sirchia G (2000) Interleukin-6 and interleukin-11 act synergistically with thrombopoietin and stem cell factor to modulate ex vivo expansion of human CD41+ and CD61+ megakaryocytic cells. Haematologica 85:25–30
158. Boiron JM, Dazey B, Cailliot C et al (2006) Large-scale expansion and transplantation of CD34(+) hematopoietic cells: in vitro and in vivo confirmation of neutropenia abrogation related to the expansion process without impairment of the long-term engraftment capacity. Transfusion 46:1934–1942
159. Watts KL, Delaney C, Humphries RK, Bernstein ID, Kiem HP (2010) Combination of HOXB4 and Delta-1 ligand improves expansion of cord blood cells. Blood 116:5859–5866
160. Schmitz N, Beksac M, Bacigalupo A et al (2005) Filgrastim-mobilized peripheral blood progenitor cells versus bone marrow transplantation for treating leukemia: 3-year results from the EBMT randomized trial. Haematologica 90:643–648
161. Dey BR, Shaffer J, Yee AJ et al (2007) Comparison of outcomes after transplantation of peripheral blood stem cells versus bone marrow following an identical nonmyeloablative conditioning regimen. Bone Marrow Transplant 40(40):19–27
162. Mohty M, Bilger K, Jourdan E et al (2003) Higher doses of CD34+ peripheral blood stem cells are associated with increased mortality from chronic graft-versus-host disease after allogeneic HLA-identical sibling transplantation. Leukemia 17:869–875
163. Joshi SS, Lynch JC, Pavletic SZ et al (2001) Decreased immune functions of blood cells following mobilization with granulocyte colony-stimulating factor: association with donor characteristics. Blood 98:1963–1970

164. Miller JS, Prosper F, McCullar V (1997) Natural killer (NK) cells are functionally abnormal and NK cell progenitors are diminished in granulocyte colony-stimulating factor-mobilized peripheral blood progenitor cell collections. Blood 90:3098–3105
165. Rondelli D, Raspadori D, Anasetti C et al (1998) Alloantigen presenting capacity, T cell alloreactivity and NK function of G-CSF-mobilized peripheral blood cells. Bone Marrow Transplant 22:631–637
166. Yamasaki S, Henzan H, Ohno Y et al (2003) Influence of transplanted dose of CD56+ cells on development of graft-versus-host disease in patients receiving G-CSF-mobilized peripheral blood progenitor cells from HLA-identical sibling donors. Bone Marrow Transplant 32:505–510
167. Arpinati M, Green CL, Heimfeld S, Heuser JE, Anasetti C (2000) Granulocyte-colony stimulating factor mobilizes T helper 2-inducing dendritic cells. Blood 95:2484–2490
168. Sun LX, Ren HY, Shi YJ, Wang LH, Qiu ZX (2009) Recombinant human granulocyte colony-stimulating factor significantly decreases the expression of CXCR3 and CCR6 on T cells and preferentially induces T helper cells to a T helper 17 phenotype in peripheral blood harvests. Biol Blood Marrow Transplant 15:835–843
169. Hill GR, Olver SD, Kuns RD et al (2010) Stem cell mobilization with G-CSF induces Type-17 differentiation and promotes scleroderma. Blood 116(5):819–828
170. Hill GR, Kuns RD, Raffelt NC et al (2010) SOCS3 regulates graft-versus-host disease. Blood 116:287–296
171. Group SCTC (2005) Allogeneic peripheral blood stem-cell compared with bone marrow transplantation in the management of hematologic malignancies: an individual patient data meta-analysis of nine randomized trials. J Clin Oncol 23:5074–5087
172. Morris ES, MacDonald KP, Hill GR (2006) Stem cell mobilization with G-CSF analogs: a rational approach to separate GVHD and GVL? Blood 107:3430–3455
173. Hiwase DK, Hiwase S, Bailey M, Bollard G, Schwarer AP (2008) The role of stem cell mobilization regimen on lymphocyte collection yield in patients with multiple myeloma. Cytotherapy 10:507–517
174. LB, Levesque J-P, Herbert KE. (2011) How I treat patients who mobilize haematopoietic stem cells poorly. Blood. Published ahead of print August 10, 2011, doi:10.1182/blood-2011-06-318220
175. Herbert KE, Levesque J-P, Mills AK, Gottlieb DJ, Cooney J, Szer J, Rasko J, To LB. (2011) How we mobilize haematopoietic stem cells. Internal Medicine Journal 41:588–594
176. To LB, Levesque J-P, Herbert KE, Winkler I, Bendall L, Hiwase D, Antonenas V, Rice A, Gottlieb D, Mills A, Rasko J, Larsen S, Beligaswatte A, Nilsson S, Cooney J, Cambareri T, Lewis I. Mobilization strategy for normal and malignant cells. (2011) Pathology. In press

Part III
Use of rHuG-CSF in Non-Oncology Setting

rHuG-CSF for the Treatment of Severe Chronic Neutropenia

David C. Dale and Audrey Anna Bolyard

1 Introduction

Severe chronic neutropenia (SCN) is a general term used to describe patients with blood neutrophil counts either chronically or cyclically less than 500 neutrophils per cubic millimeter (i.e., $<0.5 \times 10^9$/L). Many causes for this condition are known, both hereditary and acquired diseases. The common feature is that low blood neutrophil counts predispose patients to develop fever, ulcerations along the gastrointestinal tract from the mouth to the anus, and recurrent infections. The risk of these events varies depending upon the patient's general health, the cause of the neutropenia, and the capacity of the bone marrow to produce more neutrophils when infections occur [1, 2].

2 Granulocyte Colony-Stimulating Factor

The development of recombinant human granulocyte colony stimulating factor (rHuG-CSF) dramatically changed the lives of patients with SCN. We now know that most patients with chronic neutropenia, regardless of the cause, will increase their blood neutrophil counts in response to treatment with rHuG-CSF, administered subcutaneously on a daily or alternate-day basis. For most patients, the response is sustained for the duration of therapy [3, 4]. This chapter reviews the

D.C. Dale (✉)
Division of General Internal Medicine, University of Washington, 1959 NE Pacific St, AA 522 Health Sciences Building, Box 356422, Seattle, WA 98195, USA
e-mail: dcdale@u.washington.edu

A.A. Bolyard
Division of General Internal Medicine, University of Washington, Box 356422, Seattle, WA, USA

G. Molineux et al. (eds.), *Twenty Years of G-CSF*, Milestones in Drug Therapy,
DOI 10.1007/978-3-0348-0218-5_15,

development of rHuG-CSF for treatment of SCN and provides a summary of current recommendations for its appropriate use in these patients.

3 Nonclinical Studies

Soon after G-CSF was produced by recombinant DNA technology, research began on its clinical applications, using nonclinical models. Chemotherapy-induced neutropenia served as a principal therapeutic model. Animal models were also important for the development of rHuG-CSF as a therapy for chronic neutropenia.

Beginning in the late 1980s, Hammond et al. began nonclinical studies on rG-CSF treatment of normal dogs and gray collie dogs with cyclic neutropenia [5]. The researchers observed that normal dogs given repeated doses of rHuG-CSF developed chronic neutropenia due to cross-reacting antibodies to the human growth factor. This observation was the first evidence that G-CSF is the cytokine regulating and maintaining blood neutropenia counts [5]. After the canine *G-CSF* gene was cloned and canine G-CSF was produced in sufficient quantities, long-term trials showed that G-CSF is a very effective long-term treatment for canine cyclic neutropenia [6]. Other human growth factors were either ineffective or lost their effectiveness with chronic use [6]. Subsequent studies showed that canine G-CSF delivered through gene therapy is effective as a long-term treatment strategy in canine cyclic neutropenia, but similar studies had not been performed in humans [7].

4 Phase 2 Clinical Trials

Clinical trials of rHuG-CSF to treat SCN began in 1987. In the initial phase 1 and 2 trials, patients with cyclic neutropenia, congenital neutropenia, and chronic idiopathic neutropenia were given daily rHuG-CSF first for a few days and then for longer periods. The doses were adjusted at 1- to 2-week intervals with the goal of increasing blood neutrophil counts to normal values. The studies showed that almost all patients responded with somewhat different patterns of response for each of these diseases [7–10].

4.1 Cyclic Neutropenia

Cyclic neutropenia was described as a distinct clinical entity more than 100 years ago; it is probably the first recognized immune deficiency syndrome. Cyclic neutropenia is a very rare autosomal dominant disease now known to be caused by mutations in the gene for neutrophil elastase, the *ELA2* or *ELANE* gene [11]. Characteristically, patients have 3-week cycles of blood neutrophils with no or very few neutrophils in the blood for several days at the neutrophil nadirs. After each nadir, counts then increase briskly to a peak that is often $<2.0 \times 10^9$/L. During the severe neutropenic period, patients regularly have fever, mouth ulcers, cellulites, and bacterial infections [12]. Many reports exist of deaths due to septicemia, often with clostridial infections, during the neutropenic periods [13, 14].

The phase 2 trial of G-CSF for the treatment of cyclic neutropenia began soon after rHuG-CSF became available for clinical testing. The rationale was that these patients have a serious disease needing an effective therapy. The predictable neutropenia and problems with infections also made this disease a good model for testing the effects of this new agent [7]. Originally, rHuG-CSF was administered intravenously on a daily basis, until it was learned that subcutaneous administration was equally efficacious. A nonrandomized trial in six patients demonstrated that rHuG-CSF increased the amplitude of blood neutrophil cycles and shortened the periodicity of oscillations from the usual 21 days to about 14 days. Although G-CSF did not prevent cyclic oscillations in the blood counts, it clearly reduced days of severe neutropenia and prevented mouth ulcers, fevers, and serious infections [7].

4.2 Severe Congenital Neutropenia

Severe congenital neutropenia is also a very rare inherited disorder. It was first described as an autosomal recessive disorder occurring in a family in Northern Sweden and called Kostmann syndrome, named after the physician who originally described this disorder. This form of severe congenital neutropenia is now attributable to mutations in the *HAX1* gene [15, 16]. It is known that severe congenital neutropenia is more commonly inherited as an autosomal dominant condition attributable to mutations in the gene for neutrophil elastase, the *ELANE* gene [17]. Several other rare genetic causes have been identified, some with congenital anomalies which aid their diagnoses [18–22].

The original phase 2 clinical trials were performed before the diverse genetic causes of severe congenital neutropenia were known, so all of the cases were grouped together as congenital neutropenia. These initial studies showed that most patients respond, but the dose required to increase and maintain neutrophil counts in the normal range varied considerably. Some patients had a delay of several days before they responded. When blood neutrophils increased, they appeared to oscillate or cycle in some patients [23]. The clinical benefits of increasing the blood neutrophil counts were quickly obvious. Infections cleared and new infections occurred less often with normalization of neutrophil production and deployment [24].

4.3 Chronic Idiopathic Neutropenia

Chronic idiopathic neutropenia (CIN) is much more common than cyclic or congenital neutropenia. Its exact frequency is not known, but it is presumed to occur on an immune or autoimmune basis [25, 26]. Predominantly women are affected. Although the demographic features of patients with CIN and systemic lupus erythematosus (SLE) are similar, there is no good evidence that these are overlapping conditions or that CIN leads to the development of SLE or one of its variants. Blood neutrophil counts vary considerably in patients with this diagnosis, within the population, and over time in a given patient. Patients with blood neutrophil counts $<0.5 \times 10^9/L$ tend to have recurrent fevers and bacterial infections, and these problems are the greatest for those with the lowest counts.

The original clinical trials of treatment of CIN with rHuG-CSF showed that patients respond rapidly at very low doses. The time course for the response was relatively brief, and patients appeared to benefit from the increase in blood neutrophils soon after initiation of rHuG-CSF [9].

5 The Phase 3 Randomized Trial

By early 1988, all available evidence indicated that rHuG-CSF would be a very effective therapy for cyclic, congenital, and idiopathic neutropenia. A multicentered, randomized, phase 3 trial involving 123 patients then clearly demonstrated its effectiveness to increase blood neutrophil counts and prevent fever, mouth ulcers, and infections [27]. In this trial, the dose of rHuG-CSF was titrated to achieve a blood neutrophil count of $2.0 \times 10^9/L$, and the patients were maintained on this dose for a 4-month period. This dosage proved to be sufficient time to prove benefit, but the period of randomized treatment was not long enough to determine if rHuG-CSF was associated with long-term adverse effects.

6 The Severe Chronic Neutropenia International Registry

To determine the long-term effectiveness and safety of rHuG-CSF treatment, a group of investigators formed the Severe Chronic Neutropenia International Registry (SCNIR) in 1994 under the sponsorship of Amgen, the biopharmaceutical firm manufacturing rHuG-CSF for the initial studies [3, 28]. Patients from the original phase 2 and phase 3 trials plus new patients were enrolled and followed in this observational study, which has continued with sponsorship of the National Institutes of Health in the USA and with multi-governmental support in Europe for more than 16 years. This structure provided a data base for estimating the frequency of adverse events, and the enrollment of a diverse population allowed comparisons of treatment responses and treatment outcomes. In addition, the study has reached more than 60 countries and raised awareness of the clinical problem of chronic neutropenia. Through the SCNIR, the diverse genetic and acquired causes for SCN have been identified and progress in understanding these conditions has followed. Results from the SCNIR recommendations are summarized below and in other reports (Table 1).

Table 1 rHuG-CSF treatment of patients with severe chronic neutropenia

Neutropenia diagnosis	No. of pt	Baseline ANC ($\times 10^9$/L)	rHuG-CSF dose (μg/kg/day) yr 1 median (range)	ANC after 1 yr treatment ($\times 10^9$/L) median (range)	Yr treatment median (range)
Congenital (all causes)	415	0.3 (0.0–10.3)	5.0 (0.0–183.2)	2.0 (0.0–30.7)	7 (1–22)
SDS	29	0.4 (0.1–10.3)	1.0 (0.1–30.7)	2.7 (0.4–17.5)	6 (1–13)
Barth syndrome	8	0.4 (0.1–2.0)	2.3 (0.6–5.5)	1.3 (0.3–5.0)	11 (4–12)
GSD 1b	59	0.3 (0.1–7.7)	1.0 (0.0–7.2)	3.8 (0.5–63.2)	10 (2–22)
Cyclic	169	0.6 (0.1–2.0)	2.9 (0.0–14.1)	4.9 (0.0–19.8)	11 (1–23)
Idiopathic	355	0.4 (0.0–6.7)	1.9 (0.0–40.0)	2.2 (0.2–13.0)	7 (1–22)
Autoimmune	41	0.7 (0.1–7.7)	1.4 (0.0–4.3)	0.5 (0.5–2.0)	3 (1–11)
Total	980				

Data are from the Severe Chronic Neutropenia Registry, Seattle, WA, USA
ANC absolute neutrophil count; *GSD 1b* glycogen storage disease 1b; *pt* patient; *SDS* Shwachman–Diamond syndrome; *yr* year

7 Hereditary Diseases' Diagnosis and Recommendations

7.1 Cyclic Neutropenia

The diagnosis of cyclic neutropenia depends on finding regular oscillations in blood neutrophil counts at approximately 21-day intervals with intervening periods of very severe neutropenia. At least 3-times-per-week counts for 6 weeks are required to make this diagnosis, unless it has already been established in a first- or second-degree relative. Genotyping by sequencing of the *ELANE* gene is insufficient to establish the diagnosis of cyclic neutropenia because there is overlap of the genotype for cyclic and autosomal dominant severe congenital neutropenia (ADSCN) [11, 29]. It is important to make this distinction, because ADSCN has a risk of conversion to acute myeloid leukemia (AML), whereas patients with cyclic neutropenia are apparently not at risk [30].

Patients with cyclic neutropenia usually respond to treatment with rHuG-CSF (2–4 μg/kg/day) [3, 28]. They should be maintained on a dose (adjusted for body surface area or weight) to maintain a mean absolute neutrophil count (ANC) in the range of 1.0–2.0 $\times$ 10^9/L, determined by blood counts several times per year. The SCNIR does not recommend regular or repeat bone marrow examinations to follow these patients.

7.2 Severe Congenital Neutropenia

Approximately 60% of patients with the clinical diagnosis congenital neutropenia, i.e., noncyclic severe neutropenia recognized soon after birth, will have *ELANE* mutations [21]. More than 60 different mutations in the *ELANE* gene have been identified in this population [29]. The protein product of the mutated gene appears to accumulate in the cytoplasm rather than in the primary granules of neutrophils and to trigger the "unfolded protein response" and accelerated apoptosis of developing myeloid cells [31, 32], which is the apparent cause for the classic observation of "maturation arrest" of myeloid cells in the marrow in patients with severe congenital neutropenia.

The responsiveness to rHuG-CSF is similar for patients with and without *ELANE* mutations, and the risk of evolution to AML is also similar [33]. More than 90% of patients will respond to rHuG-CSF by increased blood neutrophil counts to normal or near-normal levels. The doses required vary substantially from 1.0 to >50 μg/kg/day [34]. Therefore, titration of the rHuG-CSF dose is required in each patient, generally beginning at a dose of 2–4 μg/kg/day and with dose adjustments on a weekly or biweekly basis. In general, blood neutrophil counts $>1.0 \times 10^9$/L are sufficient to prevent mouth ulcers and infections.

Mutations in the *HCLS1-associated protein X1* (*HAX1*) gene and *glucose 6 phosphatase, catalytic, 3 (G6PC3)* gene are also recognized as the cause for autosomal recessive severe congenital neutropenia [18, 35]. Patients with *HAX1* mutations often have congenital neurologic abnormalities, and patients with *G6PC3* mutations have cardiac and urogenital anomalies. Because these cases all are much less common than those attributable to *ELANE* mutations, the genotype–phenotype relationships and the risk of leukemic evolution are not well known. It appears, however, that these cases should be considered along with other patients with severe congenital neutropenia as being at risk of developing AML.

A few patients with mutations in the *Wiskott–Aldrich syndrome* (*WAS)* [36] and *growth factor independent transcription repressor (GFI 1*) [20] and the *G-CSF receptor* gene [37] have SCN. The clinical course and risk of malignant evolution for these cases are not known. About 30% of patients with the clinical diagnosis of severe congenital neutropenia will not have a mutation in *ELANE, HAX-1, G6PC3, GFI 1*, or *WAS*. As noted, these patients, however, respond to rHuG-CSF similar to patients with *ELANE* mutations, and recommendations for rHuG-CSF dose titration and schedule are the same.

At present, it is prudent to consider all patients with severe congenital neutropenia to be at similar risk of evolution to AML. Specific *ELANE* mutations do not appear to predict the therapeutic response to rHuG-CSF or the risk of leukemic evolution. The largest numbers of cases of AML have occurred in patients with the G185R mutation, but there are a higher proportion of cases evolving to leukemia for several other mutations [29]. For this reason, the SCNIR recommends that all patients with congenital neutropenia should be followed closely with clinical observations, blood cell counts, and annual bone marrow examinations, including cytogenetic testing. Patients requiring higher doses of rHuG-CSF are at greater risk of evolving to myelodysplastic syndromes (MDS)/AML, presumably because the intrinsic defect in neutrophil formation makes these patients more resistant to rHuG-CSF therapy [30, 33]. Patients requiring rHuG-CSF at doses >8–10 μg/kg/day and those developing anemia, thrombocytopenia, or refractoriness to rHuG-CSF or showing chromosomal changes, particularly monosomy 7, as signs of leukemic evolution, should be considered for hematopoietic transplantation. Transplantation is currently the only effective alternative to rHuG-CSF therapy [38].

7.3 Glycogen Storage Disease Type 1b

Glycogen storage disease 1b (GSD 1b) is caused by mutations in the *glucose 6 phosphate transporter (G6PT)* gene that moves glucose across cellular membranes, particularly into the cell's endoplasmic reticulum. Mutations in this gene cause a multiplicity of defects in neutrophil function, reflecting the critical role of glucose in neutrophil trafficking, oxidative burst, and bactericidal functions [39]. Neutropenia is attributable to poor survival of neutrophils, both in the marrow and in the blood [40].

Treatment with rHuG-CSF is particularly effective to reduce the occurrence and severity in GSD 1b because it both improves neutrophil functions and increases blood neutrophil counts. The exact mechanisms are not known, but presumably involve the anti-apoptotic effects of G-CSF as well as increase in the expression of multiple proteins related to chemotaxis and bacterial killing [41]. rHuG-CSF also reduces the occurrence and severity of the inflammatory bowel disease associated with GSD1b [42]. Splenomegaly, even massive splenomegaly, is a risk, but can generally be managed or avoided by using only the minimal dose to maintain blood neutrophil counts at or near 1.0×10^9/L [43].

7.4 Shwachman–Diamond Syndrome

Shwachman–Diamond syndrome (SDS) is characterized by neutropenia, bone marrow failure, pancreatic dysfunction, and skeletal anomalies [44]. It is attributable to mutations in the *Shwachman–Blackfan–Diamond* (*SBDS*) gene, mutations which lead to abnormalities in cell division and ribosomal functions [45]. Through the SCNIR, records are available for long-term treatment of >30 patients with SDS, approximately equal numbers of children and adults, followed for a median of 4 years (range: 2–13 years) [46]. Blood neutrophil counts were $<0.5 \times 10^9$/L in most of these patients. Treatment with rHuG-CSF increased blood neutrophil counts to the normal range in all cases. Leukemic evolution has been observed similar to that in patients with severe congenital neutropenia, both with and without rHuG-CSF therapy [47].

7.5 Barth Syndrome

Barth syndrome is an X-linked recessive disease characterized by neutropenia, cardiomyopathy, growth delay, and muscle weakness [48]. Its clinical manifestations include fatigue, hypokalemia, and congestive heart failure. It is attributable to mutations in the *tafazzin* (*TAZ)* gene. Through the SCNIR, records are available on the treatment of six patients for a median of 8 years. Patients are very responsive to rHuG-CSF therapy, which appears to reduce the occurrence and severity of their infectious complications. Leukemic evolution has not been observed in this disorder [49].

7.6 Other Neutropenia-Inherited Syndromes

WHIM syndrome/myelokathexis is a rare autosomal dominant cause for severe leukopenia and neutropenia. These patients often have total white blood cell counts $<1.0 \times 10^9$/L, attributable both to neutropenia and lymphocytopenia. Other blood cell counts are normal. This disorder is attributable to mutations in the *chemokine receptor 4* (*CXCR4)* gene, which encodes a key receptor regulating the trafficking of neutrophils and lymphocytes from the marrow to blood and tissues [50]. Patients may have severe warts and hypogammaglobulinemia. Blood neutrophil counts increase with rHuG-CSF therapy, but this treatment does not correct the lymphocytopenia or serve to improve the problem with warts [51].

In the Wiskott–Aldrich syndrome, neutropenia is associated with specific mutations, i.e., missense mutations in the Cdc42-binding domain of the Wiskott–Aldrich protein [52]. In other immunodeficiency disorders, the occurrence and severity of neutropenia is quite variable and poorly understood. Most of these patients will respond to treatment with rHuG-CSF. Treatment should be determined based on the severity of neutropenia, the propensity to infections, and a brief clinical trial in each individual patient.

8 Acquired Severe Neutropenia

8.1 Neonatal Immune and Autoimmune Neutropenia

Early in infancy, neutropenia can be caused by maternal antibodies to specific antigens expressed on neonatal neutrophils. This condition, immune neutropenia of infancy, is transient and usually resolves spontaneously within a few weeks after birth. Neutropenia can also occur in neonates of mothers with idiopathic neutropenia and as a consequence of prematurity [53]. rHuG-CSF may be effective to accelerate normalization of neutrophils in these patients, but there are no randomized trials showing the clinical effectiveness of this therapy [54].

8.2 Chronic Idiopathic Neutropenia

Chronic idiopathic neutropenia is a poorly understood condition affecting both children and adults. In children, it is often called "benign neutropenia of childhood [55]." In both groups, the level of blood neutrophils and the capacity of the marrow to generate more cells rapidly are the determinants of the clinical course. In general, patients with blood neutrophil counts $<0.5 \times 10^9$/L have recurrent fevers, and those with lower counts have recurrent episodes of cellulitis, pharyngitis, sinusitis, and occasional and unpredictable more-severe infections.

Chronic idiopathic neutropenia is difficult to distinguish from chronic autoimmune neutropenia because methods for detecting autoantibodies to neutrophil antigens are not widely available or necessarily predictive of specific disease patterns. In both idiopathic and autoimmune neutropenia, marrow neutrophils are normal or nearly normal in numbers and morphology. It is perhaps for this reason that these patients respond promptly to rHuG-CSF and can be effectively treated with 1–3 μg/kg/day or on an alternate-day basis. There is no evidence that patients with positive antibody tests are less responsive to rHuG-CSF or that therapy with the cytokine loses its effectiveness with chronic therapy [28].

8.3 Rheumatoid Arthritis and Acquired Large Granular Lymphocyte Syndrome

Neutropenia is an uncommon feature of rheumatoid arthritis, but it occurs in approximately 1–3% of patients often in association with splenomegaly, a high rheumatoid factor values, and sometimes with increased large granular lymphocytes in the circulation [56]. The neutropenia may be moderate or severe, and bacterial infections complicate the most severe cases. Fas-mediated apoptosis of neutrophils is currently thought to be a primary mechanism for neutropenia in these cases, but other mechanisms may be involved as well [57]. rHuG-CSF is a therapeutic option, but patients should be cautioned that it may trigger a worsening of their arthritic symptoms. Combinations of rHuG-CSF and methotrexate, rituximab, or other biologic agents are other therapeutic options [58].

8.4 HIV Infection and Neutropenia

Neutropenia was a common complication of HIV infection before the advent of effective antiviral therapies. In this era, it was learned that most HIV-associated neutropenia is responsive to rHuG-CSF administered chronically and in relatively low doses on a daily or alternate-day basis. Clinical trials suggested but did not prove the effectiveness of this therapy, largely because of the complexity and difficulty in the conduct of the randomized trials [59].

9 Guidelines for Treatment of Chronic Neutropenia with G-CSF

1. Treat patients based on diagnoses, frequency of fevers, and infection and not solely based on blood neutrophil counts. In cases of presumed congenital neutropenia, have patients maintain a careful daily diary of temperature, mouth ulcers, and evidence of infections – minor and major. A diary of antibiotic treatments is also helpful. This clinical information together with serial blood counts helps to make the important distinction of cyclic versus congenital neutropenia.
2. Initiate rHuG-CSF therapy at 1–2 μg/kg/day and gradually increase the dose. Initially, use daily therapy until the target ANC is reached (1.0×10^9/L); increase the dose interval to every other day or 3 days per week if the patient responds readily. Starting at a low dose minimizes bone pain and other acute adverse events.
3. If the initial therapy is insufficient to reach a target of 1.0×10^9/L, gradually increase the dose, usually double the dose, at 1- or 2-week intervals.
4. In patients with the clinical diagnosis of severe congenital neutropenia, consult with a neutropenia specialist if a dose >10 μg/kg/day is required to achieve a neutrophil count of 1.0×10^9/L.
5. In patients with idiopathic and autoimmune neutropenia, it is reasonable to try to discontinue therapy at any time, which is best achieved by reducing the dose of rHuG-CSF by 50% in a serial fashion and observing blood neutrophil counts after a few days on the reduced dose. Patients should be advised that idiopathic and autoimmune neutropenia usually lasts indefinitely, but minimizing the rHuG-CSF dose can best be done by careful dosage adjustment in this fashion.
6. rHuG-CSF is widely effective for the treatment of neutropenia and is well tolerated. Careful observation of the response to treatment with serial blood counts should guide therapy. Most responding patients can be maintained on the same dose for long periods.
7. Bone pain and headache are the common adverse events associated with starting treatment with rHuG-CSF.
8. Bone density measurements may be useful to follow patients for the development of osteopenia or osteoporosis.
9. Patients with severe congenital neutropenia and SDS are at risk of evolution to leukemia, with or without rHuG-CSF therapy. They should be followed carefully with clinical examinations and blood and bone marrow examinations at regular intervals.

Acknowledgments The authors express appreciation to the Advisory Board members and staff of the SCNIR and referring physicians, patients, and families for contributing to the data for this report. We also express appreciation to Laurie Steele and to Amgen and the NIH/NIAID (# 5R 24AI049393) for their support.

References

1. Dale DC (2008) Neutropenia. In: Crowther MA, Ginsberg J, Holger S, Meyer RM, Lottenberg R (eds) Evidence-based hematology. Blackwell Publishing Ltd, Chichester, UK, pp 215–220
2. Dale DC (2010) Neutropenia and neutrophilia. In: Lichtman MA, Kipps TJ, Kaushansky K, Beutler E, Seligsohn U, Prchal JT (eds) William's hematology, 8th edn. McGraw-Hill, New York, NY, pp 824–828
3. Dale DC, Cottle TE, Fier CJ et al (2003) Severe chronic neutropenia: treatment and follow-up of patients in the Severe Chronic Neutropenia International Registry. Am J Hematol 72:82–93
4. Hammond WP, Csiba E, Canin A, Souza LM, Dale DC (1991) Chronic neutropenia: a new canine model induced by human G-CSF. J Clin Invest 87:704–710
5. Hammond WP, Boone TC, Donahue RE, Souza LM, Dale DC (1990) A comparison of treatment of canine cyclic hematopoiesis with recombinant human GCSF, GMCSF, IL3 and canine GCSF. Blood 76:523–532
6. Yanay O, Brzezinski M, Christensen J, Liggitt D, Dale DC, Osborne W (2006) An adult dog with cyclic neutropenia treated by lentivirus-mediated delivery of granulocyte colony-stimulating factor. Hum Gene Ther 17:464–469
7. Hammond WP, Price TH, Souza LM, Dale DC (1989) Treatment of cyclic neutropenia with granulocyte colony stimulating factor. N Engl J Med 320:1306–1311
8. Bonilla MA, Gillo AP, Ruggeiro M et al (1989) Effects of recombinant human granulocyte colony-stimulating factor on neutropenia in patients with congenital agranulocytosis. N Engl J Med 320:1574–1580
9. Jakubowski AA, Souza L, Kelly F et al (1989) Effects of human granulocyte colony-stimulating factors in a patient with idiopathic neutropenia. N Engl J Med 320:38–42
10. Welte K, Zeidler C, Reiter A et al (1990) Differential effects of granulocyte-macrophage colony-stimulating factor and granulocyte colony-stimulating factor in children with severe congenital neutropenia. Blood 75:1056–1063
11. Dale DC. ELA2-related neutropenia. In: GeneReviews at GeneTests: medical genetics information resource. Initial posting June 2002. Updated July 2008. Last accessed July 2001 Copyright, University of Washington, Seattle. Available at http://www.genetests.org
12. Dale DC, Hammond WP (1988) Cyclic neutropenia: a clinical review. Blood Rev 2:178–185
13. Smith-Slatas CL, Bourque M, Salazar JC (2006) Clostridium septicum infections in children: a case report and review of the literature. Pediatrics 117:e796–e805
14. Barnes C, Gerstle JT, Freedman MH, Carcao MD (2004) Clostridium septicum myonecrosis in congenital neutropenia. Pediatrics 114:e757–e760
15. Klein C, Grudzien M, Appaswamy G et al (2007) HAX1 deficiency causes autosomal recessive severe congenital neutropenia (Kostmann disease). Nat Genet 39:86–92
16. Carlsson G, Melin M, Dahl N et al (2007) Kostmann syndrome or infantile genetic agranulocytosis, part two: Understanding the underlying genetic defects in severe congenital neutropenia. Acta Pediatr 96:813–819
17. Dale DC, Person RE, Bolyard AA et al (2000) Mutations in the gene encoding neutrophil elastase in congenital and cyclic neutropenia. Blood 96:2317–2322
18. Zeidler C, Germeshausen M, Klein C, Welte K (2009) Clinical implications of ELA2-, HAX1-, and G-CSF-receptor (CSF3R) mutations in severe congenital neutropenia. Br J Haematol 144:459–467
19. Klein C (2009) Congenital neutropenia. Hematology Am Soc Hematol Educ Program 344–350
20. Person RE, Li FQ, Duan Z et al (2003) Mutations in proto-oncogene GFI1 cause human neutropenia and target ELA2. Nat Genet 34:308–312
21. Xia J, Bolyard AA, Rodger E et al (2009) Prevalence of mutations in ELANE, GFI1, HAX1, SBDS, WAS and G6PC3 in patients with severe congenital neutropenia. Br J Haematol 147:535–542
22. Dale DC, Link DC (2009) The many causes of severe congenital neutropenia. N Engl J Med 360:3–5

23. Haurie C, Dale DC, Mackey MC (1999) Occurrence of periodic oscillations in the differential blood counts of congenital, idiopathic and cyclical neutropenic patients before and during treatment with G-CSF. Exp Hematol 27:401–409
24. Welte K, Zeidler C, Dale DC (2006) Severe congenital neutropenia. Semin Hematol 43:189–195
25. Kyle RA, Linman JW (1968) Chronic idiopathic neutropenia. A newly recognized entity? N Engl J Med 279:1015–1019
26. Dale DC, Guerry D, Werwerka JR, Bull JM, Chusid MH (1979) Chronic neutropenia. Medicine 58:128–144
27. Dale DC, Bonilla MA, Davis MW et al (1993) A randomized controlled phase III trial of recombinant human G-CSF for treatment of severe chronic neutropenia. Blood 81:2496–2502
28. Dale DC, Bolyard AA, Schwinzer BG et al (2006) The Severe Chronic Neutropenia International Registry: 10-year follow-up report. Support Cancer Ther 3:220–231
29. Dale DC, Makaryan V, Bolyard AA, et al (2010) Neutrophil elastase mutations and the risk of leukemia in patients with cyclic and congenital neutropenia. Blood (ASH Annual Meeting Abstracts) 116:3786
30. Rosenberg PS, Alter BP, Bolyard AA et al (2006) Severe Chronic Neutropenia International Registry. The incidence of leukemia and mortality from sepsis in patients with severe congenital neutropenia receiving long-term G-CSF therapy. Blood 107:4628–4635
31. Aprikyan AG, Kutyavin T, Stein S et al (2003) Cellular and molecular abnormalities in severe congenital neutropenia predisposing to leukemia. Exp Hematol 31:372–381
32. Grenda DS, Murakami M, Ghatak J et al (2007) Mutations of the ELA2 gene found in patients with severe congenital neutropenia induce the unfolded protein response and cellular apoptosis. Blood 110:4179–4187
33. Rosenberg PS, Alter BP, Link DC et al (2008) Neutrophil elastase mutations and risk of leukaemia in severe congenital neutropenia. Br J Haematol 140:210–213
34. Zeidler C, Boxer L, Dale DC, Freedman MH, Kinsey S, Welte K (2000) Management of Kostmann syndrome in the G-CSF era. Br J Haematol 109:490–495
35. Boztug K, Appaswamy G, Ashikov A et al (2009) A syndrome with congenital neutropenia and mutations in G6PC3. N Engl J Med 360:32–43
36. Beel K, Cotter MM, Blatny J et al (2009) A large kindred with X-linked neutropenia with an I294T mutation of the Wiskott–Aldrich syndrome gene. Br J Haematol 144:120–126
37. Beekman R, Touw IP (2010) G-CSF and its receptor in myeloid malignancy. Blood 115:5131–5136
38. Oshima K, Hanada R, Kobayashi R et al (2010) Hematopoietic stem cell transplantation in patients with severe congenital neutropenia: an analysis of 18 Japanese cases. Pediatr Transplant 14:657–663
39. Melis D, Fulceri R, Parenti G et al (2005) Genotype/phenotype correlation in glycogen storage disease type 1b: a multicentre study and review of the literature. Eur J Pediatr 164:501–508
40. Chou JY, Jun HS, Mansfield BC (2010) Neutropenia in type Ib glycogen storage disease. Curr Opin Hematol 17:36–42
41. Alsultan A, Sokol RJ, Lovell MA, Thurman G, Ambruso DR (2010) Long term G-CSF-induced remission of ulcerative colitis-like inflammatory bowel disease in a patient with glycogen storage disease Ib and evaluation of associated neutrophil function. Pediatr Blood Cancer 55:1410–1413
42. Dieckgraefe BK, Korzenik JR, Husain A, Dieruf L (2002) Association of glycogen storage disease 1b and Crohn disease: results of a North American survey. Eur J Pediatr 161(Suppl 1):88–92
43. Calderwood S, Kilpatrick L, Douglas SD et al (2001) Recombinant human granulocyte colony-stimulating factor therapy for patients with neutropenia and/or neutrophil dysfunction secondary to glycogen storage disease type 1b. Blood 97:376–382

44. Burroughs L, Woolfrey A, Shimamura A (2009) Shwachman-Diamond syndrome: a review of the clinical presentation, molecular pathogenesis, diagnosis, and treatment. Hematol Oncol Clin North Am 23:233–248
45. Rommens JM, Durie PR. Shwachman-Diamond syndrome. Posted July 2008. In: Pagon RA, Bird TC, Dolan CR, Stephens K (eds) GeneReviews. Seattle (WA): University of Washington, Seattle, 1993–2008 Jul 17. Available at http://www.genetests.org
46. Shimamura A, Bolyard AA, Marrero TM, et al (2009) Shwachman-Diamond syndrome: development of a North American registry to assess long-term outcomes, risk of leukemia and other complications. Blood (ASH Annual meeting Abstracts). 114:1363
47. Alter BP, Girl N, Savage SA et al (2010) Malignancies and survival patterns in the National Cancer Institute inherited bone marrow failure syndrome cohort study. Br J Haematol 150:179–188
48. Steward CG, Newbury-Ecob RA, Hastings R et al (2010) Barth syndrome: an X-linked cause of fetal cardiomyopathy and stillbirth. Prenat Diagn 30:970–976
49. Collins S, Bolyard AA, Marrero TM, Phan L Dale DC (2010) Barth syndrome and severe chronic neutropenia. Blood 116:3787
50. Kawai T, Malech HL (2009) WHIM syndrome: congenital immune deficiency disease. Curr Opin Hematol 16:20–26
51. Siedlar M, Rudzki Z, Strach M et al (2008) Familial occurrence of warts, hypogammaglobulinemia, infections, and myelokathexis (WHIM) syndrome. Arch Immunol Ther Exp (Warsz) 56:419–425
52. Ochs HD, Filipovich AH, Veys P, Cowan MJ, Kapoor N (2009) Wiskott–Aldrich syndrome: diagnosis, clinical and laboratory manifestations, and treatment. Biol Blood Marrow Transplant 15:84–90
53. Melvan JN, Bagby GJ, Welsh DA, Nelson S, Zhang P (2010) Neonatal sepsis and neutrophil insufficiencies. Int Rev Immunol 29:315–348
54. Kuhn P, Messer J, Paupe A et al (2009) A multicenter, randomized, placebo-controlled trial of prophylactic recombinant granulocyte-colony stimulating factor in preterm neonates with neutropenia. J Pediatr 155:324–330
55. Jonsson OG, Buchanan GR (1991) Chronic neutropenia during childhood. A 13-year experience in a single institution. Am J Dis Child 145:232–235
56. Bowman SJ (2002) Hematological manifestations of rheumatoid arthritis. Scand J Rheumatol 31:251–259
57. Liu JH, Wei S, Lamy T et al (2000) Chronic neutropenia mediated by fas ligand. Blood 95:3219–3222
58. Starkebaum G (2002) Chronic neutropenia associated with autoimmune disease. Semin Hematol 39:121–127
59. Kuritzkes DR (2000) Neutropenia, neutrophil dysfunction, and bacterial infection in patients with human immunodeficiency virus disease: the role of granulocyte colony-stimulating factor. Clin Infect Dis 30:256–260

Investigational Studies of rHuG-CSF to Promote the Regeneration of Nonhematopoietic Tissues

Stephen J. Szilvassy

1 Introduction

Granulocyte colony-stimulating factor (G-CSF) is a 19.6-kDa glycoprotein hormone secreted by monocytes, macrophages, fibroblasts, and endothelial cells and is the primary physiologic regulator of neutrophilic granulocyte production. It exerts its earliest actions on myeloid-restricted hematopoietic progenitor cells that reside in the bone marrow, stimulating their proliferation and differentiation through several developmental stages (myeloblasts, promyelocytes, myelocytes, and metamyelocytes), culminating in the release of terminally differentiated neutrophils into the peripheral blood. G-CSF also promotes the survival of mature neutrophils and enhances their effector functions required for a successful immune response against bacterial infection [1].

The relatively restricted actions of G-CSF on the neutrophil lineage were revealed by the phenotype of knock-out mice and nonclinical studies in which recombinant human (rHu)G-CSF was administered to rodents and nonhuman primates [2, 3]. These findings spurred the therapeutic development of rHuG-CSF as an agent to promote the regeneration of neutrophils in patients receiving cancer chemotherapy either alone or in conjunction with hematopoietic stem cell transplantation [4]. rHuG-CSF has also proven useful for the treatment of idiopathic neutropenias, and for inducing the release of primitive hematopoietic stem and progenitor cells from the bone marrow into the circulation (a process referred to as "mobilization") where they can be more readily collected for transplantation purposes (see other chapters for further information on uses of rHuG-CSF in various clinical settings). All of these actions of G-CSF are consistent with the expression of the G-CSF receptor (G-CSFR; also known as CD114) predominantly on cells of the granulocyte lineage; however, over the past 5 years in particular, an

S.J. Szilvassy (✉)
Amgen Inc., One Amgen Center Drive, MS 15-2-A, Thousand Oaks, CA, 91320, USA
e-mail: sszilvas@amgen.com

G. Molineux et al. (eds.), *Twenty Years of G-CSF*, Milestones in Drug Therapy, DOI 10.1007/978-3-0348-0218-5_16, increasing number of studies have reported expression of G-CSFR mRNA and/or protein on adult nonhematopoietic cells, including cardiomyocytes, endothelial, neural, and retinal cells [5–7]. This finding has prompted various nonclinical and clinical investigations of rHuG-CSF for the treatment of myocardial infarction, stroke, cognitive disorders, and degenerative eye diseases [5, 6, 8–12]. While the nonclinical data for the effects of G-CSF on nonhematopoietic tissues is intriguing, it is highly controversial. Among the most significant concerns with almost all of these studies is the lack of definitive evidence for expression of G-CSFR on the various nonhematopoietic cell types on which G-CSF has been claimed to act directly. Of critical importance to this issue is the recent demonstration that none of the anti-G-CSFR antibodies that are currently available from commercial sources, and which have been widely used for such studies, bind specifically to the human G-CSFR protein [13]. That is, although some of these antibodies do bind to G-CSFR, they also bind to other proteins either on the surface or within cells that do not express G-CSFR mRNA. Experiments that employ these antibodies for immunohistochemistry and/or flow cytometry to detect the presence of G-CSFR on the surface of intact cells must therefore incorporate rigorous controls (e.g., genetic ablation or siRNA-mediated knock-down of G-CSFR expression in the cells of interest to completely abrogate antibody binding) or immunoprecipitate the G-CSFR protein from cell lysates before Western analysis with a second noncompetitive anti-G-CSFR antibody, again using G-CSFR-positive and -negative cell types as controls to exclude the possibility that the staining observed might be due to antibody binding to cross-reactive proteins [13]. Unfortunately, such rigor is not routinely applied and so while this does not automatically invalidate the results of studies that also demonstrate a biologic effect of G-CSF on nonhematopoietic cells in vitro or in vivo (e.g., increased cell survival or proliferation), it does severely weaken the evidence supporting a direct mechanism of action.

The enthusiasm that spurred initiation of clinical studies of the efficacy of rHuG-CSF for promoting nonhematopoietic tissue repair is, if not scientifically well founded, somewhat understandable. The two G-CSF therapeutics that have enjoyed regulatory approval for the longest period of time, filgrastim and pegfilgrastim, the latter having an extended half-life in the circulation due to the covalent attachment of a polyethylene glycol moiety that increases the hydrodynamic size of the drug and largely eliminates its renal clearance [14], have established a favorable safety record over the course of more than 20 years during which they have been administered to millions of patients and normal individuals, in the latter for the purposes of allogeneic stem cell donation. Meanwhile, the debilitating functional impairments in many patients with advanced coronary heart disease and Alzheimer's disease, to name two, have remained a significant unmet medical need that it is predicted will only increase over the next decades as the population of elderly individuals in which such diseases typically manifest themselves continues to grow. Thus because rHuG-CSF has proven to be safe in the hematology and oncology setting, and an expanding body of nonclinical data suggested that it may be efficacious in settings that were not necessarily predicted by its biology known from the hematopoietic system, small clinical studies were initiated despite the incomplete understanding of the drugs proposed mechanism of action in these novel indications. A substantial body of evidence is now available for scientific and medical professionals to evaluate if rHuG-CSF meaningfully improves clinical parameters in the settings of acute myocardial infarction, stroke, and Alzheimer's disease. This chapter reviews some of the key nonclinical studies that ignited interest in G-CSF as a tissue repair factor, identifies some of the problems with these experiments that must be considered in interpreting the data, and reviews the work of others who have performed meta-analyses of several clinical studies of rHuG-CSF in these unconventional indications. Taken together, this literature highlights the significant challenges in translating laboratory observations into successful therapies.

2 rHuG-CSF for the Treatment of Myocardial Infarction

Myocardial infarction (commonly referred to as a heart attack) is responsible for about one-third of congestive heart failure cases and is the leading cause of cardiovascular mortality. The current treatment employs statins, beta-blockers, angiotensin-converting enzyme (ACE)-inhibitors and other drugs, together with early percutaneous coronary intervention (PCI) (commonly known as angioplasty), which while dramatically improving survival from the acute episode do not prevent left ventricular remodeling. Thus, approximately 35% of myocardial infarction patients develop heart failure within 5 years due to sustained damage that results in a chronically ischemic heart. The observation that after a myocardial infarction, increased numbers of hematopoietic, mesenchymal and endothelial stem/progenitor cells, and increased concentrations of factors that promote their homing to injured tissue (e.g., stromal cell-derived factor [SDF]-1) can be detected in the circulation prompted the idea that stem cell mobilization might represent a physiologic repair mechanism and that stem cell therapy, augmented by exogenous cytokine treatment, may represent a novel therapeutic option for these patients [15, 16]. Interest in the concept that hematopoietic stem cells might be able to regenerate the cardiomyocytes that die in the hours and days after myocardial infarction exploded with the publication of two papers demonstrating a remarkably high degree of engraftment of bone marrow-derived cells, augmented by rHuG-CSF therapy, in infarcted mouse hearts [17, 18]. The authors attributed the improved cardiac function and animal survival to "transdifferentiation," a hypothetical process widely investigated at that time, whereby hematopoietic stem cells were postulated to differentiate into nonhematopoietic cell types in response to tissue injury or inflammation. Despite considerable effort, the initial spectacular results could not be replicated [19, 20], and the transdifferentiation model was widely challenged and ultimately disproven, at least for adult stem cell types [21–23]. Nevertheless, the initial observation was sufficient to catalyze a paradigm shift and stimulated the publication of numerous nonclinical studies investigating the use of various hematopoietic growth factors and stem cell populations for improving cardiac function in mice, rats, rabbits, dogs, and pigs. With some exceptions, beneficial effects on structural and functional cardiac parameters were demonstrated [24–31]. Recombinant HuG-CSF garnered particular interest for this application for the reasons discussed and because it can be administered by subcutaneous injection without the need to collect bone marrow and manipulate cells in culture, or to transplant these into recipients by invasive routes such as intracoronary injection. The results of these nonclinical studies have been reviewed in detail and will not be discussed [32, 33]. Instead, this section highlights some emerging themes and consider potential mechanisms by which G-CSF may mediate the effects observed.

2.1 Nonclinical Studies and Potential Mechanisms for G-CSF's Action on the Heart

Three dominant hypotheses for G-CSF's mechanism of action emerged from the studies in which it was found to preserve the structure and function of the post-myocardial infarction heart, although it is important to emphasize that many other studies failed to demonstrate such effects compared with vehicle-treated controls. First, it was proposed that G-CSF promotes cardiomyocyte survival directly by binding to G-CSFR expressed on these cells, activating JAK/STAT signaling and up-regulating expression of antiapoptotic proteins such as Bcl-2 and Bcl-xL. Alternatively or in addition, bone marrow-derived hematopoietic stem cells mobilized by rHuG-CSF may home to the damaged heart where it has been proposed that they generate new cardiomyocytes. As noted, this "transdifferentiation hypothesis" was largely disposed of by subsequent studies and will not be discussed further. A third possibility, related to the second, is that G-CSF promotes the migration of hematopoietic stem cells and/or other stem cell types that reside in the bone marrow to the site of injury where they, or more specifically their progeny, mediate repair through paracrine mechanisms that might include the production of anti-inflammatory cytokines, chemokines, proangiogenic factors, or enzymes that break down scar tissue.

The first possibility was suggested by a report demonstrating G-CSFR mRNA expression in the adult mouse heart and in cultured cardiomyocytes by RT-PCR (reverse-transcription polymeriase chain reaction) analysis, and G-CSFR protein expression in cultured cardiomyocytes from normal (i.e., nonischemic) neonatal rats by immunohistochemistry and in cardiomyocytes from infarcted adult rat hearts by Western analysis [9]. A careful review of the data, however, reveals several important flaws in the experimental design. No negative control tissue was included in the RT-PCR or Western experiments, and the antibody used to detect G-CSFR protein expression is now known not to be specific for G-CSFR [13]. The negative controls used for immunohistochemistry were not adequate to exclude the possibility of nonspecific antibody binding, i.e., rather than comparing the level of staining to that observed with an irrelevant isotype-matched control antibody, the primary anti-G-CSFR antibody was simply omitted. The findings are also confused by the largely cytoplasmic localization of the staining claimed to be due to G-CSFR. Since G-CSF is well known to promote neutrophil survival by activation of the STAT3 signaling pathway, analogous signaling endpoints were also investigated in rat cardiomyocytes cultured in rHuG-CSF. Although STAT3 phosphorylation was observed, the amount of rHuG-CSF required to elicit this effect was approximately tenfold higher than that which stimulates signaling in hematopoietic cells. Pretreatment of cardiomyocytes with rHuG-CSF was found to suppress apoptosis induced by exposure to hydrogen peroxide in vitro; however, no data were presented to demonstrate that this protective effect was rHuG-CSF dose-dependent [9]. Indeed, a subsequent attempt to replicate this claimed antiapoptotic effect of G-CSF on cardiomyocytes was unsuccessful and failed to provide convincing evidence of G-CSFR protein expression in infarcted mouse hearts [29]. In this latter study, both STAT3 and Akt were shown to be phosphorylated in the hearts of infarcted mice that were treated with rHuG-CSF compared with saline. However, this finding does not allow any conclusions about whether G-CSF acts directly on cardiomyocytes. An alternative explanation, consistent with the known biology of G-CSF, is that it stimulates G-CSFR$^+$ hematopoietic cells (either in distal organs such as the bone marrow or in the heart itself) to produce other factors that also signal through these pathways (e.g., many interleukins, macrophage (M)-CSF, and platelet-derived growth factor also signal through STAT3) and which enhance heart function through indirect mechanisms (discussed later). Analogous experiments performed in excised rat hearts (i.e., in a Langendorff system) excluded the potential contribution of bone marrow cells to G-CSF-stimulated activation of Jak2, STAT3, ERK, Akt, and the endothelial nitric oxide synthase (eNOS) pathway [34]. However, the magnitude of the effects observed in this model were small; rHuG-CSF induced only a 1.6-fold increase in phosphorylated eNOS and a 10% increase in NO synthesis (a known mediator of cardioprotection [35]). Moreover, it remains difficult to exclude the possible contribution of endothelial cells and cardiac-resident monocytes in this system. Despite these technical concerns, the idea that G-CSF acts directly on G-CSFR-expressing cardiomyocytes to promote their survival and proliferation continues to be promulgated in the literature as scientific dogma. In fact, the data are not so clear-cut.

Evidence is accumulating that paracrine effects exerted by different types of infiltrating stem and progenitor cells or their progeny are the predominant mechanism underlying G-CSF's action in mediating post-infarction heart repair. Unfortunately, the numerous diverse and often conflicting studies that have been conducted to illuminate discrete components of these pathways have in contrast largely served to highlight how little is known. For example, G-CSF has been reported to induce the mobilization of endothelial progenitor cells that some investigators claim can directly differentiate into endothelial cells and promote revascularization in rodent models of ischemia and myocardial infarction [15]. It should be noted, however, that there is considerable disagreement regarding the existence and potentiality of these progenitor cells. Mesenchymal stem cells, another type of multipotent stem cell that resides in the marrow, were also reported to be mobilized by rHuG-CSF and to differentiate into cardiomyocytes after myocardial infarction [36], although a subsequent study found no improvement in cardiac function [37]. The well-documented proangiogenic effects of G-CSF may also enhance cardiac function through compensatory vasculogenesis and improved perfusion. Scatchard studies using ^{125}I-labeled G-CSF (a method that does not require use of the aforementioned nonspecific anti-G-CSFR antibodies to quantitate receptor expression) indicate that human endothelial cells express approximately 430 receptors/cell and G-CSF has been shown to promote their proliferation and migration in vitro and to stimulate neovascularization in a rat corneal angiogenesis model [38, 39]. Recently, a population of mouse CD11b^{+}Gr-1^{+} cells has been identified that includes cells of the granulocytic and monocytic lineages, immature dendritic cells, and a minor fraction of progenitor cells capable of differentiating into any of these populations. Because a fraction of these cells suppresses immune functions, they are referred to as myeloid-derived suppressor cells (MDSC) [40]. MDSCs are derived from granulocyte-macrophage progenitor cells (CFU-GM), which are well known to express G-CSFR. CFU-GM production is increased under inflammatory conditions and is augmented by G-CSF and other factors. MDSCs are a rich source of matrix metaloprotineinase (MMP)-9, which facilitates fibrinolysis and increases the bioavailability of vascular endothelial growth factor (VEGF). MDSCs also secrete Bv8 (also known as prokineticin-2), a VEGF-related protein that mobilizes hematopoietic stem and progenitor cells in vivo, stimulates granulocyte and monocyte progenitor growth in vitro, and is mitogenic for endothelial cells [41, 42]. Neutrophils themselves can secrete MMP-9 and VEGF, and they and other less well-characterized monocytic cells have been shown to exert proangiogenic effects that are augmented by exposure to rHuG-CSF in mice with peripheral ischemia [43, 44]. An analogous population of CD34^{+}CD13^{+}CD66b^{+} MDSCs has been described in humans. Their proangiogenic functions are less well characterized so it is not clear whether and how they may contribute to tissue repair associated with myocardial infarction and ischemia in a clinical setting. Regardless, it is clear that G-CSF's effects in the heart could be mediated through multiple diverse paracrine mechanisms, none of which depend on a direct receptor-dependent action on cardiomyocytes.

The profound improvement in hemodynamic parameters observed in some nonclinical models of myocardial infarction are difficult to reconcile with the relatively low proportion (typically <2 to 5% of the total heart tissue) of infiltrating hematopoietic cells, proliferating cardiomyocytes, or endothelial cells that have been measured in the heart after rHuG-CSF therapy. This finding suggests that partial replacement of the damaged myocardium plays at best a subordinate role. Another potential mechanism by which G-CSF could improve post-infarction performance is by enhancing the pathophysiologic healing process, for example, by modulating fibrinolysis and collagen synthesis. An infarct scar consists mainly of necrotic cells and fibrous tissue that is essential for preserving the structural integrity of the infarcted area. Analogous to its wound-healing effects in the hematopoietic system, rHuG-CSF treatment was shown to enhance the expression and deposition of type I and type III fibrillar collagens in the infarcted lesions of rats, which was associated with improved early ventricular expansion despite no change in overall infarct size [45]. Subsequently, expression of MMP-1 (a collagenase) and MMP-9 (a gelatinase) and infiltration of the scar by macrophages facilitate absorption of the necrotic tissue. All were found to be increased after 7 days in the ischemic region of the heart from rabbits subjected to coronary occlusion and reperfusion followed by rHuG-CSF treatment compared with saline-injected controls [24]. Recombinant HuG-CSF also significantly reduced cardiac fibrosis in a post-infarction mouse model, which was associated with increased expression of MMP-2 and -9 [29]. Combined treatment of infarcted mice with rHuG-CSF and stem cell factor (SCF) was reported to increase expression of the gap junctional protein connexin 43 that is required for intercellular coupling in the border zone of a healed myocardial infarction [10]. Slow conduction due to decreased intercellular coupling in this region is the pivotal factor for reentrant ventricular tachycardias. Increased connexin 43 was associated with a reduction in induced arrhythmias in infarcted hearts on a Langendorff apparatus, which may represent another mechanism by which G-CSF promotes repair in the post-myocardial infarction heart [10].

2.2 *Clinical Studies of rHuG-CSF for Heart Repair After Myocardial Infarction*

Inspired by the positive results demonstrated in many animal studies, but arguably prematurely due to the lack of any definitive understanding of its mechanism of action particularly at that time, rHuG-CSF was advanced to clinical testing in patients with acute myocardial infarction in 2005. Numerous trials have been completed and several excellent meta-analyses of the resulting data have been published [46–50] and will not be discussed in detail here. However, a number of important conclusions have emerged. Foremost, it is clear that despite the initial promising laboratory findings and even some statistically significant improvements in left ventricular ejection fraction (LVEF) in some of the early small nonrandomized, nonblinded safety and efficacy trials, analysis of larger, randomized, double-blinded, and placebo-controlled studies encompassing >250 patients indicates that rHuG-CSF therapy does not increase LVEF or reduce infarct size in patients (Fig. 1). Moreover, while it is acknowledged that these studies are remarkably heterogeneous in terms of study design (e.g., the reperfusion treatment and its timing after myocardial infarction, the dose of rHuG-CSF and its start time and method of administration, the duration of rHuG-CSF treatment, the study endpoints, and methods of cardiac functional assessment), they tell us nothing about potential mechanisms, and indeed have not typically been designed to do so. It is noteworthy, however, that in all of these studies $CD34^+$ cells were efficiently mobilized by rHuG-CSF treatment. The lack of any significant associated improvement in cardiac parameters thus lends additional weight to the idea that a direct cellular contribution of hematopoietic stem cells to the

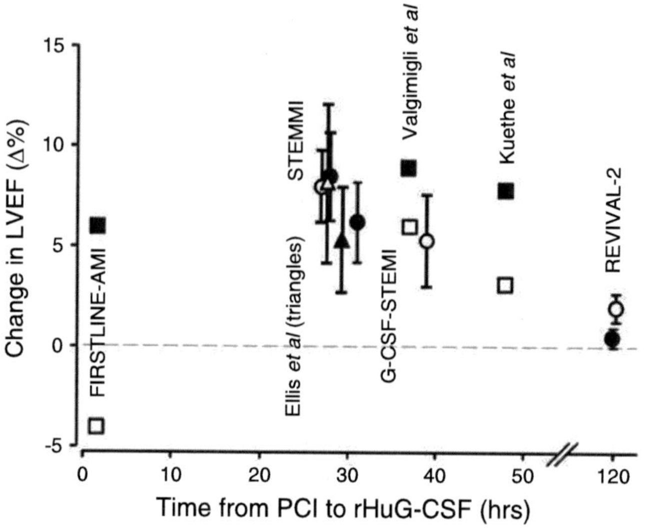

Fig. 1 Time from revascularization to initiation of rHuG-CSF therapy versus change in LVEF from baseline to follow-up. *Circles* represent data from three placebo-controlled double-blinded studies (STEMMI, G-CSF-STEMI, and REVIVAL-2) as originally shown by Ripa and Kastrup [49], with the correction of errors for the STEMMI and G-CSF-STEMI trials that were not correctly extrapolated by Ripa and Kastrup from the original papers [52, 53]. *Closed circles*: rHuG-CSF-treated patients, *open circles*: placebo groups. Data abstracted from the placebo-controlled double-blinded study of Ellis et al. [54] has been added (*closed triangles*: patients treated with 10 μg rHuG-CSF, *open triangles*: placebo group). *Squares* represent data from three nonblinded studies as described originally by Ripa and Kastrup [49] (*closed*: rHuG-CSF-treated patients, *open*: placebo groups). All values are mean ± SE. Adapted from Experimental Hematology, vol. 36, RS Ripa and J Kastrup, G-CSF therapy with mobilization of bone marrow stem cells for myocardial recovery after acute myocardial infarction – A relevant treatment?, pp. 681–686, Copyright 2008, with permission from Elsevier

cardiomyocyte population does not likely underlie the effects observed in nonclinical models.

The reason for the lack of clinical efficacy is unclear and will only be revealed through the execution of additional randomized, double-blinded, placebo-controlled, and ideally multicenter trials that are rationally designed to test mechanistic hypotheses. In the meantime, despite the lack of effect of G-CSF, some features of the four trials conducted thus far that best fit these criteria (REVIVAL-2 [51], STEMMI [52], G-CSF-STEMI [53], and the trial by Ellis et al. [54]) may be informative. In these studies, patients were administered 5 or 10 μg/kg/day rHuG-CSF for 5 or 6 days beginning from a mean of 30–114 h after PCI [47]. This timing is considerably later than in the FIRSTLINE-AMI trial completed 1 year earlier in which patients received 10 μg/kg/day rHu-G-CSF for 6 days beginning 89 ± 35 min after PCI [55]. This 50-patient trial was open label, not placebo controlled in design, but did show a statistically significant improvement in LVEF at 6 months follow-up in the rHuG-CSF treatment group (Fig. 1). Early treatment with rHuG-CSF (i.e., immediately after reperfusion) was shown to be superior to delayed treatment (beginning on day 5) in reducing ventricular dilation and left ventricular remodeling in a porcine myocardial infarction model [56]. A post-hoc analysis of data from the G-CSF-STEMI trial, however, failed to demonstrate any improvement in cardiac function in a subset of patients administered rHuG-CSF within 24 (mean: 16 ± 6) h after PCI compared with a late treatment group (mean: 49 ± 26 h) [57]. Thus the maximal functional benefits of "early" rHuG-CSF treatment that can be revealed in a controlled laboratory environment may be difficult to achieve in a clinical setting. The participants in the FIRSTLINE-AMI trial were also younger (mean: 50 years) than in the other trials (mean range: 57–60 years). Recombinant HuG-CSF was found to lack therapeutic efficacy for post-myocardial infarction remodeling in old mice because of an age-related loss in its direct ability to prevent peri-infarct apoptosis of cardiomyocytes [58]. It will be important to learn more about the impact of aging on the reparative mechanisms in the heart considering that four of five patients with coronary heart disease are aged 65 years or older.

In summary, after almost a decade of intense investigation in this area, we are still disturbingly ignorant of the mechanisms by which G-CSF may modulate physiologic repair processes and the immune system to facilitate recovery from a myocardial infarction. It is fortunate that despite some early indications that rHuG-CSF treatment might increase the rate of restenosis, subsequent analyses have demonstrated that these rare adverse events were more likely procedure- and stent-related [46], so the safety record of rHuG-CSF has remained untarnished. Undoubtedly the use of rHuG-CSF in acute myocardial infarction will remain an area of intense investigation, and perhaps the next decade will witness a more systematic approach to delineating its actions in this setting.

3 G-CSF for the Treatment of Cerebral Ischemia and Stroke

Stroke, sometimes referred to as a "brain attack," is the leading cause of adult disability and the third leading cause of death in the United States [59]. It results from a disruption of blood supply to the brain, due either to vessel occlusion or hemorrhage, and as the lack of oxygen and nutrients limit cell function and survival, an area of cell death is created, bordered by a zone of damaged cells called the penumbra. Within the total lesion area, oxidative, inflammatory, and probably also excitotoxic cascades become activated and threaten to broadly impair cerebral function. Among 30-day survivors of a first stroke, only 50% survive 5 years. Presently, the only available treatment for stroke is recombinant tissue plasminogen activator (tPA), which facilitates thrombolysis and improved blood flow, but is only approved for use up to 3 h after stroke onset and, due to restrictive patient-selection criteria, is in practice used in only about 10% of stroke patients. The availability of a trophic factor that promotes neuroprotection and regeneration in the chronic stroke setting could significantly improve patient outcomes.

3.1 *Nonclinical Studies and Proposed Mechanisms of G-CSF Actions in the Brain*

Interest in the potential use of rHuG-CSF for the treatment of cerebral ischemia as well as neurodegenerative diseases such as Alzheimer's disease was ignited in large part by a study that reported that G-CSF crosses the blood–brain barrier and reduces infarct volume in two rat models of acute stroke [8]. The G-CSFR protein was claimed to be widely expressed by neurons in the central nervous system as well as on adult neural stem cells, suggesting that G-CSF may act directly on both primitive and terminally differentiated neural cells. Proof of this hypothesis, however, requires that several key mechanistic requirements are satisfied. First, G-CSF must indeed cross the blood–brain barrier. In the first study [8] and a subsequent study [60], a temporal increase in the ratio of radioactive counts detected in the brain versus the plasma was observed after intravenous injection of ^{125}I-labeled G-CSF into intact rats. However, the twofold increase in brain:plasma radioactivity observed after 24 h by Schneider et al. was not shown to be statistically significant, and in both studies could also be explained solely by a progressive decline in plasma ^{125}I-G-CSF concentrations without any increase in cerebral accumulation. Indeed, radiolabeled G-CSF was shown in the latter study to disappear completely from the plasma by 2 h after injection, at which time the amount of radioactivity present in the brain had only increased by about twofold, and the distribution to other target tissues such as the bone marrow was not measured at all. It has been demonstrated that the blood–brain barrier is disrupted during the acute phase of cerebral ischemia, which may allow G-CSF transport in this setting, but there is no definitive evidence for G-CSF entry into the brain of mice or humans who have suffered a stroke or with Alzheimer's disease.

A second mechanistic requirement for a direct action of G-CSF in the brain is that neuronal cells must express the full-length and functional G-CSFR protein on their surface. In both of the above studies, expression of G-CSFR protein on neurons in a variety of brain regions and on brain capillaries (which could theoretically facilitate receptor-mediated transport of G-CSF across the blood–brain barrier) was evaluated by immunohistochemical analysis using a rabbit polyclonal anti-G-CSFR antibody that was subsequently shown not to be specific for G-CSR [13]. Neither study used isotype-matched primary antibodies or stained known negative tissues to control for possible nonspecific staining. In the first study [8] the anti-G-CSFR antibody was preincubated with a blocking peptide to eliminate staining, but this control cannot exclude the possibility that the epitope encoded by the control peptide might also be expressed on protein(s) other than G-CSFR to which nonspecific antibody binding would also be blocked. In the second paper, G-CSFR expression was detected by immunohistochemistry in the nuclei of endothelial cells, but not on the cell surface [60]. Finally, two bands claimed to be G-CSFR were detected by Western analysis in cell lysates prepared from the adult rat brain microvascular endothelial cell line SV-40, but the molecular weight of these proteins was not specified, negative controls were not shown, and the blots presented were cut to exclude all regions of the lane except that containing the putative G-CSFR bands so the specificity of the antibody used could not be ascertained [60].

The third requirement to definitely prove a direct action of G-CSF on neuronal cells is demonstration that G-CSF activates intracellular signaling pathways known to play a role in cell survival and/or proliferation. G-CSF was reported to protect rat cortical neurons against programmed cell death after exposure to the apoptosis inducer campothecin and nitric oxide in vitro [8]. This activity appeared to be mediated by G-CSFR as an antibody against the receptor abolished protection. Effects on cells treated with medium lacking G-CSF, and with control antibodies that bind to proteins other than G-CSFR (notwithstanding the fact that the anti-G-CSFR antibody used is not specific to this receptor), were not reported. Experiments that examined G-CSF-induced phosphorylation of STAT3, Akt, ERK1/2, ERK5, and PDK1 in cultured cortical neurons were similarly flawed, and cannot be considered to have definitively demonstrated that G-CSF activates antiapoptosis or proliferation pathways in neurons [8]. Despite these issues, this paper is widely cited as evidence for exactly these effects.

A plethora of papers have reported beneficial effects of G-CSF compared with vehicle treatment in reducing cerebral infarct volume and mortality rate, and improving locomotor function in various rodent models of cerebral ischemia [5, 61–64] (Fig. 2). Other studies have shown no benefit on long-term function, or even adverse effects of single-agent rHuG-CSF [65–67]. In assessing this literature, distinctions between the extent of blood flow restriction (i.e., global versus focal ischemia), its duration (i.e., permanent versus temporary focal ischemia), and the time after cerebral artery occlusion at which rHuG-CSF therapy is initiated (i.e., within hours in the so-called "acute" phase, or after several days or weeks in the "subacute" phase) are critical to distinguishing neuroprotective from reparative mechanisms and the general relevance of the nonclinical models to clinical conditions as these each result in different histologic and behavioral outcomes, are accompanied by variable levels of inflammation, and involve different molecular pathways [68]. A review of 19 publications incorporating 666 animals indicated that treatment with rHuG-CSF significantly reduced lesion size in models of transient but not permanent focal ischemia. Neurologic deficit and performance in limb placement tests also improved after rHuG-CSF treatment but overall locomotor activity was not improved [69]. Another meta-analysis of 277 animals from 13 studies found that for each 1 μg/kg increase in rHuG-CSF dose, infarct size was decreased by 0.8% when the cytokine was administered within the first 6 h after stroke, and 2.1% when applied at later times, up to a mean overall infarct size reduction of 42% [70]. For sensorimotor outcome, running function was significantly improved by 0.23% for each 1-μg/kg body weight increase in rHuG-CSF dose. The finding that delaying treatment with rHuG-CSF by several days might be more effective than early treatment is particularly surprising as the time window for most candidate neuroprotectants is narrow. This finding may be related to G-CSF's multimodal actions that are thought to comprise both neuroprotective and particularly proregenerative activities.

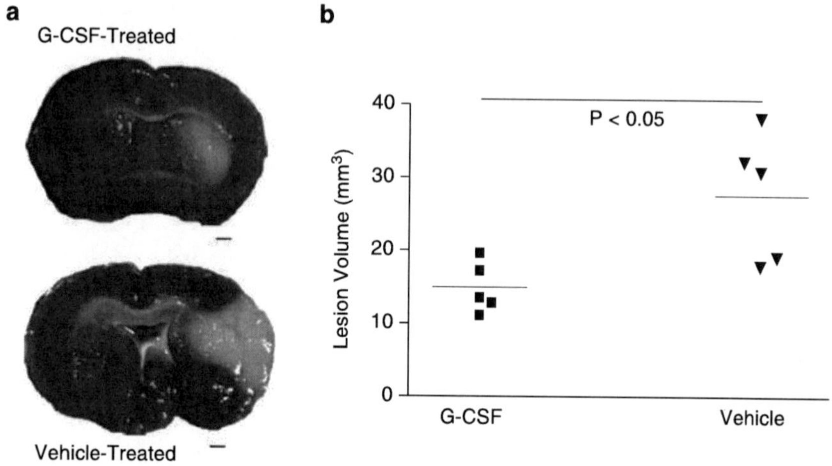

Fig. 2 G-CSF treatment reduces cerebral infarct volume. Mice were subjected to 60 min of right middle cerebral artery occlusion and then injected subcutaneously with either 50 μg/kg rHuG-CSF (*upper*) or saline (*lower*) at the time of reperfusion. Panel (**a**) 2,3,5-Triphenyltetrazolium chloride-stained brain slices depict the lesion 48 h later, as indicated by the *white* area which reflects a loss of viable tissue. The horizontal scale bars represent 1 mm. Panel (**b**) The individual lesion volumes for five mice/group (*squares* and *triangles*) together with the mean values (*horizontal lines*) are significantly different as reported [62]. Reprinted by permission from Macmillan Publishers Ltd., Journal of Cerebral Blood Flow & Metabolism 25: 431–439, 2005; Copyright 2005

Broadly speaking, the potential mechanisms that have been proposed to explain G-CSF's actions are similar to those for myocardial infarction [71]:

1. G-CSF may mobilize bone marrow-derived hematopoietic or mesenchymal stem cells that can either directly replace damaged neuronal tissue or whose progeny produce neurotrophic factors that promote the recovery of neural functions by paracrine mechanisms. As in myocardial infarction, the deposition of fibronectin in the ischemic brain appears to be a key component of the repair pathway [72] as evidenced most convincingly by the increased neuronal apoptosis and larger infarction area after induction of transient focal cerebral ischemia in fibronectin-deficient mice [73]. Although hematopoietic cells expressing cell-surface proteins such as CD34 or CXCR4, or that have incorporated BrdU (indicating proliferation), have been shown to be increased in the area surrounding a cerebral infarct after rHuG-CSF treatment [61, 66, 71], the absence (or only rare presence) of bone marrow-derived cells that coexpress markers of neural cells indicates that direct hematopoietic contribution to new brain tissue is unlikely.
2. G-CSF may directly activate antiapoptotic pathways in neurons by binding to neuronally expressed G-CSFR. As noted above, support for this hypothesis is weakened by the lack of antibodies to specifically detect G-CSFR expression on primary neurons, and inadequate controls used for immunohistochemistry and

Western analyses to quantitate changes in the expression of caspase-3/7, pSTAT3, pERK1/2 and 5, pAkt, Bcl-2, and Bcl-X_L in ischemic brains or cultured neural cell lines after exposure to rHuG-CSF [8, 74]. rHuG-CSF was reported to moderately increase the expression of the cellular inhibitor of apoptosis protein 2 (*cIAP2*; one of the genes induced by activation of the STAT3 signaling pathway) and decrease the expression of cleaved caspase-3 in the whole ischemic brain [75] but it is not possible to determine if these effects are due to a direct action of G-CSF on G-CSFR$^+$ neurons. A 30-min exposure of primary cerebellar granule cells to rHuG-CSF in vitro before addition of the excitotoxic agent glutamate was also found to dose-dependently improve their viability [5]. This system offers the advantage of excluding the effects of other cell types and soluble factors, but G-CSFR expression and signaling was not examined in this study. Jung et al. used a human cerebral-neuroblastoma hybrid cell line subjected to oxygen and glucose deprivation in vitro to examine the effects of rHuG-CSF on apoptosis [76]. As in the study from Solarglu et al. [75], rHuG-CSF improved survival and reduced apoptosis of ischemic cells, but definitive evidence that this is the result of signaling through G-CSFR was lacking as multiple bands of undefined molecular weight were identified by the anti-G-CSFR antibody used for Western blot and immunohistochemistry, and the ratio of phosphorylated to total STAT3 did not differ (despite claims to the contrary) in ischemic cells cultured with and without rHuG-CSF [76].

3. G-CSF may stimulate the differentiation of neuronal stem and progenitor cells. In a direct mechanism, this would require expression of G-CSFR on these primitive cells. G-CSFR mRNA was detected in adult neural stem cells isolated from the rat subventricular zone or hippocampal region. These cells can grow as neurospheres in vitro and double immunofluorescence labeling showed that some of the cells in these colonies appeared to coexpress G-CSFR (albeit using an antirat G-CSFR antibody of unknown specificity) with the stem cell marker nestin [8]. G-CSF also dose-dependently increased the mean expression level of various markers of neuronal and glial cells, and the absolute number of mature neurons expressing microtubule-associated protein 2 (MAP2) [8]. In intact rat brains, peripheral infusion of rHuG-CSF enhanced the migration of cells expressing doublecortin (a MAP expressed in neural progenitor cells and immature neurons) from the lateral ventricle wall into the ischemic area of the neocortex. Similar effects, however, were not observed in the striatum, another region of the brain that is close to the subventricular zone and which has also been described to exhibit some degree of progenitor activation and neurogenesis after ischemia, nor was there evidence that the doublecortin$^+$ cells which migrated to the neocortex subsequently proliferated and differentiated into mature (NeuN$^+$) neurons [8].
4. G-CSF may facilitate improved functional performance after ischemia by inhibiting local inflammation. Inflammation is an important mechanism for isolating damaged tissue from surrounding uninjured zones so that monocytes/macrophages can consume dead or dying cells and repair the extracellular matrix. Reactive oxygen species and other free radicals/oxidants that exude from

infiltrating neutrophils can also threaten the viability of adjacent tissue and adversely affect stroke outcome. While rHuG-CSF increases neutrophil numbers and augments some of these components of the inflammatory reaction, it is also well established that G-CSF limits this cascade [77]. For example, G-CSF has been shown to dampen the release of proinflammatory cytokines (tumor necrosis factor [TNF]-α, IL-1β, IL-6, 8, and 12; and interferon [IFN]-γ) and inducible nitric oxide synthase (iNOS), and to induce the release of counter regulatory molecules such as IL-1 receptor antagonist (IL-1ra) and soluble TNF receptor under several conditions [74, 78–80]. The pathophysiologic relevance of this anti-inflammatory action of G-CSF to stroke recovery is suggested by the fact that compared to wild-type mice, infarct volume is reduced by approximately 70% in IL-1α and -1β knock-out mice subjected to transient cerebral ischemia [81], and that increased expression of IL-1β mRNA and protein appear to be associated with worse neurologic outcome in ischemic stroke patients [82, 83].

5. A fifth hypothetical mechanism, that G-CSF promotes angiogenesis, will not be discussed here in the context of cerebral or myocardial ischemia as it is the subject of an entire chapter in this book (see chapter "Role of Myeloid Cells in Tumor Angiogenesis" by Ferrara).

3.2 *Clinical Studies of rHuG-CSF in Stroke*

Four small clinical studies evaluating the safety and efficacy of rHuG-CSF as a therapy for ischemic stroke were completed from 2006 to 2008 [84–87]. Together, these comprised 133 patients, but only two of these studies (Zhang et al. and Sprigg et al.) were double-blinded, placebo-controlled in design, encompassing 39 rHuG-CSF-treated and 42 placebo patients combined. All these trials examined the effects of rHuG-CSF administered only in the first 1–4 days after a stroke. No clinical studies of chronic rHuG-CSF treatment after stroke have been reported. The results of these trials were discussed elsewhere [88] and are not reviewed in detail here. Both of the fully blinded and placebo-controlled studies demonstrated increases in circulating leukocytes and $CD34^+$ cells that were rHuG-CSF-dose dependent. In the study by Sprigg et al., rHuG-CSF was administered subcutaneously at escalating doses of 1, 3, or 10 μg/kg/day for 1 or 5 days, beginning on day 7 after ictus. No differences in neurologic outcome were noted at the 90 day-follow-up [85]. Zhang et al. treated patients with 2 μg/kg subcutaneous rHuG-CSF administered daily for 5 days after stroke. Ten days after the end of treatment there was no change in neurologic score compared with pretreatment values in the rHuG-CSF group. Although a difference was noted on day 20, the improvement noted over time was not different from the placebo group [84]. The only study in which a statistically significant improvement in neurologic functioning (on four scoring scales) has been observed, in this case at 12 months after rHuG-CSF treatment, was that of Shyu et al. [86]. Interestingly, four patients who received rHuG-CSF within 1 day after stroke performed better than three patients who started therapy after 24 h.

These seven patients in the treatment group exhibited a higher initial incidence of small lacunar infarctions that occur mainly in the white matter of the brain and are known to be associated with higher recovery rates and better patient outcomes. This apparent bias and the lack of a placebo group preempts any definitive conclusions regarding the effect of rHuG-CSF from this study. Thus overall, although there were significant differences in the rHuG-CSF dose (1–60 μg/kg), duration of treatment (1–5 days), and time of initiating rHuG-CSF therapy relative to stroke (1–30 days) in all these trials, rHuG-CSF did not significantly alter functional outcome compared with placebo in stroke patients. rHuG-CSF was well tolerated and appeared to be safe, consistent with the experience established from its use in hematology and oncology.

Two additional randomized controlled trials of rHuG-CSF as a single agent in stroke are underway. The German company Sygnis Bioscience are following up their previous AXIS trial [87], which demonstrated a positive influence of intravenous filgrastim (which they refer to as “AX200”) on multiple clinical parameters compared with baseline (although not placebo), with the larger AXIS 2 trial comprising 175 patients in each group. Patients will be enrolled within 9 h after occurrence of an ischemic stroke, and rHuG-CSF will be administered as an infusion over 3 days (http://www.clinicaltrials.gov/ct2/show/NCT00927836). Final data collection for primary outcome measurement is expected in June 2011. The STEMS2 trial started in 2007 is also now complete and in the follow-up phase (http://www.strokecenter.org/trials/TrialDetail.aspx?tid=942). This phase 2 study of 60 patients administered placebo or 10 μg/kg/day rHuG-CSF subcutaneously for 5 days between 3 and 30 days after ischemic stroke will measure effects on the incidence of serious adverse events by day 90. A third study evaluating the combination of rHuG-CSF and erythropoietin (rHuEPO) on associative learning and/or motor skills in patients with chronic stroke or amyotrophic lateral sclerosis (http://www.clinicaltrials.gov/ct/show/NCT00298597) is not considered herein.

4 rHuG-CSF and Alzheimer’s Disease

Alzheimer’s disease is a progressive neurodegenerative disorder that presents clinically with dementia. The world prevalence of dementia due to Alzheimer’s disease is estimated to be between 16 and 20 million people and is increasing. The disease is most common in elderly people and is characterized by the deposition of β-amyloid peptides into neuritic plaques, the formation of neurofibrillary tangles, and ultimately the loss of synapses and destruction of neurons in regions of the brain critical for learning and memory. These processes activate microglial cells and astrocytes and trigger neuroinflammatory processes that modulate the production of many cytokines, including those whose predominant biologic actions have at least historically been described as primarily hematologic. The current therapies for Alzheimer’s disease are directed primarily toward preserving cognition and reducing the impact of other symptoms on quality of life. Medicines that modulate the underlying biologic processes are needed.

The potential efficacy of rHuG-CSF in reversing the brain amyloid burden and cognitive impairment associated with Alzheimer's disease has been tested in several models of the disease induced either by cerebral injection of aggregated β-amyloid, or expression of mutated amyloid precursor protein (APP) and presinilin 1 (PS1) in transgenic mice [12, 89]. Compared with vehicle-treated animals, a course of rHuG-CSF (250 μg/kg/day administered subcutaneously every other day for approximately 3 weeks before and throughout behavioral testing) significantly reduced the period of latency and number of errors in working memory tests to levels exhibited by intact nontransgenic littermates, as well as the size and extent of β-amyloid deposits in the hippocampus and entorhinal cortex (by about 40%) [12]. As discussed earlier in the setting of myocardial infarction and stroke, the key question is what is the mechanism for these profound effects? In the study above, human G-CSF was claimed to be "measurable" in hippocampal tissue from rHuG-CSF-treated transgenic mice, which could indicate transport of the drug across the blood–brain barrier, but no data were presented [12]. G-CSFR protein was also reported to be expressed on hippocampal neural stem cells isolated from normal adult mice (not the Alzheimer's disease rats in which the functional effects of rHuG-CSF were demonstrated), but this was determined using an anti-G-CSFR antibody and methodology (immunohistochemistry) that does not specifically identify G-CSFR [13]. Recombinant HuG-CSF appeared to increase the proliferation of normal mouse neural stem cells in vitro but the experiments were performed in the presence of serum so one cannot exclude the possibility that G-CSF synergized with another soluble factor that might predominantly account for this effect [12]. BrdU-labeled (i.e., newly generated) neurons were also detected in the brains of G-CSF-treated Alzheimer's disease mice [89], but their frequency was so low and only marginally increased relative to vehicle-treated mice that it is difficult to link this finding with the behavioral outcome observed. An indirect nonneuronal cell-mediated effect of rHu-G-CSF seems more consistent with the data and the known biologic effects of G-CSF. G-CSF increased the number of microglia around amyloid plaques in the brain of treated Alzheimer's disease rats [12]. Notably, microglial cells are derived from hematopoietic stem cells. They are the differentiated progeny of monocytes that migrate to the brain and spinal cord where they serve to eliminate damaged neurons and act as the first and predominant line of active immune defense in the central nervous system. Microglial cells have been reported to express G-CSFR [90]. Although this finding awaits confirmation using a G-CSFR-specific antibody, it is at least consistent with the myeloid hematopoietic origin of these cells and proffers the testable hypothesis that G-CSF may mediate its effects through direct stimulation of resident microglia. Sanchez-Ramos et al. demonstrated that cells that migrate from the bone marrow could infiltrate the brain and contribute to microgliosis. These experiments used chimeric transgenic Alzheimer's disease mice in which the hematopoietic system had been regenerated after total body lethal irradiation by transplantation of bone marrow cells isolated from donor animals expressing a green fluorescent protein

(GFP) transgene. Such mice exhibited increased numbers of GFP$^+$ cells in both the hippocampus and entorhinal cortex following treatment with rHuG-CSF. Technical limitations prevented quantitation and so it is not known whether this effect was statistically significant, but it is noteworthy that some GFP$^+$ cells appeared morphologically to resemble amyloid-ingesting macrophages [12]. Thus, mobilization of "new" microglial precursors may represent an additional mechanism, independent of direct stimulation of neuronal cell proliferation, for which the evidence is questionable, by which rHuG-CSF improved Alzheimer's disease parameters in this nonclinical model. Very similar findings have been reported in a mouse model of Parkinson's disease, another debilitating neurodegenerative disorder for which current therapies are not particularly effective, but wherein rHuG-CSF appears to be neuroprotective [91]. These studies are subject to the same criticisms that undermine the direct-action hypothesis as discussed, i.e., the lack of definitive evidence that G-CSF crosses the blood–brain barrier, and that the sole evidence for G-CSFR protein expression on, in this case dopaminergic neurons in the substantia nigra, are immunohistochemistry experiments performed with a nonspecific anti-G-CSFR antibody.

Notwithstanding these technical concerns, several pieces of associative evidence point to a potential involvement of G-CSF in Alzheimer's disease and strengthen the case for a link between the hematopoietic and nervous systems. Laske et al. found significantly lower amounts of G-CSF in the plasma of 50 patients with early Alzheimer's disease compared with 50 age-matched healthy controls [92]. It is noteworthy, however, that the relatively higher mean concentration of G-CSF in the normal cohort was skewed by two individuals with unusually high levels: approximately 60 and >80 pg/mL versus <30 pg/mL in most other members in this cohort and in historical controls (Fig. 3a). It is not clear if the apparent "decrease" in G-CSF in Alzheimer's disease would be significant if these outliers were excluded from the analysis. A similar concern applies to a study comparing the number of circulating CD45RO^{low}CD34$^+$ cells in 23 patients with early Alzheimer's disease versus 25 nondemented age, sex, and environmentally matched controls [93]. While the latter group appeared to exhibit higher hematopoietic stem/progenitor cell counts, the median was skewed by three individuals with excessively high numbers of these cells in the blood (Fig. 3b). Despite the caveats, these surveys are at least consistent with two molecular studies that demonstrated that reduced expression of G-CSF comprises part of a "signature" of 18 (subsequently reduced to 5) signaling proteins that can be used to classify blinded samples from Alzheimer's and control subjects and that achieves nearly 97% accuracy to predict clinical Alzheimer's disease [94, 95]. Thus, decreased expression of G-CSF in neurodegenerative disease, with concurrent reduced hematopoiesis, could theoretically be compensated for by therapeutic administration of rHuG-CSF, as has been demonstrated in nonclinical models.

Human CD34$^+$ bone marrow cells express low levels of the receptors for dopamine (a precursor of norepinephrine, the chief neurotransmitter of the sympathetic nervous system) and chatecholamines such as epinephrine and norepinephrine that are key mediators of stress responses, including hematopoietic stress.

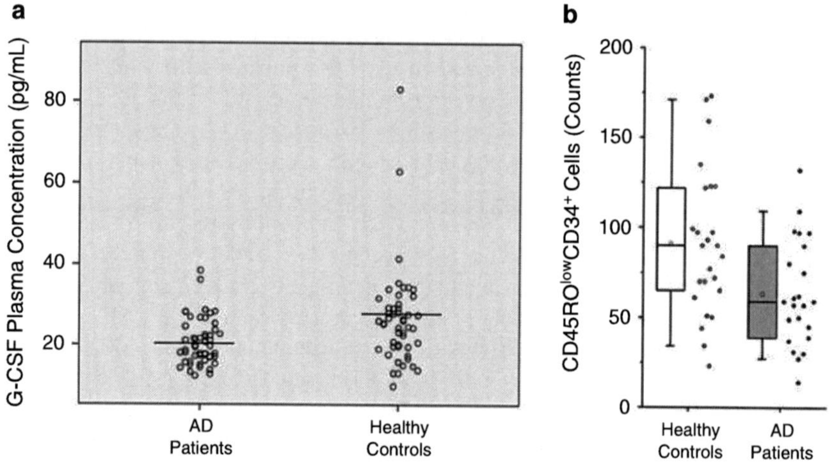

Fig. 3 Panel (**a**) G-CSF plasma concentrations are lower in patients with Alzheimer's disease than in healthy controls. The *horizontal bars* represent the mean values for each group ($n = 50$ each), which are significantly different ($P = 0.003$) as reported [92]. Reprinted from Journal of Alzheimer's Disease, Vol. 17, C Laske, K Stellos, E Stransky, T Leyhe and M Gawaz, Decreased plasma levels of granulocyte-colony stimulating factor (G-CSF) in patients with early Alzheimer's disease, pp. 115–123, Copyright 2009, with permission from IOS Press. Panel (**b**) Patients with early Alzheimer's disease ($n = 23$) have fewer circulating CD45RO^{low}CD34^{+} cells (expressed per 60 μL of whole blood) than healthy controls ($n = 25$). The mean values indicated by the small *square* symbol within each box are significantly different ($P = 0.01$) as reported [93]. The *horizontal lines* in each box mark the median. The *upper* and *lower* boundaries of the boxes demarcate the 75th and 25th percentiles, and the *upper* and *lower* whiskers indicate the 95th and 5th percentiles, respectively. Reprinted by permission from Macmillan Publishers Ltd., Molecular Psychiatry 11: 1113–1116, 2006; Copyright 2006

These receptors are expressed at slightly higher levels on the CD34^{+}CD38$^{-/low}$ subset that is more enriched in long-term repopulating hematopoietic stem cells, and expression is further increased after G-CSF stimulation [96]. Treatment of G-CSF-stimulated umbilical cord blood CD34^{+} cells with dopamine or dopamine receptor agonists in vitro has been shown to moderately increase their migration, their ability to generate colonies of mature blood cells in culture, and to repopulate the hematopoietic system after transplantation into immunodeficient mice. Similarly, treatment of mice with epinephrine (50 μg/day for 6 days) increased circulating leukocyte counts by approximately 50%, and the number of colony-forming cells and Sca-1^{+}c-kit^{+}Lin^{-} cells (highly enriched in long-term repopulating stem cells) in the peripheral blood and bone marrow, respectively, by about 2.5-fold [96]. These functional effects of neurotransmitters appeared to be mediated by activation of the Wnt pathway, a well-established determinant of hematopoietic stem cell self-renewal and proliferation. In murine models, the sympathetic nervous system (via norepinephrine signaling) has been shown to stimulate the egress from the bone marrow into the circulation of hematopoietic stem and progenitor cells both in steady state as well as after G-CSF-induced mobilization. This finding was associated with suppression of osteoblasts in stem cell "niches" and down-regulation of SDF-1, which promotes the retention of primitive hematopoietic cells within the marrow microenvironment [97, 98]. It is therefore intriguing to speculate that activation of neural signaling in response to tissue damage (stroke, ischemia) or disease (Alzheimer's disease, etc.) may stimulate the migration of bone marrow-derived hematopoietic stem cells toward the site of injury, due in part to down-regulation of SDF-1 expression in the bone marrow and its up-regulated production in these distal tissues in response to hypoxia and inflammation. As demonstrated in Alzheimer's disease models, bone marrow-derived cells might then differentiate into accessory cells (e.g., microglia) that, while hematopoietic in origin, perform highly specialized roles that enhance functional recovery. Exogenously administered rHuG-CSF could augment numerous components of this cascade, increasing stem/progenitor cell mobilization, proliferation, and expression of neurotransmitter receptors, and inhibit the production or activity of the main proinflammatory cytokines (IL-1, TNF-α, and IFN-γ) [77] to further enhance tissue repair. This model does not require G-CSF to cross the blood–brain barrier or for G-CSFR to be expressed on nonhematopoietic cell types, for which the evidence is equivocal. These proposed mechanisms will require verification in vivo through well-designed clinical studies. In the meantime, the finding that normal (i.e., nonischemic) mice in which the *csf3* gene has been deleted exhibit impaired memory formation, motor skills, reduced dendritic complexity in hippocampal neurons, and fewer neuronal precursor cells in the dentate gyrus [99], and that compared with wild-type littermates, these animals exhibit slightly higher infarct volumes and poorer motor performance after transient focal cerebral ischemia that is significantly improved by rHuG-CSF administration [100] provides compelling evidence that endogenous G-CSF plays an important role in the brain.

5 Summary

A plethora of nonclinical data accumulated over the past decade suggests that administration of exogenous G-CSF may improve functional performance in small and large animals with coronary or cerebral ischemia, and neurodegenerative diseases such as Alzheimer's disease, Parkinson's disease, and others. In stark contrast, not a single large, randomized, double-blinded and placebo-controlled clinical study has demonstrated a statistically significant improvement in clinically meaningful parameters in patients with myocardial infarction or stroke. While many of the nonclinical studies are technically flawed, these limitations apply primarily to the evidence for a hypothesized direct action of G-CSF on G-CSFR-expressing nonhematopoietic cells. The fundamental biologic observations of improved tissue recovery, more likely explained by the various indirect actions of

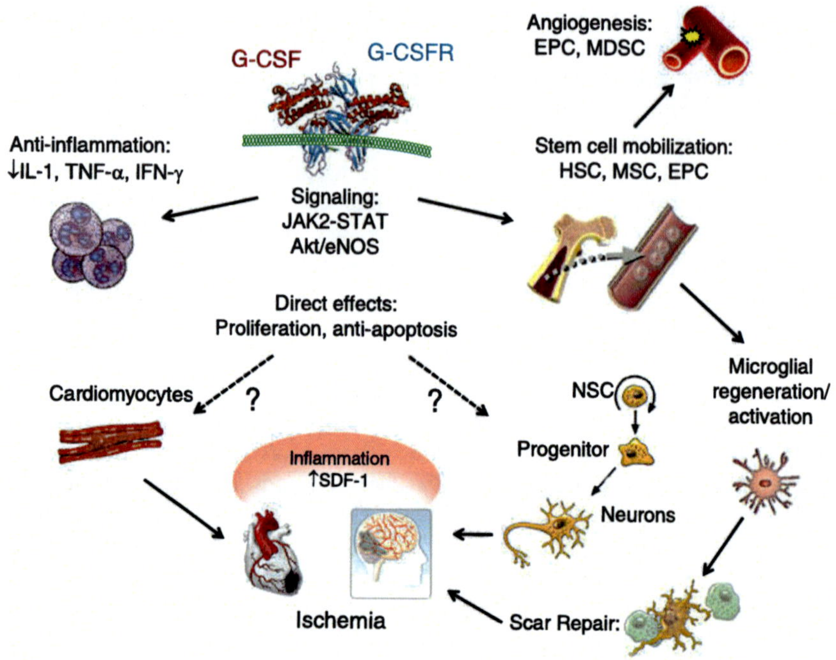

Fig. 4 Potential mechanisms by which G-CSF might facilitate improved structural and functional recovery after myocardial infarction or stroke. Binding of G-CSF to its receptor expressed on myeloid-restricted hematopoietic progenitor cells stimulates their proliferation and differentiation, resulting in the production of large numbers of mature neutrophils. As granulocytes accumulate in large numbers in the bone marrow, the enzymes they secrete cleave various adhesion molecules that retain primitive hematopoietic cells in this microenvironment. A variety of stem and progenitor cell types are thus mobilized into the circulation that can promote angiogenesis directly or indirectly (e.g., myeloid-derived suppressor cells or endothelial progenitor cells), differentiate into new mesenchymal tissue, or produce microglial precursors and macrophages that facilitate scar tissue repair. G-CSF also stimulates a cascade of anti-inflammatory processes that counter-balance the localized inflammation associated with ischemia. G-CSF has been proposed to act directly on G-CSFR$^+$ neural stem cells, mature neurons, and cardiomyocytes to activate intracellular pathways that culminate in increased proliferation and reduced cell death. However, the experimental evidence supporting such direct actions on nonhematopoietic cells is subject to challenge

G-CSF (Fig. 4), remain largely uncompromised. Thus, the challenge will be whether and how these diverse downstream effects of G-CSF observed in a nonclinical setting can be harnessed to clinical benefit in a genetically out-bred and typically aged human population whose medical profiles are obviously much more complex than are laboratory animals. The trials performed to date have been deficient in providing mechanistic insights. The next decade should be viewed as an opportunity to expand hypothesis-driven clinical study.

References

1. Molineux G, Dexter TM (1998) Biology of G-CSF. In: Morstyn G, Dexter TM, Foote M (eds) Filgrastim in clinical practice, 2nd edn. Marcel Dekker, New York, pp 1–41
2. Meisenberg BR, Davis TA, Melaragno AJ et al (1992) A comparison of therapeutic schedules for administering granulocyte colony-stimulating factor to nonhuman primates after high-dose chemotherapy. Blood 79:2267–2272
3. Lieschke GJ, Grail D, Hodgson G et al (1994) Mice lacking granulocyte colony-stimulating factor have chronic neutropenia, granulocyte and macrophage progenitor cell deficiency, and impaired neutrophil mobilization. Blood 84:1737–1746
4. Welte K, Gabrilove J, Bronchud MH et al (1996) Filgrastim (r-metHuG-CSF): the first 10 years. Blood 88:1907–1929
5. Schäbitz WR, Kollmar R, Schwaninger M et al (2003) Neuroprotective effect of granulocyte colony-stimulating factor after focal cerebral ischemia. Stroke 34:745–751
6. Oishi A, Otani A, Sasahara M et al (2008) Granulocyte colony-stimulating factor protects retinal photoreceptor cells against light-induced damage. Invest Ophthalmol Vis Sci 49:5629–5635
7. Shimoji K, Yuasa S, Onizuka T et al (2010) G-CSF promotes the proliferation of developing cardiomyocytes in vivo and in derivation from ESCs and iPSCs. Cell Stem Cell 6:227–237
8. Schneider A, Krüger C, Steigleder T et al (2005) The hematopoietic factor G-CSF is a neuronal ligand that counteracts programmed cell death and drives neurogenesis. J Clin Invest 115:2083–2098
9. Harada M, Qin Y, Takano H et al (2005) G-CSF prevents cardiac remodeling after myocardial infarction by activating the Jak-Stat pathway in cardiomyocytes. Nat Med 11:305–311
10. Kuhlmann MT, Kirchhof P, Klocke R et al (2006) G-CSF/SCF reduces inducible arrhythmias in the infracted heart potentially via increased connexin43 expression and arteriogenesis. J Exp Med 203:87–97
11. Solaroglu I, Jadhav V, Zhang JH (2007) Neuroprotective effect of granulocyte-colony stimulating factor. Front Biosci 12:712–724
12. Sanchez-Ramos J, Song S, Sava V et al (2009) Granulocyte colony stimulating factor decreases brain amyloid burden and reverses cognitive impairment in Alzheimer's Mice. Neuroscience 163:55–72
13. de Bruin C, Lincoln P, Hartley C, Shehabilen A, Van G, Szilvassy SJ (2010) Most purported antibodies to the human granulocyte colony-stimulating factor receptor are not specific. Exp Hematol 38:1025–1035
14. Lyman GH (2005) Pegfilgrastim: a granulocyte colony-stimulating factor with sustained duration of action. Expert Opin Biol Ther 5:1635–1646
15. Asahara T, Masuda H, Takahashi T et al (1999) Bone marrow origin of endothelial progenitor cells responsible for postnatal vasculogenesis in physiological and pathological neovascularization. Circ Res 85:221–228
16. Paczkowska E, Larysz B, Rzeuski R et al (2005) Human hematopoietic stem/progenitor-enriched $CD34^+$ cells are mobilized into peripheral blood during stress related to ischemic stroke or acute myocardial infarction. Eur J Haematol 75:461–467
17. Orlic D, Kajstura J, Chimenti S et al (2001) Bone marrow cells regenerate infracted myocardium. Nature 410:701–705
18. Orlic D, Kajstura J, Chimenti S et al (2001) Mobilized bone marrow cells repair the infracted heart, improving function and survival. Proc Natl Acad Sci U S A 98:10344–10349
19. Murry CE, Soonpaa MH, Reinecke H et al (2004) Haematopoietic stem cells do not transdifferentiate into cardiac myocytes in myocardial infarcts. Nature 428:664–668
20. Balsam LB, Wagers AJ, Christensen JL et al (2004) Haematopoietic stem cells adopt mature haematopoietic fates in ischaemic myocardium. Nature 428:668–673

21. Wagers AJ, Sherwood RI, Christensen JL et al (2002) Little evidence for developmental plasticity of adult hematopoietic stem cells. Science 297:2256–2259
22. Alvarez-Dolado M, Pardal R, Garcia-Verdugo JM et al (2003) Fusion of bone-marrow-derived cells with Purkinje neurons, cardiomyocytes and hepatocytes. Nature 425:968–973
23. Nygren JM, Jovinge S, Breitbach M et al (2004) Bone marrow-derived hematopoietic cells generate cardiomyocytes at a low frequency through cell fusion, but not transdifferentiation. Nat Med 10:494–501
24. Minatoguchi S, Takemura G, Chen X-H et al (2004) Acceleration of the healing process and myocardial regeneration may be important as a mechanism of improvement of cardiac function and remodeling by postinfarction granulocyte colony-stimulating factor treatment. Circulation 109:2572–2580
25. Werneck-de-Castro JPS, Costa-e-Sousa RH, de Oliveira PF et al (2006) G-CSF does not improve systolic function in a rat model of acute myocardial infarction. Basic Res Cardiol 101:494–501
26. Takahama H, Minamino T, Hirata A et al (2006) Granulocyte colony-stimulating factor mediates cardioprotection against ischemia/reperfusion injury via phosphatidylinositol-3-kinase/Akt pathway in canine hearts. Cardiovasc Drugs Ther 20:159–165
27. Li Y, Fukuda N, Yokoyama S-I et al (2006) Effects of G-CSF on cardiac remodeling and arterial hyperplasia in rats. Eur J Pharmacol 549:98–106
28. Lee SS, Naqvi TS, Forrester J et al (2007) The effect of granulocyte colony stimulating factor on regional and global myocardial function in the porcine infarct model. Int J Cardiol 116:225–230
29. Okada H, Takemura G, Li Y et al (2008) Effect of a long-term treatment with a low-dose granulocyte colony-stimulating factor on post-infarction process in the heart. J Cell Mol Med 12:1272–1283
30. Yagi T, Fukuda K, Fujita J et al (2008) G-CSF augments small vessel and cell density in canine myocardial infarction. Keio J Med 57:139–149
31. De Silva R, Raval AN, Hadi M et al (2008) Intracoronary infusion of autologous mononuclear cells from bone marrow or granulocyte colony-stimulating factor-mobilized apheresis product may not improve remodeling, contractile function, perfusion, or infarct size in a swine model of large myocardial infarction. Eur Heart J 29:1772–1782
32. Klocke R, Kuhlmann MT, Scobioala S et al (2008) Granulocyte colony-stimulating factor (G-CSF) for cardio- and cerebrovascular regenerative applications. Curr Med Chem 15:968–977
33. Vertesaljai M, Piroth Z, Fontos G et al (2008) Drugs, gene transfer, signaling factors: a bench to bedside approach to myocardial stem cell therapy. Heart Fail Rev 13:227–244
34. Ueda K, Takano H, Hasegawa H et al (2006) Granulocyte colony stimulating factor directly inhibits myocardial ischemia-reperfusion injury through Akt-endothelial NO synthase pathway. Arteroscler Thromb Vasc Biol 26:e108–e113
35. Schulz R, Kelm M, Heusch G (2004) Nitric oxide in myocardial ischemia/reperfusion injury. Cardiovasc Res 61:402–413
36. Kawada H, Fujita J, Kinjo K et al (2004) Nonhematopoietic mesenchymal stem cells can be mobilized and differentiate into cardiomyocytes after myocardial infarction. Blood 104:3581–3587
37. Cheng Z, Liu X, Ou L et al (2008) Mobilization of mesenchymal stem cells by granulocyte colony-stimulating factor in rats with acute myocardial infarction. Cardiovasc Drugs Ther 22:363–371
38. Bussolino F, Wang JM, Defilippi P et al (1989) Granulocyte- and granulocyte-macrophage-colony stimulating factors induce human endothelial cells to migrate and proliferate. Nature 337:471–473
39. Bussolino F, Ziche M, Wang JM et al (1991) In vitro and in vivo activation of endothelial cells by colony-stimulating factors. J Clin Invest 87:986–995
40. Gabrilovich DI, Nagaraj S (2009) Myeloid-derived suppressor cells as regulators of the immune system. Nat Rev Immunol 9:162–174

41. Shojaei F, Wu X, Zhong C et al (2007) Bv8 regulates myeloid-cell-dependent tumour angiogenesis. Nature 450:825–831
42. Shojaei F, Zhong C, Wu X et al (2008) Role of myeloid cells in tumor angiogenesis and growth. Trends Cell Biol 18:372–378
43. Ohki Y, Heissig B, Sato Y et al (2005) Granulocyte colony-stimulating factor promotes neovascularization by releasing vascular endothelial growth factor from neutrophils. FASEB J 19:2005–2007
44. Capoccia B, Sheperd RM, Link DC (2006) G-CSF and AMD3100 mobilize monocytes into the blood that stimulate angiogenesis in vivo through a paracrine mechanism. Blood 108:2438–2445
45. Sugano Y, Anzai T, Yoshikawa T et al (2005) Granulocyte colony-stimulating factor attenuates early ventricular expansion after experimental myocardial infarction. Cardiovasc Res 65:446–456
46. Ince H, Valgimigli M, Petzsch M et al (2008) Cardiovascular events and re-stenosis following administration of G-CSF in acute myocardial infarction: systematic review and meta-analysis. Heart 94:610–616
47. Zohlnhöfer D, Dibra A, Koppara T et al (2008) Stem cell mobilization by granulocyte colony-stimulating factor for myocardial recovery after acute myocardial infarction: a meta-analysis. J Am Coll Cardiol 51:1429–1437
48. Fan L, Chen L, Chen X et al (2008) A meta-analysis of stem cell mobilization by granulocyte colony-stimulating factor in the treatment of acute myocardial infarction. Cardiovasc Drugs Ther 22:45–54
49. Ripa RS, Kastrup J (2008) G-CSF therapy with mobilization of bone marrow stem cells for myocardial recovery after acute myocardial infarction – a relevant treatment? Exp Hematol 36:681–686
50. Abdel-Latif A, Bolli R, Zuba-Surma EK et al (2008) Granulocyte colony-stimulating factor therapy for cardiac repair after acute myocardial infarction: a systematic review and meta-analysis of randomized controlled trials. Am Heart J 156:216–226
51. Zohlnhöfer D, Ott I, Mehilli J et al (2006) Stem cell mobilization by granulocyte colony-stimulating factor in patients with acute myocardial infarction: a randomized controlled trial. JAMA 295:1003–1010
52. Ripa RS, Jørgensen E, Wang Y et al (2006) Stem cell mobilization induced by subcutaneous granulocyte-colony stimulating factor to improve cardiac regeneration after acute ST-elevation myocardial infarction: result of the double-blind, randomized, placebo-controlled stem cells in myocardial infarction (STEMMI) trial. Circulation 113:1983–1992
53. Engelmann MG, Theiss HD, Hennig-Theiss C et al (2006) Autologous bone marrow stem cell mobilization induced by granulocyte colony-stimulating factor after subacute ST-segment elevation myocardial infarction undergoing late revascularization: final results from the G-CSF-STEMI 9granulocyte colony-stimulating factor ST-segment elevation myocardial infarction) trial. J Am Coll Cardiol 48:1712–1721
54. Ellis SG, Penn MS, Bolwell B et al (2006) Granulocyte colony stimulating factor in patients with large acute myocardial infarction: results of a pilot dose-escalation randomized trial. Am Heart J 152:1051.e9–1051.e14
55. Ince H, Petzsch M, Kleine HD et al (2005) Preservation from left ventricular remodeling by front-integrated revascularization and stem cell liberation in evolving acute myocardial infarction by use of granulocyte-colony-stimulating factor (FIRSTLINE-AMI). Circulation 112:3097–3106
56. Beohar N, Flaherty JD, Davidson CJ et al (2007) Granulocyte-colony stimulating factor administration after myocardial infarction in a porcine ischemia-reperfusion model: functional and pathological effects of dose timing. Catheter Cardiovasc Interv 69:257–265
57. Engelmann MG, Theiss HD, Theiss C et al (2008) G-CSF in patients suffering from late revascularized ST elevation myocardial infarction: analysis on the timing of G-CSF administration. Exp Hematol 36:703–709

58. Lehrke S, Mazhari R, Durand DJ et al (2006) Aging impairs the beneficial effect of granulocyte colony-stimulating factor and stem cell factor on post-myocardial infarction remodeling. Circ Res 99:553–560
59. Kidd PM (2009) Integrated brain restoration after ischemic stroke – medical management, risk factors, nutrients, and other interventions for managing inflammation and enhancing brain plasticity. Alt Med Rev 14:14–35
60. Zhao L-R, Navalitloha Y, Singhal S et al (2007) Hematopoietic growth factors pass through the blood-brain barrier in intact rats. Exp Neurol 204:569–573
61. Shyu W-C, Lin S-Z, Yang H-I et al (2004) Functional recovery of stroke rats induced by granulocyte colony-stimulating factor-stimulated stem cells. Circulation 110:1847–1854
62. Gibson CL, Bath PMW, Murphy SP (2005) G-CSF reduces infarct volume and improves functional outcome after transient focal cerebral ischemia in mice. J Cerebral Blood Flow Metab 25:431–439
63. Schneider A, Wysocki R, Pitzer C et al (2006) An extended window of opportunity for G-CSF treatment in cerebral ischemia. BMC Biol 4:36–43
64. Popa-Wagner A, Stöcker K, Balseanu AT et al (2010) Effects of granulocyte-colony stimulating factor after stroke in aged rats. Stroke 41:1027–1031
65. Matchett GA, Calinisan JB, Matchett GC et al (2007) The effect of granulocyte-colony stimulating factor in global cerebral ischemia in rats. Brain Res 1136:200–207
66. Zhao L-R, Berra HH, Duan W-M et al (2007) Beneficial effects of hematopoietic growth factor therapy in chronic ischemic stroke in rats. Stroke 38:2804–2811
67. Taguchi A, Wen Z, Myojin K et al (2007) Granulocyte colony-stimulating factor has a negative effect on stroke outcome in a murine model. Eur J Neurosci 26:126–133
68. Ren JM, Finklestein SP (2005) Growth factor treatment of stroke. Curr Drug Targets CNS Neurol Dis 4:121–125
69. England TJ, Gibson CL, Bath PWM (2009) Granulocyte-colony stimulating factor in experimental stroke and its effects on infarct size and functional outcome: a systematic review. Brain Res Rev 62:71–82
70. Minnerup J, Heidrich J, Wellmann J et al (2008) Meta-analysis of the efficacy of granulocyte-colony stimulating factor in animal models of focal cerebral ischemia. Stroke 39:1855–1861
71. Lu C-Z, Xiao B-G (2006) G-CSF and neuroprotection: a therapeutic perspective in cerebral ischemia. Biochem Soc Trans 34:1327–1333
72. Yanqing Z, Yu-Min L, Jian Q et al (2006) Fibronectin and neuroprotective effect of granulocyte colony-stimulating factor in focal cerebral ischemia. Brain Res 1098:161–169
73. Sakai T, Johnson KJ, Murozono M et al (2001) Plasma fibronectin supports neuronal survival and reduces brain injury following transient focal cerebral ischemia but is not essential to skin-wound healing and hemostasis. Nat Med 7:324–330
74. Komine-Kobayashi M, Zhang N, Liu M et al (2006) Neuroprotective effect of recombinant human granulocyte colony-stimulating factor in transient focal ischemia of mice. J Cereb Blood Flow Metab 26:402–413
75. Solarglu I, Cahill J, Tsubokawa T et al (2009) Granulocyte colony-stimulating factor protects the brain against experimental stroke via inhibition of apoptosis and inflammation. Neurol Res 31:167–172
76. Jung K-H, Chu K, Lee S-T et al (2006) G-CSF protects human cerebral hybrid neurons against in vitro ischemia. Neurosci Lett 394:168–173
77. Hartung T (1998) Anti-inflammatory effects of granulocyte colony-stimulating factor. Curr Opin Hematol 5:221–225
78. Hartung T, Döcke W-D, Gantner F et al (1995) Effect of granulocyte colony-stimulating factor treatment on ex vivo blood cytokine response in human volunteers. Blood 85:2482–2489
79. So H, Fink MP (1999) Counter regulatory control of the acute inflammatory response: granulocyte colony-stimulating factor has anti-inflammatory properties. Crit Care Med 27:1019–1021

80. Zavala F, Abad S, Ezine S et al (2002) G-CSF therapy of ongoing experimental allergic encephalomyelitis via chemokine- and cytokine-based immune deviation. J Immunol 168:2011–2019
81. Boutin H, LeFeuvre RA, Horai R et al (2001) Role of IL-1alpha and IL-1beta in ischemic brain damage. J Neurosci 21:5528–5534
82. Kostulas N, Pelidou SH, Kivisäkk P et al (1999) Increased IL-1b, IL-8, and IL-17 mRNA expression in blood mononuclear cells observed in a prospective ischemic stroke study. Stroke 30:2174–2179
83. Mazzotta G, Sarchielli P, Caso V et al (2004) Different cytokine levels in thrombolysis patients as predictors for clinical outcome. Eur J Neurol 11:377–381
84. Zhang JJ, Deng M, Zhang Y et al (2006) A short-term assessment of recombinant human granulocyte colony-stimulating factor (RHG-CSF) in treatment of acute cerebral infarction. Cerebrovasc Dis 21(Suppl 4):143
85. Sprigg N, Bath PM, Zhao L et al (2006) Granulocyte-colony-stimulating factor mobilizes bone marrow stem cells in patients with subacute ischemic stroke. Stroke 37:2979–2983
86. Shyu W-C, Lin S-Z, Lee C-C et al (2006) Granulocyte colony-stimulating factor for acute ischemic stroke: a randomized controlled trial. CMAJ 174:927–933
87. Schäbitz WR, Laage R, Schwab S et al (2008) AX200 (G-CSF) for the treatment of acute ischemic stroke (AXIS). Stroke 39:561
88. Bath PMW, Sprigg N (2007) Colony stimulating factors (including erythropoietin, granulocyte colony-stimulating factor and analogues) for stroke. Cochrane Database Syst Rev 2:1–24
89. Tsai K-J, Tsai Y-C, Shen C-K (2007) G-CSF rescues the memory impairment of animal models of Alzheimer's disease. J Exp Med 204:1273–1280
90. Hasselblatt M, Jeibmann A, Riesmeier B et al (2007) Granulocyte-colony stimulating factor (G-CSF) and G-CSF receptor expression in human ischemic stroke. Acta Neuropathol 113:45–51
91. Meuer K, Pitzer C, Teismann P et al (2006) Granulocyte-colony stimulating factor is neuroprotective in a model of Parkinson's disease. J Neurochem 97:675–686
92. Laske C, Stellos K, Stransky E et al (2009) Decreased plasma levels of granulocyte-colony stimulating factor (G-CSF) in patients with early Alzheimer's disease. J Alzheimers Dis 17:115–123
93. Maler JM, Spitzer P, Lewczuk P et al (2006) Decreased circulating $CD34^+$ stem cells in early Alzheimer's disease: evidence for a deficient hematopoietic brain support? Mol Psychiatry 11:1113–1115
94. Ray S, Britschgi M, Herbert C et al (2007) Classification and prediction of clinical Alzheimer's diagnosis based on plasma signaling proteins. Nat Med 13:1359–1362
95. Ravetti MG, Moscato P (2008) Identification of a 5-protein biomarker molecular signature for predicting Alzheimer's disease. PLoS One 3:e3111. doi:10.1371/journal.pone.0003111
96. Spiegel A, Shivtiel S, Kalinkovich A et al (2007) Catecholaminergic neurotransmitters regulate migration and repopulation of immature human CD34+ cells through Wnt signaling. Nat Immunol 8:1123–1131
97. Katayama Y, Battista M, Kao W-M et al (2006) Signals from the sympathetic nervous system regulate hematopoietic stem cell egress from bone marrow. Cell 124:407–421
98. Kalinkovich A, Spiegel A, Shivtiel S et al (2009) Blood-forming stem cells are nervous: direct and indirect regulation of immature $CD34^+$ cells by the nervous system. Brain Behav Immun 23:1059–1065
99. Diederich K, Sevimli S, Dörr H et al (2009) The role of granulocyte-colony stimulating factor (G-CSF) in the healthy brain: a characterization of G-CSF-deficient mice. J Neurosci 29:11572–11581
100. Sevimli S, Diederich K, Strecker J-K et al (2009) Endogenous brain protection by granulocyte-colony stimulating factor after ischemic stroke. Exp Neurol 217:328–335

Use of rHuG-CSF in Infectious Diseases

Letizia Leone and Mario Cruciani

1 Introduction

Granulocyte-colony stimulating factor (G-CSF) is an endogenous hematopoietic growth factor that induces terminal differentiation and release of neutrophils from the bone marrow [1]. G-CSF appears to play a central role in the normal host response to infection [2] and has immunomodulatory and antibiotic enhancing functions [3]. G-CSF stimulates the growth and improves the function of both normal and defective neutrophils [4–8]. G-CSF is a member of the long-chain subtype of the class 1 cytokine superfamily, which includes growth hormone, erythropoietin (EPO), interleukin-6 (IL-6), and oncostatin M [9].

Recombinant human G-CSF (rHuG-CSF) is broadly used to accelerate bone marrow recovery after bone marrow transplantation; to increase the yield of peripheral blood stem cells collected by leukapheresis for use in peripheral blood stem-cell transplantation; to reduce the incidence and sequelae of neutropenia (low neutrophil count) in symptomatic patients with congenital, cyclic, or idiopathic neutropenia; and to shorten the duration of chemotherapy-induced neutropenia. Besides these labeled indications, rHuG-CSF has been used in other investigational settings. In its purified, cloned recombinant form, commercially approved G-CSF has been used to treat various difficult infectious problems [2, 6–8]. In non-neutropenic

L. Leone
Department of Biomedical and Surgical Science, Section of Dermatology and Venereology, University of Verona, Verona 37126, Italy
e-mail: laetitialeo@libero.it

M. Cruciani (✉)
Center of Community Medicine, Infectious Diseases Treatment and Surveillance Unit, Verona, and Infectious Diseases Service, G. Fracastoro Hospital, ULSS 20 Verona, Via Germania, 20, Verona 37135, Italy
e-mail: crucianimario@virgilio.it

G. Molineux et al. (eds.), *Twenty Years of G-CSF*, Milestones in Drug Therapy,
DOI 10.1007/978-3-0348-0218-5_17,

patients G-CSF may stimulate neutrophil production, enhancing the inflammatory response.

The safety and survival data from animal models of infection, combined with the favorable toxicity profile in humans, have led to several clinical trials of G-CSF as adjunctive therapy in a variety of conditions [9]. This chapter discusses the current status of G-CSF as adjunctive therapy in the infectious disease setting.

2 Diabetic Foot Infection

Diabetic foot infections range in severity from superficial paronychia to deep infection involving bone. Types of infection include cellulitis, myositis, abscesses, necrotizing fasciitis, septic arthritis, tendonitis, and osteomyelitis. Foot infections are among the most common and serious complications of diabetes mellitus. Foot infections in people with diabetes can be difficult to treat, and therapeutic failure often leads to a lower extremity amputation [10, 11]. Nearly all infected foot lesions require antibiotic therapy, but the penetration of antibiotics down to infected soft tissue and bone may be inadequate and the incidence of antibiotic resistance is increasing. These infections may also fail to respond to treatment because of inadequate surgical debridement, suboptimal wound care, or severe limb ischemia [12–15]. Furthermore, diabetes is associated with immunologic deficiencies, including abnormal neutrophil chemotaxis, phagocytosis, and intracellular killing [16–18]. Moreover, patients with diabetes are particularly susceptible to foot infection because of neuropathy and vascular insufficiency [19]. Peripheral neuropathy has a central role in the development of a foot infection and it occurs in about 30–50% of patients with diabetes. Patients with diabetes lose the protective sensations for temperature and pain, impairing awareness of trauma such as abrasions, blistering, or a penetrating foreign body. Motor neuropathy can result in foot deformities (e.g., claw toe) that contribute to local pressure from footwear, making skin ulceration even more likely. Once the skin is broken (typically on the plantar surface), the underlying tissues are exposed to colonization by pathogenic organisms. The resulting wound infection may begin superficially, but with delay in treatment and impaired body defense mechanisms caused by neutrophil dysfunction and vascular insufficiency, it can spread to the contiguous subcutaneous tissues and to even deeper structures. These factors help to explain reported clinical failure rates of 20–30% when treating diabetic foot infections. Thus, several investigators have sought adjunctive therapies for treating these potentially severe infections

In people with diabetes, two types of growth factors have been used in clinical practice: platelet-derived growth factors (rHuPDGF) and rHuG-CSF. However, these two growth factors are biologically different and have very different clinical roles: rHuG-CSF is used to treat infections and rHuPDGF is used for wound healing [20–22]. Although the two are potentially related (infected wounds tend not to heal until the infection is eradicated, and an open wound is at risk of becoming infected), they have different pathophysiological pathways and have very different treatment approaches. Moreover, rHuG-CSF is usually injected (subcutaneously or intravenously) and rarely topically applied, while rHuPDGF is applied directly to the wound surface. Effective management of diabetic foot infection requires appropriate antibiotic therapy, surgical drainage, debridement and resection of dead tissue, appropriate wound care and correction of metabolic abnormalities [12].

The role of rHuG-CSF as adjuvant treatment of diabetic foot infection has been the object of meta-analyses [20, 21]. The authors identified and included five eligible trials with a total of 167 patients. The investigators in primary studies administered various rHuG-CSF preparations, at different doses and for different durations. Adding rHuG-CSF did not significantly affect the likelihood of resolution of infection or wound healing, but it was associated with a significantly reduced likelihood of lower extremity surgical interventions, including amputation. Moreover, providing rHuG-CSF reduced the duration of hospital stay, but did not significantly affect the duration of systemic antibiotic therapy. Overall, in the management of diabetic foot infection, rHuG-CSF was shown to be safe and well tolerated.

Based on the results of primary studies and meta-analyses, considering that rHuG-CSF is expensive and given the lack of evidence that rHuG-CSF therapy can help cure infections or heal ulcers, one might conclude there is little reason to use it, especially for relatively mild infections. If, on the other hand, it can reduce the need for surgical interventions, especially amputations, it may be worth providing where possible. However, these data need to be confirmed in larger, well-designed clinical trials together with economic evaluation. At present, clinicians should consider using rHuG-CSF as an adjunct to other appropriate care for diabetic patients with foot infections, especially those infections that are perceived as limb threatening.

3 Bacterial Pneumonia

Pneumonia continues to be a significant cause of both morbidity and mortality, particularly in the elderly, those with significant underlying disease, or those who require mechanical ventilation. Despite causation by microorganisms that are typically sensitive to available antibiotics, mortality rates for patients with severe pneumonia requiring ICU (Intensive Care Unit) admission and mechanical ventilation remain exceedingly high (25–50%) [23].

Although the pathogenesis of bacterial pneumonia is not fully understood and is probably multifactorial, alteration of lung host defenses and the virulence of the pathogen are recognized as important factors [24]. Neutrophils, an important component of the host defense response to infection, are among the first cells to respond to the mediators released by infected tissues. Concentrations of endogenous G-CSF have been demonstrated to increase significantly during bacterial sepsis and other bacterial infections, including bacterial pneumonia [25]. In animal models of pneumonia, administration of recombinant methionyl human G-CSF (r-metHuGCSF, filgrastim) was associated both with reductions in viable bacteria counts and a significant improvement in survival [26, 27].

The clinical use of rHuG-CSF as an adjunct to antibiotics for the treatment of pneumonia in non-neutropenic adults has been the object of systematic reviews [28, 29]. The role of rHuG-CSF has been evaluated in a meta-analysis of randomized clinical trials evaluating hospitalized adult patients with either community-acquired pneumonia or hospital-acquired pneumonia [28]. Six studies with a total of 2,018 patients were identified. rHuG-CSF use appeared to be safe with no increase in the incidence of total serious adverse events or organ dysfunction. However, the use of rHuG-CSF was not associated with improved 28-day mortality. Thus, the currently available evidence seems not to support the routine use of rHuG-CSF for the treatment of non-neutropenic patients with pneumonia.

4 Bacterial Meningitis

The mortality and morbidity rates related to acute bacterial meningitis remain high, even when highly effective antibiotic therapy is given. Therefore, improvement in the outcomes of acute bacterial meningitis is unlikely to come from developments in chemotherapy but rather from measures that alleviate the damage done before the causative bacteria are killed [30]. Some of this damage is caused by bacterial toxins, but experiments in animals suggest that host inflammatory responses induced by bacterial products are also involved. Thus, strong theoretical grounds exist for believing that anti-inflammatory drugs should improve the outcomes of acute bacterial meningitis.

Some evidence from experimental pneumococcal meningitis shows that rHuG-CSF pretreatment reduces meningeal inflammation and blood bacterial concentration. [31, 32]. In experimental pneumococcal meningitis in the rabbit, rHuG-CSF pretreatment attenuates neutrophil pleocytosis and cerebrospinal fluid IL-8 levels, and delays significantly the occurrence of alterations in tumor necrosis factor (TNF)-α, IL-1β, and protein and glucose levels in the cerebrospinal fluid [32]. Furthermore, rHuG-CSF appears to attenuate the proinflammatory cytokine response in lipopolysaccharide-stimulated blood from rHuG-CSF-treated human volunteers [33]. Based on these observations, de Lalla and co-workers did a pilot study with the aim of evaluating the value of rHuG-CSF as an adjunctive agent in the standard treatment of non-neutropenic adult patients affected by *Streptococcus pneumoniae* meningitis [34]. Twenty-two non-neutropenic adult patients with *S pneumoniae* meningitis received rHuG-CSF (300–450 μg/day subcutaneously for 6 days) in addition to cefotaxime plus dexamethasone. Patients recovered without evident sequelae in all cases but one (with bilateral hearing deficit). No adverse event was recorded. Improvement of inflammation indices in the cerebrospinal fluid was rapid. The limits of this small, uncontrolled trial need to be acknowledged. Nonetheless, the safety and efficacy displayed in this pilot study by rHuG-CSF seems to justify further randomized controlled clinical trials aimed at defining precisely its therapeutic role in *S pneumoniae* meningitis.

5 Fungal Infection

Optimal regimens for the treatment of invasive fungal infections have yet to be defined, and these life-threatening conditions are one of the leading causes of treatment failure in patients with cancer [35]. The use of cytokines as immunomodulators of the multiple deficiencies involved in the progression of fungal infections in neutropenic and non-neutropenic patients with cancer is supported by a large number of nonclinical studies, but clinical experience is limited. Four cytokines (G-CSF, granulocyte-macrophage colony-stimulating factor [GM-CSF], macrophage colony-stimulating factor, and interferon [IFN]γ) show promise as adjuvant therapy for proven fungal infections in this setting [9]. Because polymorphonuclear leukocyte (PMNL) constitute the main mechanism of host defense against opportunistic fungi, including *Candida* species, these infections occur predominantly in patients with neutropenia or impaired PMNL function [36–40]. Therapeutic administration of rHuG-CSF has been investigated in several studies of invasive fungal infections in neutropenic cancer patients or in patients with chronic granulomatous disease [41–45]. However, results from randomized, well-designed studies are not available.

In a prospective, comparative study of empiric amphotericin B with or without rHuG-CSF, 59 neutropenic adults with hematologic malignancy and antibiotic-refractory fever or clinical evidence of deep-seated fungal infection were studied [41]. Thirty patients received amphotericin alone and 29 patients received amphotericin plus rHuG-CSF. Nearly twice as many patients responded to amphotericin B with concomitant administration of rHuG-CSF (62%) as responded to amphotericin alone (33%). Clinical response in patients receiving rHuG-CSF coincided with neutrophil recovery in most cases. Moreover, addition of rHuG-CSF to empiric amphotericin B significantly reduced the number of patients requiring salvage therapy with lipid-associated or liposomal formulations of amphotericin B. A pharmacoeconomic evaluation of the results of this study suggests that amphotericin B plus rHuG-CSF is cost effective compared with amphotericin B monotherapy in managing a presumed deep-seated fungal infection in neutropenic patients [42].

Another study evaluated the role of rHuG-CSF-elicited white blood cell (WBC) transfusions in patients with neutropenia-related fungal infections [43]. Adult patients with hematologic malignancies, absolute neutrophil counts (ANC) $<0.5 \times 10^9$/L and fungal infections refractory to amphotericin B, received daily transfusions of rHuG-CSF-elicited and irradiated WBC transfusions from related donors. Donors received 5 μ/kg/day rHuG-CSF subcutaneously. Eleven patients had favorable responses and eight of them remained free of infection 3 weeks after therapy.

Limited evidence from case reports of invasive aspergillosis in chronic granulomatous disease and of fusarium infections in patients with aplastic anemia suggests a possible role of rHuG-CSF as adjunctive therapy during severe invasive fungal infection [44, 45].

6 Sepsis or Septic Shock

The efficacy and safety of rHuG-CSF as an adjunct to standard care in critically ill patients with severe sepsis or septic shock has been reviewed [46]. Initial studies involving critically ill patients with severe sepsis or septic shock found mortality benefits with rHuG-CSF therapy; however, these findings are limited by factors such as small sample sizes, selection bias, and lack of an appropriate control group. Data from randomized, multicenter, double-blind studies failed to confirm the benefits in mortality for patients receiving rHuG-CSF for the treatment of severe sepsis and septic shock [47]. Among the 166 patients enrolled, the in-hospital mortality was 27% in the rHuG-CSF group and 25% in the placebo group. Secondary endpoints were not different between groups, but there was a higher rate of new organ failure in rHuG-CSF-treated patients than placebo-treated patients, most of which was accounted for by new liver dysfunction. Overall, the available data do not support the use of rHuG-CSF as an adjunct therapy to standard care for critically ill patients with severe sepsis or septic shock.

7 rHuG-CSF for the Prevention or Treatment of Infections in Neutropenic Patients

7.1 Febrile Neutropenia

rHuG-CSF has been used therapeutically to increase the PMNL count in the settings of congenital neutropenia, idiopathic neutropenia, leukemic neutropenia, and aplastic anemia [9]. Febrile neutropenia is a relatively frequent event in cancer patients treated with chemotherapy [48–50]. Febrile neutropenia is generally defined as fever (single oral temperature $\geq$38.3°C or $\geq$38.0°C for >1 h) with grade 3 or 4 neutropenia (ANC <1.0 or $<0.5 \times 10^9$/L) is associated with high morbidity and mortality and increased health-related costs.

The risk of febrile neutropenia and subsequent infection is directly related to the duration and severity of neutropenia [51]. The mortality rates associated with febrile neutropenia range from 2 to 21%, and the risk of death is increased by various factors, including patient characteristics, type of malignancy, presence of comorbidities, and infectious complications [49, 52].

rHuG-CSF or rHuGM-CSF have been evaluated in a large number of studies investigating their use as prophylaxis or treatment of febrile neutropenia, and these studies have been the object of systematic reviews and meta-analyses. rHuG-CSF or rHuGM-CSF are extensively used in clinical practice to prevent febrile neutropenia or to shorten the duration of neutropenia associated with chemotherapy, radiotherapy, or myelosuppressive drugs. Several systematic reviews and meta-analyses have examined whether or not colony-stimulating factors affect other outcomes of patients with cancer in different prophylactic settings [52–59]. These reviews, which summarized data from 6 to 34 studies, usually found that colony-stimulating factors reduced febrile neutropenia and documented infections, but statistically significant reductions in infection-related mortality and all-cause mortality was not observed. Likewise, another meta-analysis that included as many as 148 randomized controlled trials, failed to demonstrate that prophylactic colony-stimulating factors were associated with a statistically significant change in all-cause mortality [60]. Use of the factors, however, reduced infection-related mortality by a small amount, and it was found effective in reducing infections, infection-related morbidity, and resource use. The larger sample size of this meta-analysis increased the study's power compared with previously published meta-analyses, but this advantage was hampered by the substantial heterogeneity among studies. Both statistical and clinical heterogeneity were found in this meta-analysis, mostly as a consequence of the large variability in patient characteristics and trial designs.

The efficacy of adding rHuG-CSF or rHuGM-CSF to empiric broad-spectrum antibiotic treatment was examined in two meta-analyses of randomized clinical trials [61, 62]. No advantage was detected for the use of a colony-stimulating factor in terms of mortality from febrile neutropenia. In one of these meta-analyses [62], including 13 randomized clinical trials, a marginally significant result was obtained for the use of a colony-stimulating factor in reducing infection related mortality, but this result was highly driven by one study. Of relevance, the group of patients treated with a colony-stimulating factor had a shorter length of hospitalization and a shorter time to neutrophil recovery. A number of subgroup analyses were performed. The only outcome that was affected by the type of factor used was the occurrence of side effects (bone pain, joint pain, and flu-like symptoms), observed more frequently in rHuGM-CSF recipients compared with rHuG-CSF recipients. No differences for other endpoints relating to the type of factor were detected. Thus, the effectiveness of rHuG-CSF and rHuGM-CSF in the treatment of febrile neutropenic cancer patients remains controversial.

Basing on the available evidence, the European Organization for Research and Treatment of Cancer (EORTC) and the American Society of Clinical Oncology (ASCO) have produced guidelines on the use of hematopoietic colony-stimulating factors [63, 64]. According to the guidelines, prophylaxis with a colony-stimulating factor is recommended when the overall risk of chemotherapy-induced febrile neutropenia is $\geq$20%, and no other equally effective regimen that does not require colony-stimulating factors is available. Primary prophylaxis is recommended for the prevention of febrile neutropenia in patients who are at high risk based on age, medical history, disease characteristics, and myelotoxicity of the chemotherapy regimen. Moreover, colony-stimulating factor use allows a modest to moderate increase in dose-density and/or dose-intensity of chemotherapy regimens (see chapter "Practice Guidelines for the Use of rHuG-CSF in an Oncology Setting" by Saraf and Ozer for further information in guidelines for usage).

7.2 Neutropenia During HIV Infection

Neutropenia occurs frequently in the setting of HIV infection, with prevalence estimates ranging from 10 to 40% [65, 66]. The principal causes of neutropenia in this setting are thought to be HIV infection of hematopoietic progenitor cells; myelosuppressive drugs such as zidovudine, sulfamethoxazole, and ganciclovir; intercurrent myeloinfiltrative processes such as HIV-related lymphoma and infection with cytomegalovirus or *Mycobacterium avium* complex; nutritional deficiency (vitamin B12) and antibodies to gp120, which suppress bone marrow progenitors [67].

Neutrophil function is impaired in advanced HIV infection, and defects such as accelerated apoptosis, diminished fungicidal and bactericidal activity, and impaired secretion of G-CSF have been described [68, 69]. HIV-related neutropenia is clinically significant and has been shown to increase the risk for secondary infectious complications, especially when associated with a low CD4 count [70, 71].

Filgrastim was found safe and effective in preventing severe neutropenia in patients with advanced HIV infection, and may reduce the incidence and duration of bacterial infections, incidence of severe bacterial infections, duration of hospital days for infections, and days of intravenous antibacterial agents [72]. Based on these data, the 2009 Recommendations from The Centers for Disease Control, the National Institutes of Health, and the HIV Medicine Association of the Infectious Diseases Society of America state that the available evidence is not sufficient to support a recommendation for or against use of rHuG-CSF to reduce the risk of bacterial infections, including pneumonia, in neutropenic HIV-infected patients [73].

7.3 rHuG-CSF For Treating or Preventing Neonatal Sepsis

Despite significant advances in neonatal intensive care and the development of broad-spectrum antibiotics, neonatal sepsis remains a major cause of death, especially when associated with neutropenia. Sepsis in preterm newborns occurs at very high rates, ranging from 25 to 50%, and it is associated with substantial mortality and morbidity [74, 75]. Sepsis also leads to disability, including cerebral palsy through damage to white matter and other brain injury and contributes to the etiology of chronic lung disease [74–78].

Neutropenia, which is commonly observed in infants born before 32 weeks, increases significantly the risk of sepsis [79–81]. When sepsis is associated with severe neutropenia, mortality exceeds 50% [82]. Neutropenia is due to immaturity of neutrophil production [83, 84], but preterm infant neutrophils are also functionally immature, both factors contributing to infection risk and morbidity [85, 86].

Studies of infected animal and human neonates suggest that the use of rHuG-CSF or rHuGM-CSF can partially counterbalance these defects and thereby reduce morbidity and mortality. Encouraged by these data and following the two earliest phase I/II pilot studies in human neonates of rHuG-CSF [87] and rHuGM-CSF [88], a number of randomized studies have been undertaken investigating the potential of colony-stimulating factors for the treatment or prophylaxis of infection and infection-related death in preterm infants at high risk of sepsis. The results of these studies have been the object of a Cochrane review [75]. Seven treatment studies of 257 infants with suspected systemic bacterial infection and 3 prophylaxis studies comprising 359 neonates were included in the analysis. The results of the pooled analysis suggest that there is insufficient evidence to support the use of rHuG-CSF or rHuGM-CSF in neonatal practice, either as treatment of established systemic infection to reduce resulting mortality or as prophylaxis to prevent systemic infection in high-risk neonates. In the treatment studies analyzed, no evidence suggested that the addition of rHuG-CSF or rHuGM-CSF to antibiotic therapy reduces all-cause mortality (immediate and at 14 days from the start of therapy) in preterm infants with suspected systemic infection. However, a significant reduction in mortality by day 14 was found in a subgroup analysis of 97 preterm infants who, in addition to systemic infection, had clinically significant neutropenia ($<1.7 \times 10^9$/L) at trial entry. Likewise, pooled data from prophylaxis studies failed to demonstrate a significant reduction in mortality in neonates receiving rHuGM-CSF, although data from one study suggest that prophylactic rHuGM-CSF may provide protection against infection when given to preterm infants who are neutropenic or at high risk of developing postnatal neutropenia. Based on the limited available evidence, the authors of the meta-analysis concluded that colony-stimulating factor treatment deserves further well-designed clinical research only in the subgroup of infants with systemic infection and neutropenia. Overall, colony-stimulating factors were well tolerated in neonates. Specifically, there was no evidence of colony-stimulating factor-related thrombocytopenia as previously reported, and theoretical concerns that respiratory distress syndrome, chronic lung disease, and necrotizing enterocolitis of prematurity may be exacerbated by colony-stimulating factors were also not substantiated [75, 77]

A randomized, double-blind, placebo-controlled trial of premature neutropenic ($<1.5 \times 10^9$/L) infants of gestational age $\leq$32 weeks has been conducted [89]. A total of 200 infants received either rHuG-CSF (10 μg/kg/day) or placebo for 3 days. Primary outcome was survival free of infection for 4 weeks after treatment, assessed in an intention-to-treat analysis. The results of this study failed to demonstrate a significant increase in survival free of infection at 4 weeks after treatment.

Results from another randomized multicenter trial in 280 neonates receiving rHuGM-CSF or standard management showed that neutrophil counts after trial entry increased significantly more rapidly in infants treated with rHuGM-CSF than in control infants during the first 11 days, but there was no significant difference in sepsis-free survival for all infants [90]. Moreover, a pooled analysis of the results of this trial and previous published prophylactic trials showed no survival benefit. Overall, the body of clinical evidence is currently not sufficient to recommend rHuG-CSF or rHuGM-CSF administration as routine adjunctive treatment or prophylaxis for neonatal sepsis.

8 Conclusions and Perspectives

G-CSF and others cytokines play critical roles in the host defense response during infection. G-CSF is available in recombinant form, and its clinical use has received interest as therapeutic immunomodulatory agents during infection. Clinical trials have demonstrated potential utility for rHuG-CSF in the treatment of a variety of infections, including diabetic foot infection, bacterial pneumonia, bacterial meningitis, fungal infections, sepsis, and septic shock, for the prevention or treatment of infections during neutropenia, and for the prevention or treatment of neonatal sepsis. Moreover, clinical experiences have shown that rHuG-CSF therapy is generally well tolerated. Although current data have shown in part clinical benefit with rHuG-CSF use, substantial heterogeneity between available trials does not allow firm conclusions to be drawn. Based on the results of primary studies and meta-analyses, the role of rHuG-CSF as adjuvant treatment for infectious diseases is still uncertain and needs confirmation by larger-scaled clinical trials. Moreover, considering that rHuG-CSF is expensive, pharmacoeconomic endpoints to determine cost-effectiveness of rHuG-CSF therapy need to be incorporated in these trials.

At present, clinicians should consider using rHuG-CSF as an adjunct to other appropriate care for selected subgroups of patients such as those with limb-threatening diabetic foot infections, or in neutropenic cancer patients at high risk of developing febrile neutropenia.

References

1. Gough A, Clapperton M, Rolando N, Foster AV, Philpott-Howard J, Edmonds ME (1997) Randomised placebo-controlled trial of granulocyte-colony stimulating factor in diabetic foot infection. Lancet 20(350):855–859
2. Dale DC, Liles WC, Summer WR, Nelson S (1995) Review: granulocyte colony-stimulating factor–role and relationships in infectious diseases. J Infect Dis 172:1061–1075
3. Hartung T (1999) Granulocyte colony-stimulating factor: its potential role in infectious disease. AIDS 13(Suppl 2):S3–S9
4. Sato N, Kashima K, Tanaka Y, Shimizu H, Mori M (1997) Effect of granulocyte-colony stimulating factor on generation of oxygen-derived free radicals and myeloperoxidase activity in neutrophils from poorly controlled NIDDM patients. Diabetes 46:133–137

5. Nelson S, Heyder AM, Stone J et al (2000) A randomized controlled trial of filgrastim for the treatment of hospitalized patients with multilobar pneumonia. J Infect Dis 182:970–973
6. Root RK, Dale DC (1999) Granulocyte colony-stimulating factor and granulocyte-macrophage colony-stimulating factor: comparisons and potential for use in the treatment of infections in non-neutropenic patients. J Infect Dis 179(Suppl 2):S342–S352
7. Hübel K, Engert A (2003) Granulocyte transfusion therapy for treatment of infections after cytotoxic chemotherapy. Onkologie 26:73–79
8. Murata A (2003) Granulocyte colony-stimulating factor as the expecting sword for the treatment of severe sepsis. Curr Pharm Des 9:1115–1120
9. Hübel K, Dale DC, Liles WC (2002) Therapeutic use of cytokines to modulate phagocyte function for the treatment of infectious diseases: current status of granulocyte colony-stimulating factor, granulocyte-macrophage colony-stimulating factor, macrophage colony-stimulating factor, and interferon-γ. J Infect Dis 185:1490–1501
10. Reiber GE, Lipsky BA, Gibbons GW (1998) The burden of diabetic foot ulcers. Am J Surg 176(2A Suppl):5S–10S
11. Moulik PK, Mtonga R, Gill GV (2003) Amputation and mortality in new onset diabetic foot ulcers stratified by etiology. Diabetes Care 26:491–494
12. Lipsky BA, Berendt AR (2000) Principles and practice of antibiotic therapy of diabetic foot infections. Diabetes Metab Res Rev 16(suppl 1):S42–S46
13. International Working Group on the Diabetic Foot (2003) International Consensus on the Diabetic Foot. International Diabetes Federation, CD-ROM
14. Korda J, Mezo R, Balint GP (2005) Treatment of musculoskeletal infections of the foot in patients with diabetes. Therapy 2:287–300
15. Gupta S, Koirala J, Khardori R, Khardori N (2007) Infections in diabetes mellitus and hyperglycemia. Infect Dis Clin North Am 21:617–638
16. Sato N, Shimizu H (1993) Granulocyte-colony stimulating factor improves an impaired bactericidal function in neutrophils from STZ-induced diabetic rats. Diabetes 42:470–473
17. Geerlings SE, Hoepelman AI (1999) Immune dysfunction in patients with diabetes mellitus (DM). FEMS Immunol Med Microbiol 26:295–365
18. Delamaire M, Maugendre D, Moreno M, Le Goff MC, Allannic H, Genetet B (1997) Impaired leucocyte functions in diabetic patients. Diabet Med 14:29–34
19. Boulton AJ, Vileikyte L, Ragnarson-Tennvall G, Apelqvist J (2005) The global burden of diabetic foot disease. Lancet 366:1719–1724
20. Cruciani M, Lipsky BA, Mengoli C, de Lalla F (2005) Are granulocyte colony-stimulating factors beneficial in treating diabetic foot infections? A meta-analysis. Diabetes Care 28:454–460
21. Cruciani M, Lipsky BA, Mengoli C, de Lalla F (2009) Granulocyte-colony stimulating factors as adjunctive therapy for diabetic foot infections. Cochrane Database Syst Rev CD006810
22. Lipsky BA, Hoey C (2009) Topical antimicrobial therapy for treating chronic wounds. Clin Infect Dis 49:1541–1549
23. Niederman MS, Bass JB Jr, Campbell GD et al (1993) Guidelines for the initial management of adults with community acquired pneumonia: diagnosis, assessment of severity, and initial antimicrobial therapy. Am Rev Respir Dis 148:1418–1426
24. Wunderink R, Leeper K Jr, Schein R et al (2001) Filgrastim in patients with pneumonia and severe sepsis or septic shock. Chest 119:523–529
25. Cebon J, Layton JE, Maher D, Morstyn G (1994) Endogenous hemopoietic growth factors in neutropenia and infection. Br J Hematol 86:265–274
26. Hebert JC, O'Reilly M, Gamelli RL (1990) Protective effect of recombinant human granulocyte colony-stimulating factor against pneumococcal infections in splenectomized mice. Arch Surg 125:1075–1078
27. Nelson S, Summer W, Bagby G et al (1991) Granulocyte colony stimulating factor enhances pulmonary host defenses in normal and ethanol-treated rats. J Infect Dis 164:901–906

28. Cheng AC, Stephens DP, Currie BJ (2007) Granulocyte-colony stimulating factor (G-CSF) as an adjunct to antibiotics in the treatment of pneumonia in adults. Cochrane Database Syst Rev CD004400
29. Siempos II, Vardakas KZ, Kopterides P, Falagas ME (2008) Adjunctive therapies for community acquired pneumonia: a systematic review. J Antimicrob Chemother 62:661–668
30. Greenwood BM (2007) Corticosteroids for acute bacterial meningitis. N Engl J Med 357:2507–2509
31. Brandt CT, Lundgren JD, Lund SP et al (2004) Attenuation of the bacterial load in blood by pretreatment with granulocyte-colony-stimulating factor protects rats from fatal outcome and brain damage during *Streptococcus pneumoniae* meningitis. Infect Immun 72:4647–4653
32. Ostergaard C, Benfield T, Gesser B et al (1999) Pretreatment with granulocyte colony-stimulating factor attenuates the inflammatory response but not the bacterial load in cerebrospinal fluid during experimental pneumococcal meningitis in rabbits. Infect Immun 67:3430–3436
33. Hartung T (1998) Anti-inflammatory effects of granulocyte colony-stimulating factor. Curr Opin Hematol 5:221–225
34. De Lalla F, Nicolin R, Lazzarini L (2000) Safety and efficacy of recombinant granulocyte colony-stimulating factor as an adjunctive therapy for *Streptococcus pneumoniae* meningitis in non-neutropenic adult patients: a pilot study. J Antimicrob Chemother 46:843–846
35. Ruhnke M, Böhme A, Buchheidt D et al (2003) Diagnosis of invasive fungal infections in hematology and oncology – guidelines of the Infectious Diseases Working Party (AGIHO) of the German Society of Hematology and Oncology (DGHO). Ann Hematol 82(Suppl 2): S141–S148
36. Bohme A, Karthaus M (1999) Systemic fungal infections in patients with hematologic malignancies: indications and limitations of the antifungal armamentarium. Chemotherapy 45:315–324
37. Diamond RD (1993) Interactions of phagocytic cells with Candida and other opportunistic fungi. Arch Med Res 24:361–369
38. Rodriquez-Adrian LJ, Grazziutti ML, Rex JH, Anaissie EJ (1998) The potential role of cytokine therapy for fungal infection in patients with cancer: is recovery from neutropenia all that is needed? Clin Infect Dis 126:1270–1278
39. Stevens DA (1998) Combination immunotherapy and antifungal chemotherapy. Clin Infect Dis 26:1266–1269
40. Roilides E, Farmaki E (2001) Granulocyte colony-stimulating factor and other cytokines in antifungal therapy. Clin Microbiol Infect 7(Suppl 2):62–67
41. Hazel DL, Newland AC, Kelsey SM (1999) Malignancy: granulocyte colony stimulating factor increases the efficacy of conventional amphotericin in the treatment of presumed deep-seated fungal infection in neutropenic patients following intensive chemotherapy or bone marrow transplantation for haematological malignancies. Hematology 4:305–311
42. Flynn TN, Kelsey SM, Hazel DL, Guest JF (1999) Cost effectiveness of amphotericin B plus G-CSF compared with amphotericin B monotherapy. Treatment of presumed deep-seated fungal infection in neutropenic patients in the UK. Pharmacoeconomics 16:543–550
43. Dignani MC, Anaissie EJ, Hester JP et al (1997) Treatment of neutropenia-related fungal infections with granulocyte colony-stimulating factor-elicited white blood cell transfusions: a pilot study. Leukemia 11:1621–1630
44. Ozsahin H, von Planta M, Müller I et al (1998) Successful treatment of invasive aspergillosis in chronic granulomatous disease by bone marrow transplantation, granulocyte colony-stimulating factor-mobilized granulocytes, and liposomal amphotericin-B. Blood 92:2719–2724
45. Girmenia C, Iori AP, Boecklin F et al (1999) Fusarium infections in patients with severe aplastic anemia: review and implications for management. Haematologica 84:114–118
46. Mohammad RA (2010) Use of granulocyte colony-stimulating factor in patients with severe sepsis or septic shock. Am J Health Syst Pharm 67:1238–1245

47. Stephens DP, Thomas JH, Higgins A et al (2008) Randomized, double-blind, placebo-controlled trial of granulocyte colony-stimulating factor in patients with septic shock. Crit Care Med 36: 448–454
48. Aapro M, Crawford J, Kamioner D (2010) Prophylaxis of chemotherapy-induced febrile neutropenia with granulocyte colony-stimulating factors: where are we now? Support Care Cancer 18:529–541
49. Kuderer NM, Dale DC, Crawford J, Cosler LE, Lyman GH (2006) Mortality, morbidity and cost associated with febrile neutropenia in adult cancer patients. Cancer 106:2258–2266
50. Pizzo PA (1999) Fever in immunocompromised patients. N Engl J Med 341:893–900
51. Bodey GP, Buckley M, Sathe YS, Freireich EJ (1966) Quantitative relationships between circulating leukocytes and infection in patients with acute leukemia. Ann Intern Med 64:328–340
52. Herbst C, Naumann F, Kruse EB, Monsef I, Bohlius J, Schulz H, Engert A (2009) Prophylactic antibiotics or G-CSF for the prevention of infections and improvement of survival in cancer patients undergoing chemotherapy. Cochrane Database Syst Rev 21:CD007107
53. Lyman GH, Kuderer NM, Djulbegovic B (2002) Prophylactic granulocyte colony stimulating factor in patients receiving dose-intensive cancer chemotherapy: a meta-analysis. Am J Med 112:406–411
54. Bohlius J, Reiser M, Schwarzer G, Engert A (2003) Impact of granulocyte colony stimulating factor (CSF) and granulocyte-macrophage CSF in patients with malignant lymphoma: a systematic review. Br J Haematol 122:413–423
55. Bohlius J, Reiser M, Schwarzer G, Engert A (2004) Granulopoiesis-stimulating factors to prevent adverse effects in the treatment of malignant lymphoma. Cochrane Database Syst Rev CD003189
56. Berghmans T, Paesmans M, Lafitte JJ, Mascaux C, Meert AP, Sculier JP (2002) Role of granulocyte and granulocyte-macrophage colony-stimulating factors in the treatment of small-cell lung cancer: a systematic review of the literature with methodological assessment and meta-analysis. Lung Cancer 37:115–123
57. Sung L, Nathan PC, Lange B, Beyene J, Buchanan GR (2004) Prophylactic granulocyte colony-stimulating factor and granulocyte-macrophage colony-stimulating factor decrease febrile neutropenia after chemotherapy in children with cancer: a meta-analysis of randomized controlled trials. J Clin Oncol 22:3350–3356
58. Sasse EC, Sasse AD, Brandalise S, Clark OA, Richards S (2005) Colony stimulating factors for prevention of myelosupressive therapy induced febrile neutropenia in children with acute lymphoblastic leukaemia. Cochrane Database Syst Rev 3:CD004139
59. Dekker A, Bulley S, Beyene J, Dupuis LL, Doyle JJ, Sung L (2006) Meta-analysis of randomized controlled trials of prophylactic granulocyte colony-stimulating factor and granulocyte-macrophage colony-stimulating factor after autologous and allogeneic stem cell transplantation. J Clin Oncol 24:5207–5215
60. Sung L, Nathan PC, Alibhai SMH, Tomlinson GA, Beyene J (2007) Meta-analysis: effect of prophylactic hematopoietic colony-stimulating factors on mortality and outcomes of Infection. Ann Intern Med 147:400–411
61. Berghmans T, Paesmans M, Lafitte JJ et al (2002) herapeutic use of granulocyte and granulocyte-macrophage colony-stimulating factors in febrile neutropenic cancer patients. A systematic review of the literature with meta-analysis. Support Care Cancer 10:181–188
62. Clark OAC, Lyman G, Castro AA, Clark LGO, Djulbegovic B (2003) Colony stimulating factors for chemotherapy induced febrile neutropenia. Cochrane Database Syst Rev 3:CD003039
63. Aapro MS, Cameron DA, Pettengell R et al (2006) EORTC guidelines for the use of granulocyte-colony stimulating factor to reduce the incidence of chemotherapy-induced febrile neutropenia in adult patients with lymphomas and solid tumours. Eur J Cancer 42:2433–2453
64. Smith TJ, Khatcheressian J, Lyman GH et al (2006) 2006 update of recommendations for the use of white blood cell growth factors: an evidence-based clinical practice guideline. J Clin Oncol 24:3187–4205

65. Mauss S, Steinmetz HT, Willers R et al (1997) Induction of granulocyte colony-stimulating factor by acute febrile infection but not by neutropenia in HIV-seropositive individuals. J Acquir Immune Defic Syndr Hum Retrovirol 14:430–434
66. Keiser P, Rademacher S, Smith JW, Skiest D, Vadde V (1998) Granulocyte colony-stimulating factor use is associated with decreased bacteremia and increased survival in neutropenic HIV infected patients. Am J Med 104:48–55
67. Kuritzkes DR (2000) Neutropenia, neutrophil dysfunction, and bacterial infection in patients with human immunodeficiency virus disease: the role of granulocyte colony-stimulating factor. Clin Infect Dis 30:256–260
68. Pitrak DL, Tsai HC, Mullane KM, Sutton SH, Stevens P (1996) Accelerated neutrophil apoptosis in the acquired immunodeficiency syndrome. J Clin Invest 98:2714–2719
69. Laursen AL, Rungby J, Andersen PL (1885) Decreased activation of the respiratory burst in neutrophils from AIDS patients with previous *Pneumocystis carinii* pneumonia. J Infect Dis 172:497–505
70. Hermans P (1999) HIV disease-related neutropenia: an independent risk factor for severe infections. AIDS 13(Suppl 2):S11–S17
71. Moore DA, Benepal T, Portsmouth S, Gill J, Gazzard BG (2001) Etiology and natural history of neutropenia in human immunodeficiency virus disease: a prospective study. Clin Infect Dis 32:469–475
72. Kuritzkes DR, Parenti D, Ward DJ et al (1998) Filgrastim prevents severe neutropenia and reduces infective morbidity in patients with advanced HIV infection: results of a randomized, multicenter, controlled trial. AIDS 12:65–74
73. Guidelines for Prevention and Treatment of Opportunistic Infections in HIV-Infected Adults and Adolescents (2009) Recommendations from CDC, the National Institutes of Health, and the HIV Medicine Association of the Infectious Diseases Society of America. MMWR Recomm Rep 58(RR04):1–198
74. Stoll BJ, Gordon T, Korones SB et al (1996) Late-onset sepsis in very low birthweight neonates: a report from the National Institute of Child Health and Human Development neonatal research network. J Pediatr 129:63–71
75. Carr R, Modi N (1997) Haemopoietic colony stimulating factors for preterm neonates. Arch Dis Child Fetal Neonatal Ed 76:F128–F133
76. Damman O, Leviton A (1998) Infection remote from the brain, neonatal white matter damage, and cerebral palsy in the preterm infant. Semin Pediatr Neurol 5:190–201
77. Marshall DD, Kotelchuck M, Young TE, Bose CL, Kruyer L, O'Shea TM (1999) Risk factors for chronic lung disease in the surfactant era: a North Carolina population-based study of very low birth weight infants. Pediatrics 104:1345–1350
78. Bedford Russell AR, Emmerson AJ, Wilkinson N et al (2001) A trial of recombinant human granulocyte colony stimulating factor for the treatment of very low birth weight infants with presumed sepsis and neutropenia. Arch Dis Child Fetal Neonatal Ed 84:F172–F176
79. Gessler P, Lüders R, König S, Haas N, Lasch P, Kachel W (1995) Neonatal neutropenia in low birthweight premature infants. Am J Perinatol 12:34–38
80. Engle WD, Rosenfeld CR (1984) Neutropenia in high risk neonates. J Pediatr 105:982–986
81. Koenig JH, Christensen RD (1989) Incidence, neutrophil kinetics and natural history of neonatal neutropenia associated with maternal hypertension. N Engl J Med 321:557–562
82. Rodwell RL, Taylor KMCD, Tudehope DI, Gray PH (1993) Hematologic scoring system in early diagnosis of sepsis in neutropenic newborns. Pediatr Infect Dis J 12:372–376
83. Christensen RD (1989) Neutrophil kinetics in the fetus and neonate. Am J Pediatr Hematol Oncol 11:215–223
84. Carr R (2000) Neutrophil production and function in newborn infants. Br J Haematol 110:18–28
85. Hill HR (1987) Biochemical, structural, and functional abnormalities of polymorphonuclear leukocytes in the neonate. Pediatr Res 22:375–382

86. Carr R, Huizinga TWJ (2000) Low soluble FcRIII demonstrates reduced neutrophil reserves in preterm neonates. Arch Dis Child Fetal Neonatal Ed 83:F160
87. Gillan ER, Christensen RD, Suen Y, Ellis R, van de Ven C, Cairo MS (1994) A randomized, placebo-controlled trial of recombinant human granulocyte colony-stimulating factor administration in newborn infants with presumed sepsis: significant induction of peripheral and bone marrow neutrophilia. Blood 84:1427–1433
88. Cairo MS, Christensen RD, Sender LS et al (1995) Results of a phase I/II trial of recombinant human granulocyte-macrophage colony-stimulating factor in very low birth weight neonates: significant induction of circulatory neutrophils, monocytes, platelets, and bone marrow neutrophils. Blood 86:2509–2515
89. Kuhn P, Messer J, Paupe A et al (2009) A multicenter, randomized, placebo-controlled trial of prophylactic recombinant granulocyte-colony stimulating factor in preterm neonates with neutropenia. J Pediatr 155:324–330
90. Carr R, Brocklehurst P, Doré CJ, Modi N (2009) Granulocyte-macrophage colony-stimulating factor administered as prophylaxis for reduction of sepsis in extremely preterm, small for gestational age neonates (the PROGRAMS trial): a single-blind, multicentre, randomised controlled trial. Lancet 373:226–233

Use of Filgrastim (r-metHuG-CSF) in Human Immunodeficiency Virus Infection

Vagish Hemmige, W. Conrad Liles, and David L. Pitrak

1 Introduction

In 2008, an estimated 33.4 million individuals worldwide were infected with the human immunodeficiency virus (HIV) [1]. Only a few years ago, infection with HIV almost invariably culminated in the development of the acquired immunodeficiency syndrome (AIDS), characterized by severe depletion of $CD4^+$ lymphocytes leading to derangements predominantly affecting cell-mediated immunity, but affecting humoral immunity as well [2]. In the later stages of AIDS, neutropenia and neutrophil functional deficits were common sequelae of HIV infection, other opportunistic infections, or HIV- or opportunistic infection-related treatment [3]. The care of the HIV-infected patient was palliative in nature, and the possibility that use of filgrastim (rHuG-CSF) might extend survival in late-stage AIDS patients with severe neutropenia or severe opportunistic infections, or might be a treatment for HIV infection itself, was explored [4]. Subsequently, however, the development of protease inhibitors and the widespread adoption of their use in multidrug regimens of highly active antiretroviral therapy (HAART) revolutionized the care of HIV-infected patients, and the number of patients dying from HIV decreased dramatically [5]. Patients with HIV can, with current regimens, achieve prolonged survival with preservation of immunologic function, although patients infected with HIV have shortened lifespans compared with uninfected people [6] secondary to

V. Hemmige • D.L. Pitrak (✉)
Department of Medicine, Section of Infectious Diseases and Global Health, University of Chicago, 5841 S. Maryland Avenue, Chicago, IL 60637
e-mail: dpitrak@medicine.bsd.uchicago.edu

W.C. Liles
Divsion of Infectious of Diseases, University of Toronto, Toronto General Hospital, Eaton North Wing, 13th floor Room 13 EB 220, 200 Elizabeth St, Toronto, Ontario, Canada

G. Molineux et al. (eds.), *Twenty Years of G-CSF*, Milestones in Drug Therapy,
DOI 10.1007/978-3-0348-0218-5_18, © Springer Basel AG 2012

comorbid conditions and increased rates of cardiovascular [7] and oncologic death [8]. However, despite the widespread availability of HAART in the developed world, patients still present with late-stage AIDS and its attendant complications, including neutropenia, because of late diagnosis of HIV [9] or progression of HIV infection due to nonadherence to HAART [10]. Furthermore, patients with HIV face an increased risk of malignancy and the subsequent need to be treated with cytotoxic chemotherapy [11]. Accordingly, filgrastim has continued to play a role in the management of selected patients with HIV.

2 Epidemiology of Neutropenia in HIV-Infected Individuals

From the earliest days of the AIDS pandemic, neutropenia was identified as a consequence of HIV infection. In 1987, an absolute neutrophil count (ANC) of $<2.0 \times 10^9$/L was reported in 4 of 20 patients with clinically diagnosed AIDS and 15 of 59 with persistent generalized lymphadenopathy [12]. As this report predates the development of antibody testing for HIV, it was not possible to determine whether patients with persistent generalized lymphadenopathy were truly HIV-infected or not. No patient had an ANC $<1.0 \times 10^9$/L in this report, although the same author had noted an ANC of 0.4×10^9/L in one patient in a case series of homosexual men with neutropenia in 1985 [13].

Neutropenia is primarily a consequence of late-stage AIDS, with the mean $CD4^+$ lymphocyte count during the first episode of neutropenia in one large study being 85 cells/mm^3. However, 14% of patients with a neutropenic episode in this study were observed to have a $CD4^+$ lymphocyte count >200 cells/mm^3, and one patient's $CD4^+$ lymphocyte count at the time of neutropenia was 858 cells/mm^3 [14]. It is worth noting that several case reports have associated neutropenia with acute HIV infection [15–18]. Since most instances of acute HIV infection are unrecognized [19], the true incidence of neutropenia associated with acute HIV infection remains unknown.

The epidemiology of neutropenia was examined in the Women's Interagency HIV Study, a prospective cohort study of 2,059 HIV-infected women enrolled in 1994 and 1995. In this cohort, 1,729 women had a documented ANC measured. It was observed that 7% of women were neutropenic at enrollment, which reflected the prevalence of neutropenia in the pre-HAART era, and that 31% of the study cohort experienced at least one episode of neutropenia (defined as a measured ANC $<1.0 \times 10^9$/L) during the study. Low $CD4^+$ lymphocyte count and high viral load, suggestive of progression of disease, were associated with the development of neutropenia [20].

Neutropenia is often multifactorial in AIDS patients, with HIV itself, myelosuppressive drugs, opportunistic infections, and malignancy most frequently identified as the etiology. Although zidovudine is the most common antiretroviral associated with neutropenia, other nucleoside reverse transcriptase inhibitors have been implicated in drug-induced neutropenia as well [21]. Indinavir-induced neutropenia has been reported [22], as has neutropenia exacerbated by efavirenz [23].

Noteworthy also are case reports of patients receiving chemotherapy concomitantly with antiretroviral regimens containing ritonavir, a potent inhibitor of the CYP450 system, and subsequently developing neutropenia due to increased concentrations of chemotherapeutic agents [24–26]. Other prophylactic agents including dapsone [27], antifungals, and beta-lactam antibiotics [28] have also been reported to cause neutropenia. Disseminated infections with bone marrow involvement caused by fungal (e.g., endemic mycoses, cryptococcosis) [29], mycobacterial (e.g., tuberculosis or *Mycobacterium avium* complex), protozoal (e.g., leishmaniasis), or viral (e.g., cytomegaloviral) pathogens have been identified as contributors to HIV-associated neutropenia [30]. Finally, infiltration of bone marrow with lymphoma is not infrequently a cause of HIV-associated neutropenia [31], and this possibility should be considered before ascribing neutropenia to medications alone.

The proportion of neutropenic episodes attributable to each of the above causes depends on the patient population. In one Italian study including 81 neutropenic patients in the pre-HAART era, most of whom were intravenous drug users, neutropenia was attributed to HIV itself in 18.5% of the patients, to infiltration of bone marrow with lymphoma or infection in 24.6% of patients, and drugs in 56.7%. Of patients with drug-induced neutropenia, zidovudine was a contributor to 48.2% of cases, chemotherapy to 21.3% of cases, trimethoprim-sulfamethoxazole to 18.7%, amphotericin B to 17.5%, ganciclovir to 13.6%, pyrimethamine to 11.1%, and dapsone to 6.4% of cases [32]. In a different study, zidovudine therapy was a contributing factor to neutropenia in 51% of patients, trimethoprim–sulphamethoxazole treatment in 45% of patients, ganciclovir therapy in 18% of patients, and cytotoxic chemotherapy in 11.3% of patients; neutropenia was attributed to lymphoma in 6.5% of cases and HIV infection itself in 1.6% of patients [31].

The epidemiology of neutropenia in HIV patients has drastically changed with the advent of HAART. HAART prevents progression of HIV, and was found to prevent the development of neutropenia in the Women's Interagency HIV study cohort [20]. Furthermore, HAART itself is an effective long-term treatment for HIV-associated neutropenia [33, 34]. In addition, the decreased dose of zidovudine employed as part of HAART regimens may be less likely to cause neutropenia than the higher dose used in the pre-HAART era. One study where patients were routinely converted from stavudine to zidovudine after 6 months of HAART therapy reported that only 7.7% of patients developed an ANC $<0.75 \times 10^9$/L [33], and a similar conversion study of 78 HIV-infected children reported that only 6% developed an ANC $<1.0 \times 10^9$/L. In the latter study, no individual's ANC decreased to $<0.75 \times 10^9$/L [34].

3 Impaired Neutrophil Function in HIV Infection

In addition to neutropenia, there are functional defects of neutrophils, as well as monocytes and macrophages, in HIV infection and other lentivirus infections. These defects are not unique to HIV infection, and a number of viral infections have been associated with impaired neutrophil function [35]. A wide range of functional defects exist, including defects in chemotaxis, phagocytosis, the respiratory (oxidative) burst, and microbicidal capacity [36–41]. Neutrophils isolated from HIV-infected patients have a profound defect in chemotaxis in response to interleukin (IL)-8 and bacterial chemoattractant peptides, such as f-met-leu-phe (fMLP), and the degree of impairment correlates with the degree of CD4$^+$ T-cell depletion. Other physiologic functions related to recruitment to sites of infection, rolling, adhesion, and emigration are also affected, as shown in a feline leukemia virus (FIV) infection model. Evidence that this may be due to a maturation defect is decreased granularity of neutrophils from the blood and bone marrow of FIV-infected animals. The results of different studies of neutrophil function in HIV infection are at times conflicting and can be explained by different assays, different conditions, and differences in the patient populations studied. The most important clinical variable is the stage of HIV disease as evidenced by absolute CD4$^+$ T-cell count.

The respiratory burst is a very important part of the microbial killing by neutrophils, generating superoxide and other bactericidal reactive oxygen species. Our group previously examined the neutrophil respiratory burst by chemiluminescence in a cohort of 78 patients with HIV infection at different stages of disease [42]. Patients with HIV infection had altered oxidative metabolism in response to opsonin receptor-dependent stimulation with zymosan opsonized with purified human complement (C3bi) or immune globulin (IgG). Patients with early HIV infection with CD4$^+$ lymphocyte counts >500 cells/mm^3 exhibited increased neutrophil chemiluminescence in response to opsonized zymosan compared with controls, while patients with advanced disease with low CD4$^+$ lymphocyte counts showed significantly decreased chemiluminescence. Absolute CD4$^+$ lymphocyte count was the only patient variable significantly correlated with opsonin-dependent neutrophil chemiluminscence activity according to multiple regression analysis. Despite a good correlation between ANC and CD4$^+$ lymphocyte count ($R = 0.24$; $p = 0.04$), ANC was not an independent predictor of impaired neutrophil chemiluminescence by multiple regression analysis.

4 Pathogenesis of Neutropenia and Neutrophil Dysfunction

The pathophysiology of neutropenia and neutrophil dysfunction in HIV infection unrelated to drug therapy or secondary complications has been better elucidated in the past few years. A subset of neutropenic HIV-infected patients may have autoimmune neutropenia; however, this probably accounts for a small proportion of cases of neutropenia [13, 43]. A study of neutropenic children with HIV infection showed that while many children had circulating antineutrophil antibodies, the presence of these antibodies did not correlate with the ANC [44].

The various immune abnormalities observed in HIV infection occur in the setting of immune systemic activation. There has been considerable interest in immune system activation occurring as a result of T-cell depletion in gut-associated lymphoid tissue (GALT) with loss of mucosal barrier function, which results in translocation of lipopolysaccharide (LPS) and other bacterial products that activate the immune system [45]. A number of investigators have shown that neutrophils are activated in vivo throughout the course of HIV infection, even in the absence of any clinical signs of secondary infectious complications [46]. Early in the course of HIV infection in vivo, activation or priming may actually result in enhanced function. This in vivo activation continues throughout the course of HIV infection, but eventually the functional capacity of the neutrophil begins to decrease significantly. It should also be noted that immune system activation has been proposed as a mechanism for impaired neutrophil function in patients with chronic hepatitis B infection [47].

Our studies of neutrophil chemiluminescence suggest that immune system activation may contribute to neutrophil dysfunction [42]. Maximum opsonin receptor expression (MOR) is achieved by exposing neutrophil to quantities of proinflammatory mediators (primers: fMLP, complement fragment C5a, and platelet-activating factor) to induce a maximal number of CD11b (C3bi receptor) and CD35 (C3b receptor or complement receptor 1) on the neutrophil surface. Priming may also enhance oxidative responses to a second stimulus by increasing the affinity of opsonin receptors for their particular ligand or enhancing intracellular signaling. Our experiments showed increases in the ratio of the chemiluminescence unprimed neutrophil at circulating opsonin receptor expression (COR) to chemiluminescence with MOR, i.e., an increased COR/MOR ratio. By whatever mechanism, neutrophils from HIV-infected patients behave as if they have been primed or activated by proinflammatory mediators in vivo. Opsonin receptor-independent NADPH-oxidase and myeloperoxidase activities, basal and stimulated, were significantly increased for HIV-infected patients, especially those with advanced disease with $CD4^+$ lymphocyte count <200 cells/mm^3. The increase in enzyme activities of NADPH-oxidase and myeloperoxidase may also be the result of in vivo activation. The decrease in myeloperoxidase activity for patients with very advanced HIV infection with $CD4^+$ lymphocyte count <100 cells/mm^3, although significantly higher than values seen in control subjects without HIV infection, may be due to degranulation resulting from in vivo activation. This chronic in vivo activation may lead to metabolic exhaustion or "burnout," and contributes to impaired neutrophil oxidative responses in HIV-infected patients at advanced stages of disease.

In vivo activation is not the only process that contributes to neutrophil dysfunction in HIV infection. Neutrophils isolated from patients with advanced HIV disease undergo accelerated apoptosis or programmed cell death. We studied apoptosis of neutrophils isolated from ten individuals with advanced HIV infection and seven control subjects [48]. Ex vivo apoptosis was examined morphologically by fluorescent microscopy after dual staining with acridine orange and ethidium bromide, fluorescent stains that intercalate DNA. Acridine orange stains the nucleus bright green and allows visualization of the nuclear chromatin pattern, while ethidium bromide identifies nonviable cells by staining the nucleus orange. Little apoptosis was evident immediately after isolation, but over time, apoptosis was observed and

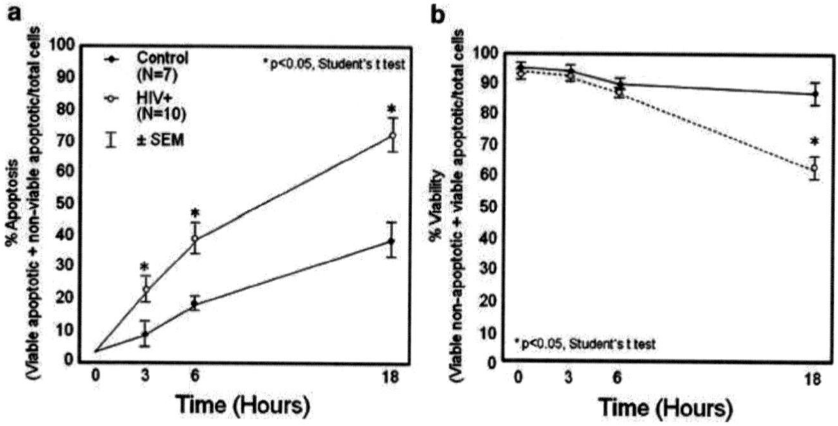

Fig. 1 Neutrophil apoptosis and viability in patients with AIDS compared with normal controls

the proportion of apoptotic cells was significantly higher for the patients with advanced HIV infection at 3, 6, and 18 h (Fig. 1). Apoptotic cells eventually die, and at 18 h there was a significant decrease in viability for the HIV patient's neutrophils due to an increased number of nonviable, apoptotic cells. As neutrophils become apoptotic, they become functionally impaired, and the accelerated apoptosis of neutrophils may in part explain the impairment of neutrophil function seen in this patient population, as well as neutropenia. A number of possible mechanisms for accelerated apoptosis exist, including increased levels of proinflammatory mediators that may accelerate the process. Other mechanisms may also be involved, including direct effects of viral proteins. HIV protease can directly induce apoptosis of a variety of leukocyte populations [49]. Alternatively, growth factor deficiency could result in accelerated apoptosis.

Granulocyte precursors can be infected by HIV, and this may affect normal proliferation and development of neutrophils, but mature cells are not targets and express very little, if any, HIV [50]. Infection of bone marrow stromal cells may affect the microenvironment for myelopoiesis. Both HIV and cytomegalovirus (CMV) can infect bone marrow stromal cells and impair production of endogenous colony-stimulating factors such as granulocyte colony-stimulating factor (G-CSF) and granulocyte-macrophage colony-stimulating factor (GM-CSF) [51, 52]. Altered cytokine secretion may result in abnormal neutrophil maturation. G-CSF is a key growth factor important in maintaining the normal number of circulating neutrophils. Although several growth factors can stimulate myelopoiesis, studies of G-CSF-deficient knock-out mice indicate G-CSF is a cytokine necessary to maintain normal neutrophil counts [53]. Evidence is accumulating that HIV infection results in a state of endogenous G-CSF deficiency [54, 55]. Patients with HIV have lower serum levels of endogenous G-CSF for a given degree of neutropenia compared with patients with neutropenia from other etiologies, such as aplastic anemia and cancer chemotherapy. The low levels of G-CSF may not only impair myelopoiesis and affect neutrophil maturation, but also reduce the circulating half-life of neutrophils. The principle way growth factors cause clonal expansion of different bone marrow precursors is by inhibiting apoptosis of these cells [56]. Our data show that accelerated neutrophil apoptosis in HIV infection is reversible in vitro and ex vivo in patients receiving filgrastim [48, 57].

Why should a decrease in growth factor(s) required for neutrophil maturation occur in a disease characterized by abnormal cell-mediated immunity? The source of G-CSF, directly or indirectly, is activated T-cells, monocytes, and macrophages, all targets for HIV infection. Depletion or dysfunction of these cells as a result of HIV infection may result in decreased production of G-CSF that is required to support normal granulopoiesis and/or neutrophilic responses to infection. We previously reported that peripheral blood mononuclear cells (PBMC) isolated from patients with AIDS produce significantly less G-CSF in response to a challenge with LPS [54]. Local G-CSF production by bone marrow stromal cells is likely to be very important in granulopoiesis, and there is a decreased G-CSF response by bone marrow stromal cells from patients with HIV infection ex vivo in response to IL-1 and LPS [51].

More is known about how $CD4^{+}$ T-helper cells recruit and activate a wide variety of cell types, including macrophages, mast cells, neutrophils, eosinophils, and basophils, in addition to adaptive immune effector cells. Several different types of $CD4^{+}$ T-cells have been described that differentiate from naive $CD4^{+}$ T-cells [55]. In addition to Th1 and Th2 cells, we now recognize regulatory T-cells (Treg) [58]. These cells not only are involved in immunosuppression, but also can differentiate into effector T-cells. More has also been discovered about the interactions between phagocytic cells and T-cells. The Th17 pathway has been characterized [59]. Th17 cells have been classified as a new lineage, distinct from Th1, Th2, and Treg cells. T17 cells develop as a result of a unique set of cytokines secreted by effector cells of the innate immune system (e.g., neutrophils) and additive effects of IL-1β and tumor-necrosis factor (TNF)-α. Transforming growth factor (TGF)-β and IL-6 are crucial factors in the generation of Th17 cells from naive T-cells. The functional capacity of Th17 cells depends on the additional cytokines IL-23 and IL-1.

Th17 cells are characterized by production of the proinflammatory cytokines IL-17A, IL-17F, IL-22, and IL-26. The activity of IL-17A and IL-17F is defined by induction of increased production of a number of proinflammatory mediators, including IL-1, IL-6, CXC chemokines, TNF-α, G-CSF, and GM-CSF by epithelial and endothelial cells, macrophages, and other stromal cells. IL-17A also promotes stem-cell factor (SCF)- and G-CSF-mediated granulopoiesis [60]. It has been controversial as to whether or not neutrophils directly respond to IL-17A or IL-17F, but the downstream production of cytokines induced by IL-17A and IL-17F leads to granulopoiesis, neutrophil recruitment, and neutrophil activation. Substantial evidence exists of cross-talk between neutrophils and Th17 $CD4^{+}$ T-cells [61]. Purified human neutrophils produce chemokines that attract Th1 and Th17 cells, while Th17 cells produce CXCL8, a potent chemoattractant for neutrophils.

Th17 cells are important in host defense against infections caused by both intracellular pathogens, such as listeriosis, salmonellosis, cryptococcosis, leishmaniasis, and tularemia, and a variety of extracellular bacterial and fungal pathogens. The role of Th17 cells in specific infections is not completely understood and, in some infections such as invasive aspergillosis, the data are conflicting. It appears, however, that Th17 cells are very important for mucosal host defense against oral candidiasis, mainly through the recruitment of neutrophils, and may also be important in disseminated disease [62]. Th17 cells are also involved in the response to *Staphylococcus aureus* infection [63].

Th17 cells express CD4, so it is not surprising that dysfunction and depletion of Th17 cells occur in HIV infection. Th17 cells are efficiently infected by HIV-1 in vitro [64], which may help explain the broad range of bacterial and fungal pathogens, particularly extracellular pathogens, that infect patients with HIV. CD4+IL-17+ populations are greatly reduced in antiretroviral-naive HIV-infected patients compared to HIV-negative controls, but this subset is greatly increased after the initiation of HAART, while IFN-α T-cells (Th1) are unchanged [65]. Depletion of Th17 cells in the gut has been associated with decreased microbial barrier function of the gut and persistent immune activation due to gut translocation [66]. Other studies have even shown a decrease in TH17 cells and the Th17:Th1 ratio with preferential depletion of Th17 cells from GALT within weeks of simian immunodeficiency virus (SIV) infection in macaques [67].

The cross-talk between Th17 cells and neutrophils and other phagocytic cells of the innate immune system may explain previous observations that rHuG-CSF may be of potential benefit for immune reconstitution in HIV infection. Previously, our group was part of a study that demonstrated that rHuG-CSF can restore IL-2 production in the blood of HIV-infected individuals [68]. At that time, it was unknown how rHuG-CSF affected T-cell function or production of a lymphocyte growth factor. In retrospect, neutrophil activation may lead to IL-6 release, subsequent Th17 differentiation, and IL-17 production. An inverse relationship exists between Treg and Th17 cells, so an increase in Th17-inducing cytokines may lead to a decrease in Tregs that mediate immunosuppression. Furthermore, T-cells are not only the source but also the target of IL-17. IL-17 can modulate Th1 cell polarization both in vitro and in vivo by directly acting on $CD4^+$ T-cells [69]. Overall, however, the effects of rHuG-CSF on $CD4^+$ lymphocyte counts have been variable [70, 71].

5 Risk of Infections in HIV-Induced Neutropenia

Initial reports in the pre-HAART era suggested that the risk of infection in patients with HIV-associated neutropenia was less than the risk of infection in patients with neutropenia due to chemotherapy or hematologic malignancy, a fact attributed to, among other factors, the lack of mucosal injury in neutropenia not due to cytotoxic chemotherapy and the relatively mild nature of HIV-associated neutropenia compared to neutropenia associated with hematologic malignancy or cytotoxic chemotherapy. Farber et al., in one of the first studies to examine the risk of infection, retrospectively examined the records of 30 HIV-infected patients with ANC $<1.0 \times 10^9$/L and $CD4^+$ lymphocyte count <200 cells/mm^3, comparing infection rates during neutropenic periods and non-neutropenic periods and comparing infection rates in these patients with 37 patients with hematologic malignancies as controls. In that study, no difference was found between infection rates in HIV patients during neutropenic periods compared with non-neutropenic periods [72]. Piliero et al., in reviewing blood culture data from 38 HIV-infected patients with ANC $<0.5 \times 10^9$/L and 1,071 non-neutropenic HIV patients, found that the presence of a central venous catheter, but not neutropenia, was a risk factor for positive blood cultures in a multiple logistic regression model [73].

A number of subsequent studies, however, have suggested that HIV-associated neutropenia is associated with an increased risk of infection. Shaunik and Bartlett, in an era when zidovudine monotherapy dosed at 1,200 mg/day was the standard of care for HIV treatment (compared with the dose of 600 mg/day used in current HAART regimens), found in a study of 30 patients that an ANC $<0.5 \times 10^9$/L was associated with an incidence of bacterial infection that was 600% higher than the incidence observed when ANC was $>1.0 \times 10^9$/L [74]. Keizer et al. performed a case–control study comparing 29 HIV-infected patients followed from 1991 to 1993 with two consecutive measured ANC $<1.0 \times 10^9$/L with 29 HIV-infected controls without history of neutropenia matched for age, sex, $CD4^+$ lymphocyte count, and month of entry into the clinic. An incidence of 12.6 episodes of bacteremia per 100 patient-months was observed in the patients with a history of neutropenia compared with an incidence of 0.87 episodes of bacteremia per 100 patient-months observed in the controls, a statistically significant difference ($p = 0.0027$). In a multiple logistic regression model, neutropenia (OR = 22.7) and the presence of a central venous catheter (OR = 8.5) were independent predictors of the development of bacteremia [75]. Hambleton et al. compared the outcomes of HIV-infected inpatients who developed neutropenia after treatment with cytotoxic chemotherapy and those who developed neutropenia for other reasons from 1987 to 1990. In their cohort, few patients received filgrastim, and most patients were white men who had sex with men. They found no statistically significant difference in the rates of bacteremia or mortality between the two groups in both bivariate analysis and in multiple logistic regression models [70]. At the same center, but examining a different time period (1992–1993), Jacobsen et al. examined the rates of hospitalization for HIV-infected outpatients stratified by ANC, using ICD-9 codes to evaluate for the presence of infection. A total of 2047 outpatients were analyzed for this study. A progressive and statistically significant increase in the risk of hospitalization was noted as the nadir ANC decreased. The number of days hospitalized per 10,000 days of risk increased from 61 for patients whose ANC was always $>0.5 \times 10^9$/L during the study period to 487 for patients with ANC nadir $<0.3 \times 10^9$/L [71].

Moore et al. performed a case–control study of patients followed from 1990 to 1994; 118 HIV-infected patients with a measured ANC $<1.0 \times 10^9$/L were matched with HIV-infected non-neutropenic controls on the basis of history of intravenous drug use, $CD4^+$ lymphocyte count, enrollment date in the clinic, and duration of follow-up. A statistically significant association between neutropenia and the development of a bacterial infection was observed. In an adjusted analysis, the relative risk for developing a bacterial infection was 2.33 times higher in patients with an ANC $<1.0 \times 10^9$/L, and 7.92 times higher for patients with an ANC $<0.5 \times 10^9$/L, compared with patients with an ANC $>1.0 \times 10^9$/L; the incidence of bacterial infections was estimated as 3.5–4.5 per 100 patient-months of neutropenia [76].

Eng et al. performed a retrospective cohort study examining the records of patients who received HIV care from 1990 to 1992. Of 930 patients, 85 experienced at least one episode of neutropenia, and of 12 patients with "severe" neutropenia (ANC $<0.3 \times 10^9$/L), 3 experienced gram-negative bacteremia, compared with four episodes of gram-negative bacteremia in 61 patients with less severe neutropenia [77].

Subsequently, Caperna et al. performed a retrospective cohort study examining patients who had at least two ANC measurements between 1991 and 1995. As the degree of neutropenia worsened, increases in the observed incidence of bacteremia with *Escherichia coli*, *Pseudomonas aeruginosa*, and *Klebsiella pneumoniae* were noted in this cohort of 1,645 patients [78]. A subsequent nested case–control study from the same cohort during the same time interval concluded that, adjusting for $CD4^+$ lymphocyte count, clinical stage of HIV infection, and other confounders, a measured ANC $<0.5 \times 10^9$/L conferred an eightfold increased risk of bacteremia with one of the three aforementioned pathogens [79].

Tumbarello et al. retrospectively examined 9 years of patient records from a single center in Italy, comparing HIV-infected patients with bacteremia with both HIV nonbacteremic and HIV-uninfected bacteremic controls. In a multivariate analysis of HIV-infected patients, most of whom were intravenous drug users, low $CD4^+$ lymphocyte count, presence of a central venous catheter, and ANC $<1.0 \times 10^9$/L (OR 3.05; $p = 0.04$) were independent risk factors for the development of bacteremia [32].

Hermans et al., from a cohort of 1,870 HIV-infected patients seen at a Belgian HIV care center from 1982 to 1993, found 1,403 whose ANC had been measured at least once. From this sample, they identified 484 patients with episodes of neutropenia, defined in this study as a measured ANC $<1.0 \times 10^9$/L. A history of neutropenia was associated with bacteremia or bacterial pneumonia. Subsequently, 177 neutropenic patients from this cohort were compared with 177 non-neutropenic controls matched for initial $CD4^+$ lymphocyte count and duration of follow-up. Although, in unadjusted analysis, neutropenia was associated with increased odds of developing infection (OR 3.29), this association disappeared in a multiple logistic regression model which also included clinical stage of AIDS and hemoglobin concentration [80].

Meynard et al. performed a single-institution prospective study that enrolled 62 HIV-infected patients with a measured ANC $<1.0 \times 10^9$/L. In their study, a higher risk of infection was noted in neutropenic patients with malignancy compared with other neutropenic patients, and logistic regression modeling identified a history of neutropenia, the presence of a central venous catheter, and trough ANC as independent predictors of developing an infection [31].

Moore et al. retrospectively identified 328 HIV-infected individuals at their center in the United Kingdom with an ANC $<1.0 \times 10^9$/L from 1994 to 1995, excluding 78 patients receiving cytotoxic chemotherapy. Bacteremias were documented in 21% of patients and were observed more commonly in patients with brief, profound neutropenias, rather than in patients with milder but more prolonged neutropenic episodes, and the degree of neutropenia correlated with the risk of infection [14]. A subsequent prospective study at the same center enrolled 87 patients in 1996 and 1997. No patients were receiving chemotherapy. Upon enrollment in the study, blood was sampled weekly to measure the duration of neutropenia. Filgrastim was only given for documented infection. The median duration of neutropenia was 13 days. Twelve subjects (17%) were diagnosed with neutropenia-associated infection, in whom six were found to have infection serious enough to warrant filgrastim therapy, and four required hospital admission. All serious infections occurred in patients with ANC $<0.5 \times 10^9$/L. A further four patients received filgrastim due to prolonged neutropenia [81].

Most of the studies on the risk of infection in neutropenia were conducted in the pre-HAART era. Relatively fewer studies have attempted to assess the link between neutropenia and infection after the widespread implementation of HAART. In the Women's Interagency HIV study, which began in the pre-HAART era but extended through the development of HAART, it was observed that, independent of $CD4^+$ lymphocyte count and viral load, which were strong predictors of mortality, the presence or absence of neutropenia was not predictive of mortality [20].

Toure et al., in examining the incidence of neutropenia in a prospective cohort of 533 African patients over 6 years taking trimethoprim–sulfamethoxazole prophylaxis in a setting where HAART was available if indicated, found that 36% of patients had at least one measured ANC $<1.0 \times 10^9$/L during the study, that developing neutropenia was associated with low initial $CD4^+$ lymphocyte count, and that the adjusted hazard ratio of developing bacterial morbidity was 1.50 for patients with a history of neutropenia to that degree, but the overall likelihood of bacterial morbidity was low (36 patients overall) [82].

Most of the preceding discussion has focused on bacterial infections. In patients with hematologic malignancy or in patients undergoing stem cell transplantation, invasive aspergillosis is a feared complication of the prolonged neutropenia that these patients experience [83]. Case reports and case series have noted the occurrence of aspergillosis in patients with advanced HIV [84]; however, aspergillosis as a complication of neutropenia was an uncommon event in HIV-infected patients even before the advent of HAART [85]. Case series suggest that neutropenia or steroid use may predispose to invasive aspergillosis in HIV-infected patients [86, 87]. Mylonakis et al. reviewed 342 reported cases of aspergillosis in HIV-infected patients obtained by MEDLINE and AIDSLINE searches through 1997, and found that 93 patients were diagnosed with "definite" invasive aspergillosis, of whom 16 were reported to have an ANC $<0.5 \times 10^9$/L [88].

6 The Use of Filgrastim in Nonmalignant Conditions in HIV Infection

Since the beginning of the HIV epidemic, several dozen case reports, case series, and clinical trials have described the use of filgrastim and other formulations of rHuG-CSF in HIV-infected individuals without malignancy and have described outcomes (summarized in Table 1). Kimura et al. were the first to publish their experience, reporting on the use of rHuG-CSF in 14 Japanese patients, 11 of whom were neutropenic. The patient cohort primarily consisted of 12 patients with hemophilia, eight of whom demonstrated concurrent infection at the time of enrollment, and two of whom were experiencing fever of unknown origin. Patients were randomly assigned to receive 100 $\mu g/m^2$ or 200 $\mu g/m^2$ rHuG-CSF daily intravenously. The dose was subsequently titrated to maintain the ANC $>3.0 \times 10^9/L$. A dose-dependent increase in ANC was observed in patients not receiving zidovudine, and escalation of rHuG-CSF dosage prevented the development of neutropenia in patients receiving zidovudine [89].

Miles published the first reports specifically evaluating the benefit of filgrastim in neutropenic patients with HIV. In the first report, 13 patients were selected to receive filgrastim, which was initially dosed at 3 μg/kg daily with subsequent dose escalation until ANC $>6.0 \times 10^9/L$. Once an adequate dose was established, it was maintained for 2 weeks; rHuEPO was subsequently given. Of the 13 patients selected, 1 died and 1 was removed from the study for noncompliance. In the other 11 patients, filgrastim therapy alone was associated with an increase in BFU-E (burst forming unit-erythron) levels, as well as a statistically significant mean hemoglobin increase of 1.04 g/dL [90]. Patients were then given zidovudine, dosed at either 1,000 or 1,500 mg/day. In the full analysis based on 22 recruited patients, 20 of whom were included in the final analysis, growth factors ensured that no patient needed to discontinue zidovudine therapy secondary to neutropenia, although eight patients developed transfusion dependence necessitating cessation of therapy [91].

Bratt et al. were among the first to describe long-term outcomes from rHuG-CSF therapy, noting in their series of 17 patients that appropriate white blood cell counts could be maintained for up to 7 months in patients with initial counts $<1.0 \times 10^9/L$ [92]. Several small case series and small pilot studies were published in the ensuing years, but a pair of studies notable due to their size are the retrospective cohort studies reported by Grutzmeier and colleagues in 1996, reflecting treatment experience in the pre-HAART era. In one cohort of gay men with $CD4^+$ lymphocyte counts <50 cells/mm^3 in Sweden, treatment with rHuG-CSF was initiated upon measuring a white blood cell count $<1.0 \times 10^9/L$ or ANC $<0.5 \times 10^9/L$. Median survival in the 60 patients who received rHuG-CSF was 658 days, compared with a median survival of 511 days in the 104 patients who did not receive rHuG-CSF ($p < 0.01$). A similar analysis by the same authors performed on data from a similar cohort in Denmark found that the 60 patients who had received rHuG-CSF lived for

Table 1 Summary of published papers

Reference no.	Type of study	No. of patients	Condition(s)	Intervention	Results
[89]	Prospective, uncontrolled	14	Neutropenic or scheduled to receive ZDV. Most pt were hemophiliacs; 8 had other infections	Randomly received 100 or 200 μg/m^2/d G-CSF initially, then titrated to keep ANC >3.0 × 10^9/L	Improvement in neutrophil count, even when pt received ZDV. Improvement in control of other infections
[90]	Prospective, uncontrolled	22	Severe HIV; ANC <2.5 × 10^9/L; Hgb <11.5 g/dL; platelet count >50 × 10^9/L	Myelosuppressive drugs stopped; G-CSF 3 μg/kg/d given and dose escalated until ANC maintained >6.0 × 10^9/L for 2 wk; rHuEPO subsequently given	Correlation between blood forming units (BFU-E) and G-CSF dose ($R = 0.65$; $p = 0.02$). Mean Hgb increase 1.04 ± 0.34 g/dL
[91]	Prospective, uncontrolled	22	ANC <2.5 × 10^9/L; Hg <12 g/dL; platelet >50 × 10^9/L	Myelosuppressive drugs stopped; G-CSF 3.6 μg/kg/d and escalated until ANC >6 × 10^9/L for first 11 pts and 1.5–5.0 × 10^9/L for subsequent 11 pts; maintained for 2 wk; Epoetin alfa subsequently given. ZDV administered and titrated to dose 1,500 mg/d as tolerated	No pt required discontinuation of ZDV for neutropenia, but 6 pts required ZDV discontinuation for transfusion-dependent anemia
[108]	Prospective, uncontrolled	12	CDC stage IV AIDS on ZDV with ANC <0.1 × 10^9/L on 2 occasions	G-CSF 0.4 μg/kg/d and titrated upward to achieve ANC >3.0 × 10^9/L. Maintained for 3 wk	Increase in mean ANC from 0.65 to 6.01 × 10^9/L after 1 wk and 5.54 × 10^9/L after 4 wk ($p < 0.01$) of G-CSF therapy; values returned to baseline after cessation of treatment
[92]	Pilot followed by prospective, uncontrolled	6 in pilot phase; 11 in prospective phase	AIDS pt with Kaposi's sarcoma (8 pt), CMV retinitis (8 subjects), or lymphoma (1 pt) and WBC <1.0 × 10^9/L	G-CSF 5 μg/kg until WBC >1.5 × 10^9/L, then maintenance	All 11 pts in prospective with ANC >1.0 × 10^9/L within 24 hr. No bacterial infections seen when WBC >1.5 × 10^9/L or ANC >1.0 × 10^9/L during the subsequent 7 mo

(continued)

Table 1 (continued)

Reference no.	Type of study	No. of patients	Condition(s)	Intervention	Results
[110]	Prospective single-blind randomized dose-varying	11	CDC stage IV AIDS with ANC 0.5–1.0 × 10^9/L on two occasions while receiving ZDV	G-CSF 0.5–10 μg/kg administered for 10 d	Mean baseline ANC 0.67 × 10^9/L; ANC increased at all doses except placebo in a dose-dependent manner and decreased rapidly after cessation of therapy
[111]	Case series	6	Pt with ZDV-associated neutropenia (ANC <1.0 × 10^9/L)	G-CSF 300 μg/d for 10 d with cessation of ZDV therapy	Increase in mean ANC from 0.89 × 10^9/L before G-CSF administration to 1.55 × 10^9/L after
[112]	Randomized placebo-controlled	8 experimental; 2 control	AIDS pt receiving ZDV with ANC <1.0 × 10^9/L	Placebo or varying doses of G-CSF administered for 10 d	Mean ANC increase from 0.99 × 10^9/L to 3.81 × 10^9/L after 8 hr, which reversed after treatment cessation
[112]	Case series	6	AIDS pt receiving ganciclovir for CMV retinitis	G-CSF 5 μg/kg given for 6–117 d	Mean ANC increase from 0.65 × 10^9/L to 2.17 × 10^9/L within 24 hr
[113]	Case series	8	Stage IV-A AIDS pt with ANC <0.75 × 10^9/L and concomitant infection	Filgrastim 3–4 μg/kg/d	Filgrastim was well-tolerated and infection was successfully treated in all pt
[114]	Case series	7	CMV end organ disease and ANC <0.58 × 10^9/L	Filgrastim titrated to maintain trough ANC 0.5–1.5 × 10^9/L range	All 7 pts were successfully treated with a full course of ganciclovir
[115]	Prospective, uncontrolled	19	Children <18 years of age receiving ZDV and with ANC <0.8 × 10^9/L despite ZDV dose reduction	Filgrastim 1 μg/kg/d was begun and titrated to keep ANC in 1.5–5.0 × 10^9/L range	Increase in median ANC from 1.02 to 2.96 × 10^9/L ($p = 0.0006$); 17/19 pts were able to continue ZDV therapy

[116]	Prospective, uncontrolled	15	AIDS pt with baseline ANC <1.0 × 10^9/L	G-CSF 300 μg/d with 500 mg/d ZDV for 11 mo	Mean final ANC 2.54 × 10^9/L. No pt developed a new infection or sepsis during follow-up. One death due to cerebral hemorrhage and one due to lymphoma
[117]	Case series	3	HIV-infected children with leucopenia and anemia	G-CSF 5 μg/kg 3 times wk with Epoetin alfa	Cessation of transfusion requirements and decrease in hospitalizations
[93]	Retrospective cohort	60 experimental; 104 control	AIDS pt with CD4+ count< 50 cells/mm^3 given G-CSF compared with controls who were not given G-CSF	Daily G-CSF given until WBC >1.5 × 10^9/L, then several times wk	Median survival of 658 d in the G-CSF group and 511 d in controls ($p < 0.01$)
[93]	Retrospective cohort	60 experimental; 65 control	AIDS pt with WBC <1,000 cells/mm^3 or ANC <0.5 × 10^9/L given G-CSF compared with controls not given G-CSF	Daily G-CSF given until WBC >1.5 × 10^9/L, then several times wk	Median survival of 248 d in the G-CSF group vs. 145 d in controls ($p = 0.06$)
[118]	Open-label, noncomparative	20	AIDS pt with 2 ANC <1.0 × 10^9/L within 7 d on myelo-suppressive therapy	G-CSF 1 μg/kg daily, with dose escalation to maintain ANC 2.0 to 5.0 × 10^9/L	One pt died and one withdrew from study. Other pt were able to maintain myelo-suppressive therapy without neutropenia. Mean ANC increased from 0.72 × 10^9/L to 3.11 × 10^9/L
[94]	Prospective, open-label, noncomparative	200	AIDS pt with 3 measured ANC <1.0 × 10^9/L not attributable to cancer therapy	Filgrastim 1 μg/kg/d initiated and titrated to maintain ANC >2.0 × 10^9/L	Reversal of neutropenia in 98% of pt, sustainable with maintenance therapy
[119]	Case report	1	38-year-old man with AIDS, neutropenia, and oral ulcers refractory to steroids, acyclovir, and antibiotics	G-CSF 300 μg daily, titrated to maintain ANC >2.0 × 10^9/L	Resolution of neutropenia and of ulceration. Subsequent death from other AIDS-related causes

(continued)

Table 1 (continued)

Reference no.	Type of study	No. of patients	Condition(s)	Intervention	Results
[120]	Randomized, controlled, double-blinded	10 experimental; 10 control	Asymptomatic HIV infection with $CD4^+$ count <500 cells/mm^3	ZDV (500 mg/d) with G-CSF (10 μg/kg biweekly) and Epoetin alfa in trial group; ZDV alone in controls	Significant increase in experimental group in Hgb, ANC, and $CD4^+$ count compared with controls ($p < 0.01$)
[121]	Prospective, uncontrolled	6	Class IV AIDS with lymphocyte count <4,300 cells/mm^3 and ANC $<1.9 \times 10^9$/L	Filgrastim 3–5 μg/kg 3 times wk for 2 wk or until recovery of counts	Statistically significant increase in WBC, total lymphocyte, $CD3^+$, $CD4^+$, $CD8^+$, and NK^+ counts
[122]	Prospective, uncontrolled	11	HIV-infected children	G-CSF 5 μg/kg 2–3 times wk with concomitant Epoetin alfa	Increased leukocyte ($p = 0.003$) and neutrophil ($p = 0.009$) counts
[123]	Retrospective cohort	71 experimental; 157 control	HIV-infected pt with ANC $<1.0 \times 10^9$/L given G-CSF or GM-CSF compared with controls not given G-CSF or GM-CSF	G-CSF 300 μg or GM-CSF 250 μg administered per physician discretion	Improved survival in multiple logistic regression model associated with G-CSF or GM-CSF administration (OR 4.2)
[98]	Retrospective cohort	25 experimental; 66 control	HIV-infected pt with disseminated MAI infection	G-CSF as per physician discretion	G-CSF usage associated with improved survival in Cox proportional hazards model (HR 0.22; $p = 0.01$)
[124]	Retrospective cohort	71 experimental; 81 control	HIV-infected pt with ANC $<1.0 \times 10^9$/L	G-CSF 300 μg 5 times wk initiated when ANC$<0.5 \times 10^9$/L and titrated to maintain ANC $>1.0 \times 10^9$/L	In both bivariate and multivariate analysis, G-CSF administration was associated with a decreased risk of bacteremia and death

[95]	Prospective, randomized, unblinded	172 experimental; 86 control	HIV-infected pt with CD4$^+$ count <200 cells/mm^3, ANC 0.75–1 0 × 10^9/L, and platelet >50 × 10^9/L	Daily or intermittent filgrastim titrated to maintain ANC between 2.0 and 10.0 × 10^9/L in intervention group. Controls received filgrastim only if ANC <0.5 × 10^9/L persistently	Decreased incidence of severe neutropenia and severe bacterial infections (RR 0.46; $p < 0.01$) associated with filgrastim usage
[125]	Prospective, randomized controlled	5 experimental, 5 control	HIV-infected pt with disseminated MAI infection	Intermittent filgrastim titrated to maintain WBC >5.0 × 10^9/L	Clearance of infection and survival in all 5 pts in experimental group; death in all 5 pts in control group
[126]	Randomized, single-blind, controlled	55 experimental; 15 control	HIV-infected pt with CMV infection and ANC <1.0 × 10^9/L during first 12 d ganciclovir treatment	Lenograstim 25–150 μg/m^2 daily or placebo administered concomitantly with ganciclovir 5 mg/kg every 12 hr	Higher median ANC was noted in the intervention groups compared with the control group. No additional effect was noted for doses >50 μg/m^2/d
[109]	Pre- and post-intervention	71 pre-intervention; 81 post-intervention	HIV-infected pt who met detailed guidelines for initiating G-CSF therapy	Pharmacist-driven protocol for G-CSF dosing in the intervention group vs. standard dosing in the control group	Decrease in mean number of daily G-CSF doses given (0.51 to 0.29) and quarterly cost of G-CSF to HIV-infected pt ($90,000 to $22,000) without increase in number of neutropenic days
[127]	Prospective randomized, double-blinded, controlled	6 experimental, 5 control	HIV-infected pt with CD4$^+$ count <350 cells/mm^3 initiating HAART	Filgrastim 300 μg 3 times wk or placebo with HAART for 12 wk	Trial terminated due to adverse event. HIV viral load decreased less rapidly in filgrastim group compared with placebo group

(continued)

Table 1 (continued)

Reference no.	Type of study	No. of patients	Condition(s)	Intervention	Results
[128]	Prospective randomized, double-blinded, controlled	12 experimental; 15 control	HIV-infected pt with $CD4^+$ count<350 cells/mm^3 on stable HAART regimen	G-CSF 300 μg 3 times wk or placebo for 12 wk	Significant increases in lymphocyte count ($p = 0.002$), $CD4^+$ count ($p = 0.03$), $CD8^+$ count ($p = 0.004$), and NK cell count ($p = 0.001$) in the G-CSF group. Increases were reversed after termination of study
[129]	Prospective, uncontrolled	18	HIV-infected pt on no or stable HAART regimen	Filgrastim 10 μg/kg daily for 7 d to stimulate mobilization for stem cell transplantation	9/18 pts experienced at least one significant increase in HIV-1 RNA viral load
[130]	Prospective randomized, double-blinded, controlled	12 experimental; 15 control	HIV-infected pt with $CD4^+$ count <350 cells/mm^3 on stable HAART regimen	G-CSF 300 μg 3 times wk or placebo for 12 wk	Significant increase in $CD34^+$ ($p = 0.006$) and $CD4^+$ counts
[96]	Retrospective cohort study of data from multiple prospective clinical trials	398 experimental; 311 control	HIV-infected pt receiving anti-CMV therapy who received filgrastim compared with controls who did not	Filgrastim with dose varying between studies	56% reduction in risk of death associated with filgrastim use in regression model ($p < 0.001$). Marginal benefit in same model on risk of bacterial infection (OR 0.48; $p = 0.062$)
[131]	Retrospective cohort study of data from prospective clinical trial	36 experimental; 36 control	HIV-infected pt receiving anti-CMV therapy who received filgrastim compared with controls who did not	Filgrastim given at physician discretion	Viral load 1.7-fold higher in pt receiving filgrastim ($p = 0.12$) after adjustment for confounders
[132]	Secondary analysis of prospective, uncontrolled trial	18	HIV-infected pt undergoing stem cell mobilization	Filgrastim 10 μg/kg daily for 7 d	In 7 pts whose HIV RNA viral load increased with filgrastim usage, an increase in quasispecies noted

[133]	Secondary analysis of prospective, uncontrolled trial	6	HIV-infected pt undergoing stem cell mobilization	Filgrastim 10 μg/kg daily for 7 d	2/3 pts with HHV-8 antibodies experienced reactivation of HHV-8 viremia after receiving filgrastim, compared with 0/3 pts without HHV-8 antibodies
[134]	Prospective, uncontrolled	32	HIV/HCV coinfected pt receiving IFNα 2b and ribavirin	Filgrastim 300 μg/week initiated when ANC $<0.75 \times 10^9$/L and titrated to keep ANC $>0.75 \times 10^9$/L	50% of pt required filgrastim during therapy. Baseline ANC $<2.25 \times 10^9$/L was predictive of requiring filgrastim usage while receiving anti-HCV therapy

ANC absolute neutrophil count; *d* day; *HAART* highly active antiretroviral therapy; *HCV* hepatitis C virus; *Hgb* hemoglobin; *hr* hour; *IFN* interferon; *MAI* *Mycobacterium avium* infection; *mo* month; *NK* natural killer cell; *pt* patients; *wk* week; *WBC* white blood cell; *ZDV* zidovudine a median of 248 days, compared with a median survival of 145 days in the 65 patients who did not receive rHuG-CSF [93].

Also notable for both its size and its prospective nature is the multicenter trial conducted by the G-CSF 92105 group and published in 1996. Patients were eligible if, on at least three occasions within a 2-week period, their measured ANC was found to be $<1.0 \times 10^9$/L. Filgrastim dose was titrated to maintain ANC $>2.0 \times 10^9$/L, a target obtained in 98% of patients. Patients received filgrastim for at least 28 days, but could receive it for a longer duration at the discretion of their physicians. The number of patients receiving what were considered to be the drugs most associated with neutropenia (zidovudine, trimethoprim-sulfamethoxazole, ganciclovir, and pyrimethamine) increased 20% over the course of the study [94].

The results of the largest prospective randomized control trial to investigate the outcome of filgrastim administration to neutropenic patients with HIV were published by Kuritzkes et al. in 1998. Of note, patients were accrued in the pre-HAART era. Eligibility requirements for this multicenter, nonblinded study included at least one measured ANC between 0.75 and 1.0×10^9/L, $CD4^+$ lymphocyte count <200 cells/mm^3, platelet count $>50.0 \times 10^9$/L, Karnovsky score >50%, and life expectancy >6 months. Malignancy was the most notable exclusion criterion. In all, 258 patients were randomly assigned to daily, intermittent, or no filgrastim administration for 24 weeks, with filgrastim dose subsequently titrated to maintain white cell count in the $2.0–10.0 \times 10^9$/L range. Control patients receiving no filgrastim who developed an ANC $<0.5 \times 10^9$/L were censored and subsequently re-randomized to one of the two treatment groups, an event which occurred in 18 patients. In all, 34.1% of patients who were assigned to receive no filgrastim developed the primary endpoint of severe neutropenia or death, compared with 12.8% of patients in the intermittent filgrastim group and 8.2% of patients in the daily filgrastim group. Patients who received filgrastim were 54% less likely than controls to develop a serious bacterial infection ($p = 0.005$) [95].

The most recent large study to examine the effect of filgrastim administration on outcomes in patients with HIV was the retrospective cohort study published by Davidson et al. in 2002. This study was a retrospective analysis of data from several prospectively conducted studies of HIV-infected patients with CMV retinitis. Of note, significant heterogeneity in the use of filgrastim was present in the original trials. In an analysis made complex by the need to correct for the inherent biases of a retrospective analysis, filgrastim use was associated with a 56% reduction in death in a multivariate model ($p < 0.01$). Associations between filgrastim use and such outcomes as catheter-associated bacteremia and repeat bacterial infection were significant in unadjusted analysis but not after adjustment for confounders [96].

Other interesting uses for rHuG-CSF which single reports have explored include the use of filgrastim as an adjunct to a tagged white blood cell scan in a neutropenic HIV-infected patient in the workup of postoperative fever [97], and the use of filgrastim concomitantly with antimicrobials in the treatment of disseminated mycobacterial infection [98].

7 The Use of Filgrastim in Malignant Conditions in HIV Infection

The benefits of more widespread use of colony-stimulating factors in HIV-negative patients undergoing chemotherapy are now widely accepted, as demonstrated by recently written clinical guidelines which expand the criteria for their use [99]. This change in practice applies to HIV-infected patients as well. Although no randomized trial specifically has addressed the use of filgrastim in HIV-associated malignancy, several nonrandomized comparative studies looking at the benefits of adding rHuG-CSF to a chemotherapy regimen in patients with HIV-associated lymphoma have been published. In a pre- and postintervention trial where 65 HIV-infected patients with non-Hodgkin's lymphoma (NHL) either received a dose-reduced variant of the CHOP regimen (chemotherapy with cyclophosphamide, doxorubicin, vincristine, and prednisone) or full-dose CHOP augmented by filgrastim, grade 3 or 4 neutropenia occurred in fewer patients receiving full-dose CHOP with filgrastim (13% vs. 25%), a nonsignificant difference. The study was neither intended to nor powered to measure the effect of filgrastim independently [100]. A similar, nonrandomized study by Rossi found that HIV-positive patients receiving rHuG-CSF during chemotherapy were more likely than those who did not to receive full doses of chemotherapy and to receive cycles without delays, although the likelihood of complete response was actually lower in those who received rHuG-CSF [101]. A third study, also nonrandomized, compared the outcomes of HIV-positive patients undergoing chemotherapy for NHL, finding a decrease in mean duration of treatment delays between cycles from 9d to 4d ($p = 0.01$). Again, however, mean duration of survival was worse in the group receiving rHuG-CSF, which also trended toward having more advanced HIV [102]. Further discussion of the role of rHuG-CSF in HIV-associated malignancies is beyond the scope of this chapter; a review of the use of hematopoietic growth factors in HIV-associated malignancies has been previously published [103].

8 Adverse Effects of Use

The most common side effect of filgrastim is bone pain, with headache, nausea, and vomiting also frequently reported [104]. Although generally safe, filgrastim usage in HIV patients has been associated with disseminated intravascular coagulation [105], hepatitis, and pancreatitis [106]. In non-HIV patients, a review of case reports of adverse effects of filgrastim noted a number of rare adverse events, many of which were attributed to increased inflammation (ARDS, shock, worsening autoimmune disease) or leukostasis (arterial thrombosis or myocardial infarction, interstitial nephritis, bone marrow necrosis) [107].

Nevertheless, the adverse effect of filgrastim that serves as the biggest barrier to its use is its cost. However, filgrastim may be effective in HIV-infected patients at doses lower than those traditionally used for patients with chemotherapy-induced neutropenia [108]. One center developed a clinical pharmacist-driven protocol for dosing filgrastim in HIV-infected inpatients based on ANC; subsequently, the quarterly cost for filgrastim for the HIV service decreased from US$90,000 to US$22,000 [109].

9 Conclusion

The incidence of neutropenia in HIV-infected patients has decreased with the advent of therapies that arrest and reverse the progression of AIDS. In the future, most patients with HIV who develop neutropenia will likely do so from oncologic disease and its treatment, not HIV itself. However, the inability of all HIV-infected patients to benefit from the advances in HIV treatment means that filgrastim will continue to play a role in the management of complications of advanced HIV and its treatment.

References

1. UNAIDS (2009) AIDS epidemic update: November 2009. UNAIDS, Geneva, Switzerland
2. Moir S, Chun TW, Fauci AS (2011) Pathogenic mechanisms of HIV disease. Annu Rev Pathol 6:223–248
3. Israel DS, Plaisance KI (1991) Neutropenia in patients infected with human immunodeficiency virus. Clin Pharm 10:268–279
4. Pitrak DL (1998) Use of filgrastim (r-metHuG-CSF) in human immunodeficiency virus infection. In: Morstyn G, Dexter TM, Foote M (eds) Filgrastim (r-metHuG-CSF) in clinical practice, 2nd edn. Marcel Dekker, New York, pp 441–468
5. Palella FJJ, Baker RK, Moorman AC et al (2006) Mortality in the highly active antiretroviral therapy era: changing causes of death and disease in the HIV outpatient study. J Acquir Immune Defic Syndr 43:27–34
6. Losina E, Schackman Bruce R, Sadownik Sara N et al (2009) Racial and sex disparities in life expectancy losses among HIV-infected persons in the United States: impact of risk behavior, late initiation, and early discontinuation of antiretroviral therapy. Clin Infect Dis 49:1570–1578
7. Isabella S, Lukas ES, Georg N, Roberto C, Rainer W, Thomas FL (2006) Cardiovascular disease in HIV infection. Am Heart J 151:1147–1155
8. Sigel K, Wisnivesky J, Justice A, et al (2010) HIV infection is an independent risk factor for lung cancer. 7th Conference on retroviruses and opportunistic infections (CROI 2010). San Francisco
9. Mugavero MJ, Castellano C, Edelman D, Hicks C (2007) Late diagnosis of HIV infection: the role of age and sex. Am J Med 120:370–373
10. Wood E, Hogg RS, Yip B, Moore D, Harrigan PR, Montaner JS (2006) Impact of baseline viral load and adherence on survival of HIV-infected adults with baseline CD4 cell counts $\geq$200 cells/μl. AIDS 20:1117–1123

11. Crum-Cianflone N, Hullsiek KH, Marconi V et al (2009) Trends in the incidence of cancers among HIV-infected persons and the impact of antiretroviral therapy: a 20-year cohort study. AIDS 23:41–50
12. Murphy MF, Metcalfe P, Waters AH et al (1987) Incidence and mechanism of neutropenia and thrombocytopenia in patients with human immunodeficiency virus infection. Br J Haematol 66:337–340
13. Murphy MF, Metcalfe P, Waters AH et al (1985) Immune neutropenia in homosexual men. Lancet 325:217–218
14. Moore DA, Sullivan A, Hilstead P, Gazzard BG (2000) A retrospective study of neutropenia in HIV disease. Int J STD AIDS 11:8–14
15. Ben-Galim P, Shaked Y, Garty M, Vonsover A (1996) Immediate immunosuppression caused by acute HIV-1 infection: a fulminant multisystemic disease 2 days post infection. Infection 24:332–335
16. Philippe C, Cédric F, Malika M, Catherine T (2005) Severe transient neutropenia associated with acute human immunodeficiency virus type 1 infection. Eur J Intern Med 16:120–122
17. Ribera E, Ocaña I, Almirante B, Gómez J, Monreal P, Martinez Vázquez JM (1989) Autoimmune neutropenia and thrombocytopenia associated with development of antibodies to human immunodeficiency virus. J Infect 18:167–170
18. Skiest DJ, King ME (1994) Granulocytopenia secondary to acute infection with the human immunodeficiency virus. J Infect 28:315–318
19. Schacker T, Collier AC, Hughes J, Shea T, Corey L (1996) Clinical and epidemiologic features of primary HIV infection. Ann Intern Med 125:257–264
20. Levine AM, Karim R, Mack W et al (2006) Neutropenia in human immunodeficiency virus infection: data from the women's interagency HIV study. Arch Intern Med 166:405–410
21. Benson CA, van der Horst C, LaMarca A et al (2004) A randomized study of emtricitabine and lamivudine in stably suppressed patients with HIV. AIDS 18:2269–2276
22. Albrecht D, Vieler T, Horst HA (2002) Rash-associated severe neutropenia as a side-effect of indinavir in HIV postexposure prophylaxis. AIDS 16:2098–2099
23. Healy BJ, Freedman AR (2006) HIV-related neutropenia exacerbated by efavirenz. HIV Med 7:129–131
24. Makinson A, Martelli N, Peyriere H, Turriere C, Le Moing V, Reynes J (2007) Profound neutropenia resulting from interaction between antiretroviral therapy and vinblastine in a patient with HIV-associated Hodgkin's disease. Eur J Haematol 78:358–360
25. Cingolani A, Torti L, Pinnetti C et al (2010) Detrimental clinical interaction between ritonavir-boosted protease inhibitors and vinblastine in HIV-infected patients with Hodgkin's lymphoma. AIDS 24:2408–2412
26. Mir O, Dessard-Diana B, Louet AL et al (2010) Severe toxicity related to a pharmacokinetic interaction between docetaxel and ritonavir in HIV-infected patients. Br J Clin Pharmacol 69:99–101
27. Coleman MD (2001) Dapsone-mediated agranulocytosis: risks, possible mechanisms and prevention. Toxicology 162:53–60
28. Olaison L, Belin L, Hogevik H, Alestig K (1999) Incidence of {beta}-lactam-induced delayed hypersensitivity and neutropenia during treatment of infective endocarditis. Arch Intern Med 159:607–615
29. Zuger A (1996) Profound neutropenia in an HIV-infected man. AIDS Clin Care 8(67):69–70
30. Lim PL (2009) HIV-related haematological conditions. In: Menon A, Kamarulzaman A (eds) Is it HIV? A handbook for health care providers. Australasian Society for HIV Medicine, Darlinghurst, NSW, pp 52–56
31. Meynard J-L, Guiguet M, Arsac S, Frottier J, Meyohas MC (1997) Frequency and risk factors of infectious complications in neutropenic patients infected with HIV. AIDS 11:995–998
32. Tumbarello M, Tacconelli E, Caponera S, Cauda R, Ortona L (1995) The impact of bacteraemia on HIV infection. Nine years experience in a large Italian University Hospital. J Infect 31:123–131

33. Toeung P, Pouv S, Chel S, et al (2007) Routine switch after 6 months from d4t to AZT containing antiretroviral therapy, at an outpatient HIV clinic in Phnom Penh, Cambodia. International AIDS society conference on HIV pathogenesis, treatment, and prevention. Sydney
34. Aurpibul L, Puthanakit T, Sirisanthana T, Sirisanthana V (2008) Haematological changes after switching from stavudine to zidovudine in HIV-infected children receiving highly active antiretroviral therapy. HIV Med 9:317–321
35. Abramson JS, Mills EL (1988) Depression of neutrophil function induced by viruses and its role in secondary microbial infections. Rev Infect Dis 10:326–341
36. Valone FH, Payan DG, Abrams DI, Goetzl EJ (1984) Defective polymorphonuclear leukocyte chemotaxis in homosexual men with persistent lymph node syndrome. J Infect Dis 150:267–271
37. Lazzarin A, Uberti Foppa C, Galli M et al (1986) Impairment of polymorphonuclear leucocyte function in patients with acquired immunodeficiency syndrome and with lymphadenopathy syndrome. Clin Exp Immunol 65:105–111
38. Nielsen H, Kharazmi A, Faber V (1986) Blood monocyte and neutrophil functions in the acquired immune deficiency syndrome. Scand J Immunol 24:291–296
39. Ellis M, Gupta S, Galant S et al (1988) Impaired neutrophil function in patients with AIDS or AIDS-related complex: a comprehensive evaluation. J Infect Dis 158:1268–1276
40. Murphy PM, Lane HC, Fauci AS, Gallin JI (1988) Impairment of neutrophil bactericidal capacity in patients with AIDS. J Infect Dis 158:627–630
41. Pitrak DL, Bak PM, DeMarais P, Novak RM, Burton RA (1993) Depressed neutrophil superoxide production in human immunodeficiency virus infection. J Infect Dis 167:1406–1410
42. Mullane K, Pitrak D, Bilek M, Novak R, Allen R, Stevens P (1994) In vivo neutrophil activation and burnout in HIV infection. Clin Res 42:155a
43. Outwater E, McCutchan J (1985) Neutrophil-associated antibodies and granulocytopenia in AIDS. Clin Res 33:413a
44. Weinberg GA, Gigliotti F, Stroncek DF et al (1997) Lack of relation of granulocyte antibodies (antineutrophil antibodies) to neutropenia in children with human immunodeficiency virus infection. Pediatr Infect Dis J 16:881–884
45. Nilsson J, Kinloch de Loes S, Granath A, Sönnerborg A, Goh LE, Andersson J (2007) Early immune activation in gut-associated and peripheral lymphoid tissue during acute HIV infection. AIDS 21:565–574
46. Elbim C, Prevot M, Bouscarat F et al (1994) Polymorphonuclear neutrophils from human immunodeficiency virus-infected patients show enhanced activation, diminished fMLP-induced L-selectin shedding, and an impaired oxidative burst after cytokine priming. Blood 84:2759–2766
47. Vierucci A, De Martino M, Graziani E, Rossi ME, London WT, Blumberg BS (1983) A mechanism for liver cell injury in viral hepatitis: effects of hepatitis B virus on neutrophil function in vitro and in children with chronic active hepatitis. Pediatr Res 17:814–820
48. Pitrak DL, Tsai HC, Mullane KM, Sutton SH, Stevens P (1996) Accelerated neutrophil apoptosis in the acquired immunodeficiency syndrome. J Clin Invest 98:2714–2719
49. Nie Z, Phenix BN, Lum JJ et al (2002) HIV-1 protease processes procaspase 8 to cause mitochondrial release of cytochrome c, caspase cleavage and nuclear fragmentation. Cell Death Differ 9:1172–1184
50. Busch M, Beckstead J, Gantz D, Vyas G (1986) Detection of human immunodeficiency virus infection of myeloid precursors in bone marrow samples from AIDS patients. Blood 68:122a
51. Moses A, Heneveld M, Nelson J, Williams S, Rarick M, Bagby G (1995) CD34+ bone marrow microvascular endothelial cells (MVEC) are consistently infected by HIV-1 in patients with AIDS: suboptimal release of IL-6 and G-CSF by infected stromal cells does not depend on release of soluble inhibitory factors. Blood 86:287a

52. Lagneaux L, Delforge A, Snoeck R et al (1996) Imbalance in production of cytokines by bone marrowstromal cells following cytomegalovirus infection. J Infect Dis 174:913–919
53. Lieschke G, Grail D, Hodgson G et al (1994) Mice lacking granulocyte colony-stimulating factor have chronic neutropenia, granulocyte and macrophage progenitor cell deficiency, and impaired neutrophil mobilization. Blood 84:1737–1746
54. Pitrak D, Mullane K, Bilek M, et al (1996) Filgrastim (r-metHuG-CSF) treatment of HIV-infected patients improves neutrophil function. International conference on AIDS. Vancouver
55. Zhu J, Paul WE (2009) Heterogeneity and plasticity of T helper cells. Cell Res 20:4–12
56. Williams GT, Smith CA, Spooncer E, Dexter TM, Taylor DR (1990) Haemopoietic colony stimulating factors promote cell survival by suppressing apoptosis. Nature 343:76–79
57. Pitrak DL, Sutton SH, Tsai HC, Mullane KM, Pau AK (1999) Reversal of accelerated neutrophil apoptosis and restoration of respiratory burst activity with r-metHuG-CSF (Filgrastim therapy in patients with AIDS. AIDS 13:427–429
58. Wan YY (2010) T regulatory cells: immune suppression and beyond. Cell Mol Immunol 7:204–210
59. Korn T, Bettelli E, Oukka M, Kuchroo VK (2009) IL-17 and Th17 Cells. Annu Rev Immunol 27:485–517
60. Schwarzenberger P, Huang W, Ye P et al (2000) Requirement of endogenous stem cell factor and granulocyte colony-stimulating factor for IL-17-mediated granulopoiesis. J Immunol 164:4783–4789
61. Pelletier M, Maggi L, Micheletti A et al (2010) Evidence for a cross-talk between human neutrophils and Th17 cells. Blood 115:335–343
62. Conti HR, Shen F, Nayyar N et al (2009) Th17 cells and IL-17 receptor signaling are essential for mucosal host defense against oral candidiasis. J Exp Med 206:299–311
63. Lin L, Ibrahim AS, Xu X et al (2009) Th1-Th17 cells mediate protective adaptive immunity against *Staphylococcus aureus* and *Candida albicans* infection in mice. PLoS Pathog 5: e1000703
64. ElHed A, Unutmaz D (2010) Th17 cells and HIV infection. Curr Opin HIV AIDS 5:146–150
65. Peng QL, Zhang MX, Li GY, Liu YX, Zhou BP, Wang H (2010) [Loss of the balance between Th17 and Th1 populations in HIV/AIDS patients]. Zhonghua Shi Yan He Lin Chuang Bing Du Xue Za Zhi 24.17–20
66. Hunt PW (2010) Th17, gut, and HIV: therapeutic implications. Curr Opin HIV AIDS 5:189–193
67. Milner JD, Sandler NG, Douek DC (2010) Th17 cells, Job's syndrome and HIV: opportunities for bacterial and fungal infections. Curr Opin HIV AIDS 5:179–183
68. Hartung T, Pitrak DL, Foote MA, Shatzen EM, Verral SC, Wendel A (1998) Filgrastim restores interleukin-2 production in blood from patients with advanced human immunodeficiency virus infection. J Infect Dis 178:686–692
69. O'Connor W Jr, Kamanaka M, Booth CJ et al (2009) A protective function for interleukin 17A in T cell-mediated intestinal inflammation. Nat Immunol 10:603–609
70. Hambleton J, Aragon T, Modin G, Northfelt DW, Sande MA (1995) Outcome for hospitalized patients with fever and neutropenia who are infected with the human immunodeficiency virus. Clin Infect Dis 20:363–371
71. Jacobson MA, Liu RC, Davies D, Cohen PT (1997) Human immunodeficiency virus disease-related neutropenia and the risk of hospitalization for bacterial infection. Arch Intern Med 157:1825–1831
72. Farber BF, Lesser M, Kaplan MH, Woltmann J, Napolitano B, Armellino D (1991) Clinical significance of neutropenia in patients with human immunodeficiency virus infection. Infect Control Hosp Epidemiol 12:429–434
73. Piliero P, Currier J, Barlam T, Ives D (1995) Risk of bacteremia in HIV-infected person with neutropenia. Second national conference on human retroviruses and related infections. Washington, DC

74. Shaunak S, Bartlett J (1989) Zidovudine-induced neutropenia. Are we too cautious? Lancet 334:91–92
75. Keiser P, Higgs E, Smith J (1996) Neutropenia is associated with bacteremia in patients infected with the human immunodeficiency virus. Am J Med Sci 312:118–122
76. Moore RD, Keruly JC, Chaisson RE (1995) Neutropenia and bacterial infection in acquired immunodeficiency syndrome. Arch Intern Med 155:1965–1970
77. Eng RHK, Yen K, Tecson-Tumang F, Smith SM, Akgun K (1994) Risk of gram-negative bacteremia during neutropenia in patients with AIDS. Infect Dis Clin Pract 3:373–375
78. Caperna J, Barber RE, Toerner JG, Mathews WC (1998) Estimation of the effect of neutropenia on rates of clinical bacteraemia in HIV-infected patients. Epidemiol Infect 120:71–80
79. Mathews WC, Caperna J, Toerner JG, Barber RE, Morgenstern H (1998) Neutropenia is a risk factor for gram-negative bacillus bacteremia in human immunodeficiency virus-infected patients: results of a nested case-control study. Am J Epidemiol 148:1175–1183
80. Hermans P, Sommereijns B, Van Cutsem N, Clumeck N (1999) Neutropenia in patients with HIV infection: a case control study in a cohort of 1403 patients between 1982 and 1993. J Hematother Stem Cell Res 8:S23–S32
81. Moore DA, Benepal T, Portsmouth S, Gill J, Gazzard BG (2001) Etiology and natural history of neutropenia in human immunodeficiency virus disease: a prospective study. Clin Infect Dis 32:469–475
82. Toure S, Gabillard D, Inwoley A, Seyler C, Gourvellec G, Anglaret X (2006) Incidence of neutropenia in HIV-infected African adults receiving co-trimoxazole prophylaxis: a 6-year cohort study in Abidjan, Cote d'Ivoire. Trans R Soc Trop Med Hyg 100:785–790
83. Vehreschild JJ, Rüping MJ, Steinbach A, Cornely OA (2010) Diagnosis and treatment of fungal infections in allogeneic stem cell and solid organ transplant recipients. Expert Opin Pharmacother 11:95–113
84. Libanore M, Sighinolfi L, Ghinelli F (2010) Invasive Aspergillosis and HIV infection. In: Comarú Pasqualotto A (ed) Aspergillosis: from diagnosis to prevention. Springer, Netherlands, pp 559–566
85. Khoo SH, Denning DW (1994) Invasive aspergillosis in patients with AIDS. Clin Infect Dis 19:S41–S48
86. Pursell KJ, Telzak EE, Armstrong D (1992) Aspergillus species colonization and invasive disease in patients with AIDS. Clin Infect Dis 14:141–148
87. Libanore M, Prini E, Mazzetti M et al (2002) Invasive aspergillosis in Italian AIDS patients. Infection 30:341–345
88. Mylonakis E, Flanigan T, Rich JD, Barlam TF (1998) Pulmonary aspergillosis and invasive disease in AIDS. Chest 114:251–262
89. Kimura S, Matsuda J, Ikematsu S et al (1990) Efficacy of recombinant human granulocyte colony-stimulating factor on neutropenia in patients with AIDS. AIDS 4:1251–1255
90. Miles S, Mitsuyasu R, Lee K et al (1990) Recombinant human granulocyte colony-stimulating factor increases circulating burst forming unit-erythron and red blood cell production in patients with severe human immunodeficiency virus infection. Blood 75:2137–2142
91. Miles S, Mitsuyasu R, Moreno J et al (1991) Combined therapy with recombinant granulocyte colony-stimulating factor and erythropoietin decreases hematologic toxicity from zidovudine. Blood 77:2109–2117
92. Bratt G, Grutzmeier S, Lund B, Sandstrom E. (1992) Granulocyte-colony stimulating factor G-CSF in AIDS-patients with leukocytes less than or equal to 1.0 × 10(9)/L. International Conference on AIDS. Amsterdam Jul 19–24; 8:Mo6 (abstract M0A 0007)
93. Grutzmeier S, Gerstoft J, Boje Hansen P, Sandstrom E (1996) Filgrastim (G-CSF) use is associated with prolonged survival in AIDS-patients with leukopenia and CD4 cells less than 50 x 10^6/l. International Conference on AIDS. Vancouver, Canada
94. Hermans P, Rozenbaum W, Jou A et al (1996) Filgrastim to treat neutropenia and support myelosuppressive medication dosing in HIV infection. AIDS 10:1627–1633

95. Kuritzkes DR, Parenti D, Ward DJ et al (1998) Filgrastim prevents severe neutropenia and reduces infective morbidity in patients with advanced HIV infection: results of a randomized, multicenter, controlled trial. AIDS 12:65–74
96. Davidson M, Min Y-I, Holbrook JT et al (2002) Use of filgrastim as adjuvant therapy in patients with AIDS-related cytomegalovirus retinitis. AIDS 16:757–765
97. Lin S, Marcus CS (1996) Augmentation of leukocyte count with G-CSF in a patient with neutropenia before In-111 WBC scintigraphy. Clin Nucl Med 21:544–546
98. Keiser P, Rademacher S, Smith J, Skiest D (1998) G CSF association with prolonged survival in HIV infected patients with disseminated *Mycobacterium avium* complex infection. Int J STD AIDS 9:394–399
99. Aapro MS, Cameron DA, Pettengell R et al (2006) EORTC guidelines for the use of granulocyte-colony stimulating factor to reduce the incidence of chemotherapy-induced febrile neutropenia in adult patients with lymphomas and solid tumours. Eur J Cancer 42:2433–2453
100. Ratner L, Lee J, Tang S et al (2001) Chemotherapy for human immunodeficiency virus-associated non-Hodgkin's lymphoma in combination with highly active antiretroviral therapy. J Clin Oncol 19:2171–2178
101. Rossi G, Donisi A, Casari S et al (1998) Effects of recombinant granulocyte colony-stimulating factor (G-CSF) in patients treated with ProMACE-CytaBOM for HIV-related non-Hodgkin's lymphoma (NHL). Haematologica 83:317–322
102. Tirelli U, Vaccher E (1994) Economic and clinical evaluation of therapy of HIV-related non-Hodgkin's lymphoma with chemotherapy and granulocyte colony-stimulating factor (G-CSF). Eur J Cancer 30:1589–1590
103. Foote M (2004) Use of hematopoietic growth factors in AIDS-related malignancies. In: Morstyn G, Foote M, Lieschke G (eds) Hematopoietic growth factors in oncology. Humana Press, Totowa, NJ, pp 357–371
104. Anderlini P, Donato M, Chan KW et al (1999) Allogeneic blood progenitor cell collection in normal donors after mobilization with filgrastim: the M.D. Anderson Cancer Center experience. Transfusion 39:555–560
105. Mueller BU, Burt R, Gulick L, Jacobsen F, Pizzo PA, Horne M (1995) Disseminated intravascular coagulation associated with granulocyte colony-stimulating factor therapy in a child with human immunodeficiency virus infection. J Pediatr 126:749–752
106. Hervé Z, Louis Z, Hervé H, Stéphane H, Isabelle G, Michel C (1995) Probable G-CSF-induced hepatitis and pancreatitis in an HIV-seropositive patient. J Hepatol 22:596–597
107. D'Souza A, Jaiyesimi I, Trainor L, Venuturumili P (2008) Granulocyte colony-stimulating factor administration: adverse events. Transfus Med Rev 22:280–290
108. van der Wouw PA, van Leeuwen R, van Oers RH, Lange JM, Danner SA (1991) Effects of recombinant human granulocyte colony-stimulating factor on leucopenia in zidovudine-treated patients with AIDS and AIDS related complex, a phase I/II study. Br J Haematol 78:319–324
109. Engles-Horton LL, Skowronski C, Mostashari F, Altice FL (1999) Clinical guidelines and pharmacist intervention program for HIV-infected patients requiring granulocyte colony-stimulating factor therapy. Pharmacotherapy 19:356–362
110. Clumeck N, Wit SD, Hermans P, Franchioly P, Sommereijns B (1992) Recombinant granulocyte colony-stimulating factor (rG-CSF) in HIV patients with zidovudine related neutropenia. J Nutr Sci Vitaminol (Tokyo) Spec no.: 353–356
111. Garavelli PL (1992) Efficacy of granulocyte colony-stimulating factor (G-CSF) on neutropenia in zidovudine-treated patients with AIDS and ARC: a preliminary report. Haematologica 77:293–294
112. Goebel FD, Bogner JR, Matuschke A, Heinrich B, Kronawitter U (1992) Effects of granulocyte-colony-stimulating factor in neutropenic AIDS-patients. J Nutr Sci Vitaminol (Tokyo) Spec No. : 357–360

113. Hengge UR, Brockmeyer NH, Goos M (1992) Granulocyte colony-stimulating factor treatment in AIDS patients. Clin Investig 70:922–926
114. Jacobson MA, Heard SE (1992) Ganciclovir with recombinant methionyl human granulocyte colony-stimulating factor for treatment of cytomegalovirus disease in AIDS patients. AIDS 6:515–516
115. Mueller BU, Jacobsen F, Butler KM, Husson RN, Lewis LL, Pizzo PA (1992) Combination treatment with azidothymidine and granulocyte colony-stimulating factor in children with human immunodeficiency virus infection. J Pediatr 121:797–802
116. Garavelli PL, Berti P (1993) Efficacy of recombinant granulocyte colony-stimulating factor in the long-term treatment of AIDS-related neutropenia. AIDS 7:589–590
117. Zuccotti GV, Flumine P, Locatelli V, Banderali G, Riva E (1993) Growth factors and HIV-infection in children. J Int Med Res 21:342–345
118. Hermans P, Franchioly P, Thioux C, Gray SJ, Vannerom H, Clumeck N (1996) Minimum effective dose and duration to reverse neutropenia in non-cancer patients with advanced HIV disease. AIDS 10:1050–1051
119. Luzzi GA, Jones BJ (1996) Treatment of neutropenic oral ulceration in human immunodeficiency virus infection with G-CSF. Oral Surg Oral Med Oral Pathol Oral Radiol Endod 81:53–54
120. Perrella O, Finelli E, Perrella A, Tartaglia G, Scognamiglio P, Scalera G (1996) Combined therapy with zidovudine, recombinant granulocyte colony stimulating factors and erythropoietin in asymptomatic HIV patients. J Chemother 8:63–66
121. Stricker RB, Goldberg B (1996) Increase in lymphocyte subsets following treatment of HIV-associated neutropenia with granulocyte colony-stimulating factor. Clin Immunol Immunopathol 79:194–196
122. Zuccotti GV, Plebani A, Biasucci G et al (1996) Granulocyte-colony stimulating factor and erythropoietin therapy in children with human immunodeficiency virus infection. J Int Med Res 24:115–121
123. Ambati BK, Perlman DC, Salomon N (1998) Outcomes of granulocyte colony-stimulating factor or granulocyte-macrophage colony-stimulating factor use in neutropenic patients infected with human immunodeficiency virus. Int J Infect Dis 3:70–75
124. Keiser P, Rademacher S, Smith JW, Skiest D, Vadde V (1998) Granulocyte colony-stimulating factor use is associated with decreased bacteremia and increased survival in neutropenic HIV-infected patients. Am J Med 104:48–55
125. Mullane KM. (1998) Rapid clearance of *Mycobacterium avium* from the blood of nonneutropenic AIDS patients treated with G-CSF (Filgrastim) and antimycobacterial therapy. International congress on infectious diseases. Boston
126. Dubreuil-Lemaire ML, Gori A, Vittecoq D et al (2000) Lenograstim for the treatment of neutropenia in patients receiving ganciclovir for cytomegalovirus infection: a randomised, placebo-controlled trial in AIDS patients. Eur J Haematol 65:337–343
127. Aladdin H, Ullum H, Katzenstein T, Gerstoft J, Skinhoslash P, Klarlund B (2005) Immunological and virological changes in antiretroviral naïve human immunodeficiency virus infected patients randomized to G-CSF or placebo simultaneously with initiation of HAART. Scand J Immunol 51:520–525
128. Aladdin H, Ullum H, Nielsen SD et al (2000) Granulocyte colony-stimulating factor increases $CD4^+$ T cell counts of human immunodeficiency virus-infected patients receiving stable, highly active antiretroviral therapy: results from a randomized, placebo-controlled trial. J Infect Dis 181:1148–1152
129. Campbell TB, Sevin A, Coombs RW et al (2000) Changes in human immunodeficiency virus type 1 virus load during mobilization and harvesting of hemopoietic progenitor cells. Blood 95:48–55
130. Nielsen, SøRensen, Aladdin et al (2000) The effect of long-term treatment with granulocyte colony-stimulating factor on hematopoiesis in HIV-infected individuals. Scand J Immunol 52:298–303

131. Davidson M, Min Y-I, Holbrook JT et al (2002) Influence of filgrastim (granulocyte colony-stimulating factor) on human immunodeficiency virus type 1 RNA in patients with cytomegalovirus retinitis. J Infect Dis 186:1013–1018
132. Campbell TB, Rapaport E, Schooley RT, Kuritzkes DR (2004) Increased replication of HIV-1 minor variants during hematopoietic stem-cell mobilization with filgrastim. J Infect Dis 190:257–266
133. Neid JM, Schooley RT, Campbell TB (2004) Stimulation of Kaposi's sarcoma-associated herpesvirus viremia during hematopoietic stem cell mobilization with filgrastim. Am J Hematol 77:410–412
134. Pau AK, McLaughlin MM, Hu Z, Agyemang AF, Polis MA, Kottilil S (2006) Predictors for hematopoietic growth factors use in HIV/HCV-coinfected patients treated with peginterferon alfa 2b and ribavirin. AIDS Patient Care STDS 20:612–619

Recombinant Human G-CSF Enhances Recovery and Improves Survival from Severe Radiation-Induced Myelosuppression

Ann M. Farese, Melanie V. Cohen, and Thomas J. MacVittie

1 Introduction

Granulocyte colony stimulating factor (G-CSF): Its demonstrated efficacy in animal models of radiation-induced hematopoietic injury sets the stage for its use as a key medical countermeasure (MCM) in the event of nuclear terrorism.

It is well known that recombinant human granulocyte colony-stimulating factor (rHuG-CSF) is approved by the United States Food and Drug Administration (FDA) to treat chemotherapy-induced neutropenia. The FDA relied on the large literature database of nonclinical and clinical studies, as well as successful clinical trials to support approval of rHuG-CSF for this condition. rHuG-CSF also has utility for another indication: radiation-induced myelosuppression and treatment of potentially lethally irradiated personnel. rHuG-CSF showed significant efficacy in ameliorating the radiation-induced effects of myelosuppression. It enhanced survival from lethal radiation exposure within the hematopoietic subsyndrome (H) of the acute radiation syndrome (H-ARS) in all animal models evaluated. The database is substantial and consistent in demonstrating efficacy in mouse, canine, and nonhuman primate models of radiation-induced myelosuppression and lethality [1–10]. rHuG-CSF, when administered early after high-dose, total-body irradiation (TBI) enhanced the requisite recovery of bone marrow-derived myeloid cell progenitors and neutrophils within a critical, clinically manageable period of time. The recovery of the myeloid progenitors coupled with the reduced maturation time, increased survival, and early release of neutrophils to the peripheral

A.M. Farese • M.V. Cohen
Department of Radiation Oncology, School of Medicine, University of Maryland, 10 South Pine Street, MSTF 6-34D, Baltimore, MD 21201, USA
e-mail: afarese@som.umaryland.edu; melveirs@yahoo.com

T.J. MacVittie (✉)
Department of Radiation Oncology, School of Medicine, University of Maryland, 10 South Pine Street, MSTF 6-34E, Baltimore, MD 21201, USA
e-mail: tmacvittie@som.umaryland.edu

G. Molineux et al. (eds.), *Twenty Years of G-CSF*, Milestones in Drug Therapy, DOI 10.1007/978-3-0348-0218-5_19, circulation reduced the duration of severe neutropenia and minimized the risk of bacteremia and sepsis [11–14].

World events over the past decade have emphasized the increasing threat of nuclear terrorism. This threat underscores the requirement for a medical countermeasure(s) against the myelosuppressive and potentially lethal effects consequent to high-dose radiation exposure in this scenario. Treatment strategies for personnel exposed to moderate or lethal doses of radiation have been the subject of several international conferences during the past 20 years [15–19]. A consensus on treatment of severely irradiated personnel was presented in a 1993 Conference on Treatment of Radiation Injuries and again in a more recent publication [18, 20]. Despite the noted threat and the substantial database demonstrating the efficacy of rHuG-CSF in lethally irradiated animal models, rHuG-CSF has yet to be approved by the FDA as a medical countermeasure. Due to the ethical constraints of conducting a clinical trial in lethally irradiated people, approval of a medical countermeasure by the FDA in these cases will only be granted under the requirements set forth by the "Animal Rule" [21].

The need for an FDA-approved medical countermeasure (MCM) for the treatment of severely irradiated personnel (as well as those exposed to chemical or biologic agents) has resulted in the FDA generating published guidelines that define a path to MCM licensure when efficacy studies in humans are not feasible or ethical. These guidelines are known as the "Animal Rule" [21]. Under the Animal Rule, the FDA will rely on data from relevant animal species to provide evidence of treatment effectiveness. The FDA also published a draft document entitled "Guidance for Industry, Animal Models – Essential Elements to Address Efficacy under the Animal Rule" [22]. Together, these documents provide an outline and description of the criteria to be satisfied for successful use of the FDA Animal Rule to gain approval for candidate medical countermeasure to treat lethally irradiated personnel.

rHuG-CSF is positioned as the primary medical countermeasure to be acquired by the Centers for Disease Control and Prevention (CDC) for placement in the Strategic National Stockpile (SNS). rHuG-CSF is considered a leading MCM for the SNS due to the substantial database in nonclinical models of radiation-induced myelosuppression and lethality, as well as the extensive database of clinical studies in normal and myelosuppressed subjects demonstrating a successful safety profile spanning 2 decades.

In light of rHuG-CSF's status as a desired MCM for the SNS, this review first examines the pathophysiology of lethal radiation exposure and summarizes the nonclinical model database for the rodent, canine, and nonhuman primate. Also, the human radiation experience is discussed, as well as the potential use of rHuG-CSF as an MCM for treating individuals who may have been lethally irradiated in a nuclear terrorist event.

2 The Pathophysiology of Lethal Radiation Exposure: Radiation-Induced Neutropenia and Host Resistance to Sepsis

The bone marrow is the most radiation-sensitive organ in the body. Radiation-induced pathophysiology after a single, acute dose of ionizing radiation in the species-dependent, lethal range of 2.00–10.00 Gy results in H-ARS [23–25].

The lethal H-ARS is characterized by the frequent incidence and severity of neutropenia, thrombocytopenia, and lymphopenia that are dose and time dependent. The respective lineage-specific cytopenia is evident within days to weeks after exposure and is a result of the marked loss of bone marrow-derived myeloid progenitor cells within the initial 24 hours after irradiation as well as the circulating half-life of the respective lineage-specific cells, e.g., neutrophils, platelets, and lymphocytes.

Recovery from the lethal effects of H-ARS requires at least three key events to occur. First, hematopoietic stem and progenitor cells must initiate self-renewal required for short- and long-term marrow reconstitution and production of functional neutrophils and platelets. Second, regeneration of functional neutrophils and platelets, which are required for prevention of morbidity and mortality associated with consequent sepsis and spontaneous hemorrhage, must occur. Third, the initiation of progenitor self-renewal and production of functional end cells must occur within a critical, clinically manageable period of time, which is defined by the duration of severe neutropenia and thrombocytopenia and the capacity of health facilities and professionals to medically manage lethally irradiated personnel.

The neutrophil is the key cellular component in nonspecific host resistance against opportunistic pathogens. Neutropenia increases the risk of life-threatening infection and the relationship between the degree and duration of neutropenia and mortality is well documented [26–31]. The incidence of infection is inversely proportional to the absolute neutrophil count (ANC). The survival of neutropenic patients with presumed bacterial infection is greatest after resolution of neutropenia and restoration of neutrophil numbers and function [26, 32, 33].

3 Severe Neutropenia Consequent to Lethal Doses of TBI Determine the Degree of Medical Management Required

Severe neutropenia, an ANC $< 0.5 \times 10^9$/L, is predictive of impending infection. The most severe infections and all bacteremias occur when the patient has an ANC $< 0.1 \times 10^9$/L, and are most prevalent in patients with prolonged periods of bone marrow aplasia [26, 28, 30]. Lethal radiation exposure induces significant bone marrow myelosuppression with subsequent profound and prolonged neutropenia. These conditions predispose the patient to infection and thereby necessitate the prophylactic administration of antibiotics in afebrile but severely neutropenic, lethally irradiated personnel. Several organizations and a meta-analysis suggest that afebrile patients who are neutropenic and are expected to be profoundly neutropenic (ANC $< 0.1 \times 10^9$/L) should have prompt empirical antibiotic therapy [28, 29, 31, 34–36].

Further evidence for initiating antibiotic therapy in exposed personnel comes from examining the data after several cases of accidental exposure, in addition to the reactor explosion at the Chernobyl Nuclear Power Plant in the former Soviet

Union. These cases provide a relevant database for assessing the effects of approximate uniform, sublethal, and lethal doses of radiation on healthy young adults [25, 37]. The most relevant parameter that correlates with radiation dose causing severe H-ARS in humans is the day on which peripheral neutrophils decreased to 0.5×10^9/L, called the "d500" by the authors [37, 38]. If the patient's "d500" score was <14, it corresponded to a total body exposure of 5.00–6.00 Gy, which is an approximate lethal dose (LD)50/60 (50% LD within 60 days) of the exposed population administered medical management [23, 24, 39].

Further evaluation of the neutrophil kinetics after irradiation from 18 patients who received an estimated TBI of 4.70–8.30 Gy showed that each person experienced an ANC < 0.1×10^9/L within a range of 1–4 days after their respective "d500" occurred [37]. It was reported that although the incidence of fever and infection coincided exactly with the duration of neutropenia in the radiation exposure range of 4.00–5.00 Gy, the correlation was even more consistent in the dose range >5.00 Gy. These data, in addition to effects noted in lethal-radiation exposure of canine and nonhuman primate models, justifies the use of prophylactic administration of antibiotics in afebrile but neutropenic personnel that are estimated to have received lethal doses of TBI [28, 30, 34–36].

4 The Value of rHuG-CSF as a Medical Countermeasure: G-CSF and Neutrophil Function

A critical property of an effective medical countermeasure against H-ARS is its ability to enhance the production of myeloid progenitors to produce mature, functional neutrophils. Because receptors for G-CSF exist on all cells of the granulocyte lineage, it is an ideal candidate for MCM. The value of rHuG-CSF as a countermeasure against radiation-induced neutropenia extends beyond its ability to increase the survival, self-renewal, differentiation, and amplification of myeloid progenitor cells to produce mature neutrophils.

G-CSF has been shown to promote enhanced cellular functions critical to host defense against bacteria. For example, when rHuG-CSF is administered either as a single agent or in combination with antibiotic therapy, bactericidal activity was improved [40–43]. The augmentation of neutrophil function after rHuG-CSF administration has been demonstrated in several models of normal and radiation- or chemotherapy-induced-neutropenic animals. rHuG-CSF administration stimulated an increase in phagocytosis and hydrogen peroxide formation and protected cyclophosphamide-induced neutropenic mice from systemic challenge with numerous strains of bacteria [44, 45]. The treatment efficacy of rHuG-CSF has been demonstrated in a mouse model of intra-abdominal infection via cecal ligation and puncture [46]. Similar results were obtained in other studies when rHuG-CSF was administered to neutropenic mice experiencing bacterial and fungal infections [40, 43, 47, 48].

Furthermore, the neutrophil is an interactive cell and may have a regulatory role within the irradiated-tissue microenvironment, rather than one focused only on the innate microbial host defense. Neutrophils produce cytokines, cytokine antagonists, or binding proteins in response to stimulatory cytokines or endotoxin [49–51]. The neutrophil is also capable of producing and shedding both tumor-necrosis factor (TNF)-α and -β receptors, the (interleukin) IL-1 receptor, and the IL-1 receptor antagonist [52–55]. TNF and IL-1 are known proinflammatory cytokines that may be intimately involved in multiple organ system failure [56–59]. These properties underscore the autocrine and paracrine nature of the neutrophil and stress the potential ability of the neutrophil to orchestrate host defense and inflammation within the tissue site.

In summary, potentially lethal doses of TBI cause severe and prolonged neutropenia. Adequate neutrophil production within a critical, medically manageable period of time is essential for survival after lethal TBI. The risk of infection is inversely related to the neutrophil count; therefore, antibiotics should be administered to afebrile, neutropenic personnel exposed to potentially lethal doses of TBI. Because it takes 14 days for the bone marrow to produce a mature neutrophil, the use of prophylactic antibiotics is warranted to allow the patient to competently protect against bacterial infections until their ANC recovers to a sufficient level. rHuG-CSF will stimulate lineage-specific granulopoietic progenitors and increase production of neutrophils within this specified timeframe to reduce the duration of severe neutropenia and increase survival.

5 rHUG-CSF Enhances Recovery and Survival in Animal Models of Severe Radiation-Induced Bone Marrow Suppression and Neutropenia

The value of rHuG-CSF in the treatment of severe, radiation-induced bone marrow myelosuppression lies in its ability to modulate the expansion of marrow granulopoiesis and the consequent production and function of the neutrophil, the key cell in the hosts' nonspecific response to opportunistic infection. The database is consistent and substantial in animal models of radiation-induced myelosuppression. rHuG-CSF has been shown to effectively stimulate marrow granulopoiesis and decrease the duration of neutropenia and recovery time of neutrophils in irradiated mice, canines, and nonhuman primates [1–8, 10, 60–64].

5.1 Rodent Studies

rHuG-CSF has been shown to enhance recovery from severe neutropenia and increase survival in mice when administered early, after high-dose TBI [2, 4, 6, 7].

When administered within a range of 1 hour to 3 days after high doses of TBI, rHuG-CSF significantly enhanced the survival and recovery of bone marrow, splenic cellularity, and marrow-derived granulocyte progenitor cells relative to the control cohort. In studies using 8.00 Gy TBI, control mice experienced a 27% survival, whereas the cohorts administered rHuG-CSF after TBI either within hours through day 12, or day 1 through day 12, or day 3 through day 12 after TBI, had 95%, 70%, or 57% survival, respectively (Table 1) [2, 4, 64]. rHuG-CSF was administered on consecutive days from 2 hours to 6 days after lethal doses of TBI at 8.50 Gy, 9.00 Gy, and 9.50 Gy (Table 1). Survival was increased in the rHuG-CSF-treated mice versus the control mice from 44.0 to 83.3%, 14.8 to 54.5%, and 0 to 45.8%, respectively [7]. rHuG-CSF was ineffective when administered after 10.50 Gy TBI (Table 1). The 10.50 Gy total-body exposure likely encroached on the gastrointestinal subsyndrome in this mouse strain, for which rHuG-CSF will be ineffective. These results provide insight into the capability of rHuG-CSF to stimulate recovery after dose-dependent destruction of granulocyte progenitor cells and the limited, clinically relevant time available to produce threshold numbers of functional neutrophils [6, 7]. Results from rodent models of lethal TBI consistently demonstrate that therapeutic administration of rHuG-CSF accelerates hematopoietic regeneration, enhances survival, and is effective when administered within hours to days after exposure.

Table 1 Summary of survival in mice at 30 days after irradiation

Exposure (Gy)	Treatment	Schedule	Percent survival
8.00	Control		27
	rHuG-CSF	1 hour through day 12	95
		Day 1 through day 12	70
		Day 3 through day 12	57
8.50	Control		44.0
	rHuG-CSF	2 hours through day 6	83.3
9.00	Control		14.8
	rHuG-CSF	2 hours through day 6	54.5
9.50	Control		0
	rHuG-CSF	2 hours through day 6	45.8
10.50	Control		0
	rHuG-CSF	2 hours through day 6	0

The data provide evidence that rHuG-CSF offers a survival benefit for mice exposed to lethal levels of radiation. The table summarizes the survival data for mice (C3H/HeN) exposed to 8.00 Gy and administered rHuG-CSF daily, through day 12 post-irradiation when rHuG-CSF was initiated either within hours of irradiation, or beginning on either day 1 or day 3 post-irradiation [2, 4, 64]. Additionally, a radiation dose response is summarized. Mice (BDF_1) were irradiated at 8.50, 9.00, 9.50, and 10.50 Gy and rHuG-CSF was administered to all radiation cohorts beginning at 2 hours after irradiation and then daily through day 6 post-irradiation [4, 5]

5.2 Canine Studies

Experiments in large animal models, such as the canine and nonhuman primate, permit the effective use of clinically relevant medical management (supportive care) consisting of sequential treatment of each animal with the appropriate administration of antibiotics, blood products, fluids, analgesics, and antiemetics. The administration of supportive care in the neutropenic and thrombocytopenic animal provides the host with additional survival time required for spontaneous endogenous and rHuG-CSF-induced regeneration of hematopoietic progenitor cells. All canine models reported herein, of either sublethal or lethal TBI, were treated with supportive care as defined by the respective experimental protocol. Extensive investigation of rHuG-CSF efficacy in enhancing recovery from TBI was conducted primarily in two laboratories. These studies provided significant evidence that when rHuG-CSF was administered early (4–24 hours) after lethal TBI and continued once daily through the desired effect on neutrophil recovery, a survival benefit was observed. One set of studies investigated the therapeutic administration of rHuG-CSF in canines exposed to TBI over a range of lethal doses from 3.50 to 6.00 Gy cobalt-60 gamma TBI (Table 2). All animals received supportive care as per proscribed study protocols. rHuG-CSF was administered once daily from day 1 to day 14 to 23 after TBI depending on the dose of TBI and the response to treatment (e.g., recovery time of neutrophils). The LD50/30 for the control cohorts, administered supportive care alone was 3.38 Gy [10, 65]. Treatment with rHuG-CSF early, within 20 hours after TBI and daily through recovery of neutrophils to 1.0×10^9/L plus supportive care, increased the LD50/30 to 4.88 Gy resulting in a dose reduction factor

Table 2 Summary of survival at 60 days after irradiation

Exposure (Gy)	Treatment	Survival at 60 days	
		n/*N*	%
3.50	Control	12/19	63
	Filgrastim 10 μg/kg, BID	1/2	50
	Filgrastim 10 μg/kg, QD	6/7	85
	rcG-CSF 5 μg/kg, QD	4/6	67
	rcG-CSF 1 μg/kg, QD	2/3	67
4.00	Control	1/16	6
	Filgrastim 10 μg/kg, QD	8/8	100
4.50	Control	0/5	0
	Filgrastim 10 μg/kg, QD	3/7	43
5.00	Control	0/13	0
	Filgrastim 10 μg/kg, QD	6/8	75
	rcG-CSF 5 μg/kg, QD	6/9	63
6.00	Control	0/4	0
	Filgrastim 10 μg/kg, QD	2/10	20

The data presented in **Table 2** provide evidence that G-CSF (either Filgrastim or recombinant canine [rc] G-CSF) offers a survival benefit for canines exposed to lethal levels of radiation. **Table 2** summarizes the survival data for dogs exposed to 3.50 Gy, 4.00 Gy, 4.50 Gy, 5.00 Gy, and 6.00 Gy. Improved survival was observed at all exposure levels from 3.50 to 6.00 Gy[10, 66].

of 1.44 relative to the LD50/30 of 3.38 Gy of control cohorts treated with supportive care alone [10]. Canines were exposed to TBI and both treatment groups, control and rHuG-CSF, received supportive care per the experimental protocol. Cohorts of control animals were exposed to 3.50, 4.00, 4.50, 5.00, or 6.00 Gy and experienced 63.2%, 6.3%, 0%, 0%, and 0% survival, respectively (Table 2). The rHuG-CSF-treated cohorts exposed to the same radiation doses experienced survival to 77.8%, 100%, 42.8%, 75.0%, and 20.0% relative to the respective control cohorts.

These data are concordant with contemporary studies [1, 5, 66]. These authors noted that 3.6% of canines survived when exposed to 4.00 Gy TBI plus supportive care alone, whereas treatment with rHuG-CSF plus supportive care increased survival to 80.0%. Treatment of canines exposed to the supralethal (100% lethal) dose of 5.00 Gy increased survival to 30% but was not effective above this radiation dose, in contrast to that noted in the previous study.

Collectively, these data suggest that survival could be enhanced by administration of rHuG-CSF plus supportive care after 100% lethal doses of TBI when hematopoietic progenitor cells have been reduced to levels otherwise incompatible with survival. rHuG-CSF was necessary to stimulate regeneration of the lineage-specific progenitor cell compartment, as well as stimulate the production of functional neutrophils within the critical, clinically manageable period of time dictated by the efficacy of supportive care. As with the mouse experiments, there is a threshold dose of TBI (approximately 6.00 Gy) at which the administration of rHu G-CSF in concert with supportive care will not stimulate granulopoiesis within the critical time required for survival of the lethally irradiated canine.

5.3 Nonhuman Primate Studies

rHuG-CSF administration has demonstrated efficacy in enhancing the recovery of neutrophils and survival in nonhuman primates exposed to sublethal (low- to mid-lethal) doses of TBI [8, 9, 67–69]. Several studies were conducted at three research sites by two primary groups of investigators. All of the studies administered supportive care equivalent to that described in the canine studies. Specifically, all control cohorts were administered supportive care alone while the treatment cohorts were administered supportive care plus rHuG-CSF. Farese et al. [8, 9, 70] used three nonhuman primate models of high-dose TBI:

- 7.00 Gy of TBI with Co-60 gamma radiation
- 4.50 Gy of TBI with a 0.6:1 mixed fission neutron: gamma radiation from a TRIGA reactor operated at steady state dose rate of 20 rad/min
- 6.00 Gy of TBI with 250 kVp X-radiation to investigate the ability of rHuG-CSF to enhance recovery from radiation-induced myelosuppression

rHuG-CSF (10 μg/kg) was administered by subcutaneous injection within 20 hours after TBI and continued daily until circulating neutrophils recovered to a predetermined level, e.g., 0.5×10^9/L, 1.0×10^9/L, or 2.0×10^9/L. In nonhuman primates exposed to 6.00 Gy of X-radiation, the administration of rHuG-CSF significantly decreased the mean duration of neutropenia and recovery time to an ANC $>0.5 \times 10^9$/L or 2.0×10^9/L. The duration of neutropenia (ANC $<0.5 \times 10^9$/L) was significantly reduced from 15.5 days to 12.3 days by the administration of rHuG-CSF compared with the control cohort [8]. The ANC nadir was not significantly modified with administration of rHuG-CSF, whereas the time to recovery of ANC $>0.5 \times 10^9$/L was reduced from 24.0 days in the control cohort to 18.0 days with rHuG-CSF treatment. In a similar study, nonhuman primates were exposed to 7.00 Gy of TBI with Co-60 gamma radiation. rHuG-CSF administration significantly reduced the duration of neutropenia compared with the control from 18.8 days to 9.8 days, the recovery time for ANC to reach 0.5×10^9/L was reduced from 24.4 days in the control cohort to 16.8 days in the rHuG-CSF-treated cohort [69]. In this case, the ANC nadir was significantly increased to an ANC of 0.53×10^9/L relative to 0.19×10^9/L for the control cohort. The third previously unpublished study compared the efficacy of rHuG-CSF administered to nonhuman primates after exposure to 4.50 Gy of mixed fission neutron:gamma radiation. Administration of rHuG-CSF reduced the duration of neutropenia from 14.5 days in the control cohort to 12.0 days; the respective nadirs were not different from each other while the recovery of ANC to $>0.5 \times 10^9$/L or $>2.0 \times 10^9$/L was reduced from the control levels of 20.9 days and 23.7 days to 17.0 days and 19.3 days after rHuG-CSF administration, respectively.

A set of studies assessed the treatment efficacy of rHuG-CSF administered to nonhuman primates exposed to 5.00 Gy TBI 250 kVp X-radiation [67, 68]. The animals were divided into two cohorts that received rHuG-CSF at either 5 or 10 μg/kg/day initiated 1 day after TBI and continued for 21 or 14 consecutive days, respectively. The control cohort required approximately 22 days to reach an ANC $>0.5 \times 10^9$/L. rHuG-CSF treatment irrespective of dose or schedule, significantly improved the ANC recovery time to $>0.5 \times 10^9$/L by 3–5 days. Irradiated nonhuman primates, in all studies, experienced severe neutropenia (ANC $< 0.5 \times 10^9$/L) that ranged from 12 to 28 days duration after TBI. rHuG-CSF, when administered within 1 day after TBI and continued daily until evidence of neutrophil recovery, significantly enhanced recovery of neutrophil-related parameters in rhesus macaques exposed to high, myelosuppressive doses of TBI.

6 Delayed Administration of rHuG-CSF After High-Dose Irradiation in Mouse, Canine, and Nonhuman Primate Models

The common protocol for rHuG-CSF therapy is to initiate treatment within several hours up to 24 hours after TBI or chemotherapy-induced myelosuppression. Therapy is continued with daily administration to ensure regeneration of marrow-derived progenitor cells and earlier production of neutrophils. The database in all three animal species does not provide a definitive answer relative to whether treatment efficacy is diminished as the interval between TBI and initiation of treatment increases. The mouse models provide a more consistent conclusion relative to efficacy of a treatment interval relevant to a terrorist scenario. Studies suggested that the earlier and more frequently rHuG-CSF is administered, the earlier the recovery of hematopoiesis [2, 7].

Testing the delayed administration of rHuG-CSF after lethal doses of TBI in canines has led to conflicting results [1, 70]. Schuening et al delayed administration of rHuG-CSF for 7 days to otherwise lethally irradiated canines (LD99/30) compared with initiating daily treatment within 2 hours of TBI [1]. All canines in the "delayed" rHuG-CSF protocol succumbed to hematopoietic aplasia compared with one of five canines in the "early" rHuG-CSF protocol. In contrast, two other studies, one in canines and one in nonhuman primates, have shown that a similarly delayed protocol did not alter the hematopoietic response or survival relative to the "early" rHuG-CSF protocol [70, 71]. The lack of a database on the efficacy of delayed administration of rHuG-CSF indicates further studies are required to test alternate treatment schedules where initiation of rHuG-CSF therapy occurs 48 or 72 hours after TBI. These delayed administration studies will provide critical information in support of rHuG-CSF approval as an MCM under the FDA-Animal Rule for use in a nuclear terrorist scenario.

7 The Human Experience: Accidental, Acute Radiation Exposure and Use of Medical Management and Hematopoietic Growth Factors for Treatment

Data generated from accidental exposure of humans to ionizing radiation has served as the source of information to determine the human radiation dose response relationship for the occurrence of the major sequelae of the acute radiation syndrome and its modification by medical management and administration of hematopoietic growth factors [20, 23–25, 38, 72, 73].

The lethality after TBI is directly proportional to the dose received. Reviews of the cumulative data on human radiation exposure suggests that the LD50/60 is approximately 3.26–4.50 Gy in the absence of medical management [23, 24, 39, 72]. The LD50/60 increases to approximately 6.00–7.00 Gy when medical management (consisting of fluids, antibiotics, antiviral agents, and blood products) is provided [23–25, 37]. Note that antiviral agents were not used in treatment protocols in above-mentioned animal models.

8 The Reality of the Radiation Exposure Scenario in an Accident or Terrorist Scenario: Its Relationship to Effective Treatment with rHuG-CSF

The reality of the accidental or nuclear terrorist scenario is that the exposure environment will be ill defined and uncontrolled. Body position, affordable shielding, and distance from the source will result in a nonuniform and heterogeneous exposure to any group of individuals. Additionally, the time interval between exposure and initiation of treatment protocols will be less than optimal (e.g., >24 hours). However, a hopeful feature of the accident or terrorist scenario is that the above characteristics forecast a highly variable dose distribution to the body with likely sparing of bone marrow, and in turn hematopoietic stem and progenitor cells. This feature is an encouraging aspect relative to treatment with supportive care and hematopoietic growth factors. Effective medical management will provide additional time after the exposure for endogenous regeneration of the surviving hematopoietic progenitor cells. Hematopoietic progenitors will initiate spontaneous self-renewal within their niche in response to endogenous hematopoietic growth factors as well as be responsive to treatment with rHuG-CSF.

The database from the Chernobyl accident and other accidents is invaluable in demonstrating the efficacy of medical management administered to lethally irradiated personnel with an estimated nearly uniform exposure [25, 37, 74–76]. The enhanced survival of personnel exposed to potentially high, lethal doses of radiation with medical management alone, suggests a marked increase in potential for treating lethally irradiated personnel with rHuG-CSF and other MCM. Effective medical management provides valuable replacement therapy, e.g., antibiotics, blood transfusions, and fluids during periods of neutropenia, thrombocytopenia, and dehydration, respectively. Bacteremia, sepsis, and hemorrhage can be delayed or averted depending on radiation dose and severity and duration of consequent myelosuppression. The data provided for the Chernobyl victims clearly emphasized the positive role of medical management in patients with severe acute radiation syndrome and the minimal role to be played by bone marrow or stem cell transplantation [38, 77–80].

Before Chernobyl, only six known patients with acute radiation syndrome with severe myelosuppression and medical management demonstrated early recovery of hematopoiesis. These survivors were estimated to have received uniform irradiation of no more than 6.00–8.00 Gy. In 28 Chernobyl patients with very severe myelosuppression, 14 demonstrated spontaneous recovery of hematopoiesis. These data suggested that hematopoiesis could recover after total-body exposure up to an estimated 8.00 Gy, well above the calculated LD50/60 value [23, 25, 81]. It was also noted that fever and infection coincided with the initiation and duration of the neutropenic periods in exposure doses of 4.00–5.00 Gy, and that these signs were more aggressive in all patients in an estimated 5.00–6.00 Gy range. It is highly probable that the use of hematopoietic growth factors, such as rHuG-CSF, in these patients would have decreased the severity of myelosuppression and further diminished the risk of infection and morbidity.

9 Conclusion

This review has linked the lethal doses of TBI to severe hematopoietic myelosuppression, consequent prolonged and severe neutropenia, increased risk for lethal bacteremia and sepsis, and the beneficial effect of administering rHuG-CSF to enhance neutrophil recovery and survival of lethally irradiated personnel. These effects are similar to those of high-dose chemotherapy except in degree of myelosuppression and lethal risk. We note that the American Society of Clinical Oncology (ASCO) and the European Organization for Research and Treatment of Cancer (EORTC) have published evidence-based clinical practice guidelines on the use of hematopoietic growth factors for chemotherapy-induced myelosuppression as a primary risk factor for infection-related morbidity and mortality, as well as dose-limiting toxicity and risk of developing severe neutropenia [82, 83] (see chapter "Practice Guidelines for the Use of rHuG-CSF in an Oncology Setting" by Saraf and Ozer.). ASCO extended their recommendation for the management of patients exposed to lethal doses of total-body radiotherapy, including the prompt use of rHuG-CSF or pegylated rHuG-CSF. The EORTC guidelines recommended the use of rHuG-CSF and pegylated rHuG-CSF to prevent febrile neutropenia and febrile neutropenia-related complications. The CDC also acknowledges the potential benefit of filgrastim administration to people exposed to very high radiation doses [84]. The United States Department of Health and Human Services has published a guidance entitled "Radiation Emergency Medical Management" [85]. This document recommends treatment with filgrastim at 5 μg/kg/day, administered subcutaneously in combination with medical management for individuals exposed to TBI $\geq$2.00 Gy or those who have a diagnosis of the hematopoietic syndrome as manifest by neutropenia (ANC $<0.5 \times 10^9$/L).

References

1. Schuening FG, Storb R, Goehle S et al (1989) Effect of recombinant human granulocyte colony-stimulating factor on hematopoiesis of normal dogs and on hematopoietic recovery after otherwise lethal total body irradiation. Blood 74:1308–1313
2. Patchen ML, MacVittie TJ, Solberg BD, Souza LM (1990) Therapeutic administration of recombinant human granulocyte colony stimulating factor accelerated hemopoietic regeneration and enhances survival in a murine model of radiation-induced myelosuppression. Int J Cell Cloning 8:107–122
3. MacVittie TJ, Monroy RL, Patchen ML, Souza LM (1990) Therapeutic use of recombinant human G-CSF in a canine model of sublethal and lethal whole-body irradiation. Int J Radiat Biol 57:723–736
4. Patchen ML, Fischer R, MacVittie TJ (1993) Effects of combined administration of IL-6 and G-CSF on recovery from radiation-induced hemopoietic aplasia. Exp Hematol 21:338–344
5. Schuening FG, Appelbaum FR, Deeg HJ et al (1993) Effects of recombinant canine stem cell factor, a c-kit ligand and recombinant granulocyte colony stimulating factor on hematopoietic recovery after otherwise lethal total body irradiation. Blood 81:20–26
6. Tanikawa S, Nakao I, Tsuneska K, Nobio N (1989) Effects of recombinant granuloctye colony-stimulating factor (rG- CSF) and recombinant granuloctye-macrophage colony-stimulating factor (rGM-CSF) on acute radiation hematopoietic injury in mice. Exp Hematol 17:883–888
7. Tanikawa S, Nose M, Yoshiro A, Tsuneoka K, Shikita M, Nara N (1990) Effects of recombinant human granuloctye colony-stimulating factor on the hematologic recovery and survival of irradiated mice. Blood 76:445–449

8. Farese AM, Hunt P, Grab LB, MacVittie TJ (1996) Combined administration of recombinant human megakaryocyte growth and development factor and granulocyte colony-stimulating factor enhances multilineage hematopoietic reconstitution in nonhuman primates after radiation-induced marrow aplasia. J Clin Invest 97:2145–2151
9. MacVittie TJ, Farese AM, Herodin F, Grab LB, Baum CM, McKearn JP (1996) Combination therapy of radiation-induced bone marrow aplasia in nonhuman primates using synthokine SC-55494 and recombinant human granulocyte colony-stimulating factor. Blood 87:4129–4135
10. MacVittie TJ, Farese AM, Jackson WI (2005) Defining the full therapeutic potential of recombinant growth factors in the post radiation-accident environment: the effect of supportive care plus administration of G-CSF. Health Phys 89:546–555
11. Mackey M, Aprikyan AAG, Dale DC (2003) The rate of apoptosis in post mitotic neutrophil precursors of normal and neutropenic humans. Cell Prolif 36:27–34
12. Lord BI, Woolford LB, Molineux G (2001) Kinetics of neutrophil production in normal and neutropenic animals during the response to filgrastim (r-metHu G-CSF) or filgrastim SD/01 (peg-r-metHu G-CSF). Clin Cancer Res 7:2085–2090
13. Lord BI, Bronchud MH, Owens S et al (1989) The kinetics of human granulopoiesis following treatment with granulocyte colony-stimulating factor in vivo. Proc Natl Acad Sci U S A 86:9499–9503
14. Rieger MA, Hoppe PS, Smejkal BM, Eitelhuber AC, Schroeder T (2009) Hematopoietic cytokines can instruct lineage choice. Science 325:217–218
15. Dainiak N (1997) Biological effects of radiation injury. 12(Suppl 2), iii–iv. In: Dainiak N, Schull WJ, Karkanitsa L, Aleinikova OA (eds) Radiation injury and the Chernobyl catastrophe. Miamisburg, OH, AlphaMed Press, 3-22-1996
16. Ricks RC, Fry SA (1990) The medical basis for radiation accident preparedness II: clinical experience and follow-up since, 1979. Elsevier Science, New York, NY
17. Browne D, Weiss JF, MacVittie TJ, Pillai MV (1990) Treatment of radiation injuries. Proceedings of the first consensus development conference on the treatment of radiation injuries, Plenum Press, New York
18. MacVittie TJ, Weiss JF, Browne D (1996) Advances in the treatment of radiation injuries. Proceedings of second consensus development conference on the treatment of radiation injuries, Pergamon, Elsevier Science Inc, Terrytown, NY
19. Ricks RC, Berger ME, O'Hara F (2001) The medical basis for radiation accident preparedness IV: clinical care of victims. Parthenon Publishers, Washington, DC
20. Waselenko JK, MacVittie TJ, Blakely WF et al (2004) Medical management of the acute radiation syndrome: recommendations of the Strategic National Stockpile Radiation Working Group. Ann Intern Med 140:1037–1051
21. Crawford LM (2002) New drug and biological drug products; evidence needed to demonstrate effectiveness of new drugs when human efficacy studies are not ethical or feasible. Fed Regist 67:37988–37998. 5-31-2002. 21 CFR parts 314 and 601, FDA, HHS; ACTION: Final Rule
22. Food and Drug Administration, Center for Drug Evaluation and Research and Food and Drug Administration, Center for Biologics Evaluation and Research (2009) Guidance for industry: animal models-essential elements to address efficacy under the animal rule, 1–19, 1-13-2009
23. Anno GH, Young RW, Bloom RM, Mercier JR (2003) Dose response relationships for acute ionizing-radiation lethality. Health Phys 84:565–575
24. Anno GH, Baum SJ, Withers HR, Young RW (1989) Symptomatology of acute radiation effects in humans after exposure to doses of 0.5-30 Gy. Health Phys 56:821–838
25. Baranov AE, Guskova AK (1990) Acute radiation disease in Chernobyl accident victims. In: Ricks RC et al (eds) The medical basis for radiation accident preparedness II; clinical experience and follow-up since 1979. Elsevier, New York, pp 79–87
26. Bodey GP, Buckley M, Sathe YS, Freireich EJ (1966) Quantitative relationships between circulating leukocytes and infection in patients with acute leukemia. Ann Intern Med 64:328–340

27. Hughes WT, Armstrong D, Bodey GP et al (1990) Guidelines for the use of antimicrobial agents in neutropenic patients with unexplained fever. J Infect Dis 161:381–396
28. Hughes WT (2002) Use of antimicrobial agents for treatment of infection in the neutropenic immunocompromised patient. In: Ricks R et al (eds) The medical basis for radiation-accident preparedness. The clinical care of victims. The Parthenon Publishing Group, Washington, DC, pp 117–129
29. Gafter-Gvili A, Fraser A, Mical P, Leibovici L (2005) Meta-analysis: antibiotic prophylaxis reduces mortality in neutropenic patients. Ann Intern Med 142:979–995
30. Schimpff SC (1990) Infections in radiation accidents. In: Browne D et al (eds) Treatment of radiation injuries. Plenum, New York, pp 75–85
31. Timmer-Bonte JN, de Boo TM, Smit HJ et al (2005) Prevention of chemotherapy-induced febrile neutropenia by prophylatic antibiotics plus or minus granulocyte colony-stimulating factor in small cell lung cancer: a Dutch randomized phase III study. J Clin Oncol 23:7974–7984
32. Pizzo PA (1984) Granulocytopenia and cancer therapy. Cancer 54:2649–2661
33. Pizzo PA (1993) Management of fever in patients with cancer and treatment-induced neutropenia. N Engl J Med 328:1323–1332
34. Hughes WT, Armstrong D, Bodey GP, Bow EJ, Brown AE, Calandra T (2002) 2002 guidelines for the use of antimicrobial agents in neutropenic patients with cancer. Clin Infect Dis 34:730–751
35. National Comprehensive Cancer Network (2005) Clinical practice guidelines in oncology: fever and neutropenia. Version 1.2005. 8-31-2005
36. Hughes WT, Armstrong D, Bodey GP et al (1997) 1997 Guidelines for the use of antimicrobial agents in neutropenic patients with unexplained fever. Clin Infect Dis 25:551–573
37. Baranov AE, Konchalovski MV, Soloviev WY, Guskova AK (1998) Use of blood cell count changes after radiation exposure in dose assessment and evaluation of bone marrow function. In: Ricks RC et al (eds) The medical basis for radiation accident preparedness II. Elsevier, New York, pp 427–443
38. Baranov AE (1996) Allogenic bone marrow transplantation after severe, uniform total-body irradiation: experience from recent (Nyasvizh, Belarus) and previous radiation accidents. In: MacVittie TJ et al (eds) Advances in the treatment of radiation injuries: advances in the bioscience, vol 94. Pergamon, Elsevier Science Ltd, Tarrytown, NY, pp 281–293
39. Fliedner TM et al (1988) Hematological indicators to predict patient recovery after whole-body irradiation as a basis for clinical management. In: Ricks RC (ed) The medical basis for radiation accident preparedness. Elsevier, New York, pp 445–459
40. Toda H, Murata A, Matsuura N et al (1993) Therapeutic efficacy of granulocyte colony-stimulating factor against rat cecal ligation and puncture model. Stem Cells 11:228–234
41. Weisbart RH, Gasson JC, Golde DW (1989) Colony-stimulating factors and neutrophils. Colony-stimulating factors and host defense. Ann Intern Med 110:297–303
42. Cohen AM, Zsebo KM, Inoue H et al (1987) In vivo stimulation of granulopoiesis by recombinant human granulocyte colony-stimulating factor. Proc Natl Acad Sci USA 84: 2484–2488
43. Matsumoto M, Matsubara S, Matsuno T et al (1987) Protective effect of human granulocyte colony-stimulating factor on microbial infection in neutropenic mice. Infect Immunol 55:2715–2720
44. Shinomiya N, Tsuru S, Katsura Y, Kayashima S, Nomoto K (1991) Enhanced resistance against *Listeria monocytogenes* achieved by pretreatment with granulocyte colony-stimulating factor. Infect Immunol 59:4740–4743
45. Wakiyama H, Tsuru S, Hata N et al (1993) Therapeutic effect of granulocyte colony-stimulating factor and cephem antibiotics against experimental infections in neutropenic mice induced by cyclophosphamide. Clin Exp Immunol 92:218–224
46. O'Reilly M, Silver GM, Greenhalgh DG, Gamelli RL, Davis JH, Hebert JC (1992) Treatment of intra-abdominal infection with granulocyte colony-stimulating factor. J Trauma 33:679–682

47. Matsumoto M, Matsubara S, Yokota T (1991) Effect of combination therapy with recombinant granulocyte colony-stimulating factor (rG-CSF) and antibiotics in neutropenic mice unresponsive to antibiotics alone. J Antimicrob Chemother 28:447–453
48. Ono M, Matsumoto M, Matsubara S, Tomioka S, Asano S (1988) Protective effect of human granulocyte colony-stimulating factor on bacterial and fungal infections in neutropenic mice. Behring Inst Mitt 83:216–221
49. Lindemann A, Riedel D, Oster W et al (1988) Granulocyte/macrophage colony-stimulating factor induces interleukin-1 production by human polymorphonuclear neutrophils. J Immunol 140:837–839
50. Lindemann A, Riedel D, Oster W, Ziegler-Heitbrock HW (1989) Granulocyte-macrophage colony-stimulating factor induces cytokine secretion by human polymorphonuclear leukocytes. J Clin Invest 83:1308–1312
51. Lloyd AR, Oppenheim JJ (1992) Poly's lament: the neglected role of the polymorphonuclear neutrophil in the afferent limb of the immune response. Immunol Today 13:169–172
52. Fasano MB, Cousart S, Neal S, McCall CE (1991) Increased expression of the interleukin-1 receptor on blood neutrophils of human with the sepsis syndrome. J Clin Invest 88:1452–1459
53. McColl SR, Paquin R, Manard C, Beaulieu AD (1992) Human neutrophils produce high levels of the interleukin 1 receptor antagonist in response to granulocyte/macrophage colony-stimulating factor and tumor necrosis factor. J Exp Med 176:593–598
54. Porteu F, Nathan C (1990) Shedding of tumor necrosis factor receptors by activated human neutrophils. J Exp Med 172:599–607
55. Re F, Megozzi M, Muzio M, Dinarello CA, Mantovani A, Colotta F (1993) Expression of interleukin-1 receptor antagonist (IL-1ra) by human circulating polymorphonuclear cells. Eur J Immunol 23:570–573
56. Dinarello CA (2010) IL-1: discoveries, controversies and future directions. Eur J Immunol 40:606
57. Dinarello CA, Wolff RL (1974) Human leukocytic pyrogen: purification and development of a radioimmunoassay. Proc Natl Acad Sci U S A 74:4624–4627
58. Strieter RM, Kunkel SL, Bone RC (1993) Role of tumor necrosis factor-alpha in disease states and inflammation. Crit Care Med 21:S447–S463
59. Beutler B, Grau G (1993) Tumor necrosis factor in the pathogenesis of infectious diseases. Crit Care Med 21:S423–S435
60. Kobayashi Y, Okabe T, Urabe A, Suzukci N, Takaku F (1987) Human granulocyte colony stimulating factor produced by *Escherichia coli* shortens the period of granulocytopenia induced by irradiation in mice. Jpn J Cancer Res 78:763–768
61. Fushiki M, Ono K, Sasai K et al (1990) Effect of recombinant human granulocyte colony stimulating factor on granulocytopenia in mice induced by irradiation. Int J Radiat Oncol Biol Phys 18:353–357
62. Farese AM, Roskos L, Cheung E, Stead RB, Yin SM, MacVittie TJ (1998) A single administration of r-metHuG-SD/01 (SD01) significantly improves neutrophil recovery following autologous bone marrow transplantation. Blood 92:112
63. MacVittie TJ, Monroy RL (1990) Rescue of lethally irradiated animals: therapeutic use of rhG-CSF and rhGM-CSF in preclinical models of radiation-induced marrow aplasia. In: Browne D et al (eds) Treatment of radiation injuries. Plenum, New York, pp 35–49
64. Patchen ML, MacVittie T (1994) Granulocyte colony-stimulating factor and amifostine (Ethyol) synergize to enhance hemopoietic reconstitution and increase survival in irradiated animals. Semin Oncol 21:26–32
65. MacVittie TJ, Monroy R, Vigneulle RM, Zeman GH, Jackson WE (1991) The relative biological effectiveness of mixed fission-neutron:gamma radiation on the hematopoietic syndrome in the canine: effect of therapy on survival. Radiat Res 128:S29–S36
66. Nash RA, Schuening FG, Seidel K et al (1994) Effect of recombinant canine granulocyte-macrophage colony-stimulating factor of hematopoietic recovery after otherwise lethal total body irradiation. Blood 83:1963–1970

67. Neelis KJ, Dubbelman YD, Qingliang L, Thomas GR, Eaton DL, Wagemaker G (1997) Simultaneous administration of TPO and G-CSF after cytoreductive treatment of rhesus monkeys prevents thrombocytopenia, accelerates platelet and red cell reconstitution, alleviates neutropenia, and promotes the recovery of immature bone marrow cells. Exp Hematol 25:1084–1093
68. Neelis KJ, Hartong SCC, Egeland T, Thomas GR, Eaton DL, Wagemaker G (1997) The efficacy of single-dose administration of thrombopoietin with coadministration of either granulocyte/macrophage or granulocyte colony-stimulating factor in myelosuppressed rhesus monkeys. Blood 90:2565–2573
69. Farese AM, Casey DB, Smith WG, Vigneulle RM, McKearn JP, MacVittie TJ (2001) Leridistim, a chimeric dual G-CSF and IL-3 receptor agonist, enhances multilineage hematopoietic recovery in nonhuman primate model of radiation-induced myelosuppression: effect of schedule, dose, and route of administration. Stem Cells 19:522–533
70. Farese AM, Kirschner KF, Patchen ML, Zsebo KM, MacVittie TJ (1993) The effect of recombinant canine stem cell factor and/or recombinant canine granulocyte colony stimulating factor on marrow aplasia recovery in lethally irradiated canines. Exp Hematol 21:1169
71. Meisenberg BR, Davis TA, Melaragno AJ, Stead R, Monroy RL (1992) A comparison of therapeutic schedules for administering granulocyte colony-stimulating factor to nonhuman primates after high-dose chemotherapy. Blood 79:2267–2272
72. Barrett A (1984) Total body irradiation and LD 50 in man. In: Roerse JJ et al (eds) Response of different species to total body irradiation. Martinus Nijhoff Publishers, Dordrecht, The Netherlands, pp 205–208
73. Barabanova AV (1995) Acute radiation syndrome with cutaneous syndrome. In: Lord BI et al (eds) Radiation toxicology: bone marrow and leukaemia. Taylor and Francis, Bristol, PA, p 217
74. Konchalovsky MV, Baranov AE, Kolganov AV (2005) Multiple organ involvement and failure: selected Russian radiation accident cases re-visited. BJR Suppl. 2005; 27:26–29
75. Uozaki H, Fukayama M, Nakagawa K et al (2005) The pathology of multi-organ involvement: two autopsy cases from the Tokai-mura criticality accident. Br J Radiol 27:13–16
76. Asano S (2005) Multi-organ involvement: lessons from the experience of one victim of the Tokai-mura criticality accident. Br J Radiol 27:9–12
77. Densow D, Kindler H, Baranov AE, Tibken B, Hofer EP, Fliedner TM (1997) Criteria for the selection of radiation accident victims for stem cell transplantation. Stem Cells 15(suppl 2):287–297
78. Georges G, Storb R (2001) Experimental and clinical experience with hematopoietic stem cell transplants. In: Ricks R et al (eds) The medical basis for radiation-accident preparedness. Parthenon Publishing Group, Boca Raton, pp 73–93
79. Georges GE, Storb RF (1997) Experimental and clinical experience with hematopoietic stem cell transplants. Stem Cells 15:73–94
80. Fliedner TM, Kindler H, Densow D, Baranov AE, Guskova AK, Szepesi T (1996) The Moscow-Ulm radiation accident clinical history data base. In: MacVittie T et al (eds) Advances in the treatment of radiation injuries. Pergamon, New York, pp 271–279
81. Baranov AE, Selidovkin GD, Butturini A, Gale RP (1994) Hematopoietic recovery after 10-Gy acute total body radiation. Blood 83:596–599
82. Smith TJ, Khatcheressian J, Lyman GH et al (2006) 2006 update of recommendations for the use of white blood cell growth factors: an evidence-based clinical practice guideline. J Clin Oncol 24:3187–3205
83. Aapro MS, Cameron DA, Pettengell R et al (2006) EORTC guidelines for the use of granulocyte-colony stimulating factor to reduce the incidents of chemotherapy-induced febrile neutropenia in adult patients with lymphomas and solid tumours. Eur J Cancer 42:2433–2453
84. United States Centers for Disease Control and Prevention (2005) Facts about Neupogen. Available at: http://emergency.cdc.gov/radiation/neupogenfacts.asp. Accessed 23 Nov 2010
85. United States Department of Health and Human Services (2010) Radiation emergency medical management; Hematopoietic subsyndromes; Hematopoeitic countermeasures. http://www.remm.nlm.gov/cytokines.htm. Accessed 22 Nov 2010

Neutropenias in Felty's Syndrome and Systemic Lupus Erythematosus

Mojtaba Akhtari and Edmund K. Waller

1 Introduction

Autoimmune neutropenia is characterized by the presence of autoantibodies produced by the patient's immune system against antigens on neutrophils, and it can be divided into primary and secondary forms. Primary autoimmune neutropenia is not a common condition in the adult population, but secondary autoimmune neutropenia may be as common as autoimmune disorders of red blood cells and platelets. Secondary autoimmune neutropenia often is not diagnosed because symptoms are less overt with no reliable markers for in vivo neutrophil lysis in contrast with hemoglobin in hemolytic anemia. Autoimmune neutropenia is usually associated with autoimmune disorders, hematologic and nonhematologic malignancies, and drug exposure (Table 1) [1].

The number of neutrophils in circulation is determined by the rate of their production and destruction. Neutrophils have a very high turnover rate (10^9 cells/day), and bone marrow is the exclusive site for their production under normal conditions. Normally, most neutrophils are in the bone marrow storage pool where they spend 6–7 days, and then they are released into the blood circulation where their half-life is 6.7 h. Circulating neutrophils comprise <5% of the total body mass

M. Akhtari (✉)
Division of Hematology and Oncology, Department of Internal Medicine, University of Nebraska Medical Center, 987680 Nebraska Medical Center, Omaha, NE 68198-7680, USA
e-mail: Mojtaba.akhtari@unmc.edu

E.K. Waller
Department of Hematology and Medical Oncology, Winship Cancer Institute, Emory University School of Medicine, Atlanta, GA 30322, USA
e-mail: ewaller@emory.edu

G. Molineux et al. (eds.), *Twenty Years of G-CSF*, Milestones in Drug Therapy,
DOI 10.1007/978-3-0348-0218-5_20,

Table 1 Causes of autoimmune neutropenia in adults [1]

Primary autoimmune neutropenia
Secondary autoimmune neutropenia
Alternating autoimmune hemocytopenia
Autoimmune diseases
Systemic lupus erythematosus
Rheumatoid arthritis (Felty's syndrome)
Sjögren's syndrome
Scleroderma
Primary biliary cirrhosis
Graves' disease
Immunologic diseases
X-linked autoimmune lymphoproliferative syndrome
Common variable immune deficiency
Infectious diseases
Viral
Human immunodeficiency virus
Infectious mononucleosis
Parvovirus B19
Hepatitis B virus
Hepatitis C virus
Bacterial
Helicobacter pylori
Mycoplasma pneumonia
Malignancy
Large granular lymphocyte leukemia
Hairy cell leukemia
Hodgkin's lymphoma
Non-Hodgkin's lymphoma
Chronic lymphocytic leukemia
Bone marrow sea-blue histiocytosis
Thymoma
Melanoma
Neurological diseases
Multiple sclerosis
Transplantation
Hematopoietic stem cell transplantation (autologous and allogeneic)
Kidney transplant
Liver transplant
Drugs

of neutrophils. After bone marrow, the bulk of mature neutrophils are located in the tissues where their half-life may extend to 24–48 h, and they offer local defense. In addition to neutrophils in circulation and tissues, there is a large marginated pool that allows for rapid recruitment. Chronic neutropenia is defined by an absolute neutrophil count (ANC) $<1.5 \times 10^9$/L lasting for at least 6 months. Neutrophil counts can be lower than average in Africans, black Americans, and Yemenite Jews. Neutropenia is classified based on the ANC as mild (1.0–1.5×10^9/L), moderate (0.5–1.0×10^9/L), and severe ($<0.5 \times 10^9$/L). Neutropenia is associated with an increased risk of infection, particularly if the ANC is $<0.5 \times 10^9$/L. Bacterial infections, especially intermittent stomatitis and gingivitis, perirectal abscess, and cellulites are more common than pneumonia or septicemia. Fungal infections are unusual, but oral candidiasis is common, but the risk for viral or parasitic infections is generally not increased. Chronic fatigue is a frequent symptom among patients with neutropenia.

Early reports on autoimmune neutropenia were published in the 1950s when it was shown that infusion of plasma from patients with neutropenia to normal recipients caused leucopenia [2]. In 1960, it was shown that fetal–maternal neutrophil incompatibility causes neonatal neutropenia [3], a study that led to the discovery of neutrophil-specific antigens [4]. In 1975, chronic neutropenia was shown to be caused by autoantibodies against neutrophil-specific antigens [5].

2 Pathogenesis and Pathophysiology of Autoimmune Neutropenia

Two major mechanisms have been proposed for antibody formation against autologous antigens [6]. The first is the formation of antibodies against foreign antigens, e.g., viruses or bacteria that crossreact with autologous antigens. Antibodies in these conditions are usually polyclonal and in a normally developed immune system, this type of immune reaction is self-limited. The second mechanism involves the loss of suppression by a clone of cells that normally protect reaction of the immune system against autologous antigens. The concept of clonality of the antibodies in autoimmune neutropenia is not well understood. In secondary autoimmune neutropenia, the development of other autoimmune diseases suggests an etiology in which disturbances of self-tolerance play an important role.

Our current knowledge about mechanisms of neutrophil destruction is based primarily on in vitro observations. Agglutination is the major form of neutrophil response to antineutrophil antibodies. Complement-induced neutrophil agglutination due to antineutrophil autoantibodies is another suggested mechanism for neutrophil destruction. Evidence exists that activated complement system can cause neutrophil aggregation and adherence of neutrophils to endothelial cells [2]. Medical devices with filtration membranes such as cardiopulmonary bypass [7] and hemodialysis [8] can cause complement-activation neutropenia. C5a is usually activated by the filtration membranes causing neutrophil aggregation. Phagocytosis of the neutrophils, which are coated with neutrophil antibodies, is another probable mechanism for neutrophil destruction. It has been shown that neutrophil antibodies can cause opsonization of neutrophils resulting in clearance by phagocytic cells [2]. Evidence suggests that opsonized neutrophils are cleared from the circulation through phagocytosis in the spleen; however, the spleen is not the only site for phagocytosis of the sensitized neutrophils and it has been shown by ^{111}In radiolabeling techniques that they can be removed by various tissues that contain phagocytotic cells [9]. Lymphocytes from neutropenic patients with Felty's syndrome and systemic lupus erythematosus (SLE) have been shown to inhibit cell colony-forming units (CFU-C) in bone marrow culture [10] and in some patients with autoimmune neutropenia, increased T-cell-mediated cytotoxicity and production of interferon (IFN)-γ have been demonstrated [11]. It is recognized that circulating immune complexes bind to neutrophil Fc receptors, particularly in conditions with high amounts of circulating immune complexes. The neutrophil–immunoglobulin complex can result in complement activation, which in turn, can cause neutrophil destruction. The neutrophil-immunoglobulin complex can be cleared from the circulation by phagocytosis through the reticuloendothelial system and tissue macrophages. Neutropenia, however, is not a common finding in many conditions characterized by the presence of immune complexes. Neutropenia in acquired immunodeficiency syndrome (AIDS) is complex and multifactorial. Patients with AIDS usually have high amounts of neutrophil-bound immunoglobulins. Furthermore, myelodysplasia and bone marrow suppression can contribute to neutropenia in these patients.

Antineutrophil antibodies can affect neutrophil function, causing qualitative abnormalities such as defective response to chemotaxis. It has been shown in an in vitro experiment that IgG from a patient with autoimmune neutropenia and recurrent infections had abnormal aggregation–disaggregation response to formyl peptide chemotaxic factors [12]. Antineutrophil autoantibodies can interfere with neutrophil function in patients who are not neutropenic. Inhibition of neutrophil motility by an IgG autoantibody in a patient with normal neutrophil counts and recurrent skin infections has been reported [2]. This observation can explain the severity of clinical presentations in some patients with autoimmune neutropenia, in which patients are not markedly neutropenic. Pathophysiology of autoimmune neutropenia in adults is somehow different from autoimmune neutropenia in children. In the pediatric population, autoimmune neutropenia of infancy has an early onset and is usually self-limited. Many researchers believe that the difference is in the immune system of children, particularly the suppressor T-cell function, which is quite immature at birth and does not reach full development until about 3 years of age [6]. In a series of patients with autoimmune neutropenia of infancy, neutrophil antibodies were shown in 119 of 121 patients, and 10% of the antibodies had specificity for neutrophil antigen system, NA1 or NA2 [13]. The neutrophil antibody antigens are located on the IgG-Fc receptor type IIIb (FcγRIIIb or CD16), which is expressed exclusively by neutrophils [14]. Although in autoimmune neutropenia of infancy the autoantibodies frequently target the neutrophil antigenic system with preference for the NA1 alleloform [15, 16]. In children with secondary autoimmune neutropenia, the neutrophil antibodies have pan-Fcγ-RIIIb specificity [17].

3 Clinical Manifestations of Neutropenia

The clinical findings of neutropenia usually are related to the severity and duration of neutropenia. The hematologic profile shows the absence or reduction in the number of neutrophils, while the number of eosinophils and basophils are either normal or are increased. Monocytosis is a common finding and lymphocyte numbers are normal or, occasionally, reduced. A mild anemia may be present. Platelet counts are usually normal. Hypergammaglobulinemia is a common finding that indicates a chronic infection. Bone marrow examination reveals myeloid hyperplasia with distinctly reduced numbers of mature cells that may resemble maturation arrest. Splenomegaly is a not a common finding.

4 Neutropenia in Felty's Syndrome

Felty's syndrome is a form of seropositive rheumatoid arthritis that is associated with neutropenia and splenomegaly. Felty's syndrome is an uncommon clinical entity, and the lifetime risk of developing the disease in a patient with rheumatoid arthritis is 1–3% [18]. Neutropenia in patients with Felty's syndrome is usually chronic and can be severe and complicated by recurrent bacterial infections. Skin and respiratory infections occur commonly, particularly in patients who have ANC $<1.0 \times 10^9$/L, have skin ulcerations, and receive steroid treatment [19]. The pathophysiology of neutropenia in Felty's syndrome is complex and multifactorial. Splenomegaly can cause splenic sequestration of the neutrophils, and bone marrow examination may reveal such abnormalities as maturation arrest and hypoplasia [20]. Neutrophil autoantibodies and immune complexes, when present, cause destruction and peripheral margination of neutrophils [21]. It has been shown that proimflammatory cytokines (interleukin [IL]-8, tumor-necrosis factor [TNF]-α, and IFN-γ) also affect the bone marrow, causing inhibition of granulopoiesis [22]. Low serum amounts of soluble FcγIII and serum granulocyte colony-stimulating factor (G-CSF) can be used to identify patients at risk for infectious complications [23]. The same investigators have shown the presence of antibodies to G-CSF in patients with Felty's syndrome or SLE who had neutropenia; however, in most patients, the anti-G-CSF antibodies did not have neutralizing activity. Evidence exists that the eukaryotic elongation factor 1A-1 antigen is the target for antineutrophil antibodies in patients with Felty's syndrome. It has been shown that during apoptosis, elongation factor 1A-1 antigen is translocated from the nucleus to the cell membrane, a process that may explain the mechanism by which such antibodies bind to neutrophil membranes [24]. Patients with Felty's syndrome can develop an increased number of large granular lymphocytes in circulation and in their bone marrow. Large granular lymphocytes are a heterogeneous population of lymphoid cells that includes natural killer cells and activated cytotoxic T cells [25]. The expansion of these lymphocytes may be reactive, but more often it is clonal giving rise to large granular lymphocyte syndrome that is based on detection of clonal large granular lymphocytes and positive T-cell receptor rearrangement [26]. Patients with this condition usually have moderate splenomegaly and 50% of patients have an ANC of $<0.5 \times 10^9$/L. T-cell large granular lymphocyte syndrome is associated with rheumatoid arthritis in 25% of cases. Most patients demonstrate mild to moderate absolute lymphocytosis in the peripheral blood, and the bone marrow is infiltrated with lymphocytes in most patients [27, 28] Evidence suggests that increased peripheral destruction of neutrophils secondary to immune complexes and bone marrow suppression of granulopoiesis through Fas ligand secretion are the main causes of neutropenia in patients with large granular lymphocytosis [29].

5 Secondary Autoimmune Neutropenia in SLE

Neutropenia is common in patients with SLE and can be detected in 50% to 60% of the patients during the course of their disease [30]. Detecting antineutrophil antibodies in the serum of patients with SLE is a common finding, but not every patient with positive antibody findings is neutropenic. Neutropenia is considered to be a marker of SLE activity and SLE-related neutropenia is a predisposing factor for clinical infections [31]. Pathophysiology of neutropenia in SLE is complex; however, evidence is growing that accelerated apoptosis of mature neutrophils is a major contributing factor. It has been shown that Fas (CD95) is involved in apoptosis of circulating neutrophils, monocytes, and lymphocytes of patients with SLE [32, 33]. Marrow hypoplasia and Fas-mediated apoptosis of $CD34^+$ hematopoietic progenitor cells are additional factors contributing to SLE-associated cytopenias [34]. It has been shown that in patients with SLE, serum concentrations of TNF-related apoptosis-inducing ligand (TRAIL) were inversely proportional to blood neutrophil counts [35]. Expression of TRAIL receptor 3, a decoy receptor for TRAIL, was also lower in patients with neutropenia than in controls or patients without neutropenia. In vitro, TRAIL induced apoptosis in neutrophils. Corticosteroid therapy reduced the expression of TRAIL on T cells and enhanced expression of Fas-associating protein with death domain-like IL-1^3-converting enzyme (FLICE)-inhibitory protein (FLIP), an antiapoptotic protein, by neutrophils. Approximately 66% of patients with SLE have neutrophil auto-antibodies (usually IgG) [36]. Neutropenia is more common in patients with anti-Ro autoantibodies [37]. Evidence suggests that anti-SSB/La is one of the anti-neutrophil antibodies that can bind to the surface-expressed SSB/La on neutrophils and is probably responsible for neutropenia in patients with SLE [38].

6 Treatment

It is well established that severe neutropenia (i.e., ANC $<0.5 \times 10^9$/L) is correlated with an increased risk for infections [39]. Prophylactic therapy with antibiotics with or without recombinant human granulocyte colony-stimulating factor (rHuG-CSF) is common practice for the treatment of patients who are severely neutropenic. Tissue neutrophil concentration may be the primary determinant of host defense against bacteremia [40], and it has been shown that bacteremia in patients who are neutropenic correlates more closely with total body mass of neutrophils than with blood neutrophil count [41]. Patients with severe neutropenia and fever (>38°C) may require hospitalization and antibiotics after blood cultures have been obtained. The most likely sources of infection are the lungs and the skin, and infecting organisms are usually endogenous. Treatment with rHuG-CSF at a starting dose of 3–5 μg/kg/day administered subcutaneously is reasonable and most patients respond within days, but the duration of neutrophil count recovery often is not long. The dose of rHuG-CSF should be individualized and the treatment given only for control of complications and in preparation for surgical procedures. Some patients need to be kept on a regular maintenance treatment based on their response. rHuG-CSF is usually well tolerated by most patients. The most commonly reported adverse event is bone pain. Splenic rupture and acute lung injury are considered to be very rare complications (see chapter "The Safety Profile of Filgrastim and Pegfilgrastim" by Neumann and Foote for further information). Although the association of rHuG-CSF treatment with increased risks of developing myelodysplastic syndromes (MDS) and acute myeloid leukemia (AML) in patients with severe congenital neutropenia have been reported, the causal relationship between rHuG-CSF therapy and myelodysplasias or leukemia remains uncertain [42, 43] (see chapter "rHuG-CSF for the Treatment of Severe Chronic Neutropenia" by Dale and Bolyard for further information). Endogenous G-CSF induces neutrophils to shed FcγRIIIb from their plasma membrane which, in turn, increases the amount of soluble FcγRIIIb causing sequestration of neutrophil autoantibodies [44, 45].

7 Treatment of Neutropenia in Felty's Syndrome

rHuG-CSF is effective in the management of severe neutropenia in patients with Felty's syndrome [36, 46]. Arthritic exacerbation and vasculitic skin rash have been reported in some patients after administration of rHuG-CSF; however, some patients were able to continue the treatment with reduced doses of rHuG-CSF [47–53]. rHuG-CSF dose and frequency should be adjusted at the lowest effective dose to maintain ANC $>1.0 \times 10^9$/L, a goal that can be achieved by injections once or twice a week. rHuG-CSF also has therapeutic value in the management of the preoperative period in patients who need a splenectomy. It has been suggested that long-term treatment with low doses of rHuG-CSF may be associated with clinical benefits in patients who tolerate therapy and demonstrate good hematologic response [53, 54]. Reports exist of responses to colony-stimulating factors, either rHuG-CSF or rHu granulocyte-macrophage colony-stimulating factor (rHuGM-CSF), in chronic neutropenia associated with large granular lymphocyte syndrome [53, 55–61].

8 Treatment of Neutropenia in SLE

rHuG-CSF is an effective therapeutic intervention in the management of neutropenia in patients with SLE, and there are reports of successful rHuG-CSF treatment of neutropenia and infectious complications in patients with SLE [62–64]. The therapeutic effect of rHuG-CSF, however, can be variable [65]. Treatment with rHuG-CSF has been associated with flares of SLE and leukocytoclastic vasculitis [64, 66]. As in Felty's syndrome, the management of neutropenia begins with continuing the control of underlying autoimmune disorder and adding rHuG-CSF to reverse severe neutropenia. It has been suggested that in these patients, the dose and frequency of rHuG-CSF should be adjusted to the lowest effective dose with the goal of maintaining the neutrophil count at approximately 1.0×10^9/L [36, 46].

9 Conclusions

Colony-stimulating factors, such as rHuG-CSF, have become first-line treatment for primary and secondary neutropenias. Beneficial effects of rHuG-CSF in immune neutropenias, and recent advances in our understanding of its mechanisms of action suggest that rHuG-CSF can be used as an effective and generally well-tolerated therapy in the management of patients with Felty's syndrome or SLE. To prevent disease flares, however, rHuG-CSF should be administered at the lowest effective dose.

References

1. Akhtari M, Curtis B, Waller EK (2009) Autoimmune neutropenia in adults. Autoimmun Rev 9:62–66
2. Lalezari P (2006) Neutropenia. In: Rose NR, Mackay IR (eds) The autoimmune diseases. Elsevier, St. Louis, pp 585–589
3. Lalezari P, Nussbaum MG, Gelman S, Spaet TH (1960) Neonatal neutropenia due to maternal isoimmunization. Blood 15:236–243
4. Lalezari P, Murphy GB, Allen FH (1971) NB1, a new neutrophil-specific antigen involved in the pathogenesis of neonatal neutropenia. J Clin Invest 50:1108–1115
5. Lalezari P, Jiang AF, Yegen L, Santorineou M (1975) Chronic autoimmune neutropenia due to anti-NA1 antibody. N Engl J Med 293:744–747
6. Shastri KA, Logue GL (1993) Autoimmune neutropenia. Blood 81:1984–1995
7. Chenoweth DE, Cooper SW, Hugli TE, Stewart RW, Blackstone EH, Kirklin JW (1981) Complement activation during cardiopulmonary bypass: evidence for generation of C3a and C5a anaphylatoxins. N Engl J Med 304:497–503
8. Craddock PR, Fehr J, Dalmasso AP, Brighan KL, Jacob HS (1977) Hemodialysis leukopenia. Pulmonary vascular leukostasis resulting from complement activation by dialyzer cellophane membranes. J Clin Invest 59:879–888

9. McCullough J, Press C, Clay M, Kline W (eds) (1988) Granulocyte serology: a clinical and laboratory guide. ASCP Press, Chicago, pp 90–112, 113–124, 134–135
10. Logue GL, Shastri KA, Laughlin M, Shimm DS, Ziolkowski LM, Iglehart JL (1991) Idiopathic neutropenia: antineutrophil antibodies and clinical correlations. Am J Med 90:211–216
11. Wlodarski MW, Nearman Z, Jiang Y, Lichtin A, Maciejewski JP (2008) Clonal predominance of CD8(+) T cells in patients with unexplained neutropenia. Exp Hematol 36:293–300
12. Hartman KR, Wright DG (1991) Identification of autoantibodies specific for the neutrophil adhesion glycoproteins CD11b/CD18 in patients with autoimmune neutropenia. Blood 78:1096–1104
13. Lalezari P, Khorshidi M, Petrosova M (1986) Autoimmune neutropenia of infancy. J Pediatr 109:764–769
14. Lalezari P (2002) Nomenclature for neutrophil-specific antigens Transfusion 42:1396–1397
15. Huizinga TWJ, Kleijer M, Tetteroo PA, Roos D, von dem Borne AE (1990) Biallelic neutrophil NA-antigen system is associated with a polymorphism on the phospho-inositol-linked Fc gamma receptor III (CD16). Blood 75:213–217
16. Bux J (1999) Nomenclature of granulocyte alloantigens. ISBT Working Party on Platelet and Granulocyte Serology, Granulocyte Antigen Working Party. International Society of Blood Transfusion. Transfusion 39:662–663
17. Bruin MCA KR, Tamminga RYJ, Kleije M, Buddelmeijer L, de Haas M (1999) Neutrophil antibody specificity in different types of childhood autoimmune neutropenia. Blood 94: 1797–1802
18. Sibley JT, Haga M, Visram DA, Mitchell DM (1991) The clinical course of Felty's syndrome compared to matched controls. J Rheumatol 18:1163–1167
19. Campion G, Maddison PJ, Goulding N et al (1990) The Felty syndrome: a case-matched study of clinical manifestations and outcome, serologic features, and immunogenetic associations. Medicine (Baltimore) 69:69–80
20. Dancey JT, Brubaker LH (1979) Neutrophil marrow profiles in patients with rheumatoid arthritis and neutropenia. Br J Haematol 43:607–617
21. Starkebaum G, Arend WP, Nardella FA, Gavin SE (1980) Characterization of immune complexes and immunoglobulin G antibodies reactive with neutrophils in the sera of patients with Felty's syndrome. J Lab Clin Med 96:238–251
22. Meliconi R, Uguccioni M, Chieco-Bianchi F et al (1995) The role of interleukin-8 and other cytokines in the pathogenesis of Felty's syndrome. Clin Exp Rheumatol 13:285–291
23. Hellmich B, Csernok E, de Haas M et al (2002) Low Fcγ receptor III and high granulocyte colony-stimulating factor serum levels correlate with the risk of infection in neutropenia due to Felty';s syndrome or systemic lupus erythematosus. Am J Med 113:134–139
24. Ditzel HJ, Masaki Y, Nilesen H, Farnaes L, Burton DR (2000) Cloning and expression of a novel human antibody-antigen pair associated with Felty's syndrome. Proc Natl Acad Sci U S A 97:9234–9239
25. Loughran TP (1993) Clonal diseases of large granular lymphocytes. Blood 82:1–14
26. Lima M, Almeida J, Santos AH et al (2001) Immunophenotypic analysis of the TCR-Vbeta repertoire in 98 persistent expansions of CD3(+)/TCR-alphabeta(+) large granular lymphocytes: utility in assessing clonality and insights into the pathogenesis of the disease. Am J Pathol 159:1861–1868
27. Rose MG, Berliner N (2004) T-cell granular lymphocyte leukemia and related disorders. Oncologist 9:247–258
28. Burks EJ, Loughran TP (2006) Pathogenesis of neutropenia in large granular lymphocyte leukemia and Felty syndrome. Blood Rev 20:245–266
29. Liu JH, Wei S, Lamy T et al (2000) Chronic neutropenia mediated by fas ligand. Blood 95:3219–3222
30. Budman DR, Steinberg AD (1977) Hematologic aspects of systemic lupus erythematosus: current concepts. Ann Intern Med 86:220–229

31. Hadley AG, Byran MA, Chapel H, Bunch C, Holburn AM (1987) Antigranulocyte opsonic activity in sera from patients with systemic lupus erythematosus. Br J Haematol 65:61–65
32. Courtney PA, Crockard AD, Williamson K, Irvine AE, Kennedy RJ, Bell AL (1999) Increased apoptotic peripheral blood neutrophils in systemic lupus erythematosus: relations with disease activity, antibodies to double stranded DNA, and neutropenia. Ann Rheum Dis 58:309–314
33. Ren Y, Tang J, Mok MY, Chan AW, Wu A, Lau CS (2003) Increased apoptotic neutrophils and macrophages and impaired macrophage phagocytic clearance of apoptotic neutrophils in systemic lupus erythematosus. Arthritis Rheum 48:2888–2897
34. Papadaki HA, Boumpas DT, Gibson FM et al (2001) Increased apoptosis of bone marrow CD34(+) cells and impaired function of bone marrow stromal cells in patients with systemic lupus erythematosus. Br J Haematol 115:167–174
35. Matsuyama W, Yamamoto M, Higashimoto I et al (2004) TNF-related apoptosis-inducing ligand is involved in neutropenia of systemic lupus erythematosus. Blood 104:184–191
36. Starkebaum G (2002) Chronic neutropenia associated with autoimmune disease. Semin Heamtol 39:121–127
37. Kurien BT, Newland J, Paczkowski C, Moore KI, Scofield RH (2000) Association of neutropenia in systemic lupus erythematosus with anti-Ro and binding of an immunologically cross-reactive neutrophil membrane antien. Clin Exp Immunol 120:209–217
38. Hsieh SC, Yu HS, Lin WW, Sun KH, Tsai CY, Huang DF (2003) Anti-SSB/La is on of the antineutrophil autoantibodies responsible for neutropenia and functional impairment of polymorphonuclear neutrophils in patients with systemic lupus erythematosus. Clin Exp Immunol 131:506–516
39. Finch SC (1983) Neutrophil disorders-benign, quantitative abnormalities of neutrophils. In: Williams WJ, Beutler E, Erslev AJ, Lichtman MA (eds) Hematology. McGraw-Hil, New York
40. Crosby WH (1969) How many "polys" are enough? Arch Intern Med 123:722–723
41. Koene HR, de Haas M, Kleijer M, Huizinga TW, Roos D, von dem Borne AE (1998) Clinical value of soluble IgG Fc receptor type III in plasma from patients with chronic idiopathic neutropenia. Blood 91:3962–3966
42. Tigue CC, McKoy JM, Evens AM, Trifilio SM, Tallman MS, Bennett CL (2007) Granulocyte-colony stimulating factor administration to healthy individuals and persons with chronic neutropenia or cancer: an overview of safety considerations from the Research on Adverse Drug Events and Reports project. Bone Marrow Transplant 40:185–192
43. Rosenberg PS, Alter BP, Bolyard AA et al (2006) Severe Chronic Neutropenia International Registry. The incidence of leukemia and mortality from sepsis in patients with severe congenital neutropenia receiving long-term G-CSF therapy. Blood 107:4628–4635
44. Donadieu J, Leblanc T, Bader Meunier B et al (2005) French Severe Chronic Neutropenia Study Group. Analysis of risk factors for myelodysplasias, leukemias and death from infection among patients with congenital neutropenia. Experience of the French Severe Chronic Neutropenia Study Group. Haematologica 90:45–53
45. Kerst JM, de Haas M, van der Schoot CE et al (1993) Recombinant granulocyte colony-stimulating factor administration to healthy volunteers: induction of immunophenotypically and functionally altered neutrophils via an effect on myeloid progenitor cells. Blood 82:3265–3272
46. Hellmich B, Schnabel A, Gross WL (1999) Treatment of severe neutropenia due to Felty's syndrome or systemic lupus erythematosus with granulocyte colony-stimulating factor. Semin Arthritis Rhem 29:82–89
47. Jain KK (1994) Cutaneous vasculitis associated with granulocyte colony-stimulating factor. J Am Acad Dermatol 31:213–215
48. McMullin MF, Finch MB (1995) Felty's syndrome treated with rhG-CSF associated with flare of arthritis and skin rash. Clin Rheumatol 14:204–208
49. Hayat SQ, Hearth-Holmes M, Wolf RE (1995) Flare of arthritis with successful treatment of Felty's syndrome with granulocyte colony stimulating factor (GCSF). Clin Rheumatol 14:211–212

50. Farhey YD, Herman JH (1995) Vasculitis complicating granulocyte colony stimulating factor treatment of leucopenia and infection in Felty's syndrome. J Rheumatol 22:1179–1182
51. Yasuda M, Kihara T, Wada T et al (1994) Granulocyte colony stimulating factor induction of improved leukocytopenia with inflammatory flare in a Felty's syndrome patient. Arthritis Rheum 37:145–146
52. Vidarsson B, Geirsson AJ, Onundarson PT (1995) Reactivation of rheumatoid arthritis and development of leukocytoclastic vasculitis in patient receiving granulocyte colony-stimulating factor for Felty's syndrome. Am J Med 98:589–591
53. Stanworth SJ, Bhavnani M, Chattopadhya C, Miller H, Swinson DR (1989) Treatment of Felty's syndrome with the haemopoietic growth factor granulocyte colony-stimulating factor (G-CSF). Q J Med 91:49–56
54. Krishnaswamy G, Odem C, Chi DS, Kalbfleisch J, Baker N, Smith JK (1996) Resolution of neutropenia of Felty's syndrome by longterm administration of recombinant granulocyte colony stimulating factor. J Rhematol 23:763–765
55. Kaneko T, Ogawa Y, Hirata Y et al (1990) Agranulocytosis associated with granular lymphocyte leukemia: improvement of peripheral blood granulocyte count with human recombinant granulocyte colony-stimulating factor (G-CSF). Br J Haematol 74:121–122
56. Vicers M, Stross P, Millard P, Barton C (1994) Response of T-beta $CD8^+$ lymphocytosis-associated neutropenia to G-CSF. Br J Haematol 87:431–433
57. Jakubowski A, Winton EF, Gencarelli A, Gabrilove J (1995) Treatment of chronic neutropenia associated with large granular lymphocytosis with cyclosporine A and filgrastim. Am J Hematol 50:288–291
58. Mulder A, De Wof J, Smit J, Van Oostveen J, Vellenga E (1992) Correction for neutropenia by GM-CSF in patients with a large granular lymphocyte proliferation. Ann Hematol 65:91–95
59. Krieger G, Kneba M, Vehmeyer K, Nagel GA, Welte K (1990) Use of recombinant granulocyte-macrophage colony stimulating factor in T-lymphocytosis with granulocytopenia. Eur J Haematol 44:205–206
60. Thomessen C, Nissen C, Gratwohl A, Tichelli A, Stern A (1989) Agranulocytosis associated with T-gamma-lymphocytosis: no improvement of peripheral blood granulocyte count with human recombinant granulocyte-macrophage colony stimulating factor (GM-CSF). Br J Haematol 71:157–158
61. Lamy T, LePrise PY, Amiot L et al (1995) Response to granulocyte-macrophage colony stimulating factor (GM-CSF) but not to G-CSF in a case of agranulocytosis associated large granular lymphocyte (LGL) leukemia. Blood 85:3352–3362
62. Euler HH, Schwab UM, Schroeder JO (1994) Filgrastim for lupus neutropenia. Lancet 344:1513–1514
63. Kondo H, Date Y, Sakai Y, Akimoto M (1994) Effective simultaneous rhG-CSF and methylprednisolone "pulse" therapy in agranulocytosis associated with systemic lupus erythematosus. Am J Hematol 46:157–158
64. Euler HH, Harten P, Zeuner RA, Schwab UM (1997) Recombinant human granulocyte colony stimulating factor in patients with systemic lupus erythematosus associated neutropenia and refractory infection. J Rheumatol 24:2153–2157
65. Hellmich B, Csernok E, Schatz H, Gross WL, Schanbe A (2002) Autoantibodies against granulocyte colony-stimulating factor in Felty's syndrome and neutropenic systemic lupus erythematosus. Arthritis Rheum 46:2384–2391
66. Vasiliu IM, Petri MA, Baer AN (2006) Therapy with granulocyte colony-stimulating factor in systemic lupus erythematosus may be associated with severe flares. J Rheumatol 33: 1878–1880

Part IV
Safety and Economics

The Safety Profile of Filgrastim and Pegfilgrastim

Theresa A. Neumann and MaryAnn Foote

1 Introduction

The discovery of endogenous proteins that regulate hematopoiesis led to the identification of human granulocyte colony-stimulating factor (G-CSF). With the advent of recombinant DNA technology, it became possible to manufacture bioactive recombinant proteins for medicinal use. Since the approval of recombinant human G-CSF (rHuG-CSF), such as filgrastim in 1991 and pegfilgrastim in 2002, millions of patients at risk for severe myelosuppression have received these products. Overall, filgrastim and pegfilgrastim have a high margin of safety for short-term use; however, rare severe adverse events have emerged and questions remain regarding the long-term safety and consequences of use of these products. This chapter primarily focuses on the safety and adverse event profile of the most widely used commercially available rHuG-CSF, Neupogen (filgrastim) and Neulasta [a modified (pegylated) filgrastim, pegfilgrastim]. As safety information can change rapidly, we suggest readers consult the latest package inserts for any changes that have occurred from the time of this writing. Other chapters in this volume discuss key studies in specific disease settings in greater detail than is the purview of this chapter, and we encourage the interested reader to reference them for further information.

T.A. Neumann
Menlo Park, CA, USA

M. Foote (✉)
MA Foote Associates, 4284 Par Five Ct, Westlake Village, CA 91362, USA
e-mail: fmawriter@aol.com

G. Molineux et al. (eds.), *Twenty Years of G-CSF*, Milestones in Drug Therapy,
DOI 10.1007/978-3-0348-0218-5_21,

2 Filgrastim

Filgrastim is a 175-amino acid recombinant protein expressed in *Escherichia coli*. The filgrastim peptide has the same amino acid sequence as endogenous human G-CSF with the exception that the backbone of the molecule is not glycosylated and the *N*-terminus is a methionine. Endogenous human G-CSF is a lineage-specific glycoprotein and is responsible for regulation of neutrophil production in bone marrow. The lineage specificity is an important aspect of filgrastim's safety profile in that it has been shown to have a relatively consistent safety profile. Both in vivo and ex vivo studies have demonstrated the molecule acts by binding to the G-CSF receptor (G-CSFR) and it plays a key role in neutrophil regulation and differentiation and in neutrophil functions (i.e., respiratory burst, antibody-dependent filling, and phagocytosis) [1–3]. Exogenous rHG-CSF administration has been shown to mobilize stem cells from the marrow into the peripheral system [4–6]. Techniques have been developed to isolate and harvest stem cells from the blood in a process known as peripheral blood progenitor cell (PBPC) collection.

2.1 *Overview*

The clinical utility of filgrastim in correcting and reversing low neutrophil counts and in improving the function of neutrophils led to evaluation of the molecule in the setting of cancer chemotherapy, bone marrow transplantation, systemic infections, and congenital neutropenia. As a result of extensive investigations, filgrastim is approved for prevention and treatment of severe neutropenia in patients receiving myelosuppressive chemotherapy. It is also approved for use in adult patients with acute myeloid leukemia (AML) who are undergoing induction or consolidation chemotherapy, and in patients who are receiving myeloablative chemotherapy followed by bone marrow transplantation. Filgrastim is also approved, used to treat patients with chronic forms of neutropenia such as idiopathic neutropenia, congenital neutropenia, and cyclic neutropenia.

The adverse event profile of filgrastim includes bone pain, headache, allergic reactions, rash, and splenomegaly, with bone pain the only adverse event that has been consistently reported across all patient populations. Increases in lactate dehydrogenase, alkaline phosphatase, and uric acid have been reported and may be related to increased cell turnover in chemotherapy [7, 8]. These changes are transient and not associated with any clinical sequelae. Incidental reports suggest that some patients believe that bone pain is an indication that filgrastim is "working" and do not mind the pain. Obviously, pain is not necessarily indication of filgrastim's activity.

Unlike other human proteins, filgrastim is devoid of side effects such as fever, malaise, autoimmune reactions, and fluid retention. The lack of these side effects makes filgrastim an ideal product to address bone-marrow recovery after cytotoxic chemotherapy. Of note, filgrastim is used primarily to address short-term loss of neutrophils after 6–8 cycles of chemotherapy for advanced cancer. Long-term use of filgrastim, defined as >1 year, has only been characterized in a smaller subset of patients with severe chronic neutropenia (SCN).

What is important to note is what adverse events are not reported. Filgrastim does not produce dose-limiting side effects even at 115 μg/kg, a dose that can cause marked leukocytosis (50 × 10^9/L) [9]. Unlike rHu granulocyte-macrophage colony-stimulating factor (rHuGM-CSF), indicated for use in many of the same patient populations as filgrastim, treatment with filgrastim does not appear to produce fever [3, 10–12], capillary-leak syndrome [13–19], or first-dose reaction [11, 14, 18, 20].

2.2 Patients Receiving Myelosuppressive Chemotherapy

Filgrastim is typically given 24 h after chemotherapy, an important consideration to avoid stimulation of bone marrow cell division during peak systemic cytotoxic chemotherapy levels that could accentuate bone-marrow damage. The most frequently reported adverse events attributed to filgrastim are bone pain, injection-site reaction, rash, acute neutrophilic dermatoses, allergic reaction, worsening of inflammatory conditions, and splenic enlargement. The most common adverse event associated with patients receiving filgrastim relative to patients receiving placebo appears to be bone pain. This event is dose-related and commences shortly after beginning treatment with filgrastim and may reoccur or worsen shortly before neutrophilic recovery in patients who have received chemotherapy [21].

Early in the use of filgrastim in the setting of chemotherapy, reports of possible pulmonary toxicity associated with filgrastim and bleomycin surfaced. Critical review of several randomized and nonrandomized studies suggested no increase in the known pulmonary toxicity. The studies were done in the settings of non-Hodgkin's lymphoma (NHL) [22–26], and in metastatic teratoma [27], germ-cell tumors [28], and advanced testicular cancer [29].

In a pivotal randomized, placebo-controlled, double-blind phase 3 study, 210 patients with nonsmall-cell lung cancer (NSCLC) received chemotherapy with or without filgrastim [30]. The most commonly reported adverse event was mild-to-moderate medullary bone pain that was treated with non-narcotic analgesics. A total of 6% of patients given filgrastim reported allergic reactions. None of the expected side effects of human protein administration such as fever, fluid retention, arthralgia, and malaise were reported in this double-blind study.

2.3 Patients Receiving Chemotherapy with Concomitant Thoracic Radiotherapy

The use of rHuG-CSF before chemotherapy and during thoracic radiation is not recommended due to the risk of more marrow damage. A small number of studies have shown that use of growth factors, primarily studies with rHuGM-CSF, during concomitant radiation will increase the risk of thrombocytopenia [31–33].

The use of concomitant filgrastim was evaluated in 38 patients with small-cell lung cancer (SCLC) receiving cyclical chemotherapy with concurrent mediastanal irradiation, and the authors reported no increase in pulmonary toxicity associated with concomitant use of filgrastim during thoracic radiation [34]; however, thrombocytopenia did occur and it is unclear if this side effect outweighs the use of filgrastim in this setting.

2.4 Patients with AML Receiving Induction or Consolidation Chemotherapy

Induction and consolidation therapy for treatment of acute myeloid leukemia (AML) has a high mortality rate associated with a high risk of prolonged severe neutropenia, an ideal setting for intervention with filgrastim. A randomized, placebo-controlled phase 3 study evaluated the clinical utility of adjunctive filgrastim in patients with AML receiving remission induction and consolidation chemotherapy [35]. Adverse events including allergic reactions and bone pain were slightly higher in the filgrastim group compared with the placebo group. Long-term outcome of this study confirms the earlier result that filgrastim does not have any untoward impact on survival of patients with AML and is not associated with any secondary malignancies [36]. The issue may not be fully resolved, however, as the perceived relationship between the use of any hematopoietic growth factor and the risk of developing leukemia remains controversial in some patient populations [37].

2.5 Patients with NHL Receiving Stem Cell Transplantation

Studies have been reported for patients with NHL who received stem cell transplantation – either bone marrow or PBPC. Some of the earliest work in stem cell transplantation, specifically bone marrow, included patients with NHL who were receiving filgrastim support for marrow recovery [38]. In this study, no significant toxicity was noted beyond localized erythema at 2/88 infusion sites, and no significant difference was reported in veno-occulsive disease of the liver or interstitial pneumonia between the filgrastim and placebo groups.

Several early, generally small, studies in the setting of PBPC mobilization, collection, and reinfusion enrolled patients with NHL [39–42]. No untoward adverse events were reported in these studies. A comparison of filgrastim-mobilized PBPC versus autologous bone marrow transplant was evaluated in 58 patients with NHL [43]. The group receiving filgrastim-mobilized PBPC had a lower number of platelet transfusions and a shorter duration to platelet recovery and neutrophil recovery, which led to fewer days of hospitalization compared with patients who received bone marrow. No adverse events were attributed to the filgrastim-mobilized PBPC procedure.

2.6 Peripheral Blood Progenitor Cell Collection and Therapy in Cancer Patients

While use of high-dose chemotherapy has been an important advance in the treatment of patients with cancer, the consequence of this high-dose therapy is often a temporary or permanent ablation of hematopoietic activity and an increased morbidity and mortality. The focus of research turned to discovery of ways to abrogate the neutropenia, as well as anemia and thrombocytopenia, of high-dose chemotherapy. Bone marrow transplantations, both autologous and allogeneic, were first steps, but with the advent of hematopoietic growth factors, the utility of filgrastim in mobilizing PBPC for reinfusion after ablative therapy became apparent. Collection of PBPC is inherently less hazardous than harvesting of bone marrow and there was hope that reinfusion of PBPC would be without tumor contamination.

In studies with a total of 126 patients undergoing PBPC collection, filgrastim treatment was associated with a 44% incidence of mild-to-moderate muscle or bone pain with a 7% rate of headache [44]. Reversible increases in serum alkaline phosphatase occurred in 21% of patients. All patients had an increase in neutrophil counts, with two patients experiencing significant counts $>100 \times 10^9$/L with no associated clinical sequelae. Mild-to-moderate anemia and thromobocytopenia did occur in most patients, suggesting that treatment with filgrastim for mobilization can lead to lineage steal and a temporary decrease in erythrocyte and thrombocyte production.

2.7 Patients with Severe Chronic Neutropenia

Patients with idiopathic or genetic abnormalities in neutrophil production and regulation suffer with susceptibility to chronic infections, another ideal setting for the utility of filgrastim. As filgrastim was originally approved only for short-term use, the safety of the product in chronic use in children was a serious question that needed to be answered. In a phase 3 study, 123 patients with documented severe chronic neutropenia (SCN) and an absolute neutrophil count (ANC) $<0.5 \times 10^9$/L were randomly assigned to receive filgrastim or to undergo a 4-month observation period followed by treatment with filgrastim [45]. Filgrastim treatment resulted in a correction in the ANC and reduction in infection rate. The safety profile was characterized by a 30% incidence in mild and transient headache, bone pain, skin rash and manageable thrombocytopenia. Asymptomatic splenomegaly did occur in more than half of the patients but did not result in splenectomies or significant clinical sequelae. The enlargement of the spleen in response to filgrastim is attributed to extramedullary hematopoiesis which has been observed in animal models.

Safety data of long-term use of filgrastim have been collected in the Severe Chronic Neutropenia International Registry (SCNIR) and a subset of patients receiving filgrastim for congenital neutropenia has been monitored for any emergent adverse events. Case reports suggest a possible association between long-term filgrastim and splenomegaly, osteopenia, osteoporosis, vasculitis, retarded growth, and development. With treatment of 7 years or more, there is an increased risk of malignant myeloid transformation which is associated with filgrastim given for years [46, 47]. The underlying disease of chronic neutropenia, however, can also put these patients at risk of myeloid transformation.

2.8 Patients with Active Infection

Nonclinical studies suggested that filgrastim regulated the survival, proliferation, and differentiation of precursor cells of neutrophilic granulocytes, and functionally activated mature neutrophils [1]. The functional properties of neutrophils that are enhanced by filgrastim are those related to host defenses and the concentration of endogenous G-CSF has been shown to increase in a variety of infections [48]. Taken together, these observations suggested a role for filgrastim in patients with infectious diseases.

Adjunctive filgrastim in combination with antibiotics was evaluated in a phase 1 study of non-neutropenic patients with pneumonia [49]. A total of 30 patients with community-acquired pneumonia received daily filgrastim subcutaneous doses ranging from 75 to 600 μg for 10 days or until their ANC reached or exceeded 0.75×10^9/L. Safety evaluation included vital signs, pulse oximetry, arterial blood gases, daily complete blood counts with differential, serum chemistries, coagulation profiles, electrocardiograms, and chest radiographs. The results of the study indicated no evidence of pulmonary toxicity or exacerbation of the infection. Two large, randomized phase 3 studies of patients with pneumonia compared standard antibiotics with or without the addition of filgrastim and the results confirmed the safety of filgrastim in this population and demonstrated an improved resolution of chest infection based on radiographic evidence [50, 51].

2.9 Patients with HIV Infection

The hallmark of human immunodeficiency virus (HIV) infection is defects in the production and function of $CD4^+$ helper cells; anemia, neutropenia, and thrombocytopenia are major clinical problems. Patients with HIV infection have poor hematopoietic reserves [52, 53]. Recurrent bacterial infections are recognized as criteria for the diagnosis of AIDS [54]. In clinical studies of patients with HIV infection receiving filgrastim, the incidence of adverse events was similar to that reported in cancer patients and consisted of musculoskeletal pain, predominantly mild-to-moderate bone pain and myalgia [55, 56]. Splenic enlargement has been reported to be related to filgrastim therapy in <3% of patients with HIV infection/AIDS, but the splenomegaly is mild or moderate and does not result in splenectomy [57]. As the finding of splenic enlargement is common in patients with HIV infection and also common in large number of patients with AIDS, the relationship to filgrastim treatment is unclear.

2.10 Patients with Renal or Hepatic Impairment

Studies of filgrastim in patients with severe impairment of renal or hepatic function demonstrate that it exhibits a similar pharmacokinetic and pharmacodynamic profile to that seen in normal individuals, and the safety profile is similar to that seen in patients with normal renal and hepatic function [57]. No dose adjustments are required for these special populations.

3 Pegfilgrastim

Pegfilgrastim is a chemically modified derivative of filgrastim in which a polyethylene glycol (PEG) molecule is covalently attached to the *N* terminus of the peptide. This polyethylene glycol tail has no intrinsic biologic activity but does alter the pharmacokinetics and pharmacodynamic profiles enabling the half-life to be extended from 3 h to approximately 80 h and subsequently reduced clearance from the systemic circulation [58]. This extended half-life can support one injection per cycle of chemotherapy rather than daily dosing that is required with filgrastim. Pegylation of filgrastim is advantageous in that a single injection is sufficient to produce neutrophil recovery. Pegylation of proteins is also useful for reduction in immunogenicity of the native protein; however, filgrastim has not been shown to be highly immunogenic in which neutralizing antibodies that cross-react with endogenous GCSF have been detected. The effect of pegylation of filgrastim to reduce immunogencity has not been evaluated as filgrastim is not highly immunogenic. Unlike filgrastim, pegfilgrastim is only approved for the prevention and treatment of febrile neutropenia and unlike the short-acting filgrastim, pegfilgrastim is not approved for peripheral stem cell mobilization for transplantation or for use in patients with myeloid cancers.

3.1 Overview

Pegylation is a process by which a polyethylene glycol molecule is attached to a native protein in order to stabilize the protein, reduce degradation of the protein, and reduce the immunogenic potential [59]. Polyethylene glycol is categorized by the US Food and Drug Administration (FDA) as "Generally Recognized As Safe" (i.e., the GRAS List). However, attachment of a polyethylene glycol molecule to an active molecule may reduce the potency of the therapeutic molecule due to stearic hindrance but is offset by a longer circulating half-life [60].

While single-dose per cycle pegfilgrastim has comparable efficacy compared to daily injections of filgrastim, the side-effect profile has been shown to differ from filgrastim with a greater number of warning labels. Postmarketing safety studies indicate that use of pegfilgrastim is associated with severe allergic reactions such as anaphylaxis, angioedema, or urticaria, and splenic rupture, including fatal cases, has been reported with pegfilgrastim [57]. In addition to fatal splenic rupture, rare cases of acute respiratory distress syndrome (ARDS) have occurred in patients receiving pegfilgrastim [57]. The exact mechanism of the ARDS is unknown but with a longer half-life and enhanced neutrophil function may play a role.

3.2 Patients Receiving Myelosuppressive Chemotherapy

In a phase 2 randomized study of 154 patients comparing filgrastim to pegfilgratim, the most frequently reported adverse event was mild-to-moderate medullary bone pain [61]. Despite the pegylation and longer-acting filgrastim, the duration and severity of the bone pain were similar with an overall incidence of bone pain of 35% in pegfilgrastim patients and 36% in filgrastim patients; most incidences were mild to moderate in severity. In the 30-, 60-, and 100-g/kg pegfilgrastim dose groups, the incidence of bone pain was 16%, 34% and 45%, respectively. Treatment for the bone pain included the use of non-narcotic analgesics but a few patients (7% pegfilgrastim and 12% filgrastim) did require narcotics. The safety and efficacy of pegfilgrastim compared to filgrastim has been evaluated in two phase 3 trials [61, 62]. The results show equivalent efficacy and safety with approximately 25% of patients reporting mild-to-moderate bone pain.

In one of the largest randomized study to evaluate the safety of pegfilgrastim, 928 patients with breast cancer receiving chemotherapy were randomly assigned to receive pegfilgrastim or placebo [63]. The addition of pegfilgrastim reduced the febrile neutropenia rate by 94% and the side-effect profile was consistent with other large randomized studies.

3.3 Use of Pegfilgrastim in Special Populations

3.3.1 Pediatric

To obtain marketing approval by health authorities, drug sponsors must submit clinical data in the form of results from randomized controlled trials, most often conducted in adults. Many medicines that receive marketing approval can be used and often are used in the pediatric population off-label; however, a study suggests that some physicians are unaware that they are prescribing medicines off-label [64]. Off-label use of medications to treat children may produce no therapeutic benefit, but expose the child to all potential risks [65]. By virtue of its off-label status, the pharmacokinetics and pharmacodynamics of the medicine have not been studied in the pediatric population, and inherent metabolic differences between adults and children may not be detected by extrapolation methods [65]. Off-label use may produce, years later, serious, debilitating, or fatal results. Pediatric studies are mandated by the FDA and other health authorities.

Small nonrandomized studies have been conducted to evaluate pegfilgrastim in children. A total of 28 pediatric patients were given 126 injections of pegfilgrastim [66]. Adverse events included four patients with bone pain and two patients with headache.

A randomized study comparing filgrastim to pegfilgrastim was evaluated in 44 pediatric patients undergoing myelosuppressive chemotherapy for sarcoma [67]. No differences with adverse event incidence were detected across the two treatment groups, with bone pain being the most commonly reported adverse event.

3.3.2 Geriatric

The FDA also encourages drug sponsors to include elderly patients in studies of new drugs or new indications. Elderly patients are major consumers of drug products and to neglect to study the effects of drugs in this population does not provide a complete safety profile of the product.

A total of 852 elderly patients with either solid tumor or NHL who were eligible for treatment with myelosuppressive chemotherapy were randomly assigned to receive either prophylactic pegfilgrastim or physician's choice for reactive use of pegfilgrastim [68]. Severe arthralgia was a commonly reported adverse event and was considered to be related to treatment with pegfilgrastim. Relative to other populations, the reported incidence of bone pain was low with the overall incidence ranging from 9 to 12% across all groups.

3.3.3 Renal Impairment

A phase 1 study in 30 nonneutropenic patients with varying degrees of renal function was conducted to determine if renal clearance is an important determinant in the pharmacokinetics and pharmacodynamic profiles of pegfilgrastim [69]. Patients with normal, mildly impaired, moderately impaired, severely impaired, and end-stage renal disease received a single subcutaneous injection of pegfilgrastim. The results indicate no difference in pharmacokinetics and pharmacodynamic relationships, and suggest renal impairment does not impact clearance of pegfilgrastim and therefore is not a consideration from a safety perspective.

4 Biosimilar rHuG-CSF and Next Generation of rHuG-CSF

Recently, a new filgrastim biosimilar was approved in the European Union. Nivestim is a new filgrastim available in three strengths in prefilled syringes with a needle-safe device enabling self-administration at home. In a randomized phase 3 study, Nivestim demonstrated comparable efficacy to Neupogen (the original filgrastim) in the prevention of febrile neutropenia, and was as well tolerated, with a similar adverse event profile and no unexpected or untoward side effects [70, 71].

Other drug delivery formulations of rHuG-CSF are in development. To date, no new formulation has been approved by regulatory authorities. As G-CSF is a protein and can be digested through the oral route, administration of rHuG-CSF necessitates that the product be injected. Mimetics of G-CSF are currently in early discovery development but have yet to reach the clinic [72]. The next decade will see the introduction of next generation biosimilars and new delivery modalities which will further add to the body of literature regarding the safety of short- and long-term use of rHuG-CSF.

5 Conclusion

Filgrastim and pegfilgrastim were approved initially for clinical use to address the myelosuppression in patients undergoing chemotherapy. Such a drug product must itself have a reasonably high margin of safety so as not to further add to the adverse event profile of chemotherapy. Use of rHuG-CSF for acute neutropenia and reducing duration of neutropenia has resulted in reductions in infections and hospitalizations. In rare cases, the use of these drugs has caused fatal splenic rupture and other rare but serious side effects but overall the benefit of these products has clearly outweighed the risks and has contributed to the survival of millions of cancer patients. The success of these products is an important contribution in the overall supportive care of patients receiving chemotherapy for life-threatening cancers. Long-term use of filgrastim for chronic neutropenia has also been acceptable but another decade will reveal additional data on the safety of these important products.

Acknowledgement Dedicated to the memory of Jean Willsie-Chirino, colleague and friend.

References

1. Souza LM, Boone TC, Gabrilove J et al (1986) Recombinant human granulocyte colony-stimulating factor: effects on normal and leukemic myeloid cells. Science 232:61–65
2. Morstyn G, Burgess AW (1988) Hemopoietic growth factors: a review. Cancer Res 48:5624–5637
3. Metcalf D, Morstyn M (1991) Colony-stimulating factors. In: De Vita V (ed) General biology, Biologic therapy of cancer. JB Lippincott, Philadelphia, PA, pp 417–444

4. Dührsen U, Villeval JL, Boyd J, Kannourakis G, Morstyn G, Metcalf D (1988) Effects of recombinant human granulocyte colony-stimulating factor on hematopoietic progenitor cells in cancer patients. Blood 72:2074–2081
5. Molineux A, Pojda Z, Hampson IN, Lord BI, Dexter TM (1990) Transplantation potential of peripheral blood stem cells induced by granulocyte colony-stimulating factor. Blood 76:2153–2158
6. Sheridan WP, Juttner C, Szer J et al (1990) Granulocyte colony-stimulating factor (G-CSF) in peripheral blood stem cells (PBPC) and bone marrow. Blood 76:S1
7. Welte K, Gabrilove J, Bronchud MH, Platzer E, Morstyn G (1993) Filgrastim (r-metHuG-CSF): the first 10 years. Blood 88:1907–1929
8. Murata M, Harada M, Kato S et al (1999) Peripheral blood stem cell mobilization and apheresis: analysis of adverse events in 94 normal donors. Bone Marrow Transplant 24:1065–1071
9. Leischke GJ, Morstyn G (1990) Role of G-CSF and GM-CSF in the prevention of chemotherapy-induced neutropenia. In: Mertelsman R, Herrman F (eds) Hematopoietic growth factors in clinical applications. Marcel Dekker, New York, pp 191–223
10. Groopman JE, Molina JM, Scadden DT (1989) Hematopoietic growth factors, biology and clinical applications. N Engl J Med 321:1449–1459
11. Leischke GJ, Cebon J, Morstyn G (1989) Characterization of the clinical effects after the first dose of bacterially synthesized recombinant human granulocyte-macrophage colony-stimulating factor. Blood 74:2634–2643
12. Freifeld A, Pizzo P (1991) New developments in the antimicrobial supportive care of the immunocompromised patient. Prin Pract Oncol 5:1–14
13. Antman KS, Griffin JD, Elias A et al (1988) Effect of recombinant human granulocyte-macrophage colony-stimulating factor on chemotherapy-induced myelosuppression. N Engl J Med 319:593–598
14. Champlin RE, Nimer SD, Ireland P, Oette DH, Golde DW (1989) Treatment of refractory aplastic anemia with recombinant human granulocyte-macrophage colony-stimulating factor. Blood 73:694–699
15. Barlogie B, Jagannath S, Dixon DO et al (1990) High-dose melphalan and granulocyte-macrophage colony stimulating factor for refractory multiple myeloma. Blood 76:677–680
16. Emminger W, Wmminger-Schmidmeir W, Peters C et al (1990) Capillary leak syndrome during low dose granulocyte-macrophage colony-stimulating factor (rhGM-CSF) treatment of a patient in a continuous febrile state. Blut 61:219–221
17. Guinan EC, Sieff CA, Oette DH, Nathan D (1990) A phase I/II trial of recombinant granulocyte-macrophage colony-stimulating factor for children with aplastic anemia. Blood 76:1077–1082
18. Ho AD, Del Valle F, Engelhard M et al (1990) Mitoxantrone/high-dose Ara-C and recombinant human GM-CSF in the treatment of refractory non-Hodgkin's lymphoma. A pilot study. Cancer 66:423–430
19. Arning M, Kliche KO, Schneider W (1991) GM-CSF therapy and capillary-leak syndrome. Ann Hematol 62:83
20. Steward WP, Scarffe JH, Dirix LY et al (1990) Recombinant human granulocyte-macrophage colony-stimulating factor (GM-CSF) after high-dose melphalan in patients with advanced colon cancer. Br J Cancer 61:749–754
21. Wagner LM, Furman WL (2001) Haemopoietic growth factors in paediatric oncology. Paediatr Drugs 3:195–217
22. Owaga M, Masaoka T, Mizoguchi H, Takaku F, Nakashima M (1990) A phase II study of KRN 8601 rhG-CSF on neutropenia induced by chemotherapy for malignant lymphoma – a multi-institutional placebo controlled double-blind comparative study. Gan To Kagaku Ryoho (Japan) 17:365–373
23. Pettengell R, Hurney H, Radford JA et al (1992) Granulocyte colony stimulating factor to prevent dose-limiting neutropenia in non-Hodgkin's lymphoma. A randomized clinical trial. Blood 60:1430–1436

24. Bastion Y, Reyes F, Bosly A et al (1994) Possible toxicity with the association of G-CSF and bleomycin. Lancet 343:1221–1222
25. Bastion Y, Coiffier B (1994) Pulmonary toxicity of bleomycin: is G-CSF a risk factor? Lancet 344:474; letter
26. Bertini M, Freilone R, Vitolo U et al (1994) P-VEBEC: a new 8-weekly schedule with or without rG-CSF for elderly patients with aggressive non-Hodgkin's lymphoma. Ann Oncol 5:895–900
27. Fossa S, Kaye SB, Mead BM et al (1995) An MRC/EORTC randomized trial in poor prognosis metastatic teratoma comparing treatment with/without filgrastim. Proc Am Soc Clin Oncol 14:245; abstract 656
28. Saxman SB, Nichols CR, Stephens AW, Einhorn LH (1995) Pulmonary toxicity in patients with advanced stage germ cell tumors receiving bleomycin with or without granulocyte colony stimulating factor. Proc Am Soc Clin Oncol 14:255; abstract 690
29. Blanke C, Loehrer P, Einhorn L, Nichols C (1994) A phase II study of VP-16 plus ifosfamide plus cisplatin plus vinblastine plus bleomycin (VIP/BB) with Filgrastim for advanced stage testicular cancer. Proc Am Soc Clin Oncol 13:234; abstract 723
30. Crawford J, Ozer H, Stoller R et al (1991) Reduction by granulocyte colony-stimulating factor of fever and neutropenia induced by chemotherapy in patients with small-cell lung cancer. N Engl J Med 325:164–170
31. Fushiki M, Abe M (1992) Randomized double-blind controlled study of rhG-CSF in patients with neutropenia induced by radiation therapy. Proc Am Soc Clin Oncol 11:410; abstract
32. Momin F, Kraut M, Lattin P et al (1992) Thrombocytopenia in patients receiving chemoradiotherapy and G-CSF for locally advanced non-small cell lung cancer (NSCLC). Proc Am Soc Clin Oncol 11:294; abstract
33. Bunn PA Jr, Crowley J, Kelly K et al (1995) Chemoradiotherapy with or without granulocyte-macrophage colony-stimulating factor in the treatment of limited-stage small-cell lung cancer: a prospective, phase III randomized study of the Southwest Oncology Group. J Clin Oncol 13:1632–1641
34. Sheikh H, Colaco R, Lorigan P, et al (2011) Use of G-CSF during concurrent chemotherapy and thoracic radiotherapy in patients with limited-stage small-cell lung cancer: safety data from a phase II trial. Lung Cancer 74(1):75–79
35. Heil G, Heil G, Hoelzer D, Sanz MA et al (1997) Randomized, double-blind, placebo-controlled, phase III study of filgrastim in remission induction and consolidation therapy for adults with de novo acute myeloid leukemia. Blood 90:4710–4718
36. Heil G, Hoelzer D, Sanz MA et al (2006) Long-term survival data from a phase 3 study of filgrastim as an adjunct to chemotherapy in adults with de novo acute myeloid leukemia. Leukemia 20:404–409
37. Hershman D, Neugut AI, Jacobson JS et al (2007) Acute myeloid leukemia or myelodysplastic syndrome following use of granulocyte colony-stimulating factors during breast cancer adjuvant chemotherapy. J Natl Cancer Inst 99:196–205
38. Sheridan WP, Morstyn G, Wolf M et al (1989) Granulocyte colony-stimulating factor and neutrophil recovery after high-dose chemotherapy and autologous bone marrow transplantation. Lancet 2:891–895
39. DeLuca E, Sheridan WP, Watson D, Szer J, Begley CG (1992) Prior chemotherapy does not prevent effective mobilization by G-CSF of peripheral blood progenitor cells. Br J Cancer 66:893–899
40. Fukuda M, Kojima S, Matsumoto K, Matsuyama T (1992) Autotransplantation of peripheral blood stem cells mobilized by chemotherapy and recombinant human granulocyte colony-stimulating factor in childhood neuroblastoma and non-Hodgkin's lymphoma. Br J Haematol 80:327–331
41. Bolwell BJ, Fishleder A, Andresen SW et al (1993) G-CSF primed peripheral progenitor cells in autologous bone marrow transplantation: parameters affecting bone marrow engraftment. Bone Marrow Transplant 12:609–614

42. Hohaus S, Goldschmidt H, Ehrhardt R, Haas R (1993) Successful autografting following myeloablative conditioning therapy with blood stem cells mobilized by chemotherapy plus rhG-CSF. Exp Hematol 21:508–514
43. Schmitz N, Linch DC, Dreger P et al (1996) Randomised trial of filgrastim-mobilised peripheral blood progenitor cell transplantation versus autologous bone-marrow transplantation in lymphoma patients. Lancet 347:353–357, Erratum Lancet 1996;347:914
44. Neupogen Package Insert, Amgen Inc, Thousand Oaks, CA
45. Dale DC, Bonilla MA, Davis MW et al (1993) A randomized controlled phase III trial of recombinant human G-CSF for treatment of severe chronic neutropenia. Blood 81:2496–2502
46. Dale DC, Bolyard AA, Schwinzer BG et al (2006) The Severe Chronic Neutropenia International Registry: 10-year follow-up report. Support Cancer Ther 3:220–231
47. Rosenberg PS, Alter BP, Bolyard AA et al (2006) Severe Chronic Neutropenia International Registry. The incidence of leukemia and mortality from sepsis in patients with severe congenital neutropenia receiving long-term G-CSF therapy. Blood 107:4628–4635
48. Leischke GJ, Grail D, Hodgson G et al (1994) Mice lacking granulocyte colony-stimulating factor have chronic neutropenia, granulocyte and macrophage progenitor cell deficiency, and impaired neutrophil mobilization. Blood 84:1737–1746
49. de Boisblanc BP, Mason CM, Andresen J et al (1997) Phase 1 safety trial of Filgrastim (r-metHuG-CSF) in non-neutropenic patients with severe community-acquired pneumonia. Respir Med 91:387–394
50. Nelson S, Heyder AM, Stone J et al (2000) A randomized, controlled trial of filgrastim for the treatment of hospitalized patients with multilobar pneumonia. J Infect Dis 182:970–973
51. Root RK, Lodato RF, Patrick W et al (2003) Multicenter, double-blind, placebo-controlled study of the use of filgrastim in patients hospitalized with pneumonia and severe sepsis. Crit Care Med 31:367–373
52. Richman DD, Fischl MA, Grieco MH et al (1987) The toxicity of azidothymidine (AZT) in the treatment of patients with AIDS and AIDS-related complex. A double-blind, placebo-controlled trial. N Engl J Med 117:192–217
53. Fauci AS, Pantaleo G, Stnaley S, Weissman D (1996) Immunopathogeneic mechanisms of HIV infection. Ann Intern Med 124:654–663
54. Centers for Disease Control (1987) Revision of the CDC surveillance case definitions for acquired immunodeficiency syndrome. MMWR Morb Mortal Wkly Rep 36:15
55. Miles SA, Mitsuyasu RT, Moreno J et al (1990) Recombinant human granulocyte colony-stimulating factor increases circulating burst forming unit-erythron and red blood cell production in patients with severe human immunodeficiency virus infection. Blood 75:2137–2242
56. Kuritzkes DR, Parenti D, Ward DJ et al (1998) Filgrastim prevents severe neutropenia and reduces infective morbidity in patients with advanced HIV infection: results of a randomized, multicenter, controlled trial. AIDS 12:65–74
57. EPG Online NEUPOGEN. Available at http://www.epgonline.org
58. Molineux G (2003) Pegfilgrastim: using pegylation technology to improve neutropenia support in cancer patients. Anticancer Drugs 14:259–264
59. Harris JM, Martin NE, Modi M (2001) A novel process for modifying pharmacokinetics. Clin Pharmacokinet 40:539–551
60. Fishburn CS (2008) The pharmacology of PEGylation: balancing PD with PK to generate novel therapeutics. J Pharm Sci 97:4167–4183
61. Holmes FA, Jones SE, O'Shaughnessy J et al (2002) Comparable efficacy and safety profiles of once-per-cycle pegfilgrastim and daily injection filgrastim in chemotherapy-induced neutropenia: a multicenter dose-finding study in women with breast cancer. Ann Oncol 13:903–909
62. Green MD, Koelbl H, Baselga J et al (2003) A randomized double-blind multicenter phase III study of fixed-dose single-administration pegfilgrastim versus daily filgrastim in patients receiving myelosuppressive chemotherapy. Ann Oncol 14:29–35
63. Vogel CL, Wojtukiewicz MZ, Carroll RR et al (2005) First and subsequent cycle use of pegfilgrastim prevents febrile neutropenia in patients with breast cancer: a multicenter, double-blind, placebo-controlled phase III study. J Clin Oncol 23:1178–1184

64. Chen DT, Wynia MK, Moloney RM, Alexander GC (2009) U.S. physician knowledge of the FDA-approved indications and evidence base for commonly prescribed drugs: results of a national survey. Pharmacoepidemiol Drug Saf 18:1094–1100
65. Rodriguez W, Roberts R, Murphy D (2001) Adverse drug events in children: the US Food and Drug Administration perspective. Curr Ther Res 62:711–723
66. Andre N, Milano E, Rome A, Gentet JC (2008) Safety of pegfilgrastim in children. Ann Pharmacother 42:290
67. Spunt SL, Irving H, Frost J et al (2010) Phase II, randomized, open-label study of pegfilgrastim-supported VDC/IE chemotherapy in pediatric sarcoma patients. J Clin Oncol 28:1329–1336
68. Balducci L, Al-Halawani H, Charu V et al (2007) Elderly cancer patients receiving chemotherapy benefit from first-cycle pegfilgrastim. Oncologist 12:1416–1424
69. Yang BB, Kido A, Salfi M, Swan S, Sullivan JT (2008) Pharmacokinetics and pharmacodynamics of pegfilgrastim in subjects with various degrees of renal function. J Clin Pharmacol 4:1026–1031
70. Waller CF, Bronchud M, Mair S, Challand R (2010) Comparison of the pharmacodynamic profiles of a biosimilar filgrastim and Amgen filgrastim: results from a randomized phase I trial. Ann Hematol 89:871–878
71. Waller CF, Bronchud M, Mair S, Challand R (2010) Pharmacokinetic profiles of a biosimilar filgrastim and Amgen filgrastim: results from a randomized, phase I trial. Ann Hematol 89:927–933
72. Kaushansky K (2001) Hematopoietic growth factor mimetics. Ann N Y Acad Sci 938:131–138

Economics of the Recombinant Human Granulocyte Colony-Stimulating Factors

Bradford R. Hirsch and Gary H. Lyman

1 Introduction

The cost of cancer care in the United States continues to rise quickly, outpacing both inflation and the increases seen in other areas of medicine. As a result, oncology is receiving increasing scrutiny regarding the way in which resources are utilized. Statistics from the National Institute of Health (NIH) estimate that cancer accounted for >US$200 billion in direct costs in the United States during 2007 and studies estimated the impact of indirect costs such as lost productivity to be >US$900 billion as of 2000 [1–3]. During a period of health reform, increasing focus is being placed on controlling costs and supportive care in oncology is an important area of focus.

The budget for cancer care continues to grow quickly, in part, because the cost of therapies is rising both in terms of their price and frequency of utilization. This trend is true in supportive care and is the focus of this chapter as it applies to human recombinant granulocyte colony-stimulating factor (rHuG-CSF). It is critical that the same rigor be applied to assessments of the clinical utility and cost effectiveness of their use as it is in other areas of diagnosis and treatment in oncology.

In 2006, the American Society of Clinical Oncology (ASCO) as well as the European Organization for Research and Treatment of Cancer (EORTC) and the National Comprehensive Cancer Network (NCCN) changed their recommendations regarding the prophylactic administration of rHuG-CSF to state that it should be given when the risk for febrile neutropenia is >20% for a given

B.R. Hirsch
Department of Medicine, Duke University and the Duke Cancer Institute, 2424 Erwin Road, Suite 205, 27705, Durham, NC, USA
e-mail: bradford.hirsch@duke.edu

G.H. Lyman (✉)
Duke Cancer Institute, Duke University, 2424 Erwin Road, Suite 205, Durham, NC 27705, USA
e-mail: gary.lyman@duke.edu

G. Molineux et al. (eds.), *Twenty Years of G-CSF*, Milestones in Drug Therapy,
DOI 10.1007/978-3-0348-0218-5_22, © Springer Basel AG 2012

chemotherapeutic regimen [4–6]. The use of rHuCSF was previously recommended when the risk of febrile neutropenia was >40% [7, 8]. This change was largely based on the results of cost minimization modeling that had been developed over the preceding decade. It was found that prophylaxis led to cost savings when the risk of febrile neutropenia was >20% with a given regimen. It is critical to evaluate the clinical and economic research on rHuG-CSF to understand the basis and potential impact of these recommendations.

The economic value of rHuG-CSF has been predicated on its ability to limit the complications that arise from episodes of febrile neutropenia; however, an evolving body of literature looks at the cost effectiveness and cost utility of prophylaxis based on the ability of rHuG-CSF to impact mortality and maintain the relative dose intensity of chemotherapy. Over the past few years, studies have been published which help to inform this debate. In the setting of health reform and increasing concern regarding cost controls, this area is of particular interest.

2 Background on Economic Analyses

Economic studies in medicine are often referred to as cost effectiveness analyses; however, a number of variations are included in this category. One must understand the different study types in order to interpret their meaning appropriately.

2.1 Cost Minimization

Cost minimization is often the easiest form of economic analysis to understand. In this type of study, the costs of a given medication or procedure are assessed, as are the possible savings the strategy might provide. In the case of rHuG-CSF, cost minimization studies have assessed whether prophylaxis can save money by preventing episodes of febrile neutropenia, thereby lowering overall resource utilization. If these savings outweigh the cost of providing prophylaxis, the cost of care can be minimized by providing prophylaxis. To perform the study, one must calculate the costs of the agent and its administration, as well as the costs of complications from febrile neutropenia, such as hospitalization and intravenous antibiotic use. One must also have reliable data regarding how different prophylaxis regimens alter the frequency of complications. In settings where the effect of a treatment on mortality is not clear, minimizing the cost of care is a legitimate goal.

2.2 Cost Effectiveness Analysis

Cost effectiveness analyses compare treatments, taking into account the costs incurred in each scenario as well as the benefits gained. For rHuG-CSF, a relevant study might compare the costs incurred with different agents to its ability to lower the risk of febrile neutropenia by a specified proportion. To do so, one must look at the range of associated costs of each agent and the relative efficacy of the different formulations in reducing febrile neutropenia.

These data are often reported using the incremental cost effectiveness ratio (ICER), which is calculated as:

$$\text{ICER} = \frac{\text{Cost intervention} - \text{Cost alternative}}{\text{Effect intervention} - \text{Effect alternative}} = \frac{\text{Incremental Cost}}{\text{Incremental Effect}}.$$

The units for the rHuCSF might be reported as the relative cost in dollars per percent decrease in febrile neutropenia by the two formulations.

2.3 *Cost Utility Analysis*

While similar to the cost effectiveness analysis described, cost utility is reported in measures such as life years gained (LYG) or quality-adjusted life years (QALY) gained. A cost utility analysis requires a given intervention to result in a mortality benefit as it is predicated on a difference in life years. The relative difference in mortality, as well as quality of life if known, is used to calculate the difference in LYG or QALY. As an example, if rHuG-CSF is found to result in a mortality benefit by maintaining the relative dose intensity of chemotherapy, the costs of different agents as well as their relative ability to prevent mortality could be used to report their cost utility.

This information is used in the United Kingdom to help make coverage decisions, as a cost per QALY gained of £30,000 is often quoted as the maximum amount acceptable for a new intervention to be approved. In the United States, a value of US$50,000 to US$100,000 is often mentioned as an appropriate threshold; however, no clear cutoff has ever been established or used consistently to inform coverage discussions.

3 Overview of the Pharmacoeconomics of rHuG-CSF

Increasing importance is being placed on economic studies in medicine, particularly in oncology. A review suggested that there has been a steady growth in the number of cost effectiveness publications in oncology with an increase from an average of seven papers per year on the topic in the 1990s to 25 per year between 2002 and 2007 [9]. This increase can also be seen in economic publications addressing the rHuG-CSF. While the cost minimization data are the most robust as they were the initial focus of researchers in the area, there is increasing information available on cost effectiveness and cost utility as well.

In a review of the subject matter in *Pharmacoeconomics* in 2003, a paper referenced 33 economic evaluations on the prophylactic and therapeutic use of rHuG-CSF and suggested that the studies were heterogeneous and did not clearly demonstrate cost savings outside the treatment of established febrile neutropenia and primary prophylaxis during stem cell transplantation [10]. Since the time of the publication, further studies have helped clarify the topic as there have been a number of economic questions of particular interest. The initial models looked at whether cost minimization was achieved with primary prophylaxis, independent of any mortality benefit. Positive findings in these studies led to the changes in the ASCO, EORTC, and NCCN guidelines. More recently, cost-effectiveness studies have been published comparing different formulations and therapeutic settings for the use of rHuG-CSF, such as primary versus secondary prophylaxis. Finally, cost-utility studies have begun to emerge in certain clinical scenarios, focusing on potentially curative cancers such as early-stage breast cancer and non-Hodgkin's lymphoma (NHL). While the literature is likely to continue to develop as biosimilars enter the market and as more data become available regarding the effect of rHuG-CSF on long-term mortality, an impressive body of work already exists on the subject and is reviewed here.

4 Costs of rHuG-CSF and Febrile Neutropenia

To better understand the economic models of rHuG-CSF, one must first look at the relevant costs involved. Their expense results from the price of the agents themselves and their administration. The other significant cost involved is that of the resource utilization that results from complications of febrile neutropenia, such as hospitalization and intravenous antibiotic use. While risk assessment models have been published to help stratify patients with febrile neutropenia into those who can be treated safely as an out-patient versus those requiring hospitalization, most patients are still admitted for treatment [11–13]. As such, the costs of febrile neutropenia are largely driven by hospital costs.

Costs are fairly consistent across studies, with most focusing on the direct costs from a payer's perspective. In a representative study in the United States, it was calculated that a single pegfilgrastim, a type of rHuG-CSF, injection per cycle costs US$2142 while filgrastim, another type of rHuG-CSF, cost US$266 per injection and is given daily for 7–14 days per cycle for a range of costs between US$1862 and US$3724 [14, 15]. These numbers do not include the cost of administration of the agents. In a meta-analysis, the mean and median hospital costs were found to be US $19,100 and US$8376, respectively, with an average length of stay of 8 days for patients with solid tumors and 20 days for patients with leukemia [16]. Patients admitted for >10 days accounted for 75% of the costs of treatment.

It is important to note that costs vary significantly in comparisons between Europe and the United States. The costs of hospital care are much lower in Europe compared with the United States, which can lead to substantially different cost-effectiveness calculations. Table 1 provides examples of this variation between studies. As such, economic evaluations in the United States are not always directly applicable to European settings without accounting for variations in healthcare costs.

Table 1 The costs of hospital care are much lower in Europe compared with the United States, which can lead to substantially different cost-effectiveness calculations

References	Findings
US costs	
Kuderer et al 2006 [16]	Average length of stay/episode of febrile neutropenia was 11.5 days with a mean cost/episode of $19,110
Bennet et al 2007 [42]	Direct and indirect hospital costs/febrile neutropenia episode were $17,869 and $3732 for lymphoma; $10,324 and $2832 for breast cancers; and $10,311 and $1389 for lung cancer and myeloma
Crawford et al 2009 [24]	Hospital costs/episode of febrile neutropenia in non-Hodgkin's lymphoma was $15,921 (range: $4000–$24,268) with an additional cost of 10% to account for physician fees and 40% for follow-up care after discharge
European costs	
National Institute for Health and Clinical Excellence (2008) Appendix 1: further evidence requested by appeal panel. Available at: http://www.nice.org.uk/nicemedia/pdf/LungCancerErlotinibManuRoche.pdf	Average cost/hospitalization for febrile neutropenia was £3300 across four studies in the United Kingdom
Timmer-Bonte et al 2006 [20]	Cost/episode of febrile neutropenia in the Netherlands, including hospitalization, was £3285

5 Cost Minimization Analyses

The initial economic evaluations in the field focused on the minimization of costs. In these studies, it was assumed that prophylaxis with rHuG-CSF did not lead to a mortality benefit. The authors instead assessed whether prophylaxis saves more money by decreasing utilization than it costs to provide primary prophylaxis with rHuG-CSF. Many of the initial randomized, controlled trials upon which these studies were based are from the 1990s, and the assumptions use direct costs from single institutions. All episodes of febrile neutropenia required hospitalization in the models.

These studies led to the present cutoff used by ASCO, EORTC, and NCCN of rates of febrile neutropenia >20% as the point at which the savings from decreased care utilization outweighed the cost of prophylaxis [17]. This number was derived from the two-way sensitivity analysis (Fig. 1). The costs of hospitalization were between US$1676 and US$1982 per day. At this cost, a risk of febrile neutropenia >22% made the gains from prophylaxis more robust than the expense of providing prophylaxis. When indirect costs were added in a subsequent study, the threshold was reduced to 18%, leading to the present recommendations [18].

Among patients with established febrile neutropenia, the benefit of treatment with rHuG-CSF is less straightforward. No clinical guidelines presently recommend treatment except among patients with critical illnesses. A study demonstrated a net cost savings of >US$1000 per episode through treatment with rHuG-CSF

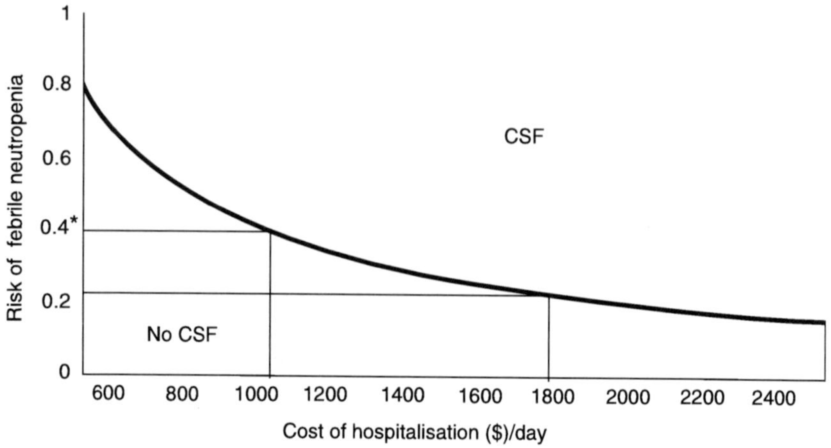

Fig. 1 Two-way sensitivity analysis, showing the relationship between the daily cost of hospitalization and the associated rate of febrile neutropenia, which are cost neutral in comparison [17, 18]

during hospitalization for febrile neutropenia as long as the anticipated length of stay was >10 days [19]. Further studies are needed to confirm these findings.

6 Cost Effectiveness and Cost Utility

Cost-effectiveness and cost-utility studies are emerging in this area. While some studies have shown an early mortality benefit from prophylaxis, the data are limited and their effect on long-term mortality is being evaluated. A few studies in the area are of particular interest.

One study looked at the addition of primary prophylaxis with rHuG-CSF to antibiotic prophylaxis [20]. Antibiotic prophylaxis is used routinely in Europe; however, concerns of antibiotic resistance prevent similar practice patterns from being adopted in the United States. In the study, the authors looked at patients with small-cell lung cancer (SCLC) who were undergoing treatment, and found a mean incremental cost of €5123 (95% CI, €3908–€6337) from the addition of rHuG-CSF to primary antibiotic prophylaxis [20]. This result was true despite a decrease in both the rate of febrile neutropenia and the complications thereof. The incremental cost to lower febrile neutropenia risk by 1% was €50 and the absolute cost per percent decrease was €240. These findings, again, must be interpreted in the context of lower healthcare costs in the Netherlands where the study was conducted, so it is unclear how representative this is of costs in the United States; however, the authors concluded that primary prophylaxis was not cost effective.

Another study comparing no prophylaxis with primary prophylaxis with filgrastim or pegfilgrastim among >26,000 patients in academic centers in the United States found that pegfilgrastim dominated [21]. With hospital costs of

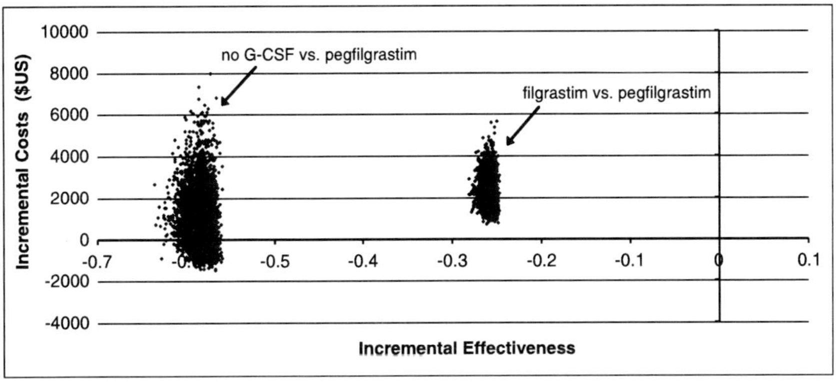

Fig. 2 Results of Monte-Carlo simulation, showing two groups of scatter plots in the cost-effectiveness plane: incremental cost effectiveness of no rHuG-CSF versus pegfilgrastim and filgrastim versus pegfilgrastim [21]

USS$1984 per day for survivors and US$3139 for patients who died, expected costs and cost effectiveness were US$3058 and 12.97 quality adjusted life-days (QALD) for pegfilgrastim, US$4203 and 2.361 QALD for no rHuG-CSF, and US$5264 and 12.698 QALD for filgrastim. This finding was true for febrile neutropenia rates >14% and held for 75% of Monte Carlo simulations (Fig. 2).

These results are consistent with a study of patients with early-stage breast cancer in which pegfilgrastim was dominant versus 11 days of filgrastim [15]. The ICER per episode of febrile neutropenia avoided was US$12,904. When mortality and quality of life were incorporated, the estimates increased to US $31,511 and US$14,415, respectively.

While few randomized controlled trials of rHuG-CSF have been powered to assess an impact on overall survival, meta-analyses of such trials have renewed interest in the potential for a reduction in both early- and long-term mortality. In a meta-analysis of randomized controlled trials of primary rHuG-CSF prophylaxis administered according to practice guidelines, significant reductions in both infection-related as well as early all-cause mortality during the course of chemotherapy were noted [22]. A systematic review and meta-analysis of 25 randomized controlled trials of systemic chemotherapy comparing primary rHuG-CSF prophylaxis to control observed a significant 10% relative risk reduction in overall all-cause mortality with an average follow-up of 5 years favoring rHuG-CSF support [23].

7 Primary Versus Secondary Prophylaxis

Present recommendations advocate the use of rHuG-CSF as primary prophylaxis, provided with the first chemotherapy cycle [4, 5, 24, 25]. Primary prophylaxis differs from secondary prophylaxis where a clinician waits for an episode of febrile neutropenia to begin prophylaxis. The guidelines are based on the fact that most patients experience febrile neutropenia during their first treatment cycle and that randomized clinical trials have shown primary prophylaxis to result in superior outcomes compared with other strategies [26–28]. Despite this fact, many clinicians use rHuG-CSF for secondary prophylaxis in practice [29, 30].

An economic analysis comparing primary to secondary prophylaxis in patients with early-stage breast cancer found an ICER of US$48,000 per febrile neutropenia episode avoided, favoring primary prophylaxis [31]. Inclusion of the possible gains of mortality benefit and an advantage realized by maintaining the relative dose intensity of chemotherapy, results in an estimate of US$110,000 per LYG or US $116,000 per QALY gained.

The benefit of adding rHuG-CSF to secondary prophylaxis with antibiotics was assessed [32]. The study did not include a direct comparison to primary prophylaxis. The authors found that the addition of rHuG-CSF increased costs from €4,496 per patient receiving antibiotic prophylaxis alone with chemotherapy to €8,998 per patient receiving antibiotics and rHuG-CSF without achieving a significant gain in benefit, leading to an unacceptably high rHuG-CSF.

8 Different Formulations of rHuG-CSF

Multiple formulations of colony-stimulating factors are available in the United States, including filgrastim, pegfilgrastim, and sargramostim. As such, studies have been published of economic comparisons among these agents. A study comparing pegfilgrastim with filgrastim in patients with NHL found an ICER of US$2167 per febrile neutropenia episode avoided, favoring pegfilgrastim [24, 33]. By including the possible mortality benefit due to the curative intent of treatment, LYG were estimated to be US$5532 while QALY gained were US$6190. A similar study in patients with early-stage breast cancer favored pegfilgrastim with an ICER of US$12,904 per febrile neutropenia episode avoided. The addition of a possible mortality benefit led to US$31,511 per LYG and US$14,415 per QALY gained [15, 24].

In a study comparing 6 days of filgrastim with 1 dose of pegfilgrastim in patients with early-stage breast cancer in the United Kingdom, the ICER was £4200 for each event avoided with a cost per percent decrease in febrile neutropenia of £42 [33]. If quality of life and mortality data were added to the calculation, the incremental cost benefit ratios favoring pegfilgrastim were £3955 per LYG and £4161 per QALY, respectively.

9 Discussion

As the focus on healthcare costs continues to increase, being able to understand the economic evaluations used to quantify the relative risks, benefits, and costs of different treatments is increasingly important. While the various forms of rHuG-CSF themselves are expensive, the literature shows that they can lead to cost savings. It also shows that it is important to utilize outcomes data from randomized clinical trials to assess cost effectiveness and utility as the results are not always intuitive.

The two-way sensitivity analysis referenced in Fig. 1 shows that primary prophylaxis with rHuG-CSF can lead to cost savings if the risk of febrile neutropenia is >20% [17, 18]. This finding was based on controlled clinical trials showing a decrease in the rate of febrile neutropenia and febrile neutropenia-related hospitalizations among those receiving primary prophylaxis, which have since been consistently reproduced [20, 22, 34–36]. This finding is independent of an effect on mortality as the model is based on a decrease in resource utilization. The effect is reproducible across a range of malignancies and chemotherapeutic regimens. The cost minimization model led to updates of the ASCO, NCCN, and EORTC recommendation over the last few years [4–6].

Cost-effectiveness analyses have helped elucidate the economic impact across agents and clinical settings. Studies comparing different formulations of rHuG-CSF have shown a cost benefit to pegfilgrastim over filgrastim, as it requires only a single injection with each cycle versus an average course of 6–14 daily injections per cycle [14, 15, 24, 33]. Studies have shown an economic benefit to primary over secondary prophylaxis [31].

Recent studies have assessed the cost utility of primary prophylaxis in terms of LYG and QALY gained. These studies evaluate the effect of factors on clinical outcomes beyond the mere avoidance of febrile neutropenia such as long-term mortality. Of particular interest is that the effect that maintaining the dose intensity of chemotherapy may have on early and overall mortality in potentially curative settings. This situation applies to malignancies such as early-stage breast cancer, NHL, and testicular cancer.

In the next few years, the economics of rHuG-CSF may change with the introduction of biosimilar agents. More than 100 biopharmaceuticals have been released in the United States since 1995, including various forms of rHu erythropoietin (rHuEPO), growth hormones, and colony-stimulating factors. Many of their patents are nearing expiration, including that of filgrastim. Filgrastim has already gone off patent in Europe and biosimilar agents are presently available there.

As biopharmaceutical agents are produced using human cells, they are not as easy to replicate as chemical pharmaceuticals and many of the publications evaluating their comparative efficacy studies are early phase trials [37, 38]. European health authorities required abbreviated testing of the compounds before approval to assist in showing comparability to their patented equivalents; however, the process was much less stringent than that imposed on novel biopharmaceuticals. The possibility of differences in immunogenicity and efficacy remains real [39–41]. The approval process in the United States is presently being established; however, the agents are expected to be introduced at low costs, potentially altering the economics of rHuG-CSF significantly.

Earlier work assessing the pharmacoeconomics of rHuG-CSF paints a convincing picture demonstrating that primary prophylaxis with pegfilgrastim can minimize costs. As the body of work continues to expand there is promise that rHuG-CSF support will provide a long-term mortality benefit in certain clinical settings further demonstrating the overall cost-effectiveness and value of appropriate G-CSF support of patients receiving cancer chemotherapy.

References

1. Cancer Trends Progress Report – 2009/2010 Update. National Cancer Institute, NIH, DHHS, Bethesda, MD April 2010; Available from: http://progressreport.cancer.gov
2. Elting LS, Shih YC (2004) The economic burden of supportive care of cancer patients. Support Care Cancer 12:219–226
3. Yabroff KR, Bradley CJ, Mariotto AB, Brown ML, Feuer EJ (2008) Estimates and projections of value of life lost from cancer deaths in the United States. J Natl Cancer Inst 100:1755–1762
4. Aapro MS, Cameron DA, Pettengell R et al (2006) EORTC guidelines for the use of granulocyte-colony stimulating factor to reduce the incidence of chemotherapy-induced febrile neutropenia in adult patients with lymphomas and solid tumours. Eur J Cancer 42:2433–2453
5. Smith TJ, Khatcheressian J, Lyman GH et al (2006) 2006 update of recommendations for the use of white blood cell growth factors: an evidence-based clinical practice guideline. J Clin Oncol 24:3187–3205
6. Lyman GH (2005) Guidelines of the National Comprehensive Cancer Network on the use of myeloid growth factors with cancer chemotherapy: a review of the evidence. J Natl Compr Canc Netw 3:557–571
7. Ozer H, Armitage JO, Bennett CL et al (2000) 2000 update of recommendations for the use of hematopoietic colony-stimulating factors: evidence-based, clinical practice guidelines. American Society of Clinical Oncology Growth Factors Expert Panel. J Clin Oncol 18: 3558–3585
8. American Society of Clinical Oncology (1996) Outcomes of cancer treatment for technology assessment and cancer treatment guidelines. J Clin Oncol 14:671–679
9. Greenberg D et al (2010) When is cancer care cost-effective? A systematic overview of cost-utility analyses in oncology. J Natl Cancer Inst 102:82–88
10. Esser M, Brunner H (2003) Economic evaluations of granulocyte colony-stimulating factor: in the prevention and treatment of chemotherapy-induced neutropenia. Pharmacoeconomics 21:1295–1313
11. Klastersky J, Paesmans M, Rubenstein EB et al (2000) The Multinational Association for Supportive Care in Cancer risk index: a multinational scoring system for identifying low-risk febrile neutropenic cancer patients. J Clin Oncol 18:3038–3051
12. Talcott JA, Siegal RD, Finberg R, Goldman L (1992) Risk assessment in cancer patients with fever and neutropenia: a prospective, two-center validation of a prediction rule. J Clin Oncol 10:316–322
13. Uys A, Rapoport BL, Anderson R (2004) Febrile neutropenia: a prospective study to validate the Multinational Association of Supportive Care of Cancer (MASCC) risk-index score. Support Care Cancer 12:555–560
14. Lyman G, Lalla A, Barron R, Dubois RW (2009) Cost-effectiveness of pegfilgrastim versus 6-day filgrastim primary prophylaxis in patients with non-Hodgkin's lymphoma receiving CHOP-21 in United States. Curr Med Res Opin 25:401–411
15. Lyman GH, Lalla A, Barron R, Dubois RW (2009) Cost-effectiveness of pegfilgrastim versus filgrastim primary prophylaxis in women with early-stage breast cancer receiving chemotherapy in the United States. Clin Ther 31:1092–1104
16. Kuderer NM, Dale DC, Crawford J, Cosler LE, Lyman GH (2006) Mortality, morbidity, and cost associated with febrile neutropenia in adult cancer patients. Cancer 106:2258–2266
17. Lyman GH, Kuderer NM, Greene J, Balducci L (1998) The economics of febrile neutropenia: implications for the use of colony-stimulating factors. Eur J Cancer 34:1857–1864

18. Cosler LE, Calhoun EA, Agboola O, Lyman GH (2004) Effects of indirect and additional direct costs on the risk threshold for prophylaxis with colony-stimulating factors in patients at risk for severe neutropenia from cancer chemotherapy. Pharmacotherapy 24:488–494
19. Cosler LE, Eldar-Lissai A, Culakova E et al (2007) Therapeutic use of granulocyte colony-stimulating factors for established febrile neutropenia: effect on costs from a hospital perspective. Pharmacoeconomics 25:343–351
20. Timmer-Bonte JN, Adang EM, Smit HJ et al (2006) Cost-effectiveness of adding granulocyte colony-stimulating factor to primary prophylaxis with antibiotics in small-cell lung cancer. J Clin Oncol 24:2991–2997
21. Eldar-Lissai A, Cosler LE, Culakova E, Lyman GH (2008) Economic analysis of prophylactic pegfilgrastim in adult cancer patients receiving chemotherapy. Value Health 11:172–1179
22. Kuderer NM, Dale DC, Crawford J, Lyman GH (2007) Impact of primary prophylaxis with granulocyte colony-stimulating factor on febrile neutropenia and mortality in adult cancer patients receiving chemotherapy: a systematic review. J Clin Oncol 25:3158–3167
23. Lyman GH, Dale DC, Wolff DA et al (2010) Acute myeloid leukemia or myelodysplastic syndrome in randomized controlled clinical trials of cancer chemotherapy with granulocyte colony-stimulating factor: a systematic review. J Clin Oncol 28:2914–2924
24. Crawford J, Armitage J, Balducci L et al (2009) Myeloid growth factors. J Natl Compr Canc Netw 7:64–83
25. Lyman GH, Kleiner JM (2007) Summary and comparison of myeloid growth factor guidelines in patients receiving cancer chemotherapy. J Natl Compr Canc Netw 5:217–228
26. Crawford J, Dale DC, Kuderer NM et al (2008) Risk and timing of neutropenic events in adult cancer patients receiving chemotherapy: the results of a prospective nationwide study of oncology practice. J Natl Compr Canc Netw 6:109–118
27. Lyman GH, Delgado DJ (2003) Risk and timing of hospitalization for febrile neutropenia in patients receiving CHOP, CHOP-R, or CNOP chemotherapy for intermediate-grade non-Hodgkin lymphoma. Cancer 98:2402–2409
28. Haithcox S, Ramnes C, Lee H, Lu J, Lyman GH (2003) The impact of frequent injections for hematopoietic growth factor support on patients receiving chemotherapy: an observational study. BMC Nurs 2:2
29. Lyman GH, Dale DC, Crawford J (2003) Incidence and predictors of low dose-intensity in adjuvant breast cancer chemotherapy: a nationwide study of community practices. J Clin Oncol 21:4524–4531
30. Lyman GH, Dale DC, Friedberg J, Crawford J, Fisher RI (2004) Incidence and predictors of low chemotherapy dose-intensity in aggressive non-Hodgkin's lymphoma: a nationwide study. J Clin Oncol 22:4302–4311
31. Ramsey SD, Liu Z, Boer R et al (2009) Cost-effectiveness of primary versus secondary prophylaxis with pegfilgrastim in women with early-stage breast cancer receiving chemotherapy. Value Health 12:217–225
32. Timmer-Bonte JN, Adang EM, Termeer E, Severens JL, Tjan-Heijnen VC (2008) Modeling the cost effectiveness of secondary febrile neutropenia prophylaxis during standard-dose chemotherapy. J Clin Oncol 26:290–296
33. Liu Z, Doan QV, Malin J, Leonard R (2009) The economic value of primary prophylaxis using pegfilgrastim compared with filgrastim in patients with breast cancer in the UK. Appl Health Econ Health Policy 7:193–205
34. Crawford J, Ozer H, Stoller R et al (1991) Reduction by granulocyte colony-stimulating factor of fever and neutropenia induced by chemotherapy in patients with small-cell lung cancer. N Engl J Med 325:164–170
35. Vogel CL, Wojtukiewicz MZ, Carroll RR et al (2005) First and subsequent cycle use of pegfilgrastim prevents febrile neutropenia in patients with breast cancer: a multicenter, double-blind, placebo-controlled phase III study. J Clin Oncol 23:1178–1184
36. Timmer-Bonte JN, de Boo TM, Smith HJ et al (2005) Prevention of chemotherapy-induced febrile neutropenia by prophylactic antibiotics plus or minus granulocyte colony-stimulating

factor in small-cell lung cancer: a Dutch Randomized Phase III Study. J Clin Oncol 23:7974–7984
37. Skrlin A, Radic I, Schwinke D et al (2010) Comparison of the physicochemical properties of a biosimilar filgrastim with those of reference filgrastim. Biologicals 38:557–566
38. Waller CF, Bronchud M, Mair S, Challand R (2010) Pharmacokinetic profiles of a biosimilar filgrastim and Amgen filgrastim: results from a randomized, phase I trial. Ann Hematol 89:927–933
39. Jelkmann W (2010) Biosimilar epoetins and other "follow-on" biologics: update on the European experiences. Am J Hematol 85:771–780
40. Mellstedt H (2010) Implications of the development of biosimilars for cancer treatment. Future Oncol 6:1065–1067
41. Mellstedt H, Niederwieser D, Ludwig H (2008) The challenge of biosimilars. Ann Oncol 19:411–419
42. Bennett CL, Calhoun EA (2007) Evaluating the total costs of chemotherapy-induced febrile neutropenia: results from a pilot study with community oncology cancer patients. Oncologist 12:478–483

Part V
The Next 20 Years

Role of Myeloid Cells in Tumor Angiogenesis

Napoleone Ferrara

1 Introduction

Angiogenesis is a fundamental process in embryonic and adult life and is also important for tissue repair [1, 2]. Normal microvessels are organized as highly ordered structures consisting of endothelial cells, pericytes, and basement membrane. Pericytes are required for vascular stabilization through establishment of contact with endothelial cells along the length of the vessels and also through paracrine signaling [3]. Angiogenesis is a relatively rare event in the adult, except in particular circumstances such as the cyclical growth of vessels in the ovarian corpus luteum [4] or during pregnancy [2]. Angiogenesis is also required for a number of pathologic processes, including tumorigenesis and intraocular neovascular diseases such as the wet form of age-related macular degeneration [5–7]. Over the last several years, clinical trials have demonstrated the clinical benefits conferred by antiangiogenic agents on patients with cancer or age-related macular degeneration (reviewed in [5, 8, 9]).

Vascular endothelial growth factor (VEGF)-A is a pivotal regulator of angiogenesis [10]. In mammals, the VEGF family comprises five members, VEGF-A (thereafter called VEGF), VEGF-B, VEGF-C, VEGF-D, and PlGF (placenta growth factor) [11]. Members of VEGF family bind to 3 tyrosine kinase receptors, VEGFR-1, VEGFR-2, and VEGFR-3. VEGFR-1 and VEGFR-2 are expressed in vascular endothelial cells, monocytes, macrophages, and hematopoietic stem cells. Early studies demonstrated that VEGF blockade with neutralizing antibodies suppresses tumor growth and angiogenesis in several models [12]. These studies prompted efforts to develop therapies to inhibit the VEGF-signaling pathway by targeting either the ligands or the receptors [6, 13, 14]. VEGF-C and VEGF-D have

N. Ferrara (✉)
Genentech Inc., 1 DNA Way, South San Francisco, CA 94080, USA
e-mail: nf@gene.com

G. Molineux et al. (eds.), *Twenty Years of G-CSF*, Milestones in Drug Therapy,
DOI 10.1007/978-3-0348-0218-5_23, © Springer Basel AG 2012

been characterized for their role in lympangiogenesis by virtue of their interaction with VEGFR-3 [15].

Bevacizumab, a humanized variant of an anti-VEGF-neutralizing monoclonal antibody [16, 17], is the first antiangiogenic agent to be approved by the US Food and Drug Administration (FDA) for cancer therapy. The drug was first approved for previously untreated metastatic colorectal cancer [18]. Subsequently it was approved for relapsed metastatic colorectal cancer, nonsmall-cell lung cancer (NSCLC) [19], renal cell carcinoma [20], and glioblastoma multiforme. In addition, a variety of small-molecule receptor tyrosine kinase inhibitors (RTKI) targeting VEGF receptor (VEGFR) signaling have been developed [13, 21]. Three VEGFR TKI have been approved by the FDA for therapy of renal cell carcinoma and other malignancies: sunitinib, sorafenib, and pazopanib.

2 Role of Myeloid Cells in Tumor Growth

Much evidence supports the notion that various bone marrow-derived cell types play important roles in regulating tumor angiogenesis and growth [22–27]. Among these cells, macrophages have been long characterized as a highly plastic cell type [28, 29]. Monocytes and macrophages have a role in physiologic angiogenesis [30], but their role in chronic diseases are especially prominent. Macrophages are present in many solid tumors, and the extent of their presence correlates with poor prognosis [31]. Indeed, gene expression profiling of stromal cells from breast cancers revealed a stromal gene expression signature associated with poor clinical outcome that included genes linked to angiogenesis, hypoxia, and tumor-associated macrophages (TAM) [32]. Recruitment of macrophages can be mediated by a number of tumor-derived chemoattractants, most notably CSF1, CCL2, CCL5, and VEGF [30, 33, 34]. In established tumors, the microenvironment often directs macrophage polarization from the M1 (classically activated) state, associated with the proinflammatory tumor and tissue-destructive response, to the M2 (alternatively activated) state, which elicits an immunosuppressive phenotype and promotes angiogenesis, tissue remodeling, and wound repair [35]. TAM, which resemble M2 macrophages phenotypically, have been shown to accumulate in avascular, necrotic regions of advanced tumors and express HIF (hypoxia inducible factor)-1α and HIF-2α resulting in the transcription of numerous genes that promote revascularization, some of which include multiple proangiogenic factors and modulators, such as VEGF, tumor-necrosis factor (TNF)-α, interleukin (IL)-8, β fibroblast growth factor (FGF), IL-1β, matrix metalloproteinase (MMP)9, and Sema4D [36].

Tie2-expressing monocytes (TEM) are a myeloid cell population that has received considerable attention [37, 38]. These cells have been reported to promote tumor growth, in part through secretion of angiogenic factors [37]. It has been suggested that tumor-derived angiopoietin (Ang)-2 plays a key role in recruiting TEM into the tumor microenvironment [39, 40]. These findings explain, at least in part, the ability of Ang-2 inhibitors to reduce tumor angiogenesis [41]. Indeed, a human antibody [42] blocking Ang-2 is undergoing clinical development. Furthermore, AMG386, a peptibody targeting Ang-1 and Ang2 [43], has shown early evidence of clinical efficacy in ovarian cancer patients (reviewed in [44]). Several observations suggest that neutrophils play an important role in tumor angiogenesis [45]. Similar to macrophages, tumor-associated neutrophils (TAN) may have different roles in tumor growth depending on their polarization status [46].

The association of mast cell accumulation with increased vasculature and poor prognosis has been observed in a wide panel of human tumors [36]. Tumor recruitment of these cells is mediated by stem cell factor (SCF), IL-3, and adrenomedullin [36]. These cells, which are preferentially localized at the tumor periphery, are a major proangiogenic source of VEGF, FGF2, IL-8, transforming growth factor (TGF)-β, TNF-α, Ang-1, and serine proteases [47].

Much research has focused on a population of myeloid cells, identified in the mouse by the expression of the cell-surface markers CD11β and Gr1 that include a variety of cell types such as neutrophils, immature dendritic cells, monocytes, and early myeloid progenitors. In cancer, interest in these cells stems from their ability to promote angiogenesis [48]. Furthermore, a subset of $CD11b^+Gr1^+$ cells, termed myeloid-derived suppressor cells (MDSC), has the ability to suppress T-cell responses and thus can promote tumor escape and progression [49–51].

3 Contribution of Myeloid Cells to VEGF-Independent Angiogenesis

To gain insight into the mechanisms of refractoriness to anti-VEGF therapy, we screened a series of murine cell lines to establish experimental tumor models that are responsive/refractory to the treatment [52]. Interestingly, refractory tumors (LLC, EL-4) were associated with a significant increase in the frequency of tumor-infiltrating $CD11b^+Gr1^+$ cells, compared to sensitive ones (B16-F1, TIB-6). Experiments in which tumor cells were mixed with $CD11b^+Gr1^+$ cells before host introduction elucidated the functional relevance of these cells. $CD11b^+Gr1^+$ cells isolated from refractory tumors, but not from sensitive tumors, were able to mediate refractoriness to anti-VEGF treatment [52]. Moreover, combination of an anti-Gr1 antibody with anti-VEGF delayed the onset of refractoriness [52]. Importantly, anti-VEGF refractoriness was observed in immunocompetent C57BL/6 or in immunocompromised XID mice, suggesting that the accumulation of $CD11b^+Gr1^+$ cells and their role in mediating refractoriness were not dependent on the adaptive immune system [52]. As previous studies implicated VEGFR-1 in recruitment of myeloid cells [53], we tested the hypothesis that the VEGFR-1 selective ligands PlGF or VEGF-B [54, 55] may mediate refractoriness to a therapy that only targets VEGF. To this end, mice harboring sensitive or refractory tumors were treated with mFlt(1-3)-IgG, a chimeric soluble VEGFR-1 variant that not only neutralizes VEGF but also PlGF and VEGF-B [56]. However, there was no difference between anti-VEGF or mFlt(1-3)-IgG-treated groups in tumor size or in the accumulation of $CD11b^+Gr1^+$ cells, arguing against a role for these ligands in VEGF-independent growth [52]. In apparent conflict with these conclusions, it was reported that targeting PlGF with neutralizing antibodies elicits significant antitumor effects in some anti-VEGFR2-refractory models, and also shows additive efficacy with an anti-VEGFR2 antibody [57]. To address such discrepancies, we developed and tested a panel of novel anti-PlGF neutralizing antibodies [58]. PlGF blockade with such novel antibodies had no significant effect on primary growth in 15 tumor models [58]. Also, neither the combination of anti-PlGF and anti-VEGF antibodies nor neutralization of all VEGFR-1 ligands with a soluble receptor further inhibited tumor angiogenesis or tumor growth compared with anti-VEGF- alone [58].

4 Role of G-CSF-Bv8 in Refractoriness to Anti-VEGF Therapy

In the course of studies aimed at identifying mediators of VEGF-independent angiogenesis, we discovered that the Bv8 protein (also known as prokineticin-2) [59–61] is upregulated in $CD11b^+Gr1^+$ cells associated with tumors refractory to anti-VEGF. Interestingly, Bv8 expression in $CD11b^+Gr1^+$ cells was strongly induced by G-CSF [62]. As extensively discussed throughout this book, G-CSF is a major regulator of granulopoiesis [63] and plays a key role in neutrophil mobilization from the bone marrow [64]. Another attractive feature of G-CSF as a potential mediator of a communication between tumor and bone marrow is its endocrine mode of action [63].

Bv8 blockade using neutralizing antibodies reduced tumor angiogenesis and growth and exhibited additive effects with anti-VEGF antibodies [62]. Additional studies indicated that production of G-CSF by tumor or stromal cells strongly correlated with refractoriness to anti-VEGF in mouse models [65]. Compared with other hematopoietic candidates (granulocyte-macrophage colony-stimulating factor [GM-CSF], stromal cell-derived factor [SDF]-1α, and PlGF), G-CSF and Bv8 were found to have preferential expression in refractory compared to sensitive tumors. Treatment with the combination anti-VEGF and anti-G-CSF antibodies reduced growth of refractory tumors compared to anti-VEGF monotherapy [65]. Importantly, anti-G-CSF treatment dramatically suppressed circulating or tumor-associated $CD11b^+Gr1^+$ cells [65]. This finding is somewhat surprising, considering that previous studies had primarily implicated other myeloid growth factors (e.g., IL-6, GM-CSF, macrophage colony-stimulating factor [M-CSF]) in the mobilization of $CD11b^+Gr1^+$ cells in tumor-bearing animals [50]. Anti-G-CSF treatment dramatically reduced Bv8 levels [65]. Conversely, G-CSF delivery to animals bearing an anti-VEGF sensitive tumor resulted in reduced responsiveness to anti-VEGF treatment, through induction of Bv8-dependent angiogenesis [65]. These data suggest that, at least in the models examined, G-CSF expression by tumor or stromal cells is an important determinant of refractoriness to anti-VEGF treatment. Furthermore, very recent studies show that neutrophils infiltrating human tumors strongly express Bv8, raising the possibility that this protein plays a pathogenic role in human malignancies [66]. Taken together these findings suggest that G-CSF or Bv8 may represent therapeutic targets.

5 Role of Myeloid Cells and G-CSF in Premetastatic Priming

Metastasis is a major cause of death from solid tumors. To metastasize, tumor cells need to degrade and invade the extracellular matrix, intravasate, be carried through blood or lymphatic vessels, extravasate at the secondary site, and finally establish secondary tumors [67]. In addition, recent evidences suggest that, at least in some circumstances, tumors are able to modify the distant microenvironment before arrival of metastatic tumor cells to create the so-called "premetastatic niche." [68] Bone marrow cells are thought to be major players in these processes [69, 70]. Although several molecules have been implicated [69–72], the mechanisms of tumor-dependent bone marrow cell mobilization and the precise identity and significance of these cells in metastasis are incompletely understood. VEGFR-1 was initially implicated as one of the key regulators of bone marrow cell mobilization and premetastatic priming, owing to its expression in a population of hematopoietic progenitor cells and the ability of anti-VEGFR-1 antibodies to reduce metastasis [69]. However, more recent studies have challenged this conclusion and reported that anti-VEGFR-1 treatment has no effect in clinically relevant models of metastasis, raising the possibility that alternative pathways mediate tissue priming for metastasis [73].

We recently analyzed several metastatic and nonmetastatic breast cancer models for their ability to trigger bone marow cell mobilization [74]. This analysis led to identification of $Ly6G^{+}Ly6C^{+}$ myeloid cells as a major cell type that accumulates in premetastatic tissues and facilitates colonization by cancer cells and subsequent metastasis. We also identified tumor-derived G-CSF as a key regulator of these processes. The metastatic 4T1, 4T07, and 66c14 mouse breast cancer cells produced high levels of G-CSF. In contract, the nonmetastatic 67NR and 168FARN cells secreted very little G-CSF. Anti-G-CSF or anti-Bv8 antibodies significantly reduced lung metastasis [74]. We found that G-CSF mobilizes a subset of $CD11b^{+}Gr1^{+}$ cells, $Ly6G^{+}Ly6C^{+}$ granulocytes, from bone marrow and also induces Bv8 expression. Bv8 in turn functions as a chemoattractant that enhances mobilization of bone marrow-derived $Ly6G^{+}Ly6C^{+}$ granulocytes and facilitates their homing into the lung before arrival of tumor cells. Once in the lungs, G-CSF-mobilized $Ly6G^{+}Ly6C^{+}$ cells may serve as a major source of Bv8, MMP9, S100A8, and S100A9. MMP-9 has been shown to enhance invasion and metastasis in lungs [75, 76]. S100A8 and S100A9 proteins have been shown to implicated in the premetastatic niche and to mediate metastasis through mobilization of myeloid cells and cancer cells to lungs [70, 77, 78]. Therefore, $Ly6G^{+}Ly6C^{+}$ cells mobilized by G-CSF create a protumorigenic microenvironment that supports extravasation, survival, and growth of secondary tumors at distant organs.

Pretreatment with recombinant G-CSF (rG-CSF) was sufficient to mimic the premetastatic environment initiated by primary tumors. rG-CSF was also able to enhance the metastatic properties of several tumors. Treatment with anti-Ly6G antibody significantly reduced the numbers of rG-CSF-induced metastases, emphasizing the important role of $Ly6G^{+}Ly6C^{+}$ cells homing in the lung in metastasis. These findings suggest that G-CSF upregulation is part of a pro-oncogenic program that confers growth and survival advantages on tumor cells.

rHuG-CSF is widely used in cancer therapy as its use has substantially reduced the risks of chemotherapy-associated neutropenia [79]. A critical question is whether rHuG-CSF administration may have protumor or prometastatic effects in patients. Our data suggest that short-term administration of rHuG-CSF, when done in conjunction with cytotoxic chemotherapy, does not increase the risk of metastasis. However, prolonged exposure to high levels of G-CSF, such as those constitutively released by some tumors, might result in enhanced metastasis. Indeed, G-CSF overexpression by a variety of tumors has been correlated with a poor prognosis [80–83]. Interestingly, patients with all solid tumors may exhibit "leukemoid reactions," characterized by extreme leukocytosis [84]. In numerous cases, the leukocytosis was secondary to a paraneoplastic syndrome linked to high G-CSF production by the tumor and, although the mechanisms remained unclear, it was associated with a particularly poor prognosis [84, 85]. This further emphasizes the deleterious effects of G-CSF overproduction by tumors.

6 Other Approaches to Target Myeloid Cells

Efforts aimed at targeting multiple myeloid growth factors or their receptors are underway. Among these, GM-CSF is being tested as a therapeutic target for inflammatory disorders and clinical efforts are ongoing [86]. It has been also reported that an anti-SDF-1 antibody reduces tumorigenesis and recruitment of $CD11b^{+}Gr1^{+}$ cells in MMTW-Wnt mammary tumors [87]. The role of M-CSF in tumorigenesis has been long object of interest [88]. M-CSF is a secreted glycoprotein that regulates growth and differentiation of macrophage lineages [89]. Unlike G-CSF and GM-CSF, which bind to receptors belonging to cytokine receptor families, M-CSF interacts with a tyrosine kinase related to PDGF/VEGF receptors, cFMS [90]. Several studies have shown that blocking M-CSF or c*fms* reduces tumor growth [91–93]. These findings support the hypothesis that blocking cFMS signaling may have therapeutic applications for cancer and inflammatory disorders. Various cFMS inhibitors are currently undergoing clinical development [88].

Given the structural homologies between cFMS and VEGFR/PDGFR receptors, a variety of small molecules developed primarily as PDGFR or VEGFR inhibitors, including sunitinib [94], ABT-869 [95], and axitinib [96] inhibit cFMS autophosphorylation. However, RTKI, especially when combined with chemotherapy, have yet to demonstrate a therapeutic advantage over more selective VEGF inhibitors. More selective blockers of M-CSF/cFMS action are required to test the hypothesis that targeting this signaling system truly confers benefits for cancer therapy.

7 Concluding Remarks

Compelling evidence suggests that both tumor and nontumor (stromal) cell types are involved in tumor angiogenesis. Stromal cell-dependent mechanisms may also contribute to reduced responsiveness to antiangiogenic treatments. This chapter focused on the role myeloid cells and myeloid growth factors in these processes. However, it is important to emphasize that additional tumor-infiltrating cell types, in particular tumor-associated fibroblasts (TAF), are also important players. TAF is increasingly being recognized as having an active role in tumorigenesis [97, 98]. Like their normal counterparts, fibroblasts from malignant and fibrotic tissues, synthesize, deposit, and remodel the extracellular matrix within the stroma. Very recently, we reported that TAF can mediate tumor refractoriness to anti-VEGF therapy and PDGF-C was identified as a key mediator [99].

In conclusion, a more complete understanding of the cellular and molecular components of the microenvironment is expected to enable advances in therapeutic strategies that may supplement current antiangiogenic therapies and also take into account tumor diversity.

References

1. Red-Horse K, Crawford Y, Shojaei F, Ferrara N (2007) Endothelium-microenvironment interactions in the developing embryo and in the adult. Dev Cell 12:181–194
2. Chung AS, Lee J, Ferrara N (2010) Targeting the tumour vasculature: insights from physiological angiogenesis. Nat Rev 10:505–514
3. Hirschi KK, D'Amore PA (1996) Pericytes in the microvasculature. Cardiovasc Res 32:687–698
4. Ferrara N, Chen H, Davis-Smyth T et al (1998) Vascular endothelial growth factor is essential for corpus luteum angiogenesis. Nat Med 4:336–340
5. Ferrara N, Mass RD, Campa C, Kim R (2007) Targeting VEGF-A to treat cancer and age-related macular degeneration. Annu Rev Med 58:491–504
6. Kerbel RS (2008) Tumor angiogenesis. N Engl J Med 358:2039–2049
7. Ferrara N, Kerbel RS (2005) Angiogenesis as a therapeutic target. Nature 438:967–974
8. Crawford Y, Ferrara N (2008) VEGF inhibition: insights from preclinical and clinical studies. Cell Tissue Res 335:261–269
9. Ferrara N (2010) Vascular endothelial growth factor and age-related macular degeneration: from basic science to therapy. Nat Med 16:1107–1111
10. Ferrara N, Carver Moore K, Chen H et al (1996) Heterozygous embryonic lethality induced by targeted inactivation of the VEGF gene. Nature 380:439–442
11. Ferrara N, Gerber HP, LeCouter J (2003) The biology of VEGF and its receptors. Nat Med 9:669–676
12. Kim KJ, Li B, Winer J et al (1993) Inhibition of vascular endothelial growth factor-induced angiogenesis suppresses tumor growth in vivo. Nature 362:841–844
13. Ellis LM, Hicklin DJ (2008) VEGF-targeted therapy: mechanisms of anti-tumour activity. Nat Rev 8:579–591
14. Ferrara N (2004) Vascular endothelial growth factor: basic science and clinical progress. Endocr Rev 25:581–611

15. Alitalo K, Tammela T, Petrova TV (2005) Lymphangiogenesis in development and human disease. Nature 438:946–953
16. Presta LG, Chen H, O'Connor SJ et al (1997) Humanization of an anti-VEGF monoclonal antibody for the therapy of solid tumors and other disorders. Cancer Res 57:4593–4599
17. Ferrara N, Hillan KJ, Gerber HP, Novotny W (2004) Discovery and development of bevacizumab, an anti-VEGF antibody for treating cancer. Nat Rev Drug Discov 3:391–400
18. Hurwitz H, Fehrenbacher L, Novotny W et al (2004) Bevacizumab plus irinotecan, fluorouracil, and leucovorin for metastatic colorectal cancer. N Engl J Med 350:2335–2342
19. Sandler A, Gray R, Perry MC et al (2006) Paclitaxel-carboplatin alone or with bevacizumab for non-small-cell lung cancer. N Engl J Med 355:2542–2550
20. Escudier B, Pluzanska A, Koralewski P et al (2008) Bevacizumab plus interferon alfa-2a for treatment of metastatic renal cell carcinoma: a randomised, double-blind phase III trial. Lancet 370:2103–2111
21. Kowanetz M, Ferrara N (2006) Vascular endothelial growth factor signaling pathways: therapeutic perspective. Clin Cancer Res 12:5018–5022
22. Shojaei F, Zhong C, Wu X, Yu L, Ferrara N (2008) Role of myeloid cells in tumor angiogenesis and growth. Trends Cell Biol 18:372–378
23. Balkwill F, Mantovani A (2001) Inflammation and cancer: back to Virchow? Lancet 357:539–545
24. Coussens LM, Werb Z (2002) Inflammation and cancer. Nature 420:860–867
25. Lin WW, Karin M (2007) A cytokine-mediated link between innate immunity, inflammation, and cancer. J Clin Invest 117:1175–1183
26. Joyce JA, Pollard JW (2009) Microenvironmental regulation of metastasis. Nat Rev 9:239–252
27. Ahn GO, Brown JM (2008) Matrix metalloproteinase-9 is required for tumor vasculogenesis but not for angiogenesis: role of bone marrow-derived myelomonocytic cells. Cancer Cell 13:193–205
28. Mantovani A, Romero P, Palucka AK, Marincola FM (2008) Tumour immunity: effector response to tumour and role of the microenvironment. Lancet 371:771–7783
29. Pollard JW (2004) Tumour-educated macrophages promote tumour progression and metastasis. Nat Rev 4:71–78
30. Pollard JW (2009) Trophic macrophages in development and disease. Nat Rev Immunol 9:259–270
31. Bingle L, Brown NJ, Lewis CE (2002) The role of tumour-associated macrophages in tumour progression: implications for new anticancer therapies. J Pathol 196:254–265
32. Finak G, Bertos N, Pepin F et al (2008) Stromal gene expression predicts clinical outcome in breast cancer. Nat Med 14:518–527
33. Solinas G, Germano G, Mantovani A, Allavena P (2009) Tumor-associated macrophages (TAM) as major players of the cancer-related inflammation. J Leukoc Biol 86:1065–1073
34. Murdoch C, Giannoudis A, Lewis CE (2004) Mechanisms regulating the recruitment of macrophages into hypoxic areas of tumors and other ischemic tissues. Blood 104:2224–2234
35. Mantovani A, Sozzani S, Locati M, Allavena P, Sica A (2002) Macrophage polarization: tumor-associated macrophages as a paradigm for polarized M2 mononuclear phagocytes. Trends Immunol 23:549–555
36. Murdoch C, Muthana M, Coffelt SB, Lewis CE (2008) The role of myeloid cells in the promotion of tumour angiogenesis. Nat Rev 8:618–631
37. De Palma M, Venneri MA, Galli R et al (2005) Tie2 identifies a hematopoietic lineage of proangiogenic monocytes required for tumor vessel formation and a mesenchymal population of pericyte progenitors. Cancer Cell 8:211–226
38. De Palma M, Murdoch C, Venneri MA, Naldini L, Lewis CE (2007) Tie2-expressing monocytes: regulation of tumor angiogenesis and therapeutic implications. Trends Immunol 28:519–524
39. Murdoch C, Tazzyman S, Webster S, Lewis CE (2007) Expression of Tie-2 by human monocytes and their responses to angiopoietin-2. J Immunol 178:7405–7411

40. De Palma M, Naldini L (2009) Tie2-expressing monocytes (TEMs): novel targets and vehicles of anticancer therapy? Biochim Biophys Acta 1796:5–10
41. Oliner J, Min H, Leal J et al (2004) Suppression of angiogenesis and tumor growth by selective inhibition of angiopoietin-2. Cancer cell 6:507–516
42. Brown JL, Cao ZA, Pinzon-Ortiz M et al (2010) A human monoclonal anti-ANG2 antibody leads to broad antitumor activity in combination with VEGF inhibitors and chemotherapy agents in preclinical models. Mol Cancer Ther 9:145–156
43. Coxon A, Bready J, Min H et al (2010) Context-dependent role of angiopoietin-1 inhibition in the suppression of angiogenesis and tumor growth: implications for AMG 386, an angiopoietin-1/2-neutralizing peptibody. Mol Cancer Ther 9:2641–2651
44. Huang H, Bhat A, Woodnutt G, Lappe R (2010) Targeting the ANGPT-TIE2 pathway in malignancy. Nat Rev 10:575–585
45. Tazzyman S, Lewis CE, Murdoch C (2009) Neutrophils: key mediators of tumour angiogenesis. Int J Exp Pathol 90:222–231
46. Fridlender ZG, Sun J, Kim S et al (2009) Polarization of tumor-associated neutrophil phenotype by TGF-beta: "N1" versus "N2" TAN. Cancer Cell 16:183–194
47. Crivellato E, Nico B, Ribatti D (2008) Mast cells and tumour angiogenesis: new insight from experimental carcinogenesis. Cancer Lett 269:1–6
48. Yang L, DeBusk LM, Fukuda K et al (2004) Expansion of myeloid immune suppressor Gr +CD11b+ cells in tumor-bearing host directly promotes tumor angiogenesis. Cancer Cell 6:409–421
49. Talmadge JE (2007) Pathways mediating the expansion and immunosuppressive activity of myeloid-derived suppressor cells and their relevance to cancer therapy. Clin Cancer Res 13 (18 Pt 1):5243–5248
50. Marigo I, Dolcetti L, Serafini P, Zanovello P, Bronte V (2008) Tumor-induced tolerance and immune suppression by myeloid derived suppressor cells. Immunol Rev 222:162–179
51. Gabrilovich DI, Nagaraj S (2009) Myeloid-derived suppressor cells as regulators of the immune system. Nat Rev Immunol 9:162–174
52. Shojaei F, Wu X, Malik AK et al (2007) Tumor refractoriness to anti-VEGF treatment is mediated by CD11b+Gr1+ myeloid cells. Nat Biotechnol 25:911–920
53. Barleon B, Sozzani S, Zhou D, Weich HA, Mantovani A, Marme D (1996) Migration of human monocytes in response to vascular endothelial growth factor (VEGF) is mediated via the VEGF receptor flt-1. Blood 87:3336–3343
54. Park JE, Chen HH, Winer J, Houck KA, Ferrara N (1994) Placenta growth factor. Potentiation of vascular endothelial growth factor bioactivity, in vitro and in vivo, and high affinity binding to Flt-1 but not to Flk-1/KDR. J Biol Chem 269:25646–25654
55. Olofsson B, Korpelainen E, Pepper MS et al (1998) Vascular endothelial growth factor B (VEGF-B) binds to VEGF receptor-1 and regulates plasminogen activator activity in endothelial cells. Proc Natl Acad Sci U S A 95:11709–11714
56. Davis-Smyth T, Chen H, Park J, Presta LG, Ferrara N (1996) The second immunoglobulin-like domain of the VEGF tyrosine kinase receptor Flt-1 determines ligand binding and may initiate a signal transduction cascade. EMBO J 15:4919–4927
57. Fischer C, Jonckx B, Mazzone M et al (2007) Anti-PlGF inhibits growth of VEGF(R)-inhibitor-resistant tumors without affecting healthy vessels. Cell 131:463–475
58. Bais C, Wu X, Yao J et al (2010) PlGF blockade does not inhibit angiogenesis during primary tumor growth. Cell 141:166–177
59. Mollay C, Wechselberger C, Mignogna G et al (1999) Bv8, a small protein from frog skin and its homologue from snake venom induce hyperalgesia in rats. Eur J Pharmacol 374:189–196
60. LeCouter J, Kowalski J, Foster J et al (2001) Identification of an angiogenic mitogen selective for endocrine gland endothelium. Nature 412:877–884
61. Li M, Bullock CM, Knauer DJ, Ehlert FJ, Zhou QY (2001) Identification of two Prokineticin cDNAs: recombinant proteins potently contract gestrointestinal smooth muscle. Mol Pharmacol 59:692–698

62. Shojaei F, Wu X, Zhong C et al (2007) Bv8 regulates myeloid-cell-dependent tumour angiogenesis. Nature 450:825–831
63. Metcalf D (1989) The molecular control of cell division, differentiation commitment and maturation in haemopoietic cells. Nature 339:27–30
64. Christopher MJ, Link DC (2007) Regulation of neutrophil homeostasis. Curr Opin Hematol 14:3–8
65. Shojaei F, Wu X, Qu X et al (2009) G-CSF-initiated myeloid cell mobilization and angiogenesis mediate tumor refractoriness to anti-VEGF therapy in mouse models. Proc Natl Acad Sci U S A 106:6742–6747
66. Zhong C, Qu X, Tan M, Meng YG, Ferrara N (2009) Characterization and regulation of Bv8 in human blood cells. Clin Cancer Res 15:2675–2684
67. Nguyen DX, Bos PD, Massague J (2009) Metastasis: from dissemination to organ-specific colonization. Nat Rev 9:274–284
68. Psaila B, Lyden D (2009) The metastatic niche: adapting the foreign soil. Nat Rev 9:285–293
69. Kaplan RN, Riba RD, Zacharoulis S et al (2005) VEGFR1-positive haematopoietic bone marrow progenitors initiate the pre-metastatic niche. Nature 438:820–827
70. Hiratsuka S, Watanabe A, Aburatani H, Maru Y (2006) Tumour-mediated upregulation of chemoattractants and recruitment of myeloid cells predetermines lung metastasis. Nat Cell Biol 8:1369–1375
71. Kim S, Takahashi H, Lin WW et al (2009) Carcinoma-produced factors activate myeloid cells through TLR2 to stimulate metastasis. Nature 457:102–106
72. Erler JT, Bennewith KL, Cox TR et al (2009) Hypoxia-induced lysyl oxidase is a critical mediator of bone marrow cell recruitment to form the premetastatic niche. Cancer Cell 15:35–44
73. Dawson MR, Duda DG, Fukumura D, Jain RK (2009) VEGFR1-activity-independent metastasis formation. Nature 461:E4
74. Kowanetz M, Wu X, Lee J et al (2010) Granulocyte-colony stimulating factor promotes lung metastasis through mobilization of Ly6G+Ly6C+ granulocytes. Proc Acad Natl Sci U S A 107:21248–22155
75. Hiratsuka S, Nakamura K, Iwai S et al (2002) MMP9 induction by vascular endothelial growth factor receptor-1 is involved in lung-specific metastasis. Cancer Cell 2:289–300
76. Acuff HB, Carter KJ, Fingleton B, Gorden DL, Matrisian LM (2006) Matrix metalloproteinase-9 from bone marrow-derived cells contributes to survival but not growth of tumor cells in the lung microenvironment. Cancer Res 66:259–266
77. Ryckman C, McColl SR, Vandal K et al (2003) Role of S100A8 and S100A9 in neutrophil recruitment in response to monosodium urate monohydrate crystals in the air-pouch model of acute gouty arthritis. Arthritis Rheum 48:2310–2320
78. Vandal K, Rouleau P, Boivin A, Ryckman C, Talbot M, Tessier PA (2003) Blockade of S100A8 and S100A9 suppresses neutrophil migration in response to lipopolysaccharide. J Immunol 171:2602–2609
79. Crawford J, Ozer H, Stoller R et al (1991) Reduction by granulocyte colony-stimulating factor of fever and neutropenia induced by chemotherapy in patients with small-cell lung cancer. N Engl J Med 325:164–170
80. Hirasawa K, Kitamura T, Oka T, Matsushita H (2002) Bladder tumor producing granulocyte colony-stimulating factor and parathyroid hormone related protein. J Urol 167:2130
81. Hasegawa S, Suda T, Negi K, Hattori Y (2007) Lung large cell carcinoma producing granulocyte-colony-stimulating factor. Ann Thorac Surg 83:308–310
82. Yamamoto S, Takashima S, Ogawa H et al (1999) Granulocyte-colony-stimulating-factor-producing hepatocellular carcinoma. J Gastroenterol 34:640–644
83. Mabuchi S, Matsumoto Y, Morii E, Morishige K, Kimura T (2010) The first 2 cases of granulocyte colony-stimulating factor producing adenocarcinoma of the uterine cervix. Int J Gynecol Pathol 29:483–487

84. Granger JM, Kontoyiannis DP (2009) Etiology and outcome of extreme leukocytosis in 758 nonhematologic cancer patients: a retrospective, single-institution study. Cancer 115: 3919–3923
85. Perez FA, Fligner CL, Yu EY (2009) Rapid clinical deterioration and leukemoid reaction after treatment of urothelial carcinoma of the bladder: possible effect of granulocyte colony-stimulating factor. J Clin Oncol 27:215–217
86. Hamilton JA (2008) Colony-stimulating factors in inflammation and autoimmunity. Nat Rev Immunol 8:533–544
87. Liu BY, Soloviev I, Chang P et al (2010) Stromal cell-derived factor-1/CXCL12 contributes to MMTV-Wnt1 tumor growth involving Gr1+CD11b+ cells. PLoS One 5:e8611
88. Patel S, Player MR (2009) Colony-stimulating factor-1 receptor inhibitors for the treatment of cancer and inflammatory disease. Curr Top Med Chem 9:599–610
89. Stanley ER, Berg KL, Einstein DB et al (1997) Biology and action of colony-stimulating factor-1. Mol Reprod Dev 46:4–10
90. Sherr CJ, Rettenmier CW, Sacca R, Roussel MF, Look AT, Stanley ER (1985) The c-fms proto-oncogene product is related to the receptor for the mononuclear phagocyte growth factor, CSF-1. Cell 41:665–676
91. Aharinejad S, Paulus P, Sioud M et al (2004) Colony-stimulating factor-1 blockade by antisense oligonucleotides and small interfering RNAs suppresses growth of human mammary tumor xenografts in mice. Cancer Res 64:5378–5384
92. Paulus P, Stanley ER, Schafer R, Abraham D, Aharinejad S (2006) Colony-stimulating factor-1 antibody reverses chemoresistance in human MCF-7 breast cancer xenografts. Cancer Res 66:4349–4356
93. Kubota Y, Takubo K, Shimizu T et al (2009) M-CSF inhibition selectively targets pathological angiogenesis and lymphangiogenesis. J Exp Med 206:1089–1102
94. Murray LJ, Abrams TJ, Long KR et al (2003) SU11248 inhibits tumor growth and CSF-1R-dependent osteolysis in an experimental breast cancer bone metastasis model. Clin Exp Metastasis 20:757–766
95. Guo J, Marcotte PA, McCall JO et al (2006) Inhibition of phosphorylation of the colony-stimulating factor-1 receptor (c-Fms) tyrosine kinase in transfected cells by ABT-869 and other tyrosine kinase inhibitors. Mol Cancer Ther 5:1007–1013
96. Sonpavde G, Hutson TE, Rini BI (2008) Axitinib for renal cell carcinoma. Expert Opin Investig Drugs 17:741–748
97. Kalluri R, Zeisberg M (2006) Fibroblasts in cancer. Nat Rev 6:392–401
98. Bhowmick NA, Neilson EG, Moses HL (2004) Stromal fibroblasts in cancer initiation and progression. Nature 432:332–337
99. Crawford Y, Kasman I, Yu L et al (2009) PDGF-C mediates the angiogenic and tumorigenic properties of fibroblasts associated with tumors refractory to anti-VEGF treatment. Cancer Cell 15:21–34

G-CSF-Induced Mobilization of Bone Marrow Stem Cells and Cardiac Repair

Buddhadeb Dawn, Santosh K. Sanganalmath, and Roberto Bolli

1 Introduction

Granulocyte colony-stimulating factor (G-CSF) plays a critical role in proliferation, maturation, survival, and activation of granulocytes and various hematopoietic progenitors [1, 2]. G-CSF is also an effective mobilizer of bone marrow cells [3, 4], and this property is used clinically for the harvest of hematopoietic stem/progenitor cells from the peripheral blood before bone marrow transplantation [5]. Emerging evidence suggests a much broader role of G-CSF signaling in the repair of a range of tissues outside the hematopoietic system.

About a decade ago, Orlic and colleagues first reported that mobilization of bone marrow cells with G-CSF and stem cell factor (SCF) administered before and after a permanent coronary occlusion improved survival, reduced infarct size, and improved cardiac structure and function in mice after an acute myocardial infarction [6]. Since this report, a number of studies in animal models of myocardial infarction have shown that cytokine-induced mobilization leads to homing of bone marrow cells to the injured myocardium with improvement in myocardial anatomy, vascularity, and function [7–12]. These promising results from animal studies were quickly translated into small-scale clinical trials of G-CSF therapy in the setting of acute myocardial infarction, but the results have been discordant and generally disappointing [13, 14]. Furthermore, recent larger randomized controlled trials have failed to show any significant benefit on infarct size or left ventricular function with G-CSF treatment in patients with acute myocardial infarction after successful

B. Dawn (✉) • S.K. Sanganalmath
Division of Cardiovascular Diseases, Cardiovascular Research Institute, University of Kansas Medical Center, 3901 Rainbow Blvd, Room 1001 Eaton Hall, MS 3006, Kansas City, KS, 66160, USA
e-mail: bdawn@kumc.edu

R. Bolli
Institute of Molecular Cardiology, University of Louisville, Louisville, KY, 40292, USA

G. Molineux et al. (eds.), *Twenty Years of G-CSF*, Milestones in Drug Therapy,
DOI 10.1007/978-3-0348-0218-5_24, © Springer Basel AG 2012

revascularization [15, 16]. Although the reasons for these negative observations remain speculative, it is plausible that the lack of benefit in clinical trials was due, at least in part, to myocardial homing of a relatively small number of stem/progenitor cells. Other factors, including patient characteristics and timing of treatment initiation, might also have impacted the reparative process [13]. It is likely that further mechanistic insights from basic studies will lead to further modification of the G-CSF regimen and better selection of suitable patients with resultant improvement in outcomes. In this chapter, we focus on the biological basis and preclinical evidence of bone marrow cell mobilization and infarct repair with G-CSF therapy, and on the results of clinical trials that have tested this approach in humans.

2 G-CSF Signaling

Produced largely by endothelial cells, macrophages, monocytes, and fibroblasts, G-CSF exerts its biological actions primarily by activating the G-CSF receptor (G-CSFR), which is expressed widely in diverse cell types [17]. The G-CSFR consists of a transmembrane region connecting an extracellular domain and a cytoplasmic domain with conserved amino acid sequences. Activation of G-CSFR plays an important role in the production, survival, and activation of granulocytes during both normal and accelerated hematopoiesis. Signaling through G-CSFR also participates in the development of other myeloid lineages, mobilization of hematopoietic stem cells, and myeloid cell migration [17, 18]. In bone marrow progenitors, G-CSFR activation leads to phosphorylation of Janus kinase (Jak)1, Jak2, and Tyk2, with subsequent activation of signal transducers and activators of transcription (STAT)3 and STAT5 [17, 18]. G-CSFR-mediated cell proliferation and survival also depend on the activation of the p21Ras/MAP kinase and PI-3K/Akt pathways [17, 18]. Although the molecular links remain to be elucidated fully, recent discoveries have identified additional important roles of G-CSFR in the protection and repair of a diverse range of tissues, including heart, muscle, liver, and neural tissues.

3 Mobilization of Bone Marrow Cells by G-CSF

Although G-CSF therapy was initially directed at improving neutropenia after chemotherapy, analysis of peripheral blood in G-CSF-treated patients revealed an increase in the number of circulating progenitor cells from granulocyte–macrophage, erythroid, and megakaryocyte lineages [19]. Further studies in radiation-ablated mice documented the ability of G-CSF-mobilized primitive stem cells to reconstitute the hematopoietic system and lymphoid populations of the thymus, suggesting a potential use of these cells for bone marrow transplantation [20]. Currently, G-CSF is used extensively for the mobilization and harvest of stem/progenitor cells from the peripheral blood with a view to curative transplantation for various pathologies [5].

However, as is not uncommon with agents used in clinical medicine, the mechanistic basis of these therapeutic actions of G-CSF remains unclear. Improved understanding of the precise molecular events that govern bone marrow cell mobilization is needed to develop optimal mobilization strategies, especially for organ repair.

Although signaling via the G-CSFR is necessary for hematopoietic progenitor cell mobilization by cyclophosphamide and interleukin (IL)-8, it is not necessary for Flt3-induced mobilization [21]. G-CSFR expression on hematopoietic progenitors is also not required for their mobilization by G-CSF [22]. Interestingly, as "mobilization" essentially means dislodging cells from their bone marrow niches into the bloodstream, molecules that anchor bone marrow cells to the stroma are thought to play critical roles in this process [23, 24]. The bone marrow stroma expresses various ligands and adhesion molecules (hyaluronic acid, vascular cell adhesion molecule-1 [VCAM-1], stromal cell-derived factor-1 [SDF-1], CD62, and others) that interact with the respective counterparts (CD44, very late antigen-4 [VLA-4], CXCR4, and P-selectin glycoprotein ligand-1, respectively, besides others) to retain stem/progenitor cells in their niches [23, 24]. G-CSF-induced mobilization involves an alteration in the bone marrow milieu that promotes the dissolution of anchorage leading to the egress of stem/progenitor cells [3, 24]. This complex process appears to utilize at least two interdependent mechanisms: the release of proteases that cleave and inactivate peptides that anchor cells to the stroma; and the disruption of the interaction between SDF-1 and its receptor CXCR4 (summarized in Fig. 1).

3.1 Role of Proteases

Several reports have documented the role of serine proteases released by activated neutrophils in G-CSF-induced hematopoietic progenitor mobilization. In two studies [25, 26], G-CSF administration was associated with a reduction in VCAM-1 expression in the bone marrow and production of proteases (neutrophil elastase and cathepsin G) that are able to cleave VCAM-1 by neutrophils. In vitro findings showed that neutrophil elastase and cathepsin G could release $CD34^+$ hematopoietic progenitor cells attached on immobilized VCAM-1 or marrow stromal monolayers [25]. Other studies have shown the ability of serine proteases to cleave additional molecules important for bone marrow cell retention, including SDF-1 [27] and c-kit [28] and CXCR4 on hematopoietic progenitor cells [29]. Consistent with this, inhibition of neutrophil elastase prevented SDF-1 degradation and reduced stem cell mobilization by G-CSF [27]. However, subsequent studies in genetic models of protease deficiency have shown normal hematopoietic progenitor cell mobilization by G-CSF in mice lacking both neutrophil elastase and cathepsin G, suggesting that other proteases may play a compensatory role [30]. These results also suggest the existence of both protease-dependent and protease-independent mechanisms for bone marrow cell mobilization.

Matrix metalloproteinase (MMP)-9, a protease with important regulatory functions in matrix homeostasis in various tissues, has also been implicated in G-CSF-induced mobilization. MMP-9 is secreted by bone marrow cell as well as neutrophils upon

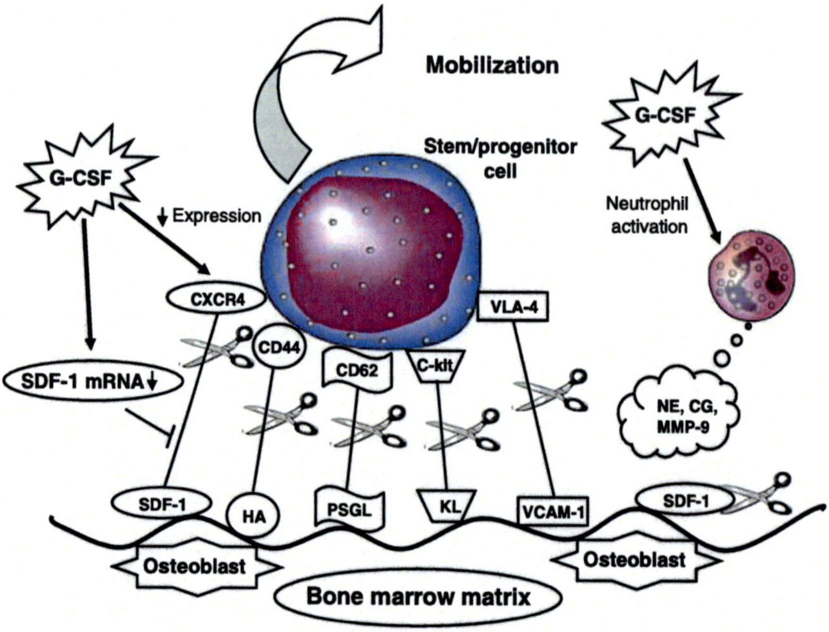

Fig. 1 Schematic representation of molecular events involved in G-CSF-induced mobilization of stem/progenitor cells from the bone marrow. The activation of neutrophils by G-CSF results in release of proteases that cleave bonds between molecules expressed in the bone marrow matrix and on stem/progenitor cells, leading to the egress of cells into the peripheral circulation. G-CSF therapy also impairs the SDF-1/CXCR4 interactions by decreasing SDF-1 levels in the bone marrow by reducing SDF-1 mRNA expression, and by reducing CXCR4 expression in cells. *CG* cathepsin G; *CXCR4* CXC chemokine receptor 4; *G-CSF* granulocyte colony-stimulating factor; *HA* hyaluronic acid; *KL* kit ligand; *MMP-9* matrix metalloproteinase-9; *NE* neutrophil elastase; *PSGL* P-selectin glycoprotein ligand-1; *SDF-1* stromal cell-derived factor-1; *VCAM-1* vascular cell adhesion molecule-1; *VLA-4* very late antigen-4

activation [31, 32]. A single IL-8 injection in rhesus monkeys has been shown to trigger a dramatic and instantaneous increase in plasma levels of MMP-9, followed by an increase in circulating hematopoietic progenitor cells [33]. The IL-8-induced mobilization could be inhibited by pretreatment with a specific monoclonal anti-MMP-9 antibody. In a subsequent study, G-CSF mobilization was associated with increased bone marrow levels of neutrophil elastase, cathepsin G, and MMP-9; and synergistic mobilization with G-CSF and GROβ/GROT correlated with a synergistic increase in plasma levels of MMP-9 [34]. Synergistic mobilization was blocked by anti-MMP-9 and was absent in MMP-9-deficient mice. However, the normal G-CSF-induced mobilization in MMP-9 knockout mice [30] suggests the presence of considerable redundancy among proteases that collectively influence bone marrow cell adherence to stroma.

In addition to neutrophil proteases, several other proteases have been shown to impact G-CSF-induced bone marrow cell mobilization in recent studies.

These include the osteolytic proteinase cathepsin K, which was shown to cleave SDF-1 and SCF [35]; dipeptidyl peptidase IV (CD26), the genetic deletion of which reduced G-CSF-induced mobilization [36]; and membrane type-1-MMP, the blockade of which inhibited G-CSF-induced mobilization of human progenitors in chimeric NOD/SCID mice [37]. Together, the current evidence indicates that possibly numerous known and unknown proteases mediate the process of bone marrow cell release from the stroma in response to G-CSF.

3.2 Role of SDF-1 and CXCR4

SDF-1 (also known as CXCL12) [38], produced in the marrow by stromal cells, is a potent chemoattractant for hematopoietic progenitor cells. It is well established that the interaction between SDF-1 and CXCR4 plays a critical role in the regulation of stem cell homing as well as mobilization [3]. SDF-1 expressed on bone marrow osteoblasts and endothelial cells effectively binds CXCR4-positive hematopoietic progenitor cells within the bone marrow [39]. Interference with this SDF-1/CXCR4 interaction, therefore, results in emigration of progenitor cells into the peripheral circulation.

G-CSF perturbs the stability of the SDF-1/CXCR4 complex in the bone marrow through several potential mechanisms. First, G-CSF reduces the level of SDF-1 in the bone marrow [27], and the magnitude of this decrease correlates with the extent of hematopoietic progenitor cell mobilization [39]. Although this decrease is likely multifactorial in origin, G-CSF has been shown to reduce osteoblast number/activity and SDF-1 mRNA expression [39]. Second, bone marrow SDF-1 levels may also decrease as a result of degradation by various proteases (neutrophil elastase, cathepsin G, MMP-9) released from G-CSF-activated neutrophils [27]. However, the decrease in bone marrow SDF-1 and G-CSF-induced hematopoietic progenitor cell mobilization were unaffected in mice deficient in neutrophil elastase and cathepsin G, suggesting an additional mechanism besides proteolytic degradation [30]. Finally, G-CSF can also reduce the expression of CXCR4 in $CD34^+$ hematopoietic progenitor cells [40]. In myeloid cells, this reduction in CXCR4 is due to reduced synthesis, and decreases their responsiveness to SDF-1 [41].

4 G-CSF Therapy for Cardiac Repair

Discoveries over the past decade have identified the ability of bone marrow cells to repair unrelated organs and tissues [42, 43], opening new areas of investigation with the creation of a new field: regenerative medicine. Quite appropriately, based on its ability to mobilize bone marrow cells, the cardiac regenerative efficacy of G-CSF has been rapidly tested in both animals and humans. Although G-CSF therapy has been used for the repair of various organs and tissues, we restrict the discussion to studies related to cardiac repair. In this regard, G-CSF has been used both in the setting of an acute myocardial infarction and in chronic heart failure.

4.1 G-CSF for Acute Myocardial Infarction in Animals

Acute myocardial infarction in humans most commonly results from sudden occlusion of a coronary artery by a thrombus developing at the site of a plaque rupture. In animal models, this event is mimicked by occluding the coronary artery (externally, with a suture or the inflation of a balloon, or internally, with a balloon) with or without subsequent release that results in reperfusion. These models typically produce cell death followed by scar formation and progressive worsening of left ventricular function over time with remodeling. In a seminal study, Orlic et al. [6] first demonstrated that administration of G-CSF and SCF (starting 5 days before until 3 days after coronary occlusion) in mice with acute myocardial infarction increases survival, improves left ventricular function, and mitigates left ventricular remodeling by inducing infarct repair with new cardiomyocytes and vessels in the infarct region. A large number of studies have since evaluated the safety and efficacy of G-CSF for cardiac repair in the setting of an acute myocardial infarction in mouse, rat, dog, pig, rabbit, and nonhuman primate models (Table 1).

The benefits of G-CSF were confirmed in a rabbit model of ischemia-reperfusion injury, in which G-CSF-mobilized leucocytes played a critical role in infarct repair by regulating phagocytosis of necrotic tissue, fibroblast proliferation, and angiogenesis, thereby attenuating left ventricular remodeling and dysfunction [7]. Other studies [8, 44] examined whether administration of G-CSF alone after myocardial infarction would still confer reparative benefits in mice and pigs. The results showed attenuation of adverse left ventricular remodeling and dysfunction with G-CSF-induced promotion of angiogenesis and reduction in apoptosis in the infarct border zone. Furthermore, in a large animal model of 90-min coronary occlusion/6-h reperfusion, intravenous administration of G-CSF for 30 min from the onset of reperfusion reduced infarct size and the incidence of arrhythmias in dogs, suggesting the potential clinical utility of such intervention in humans [45].

It should be noted that despite the above reported benefits, the failure of G-CSF as monotherapy was documented in several studies in animals. In one study [46], administration of G-CSF for 7 days after a permanent coronary occlusion in rats failed to reduce infarct size and improve left ventricular function despite adequate bone marrow cell mobilization. In the setting of a reperfused myocardial infarction in mice, we found that G-CSF monotherapy did not impart significant reparative benefits [9].

Table 1 Animal studies of G-CSF therapy in acute myocardial infarction

Species	Model	G-CSF dose	Duration	Results	References
Rabbit	Ischemia/reperfusion injury	10 μg/kg/d	From 1 to 5 d	↓ LV dilation, ↓ Infarct size, ↑ LVEF, ↑ Infarct wall thickness	[7]
Swine	Permanent coronary occlusion	10 μg/kg/d	From 24 h after MI to 7 d	↓ Apoptosis, ↓ Infarct size, ↑ LVEF, Neovascularization +	[8]
C57BL/6 mouse	Permanent coronary occlusion	200 μg/kg/d	3 d before and 5 d after MI	↓ Mortality, ↓ Infarct size, BMC mobilization +	[75]
Wistar rat	Permanent coronary occlusion	20 μg/kg/d	3 h after MI and every 24 h thereafter for 7 days	↓ LVEDD, Improved hemodynamic parameters, ↔ Infarct size	[85]
Dominant negative STAT3-Tg mouse	Permanent coronary occlusion	10–100 μg/kg/d	Immediately after MI for 5 d	↓ Apoptosis, ↓ Infarct size, ↑ LVFS, Improved hemodynamic parameters	[10]
C57BL/6 mouse	Permanent coronary occlusion	100 μg/kg/d	Immediately after MI for 5 d	↓ Mortality, ↑ LVEF, Improved hemodynamic parameters, ↔ Infarct size, Neovascularization +	[12]
Dog	Open-chest ischemia/ reperfusion injury	0.33 μg/kg/min IV	For 30 min from the onset of reperfusion	↓ Arrhythmias, ↓ Infarct size	[45]
Wistar rat	Ischemia/reperfusion in isolated-perfused hearts	300 ng/mL	Started at onset of reperfusion and continued for 2 h	↓ Infarct size, ↑ LV diastolic pressure	[81]
Rabbit	Ischemia/reperfusion injury	10 μg/kg/d	From 3 d to 7 d post-MI	↓ Scar area/LV area ratio, ↑ LVEF, ↓ LV dimensions	[80]
Sprague–Dawley rat	Permanent coronary occlusion	100 μg/kg/d	Immediately after MI for 5 d	↓ Infarct size, ↑ LVEF, Improved remodeling, Reendothelialization +	[107]
Wistar rat	Permanent coronary occlusion	100 μg/kg twice daily	3 h after MI for 7 d	↔ Infarct size, ↔ LVEF, ↔ Hemodynamics, ↔ LV remodeling	[46]
Swine	Ischemia/reperfusion injury	10 μg/kg/d	*Early group*: Immediately after injury on every other day for 20 d *Delayed group*: beginning 5 d after injury for 10 d	*Early treatment*: ↔ LVEF, Improved LVEDV, ↓ Capillary density *Delayed treatment*: ↑ LVEDV, ↓ Capillary density	[108]

(continued)

Table 1 (continued)

Species	Model	G-CSF dose	Duration	Results	References
Wistar rat	Two groups: Permanent coronary occlusion Ischemia/reperfusion	100 μg/kg/d	5 d in both the groups	*Permanent occlusion*: ↔ Infarct size, ↔ LV wall thickness, ↔ LVEF *Ischemia/reperfusion*: ↓ Infarct size, ↑ LV wall thickness, ↑ LVEF	[109]
C57BL/6 mouse	Permanent coronary occlusion	300 μg/kg/d	24 h after MI for 10 d	↑ LVEF, Improved hemodynamic parameters, ↓ Mortality, ↓ Infarct size in chronic phase, Myocardial regeneration +	[86]
Sprague–Dawley rat	Permanent coronary occlusion	50 μg/kg/d	3 h after MI for 5 d	↓ LVEF, ↑ LV dilation, Worse hemodynamic parameters, ↑ Infarct size, ↑ Cardiac fibrosis, ↑ Mortality	[110]
Swine	Ischemia/reperfusion injury	Bolus: 10 μg/kg/d SC injections: 5 μg/kg/d	IV bolus at reperfusion, daily SC injections on 5–9 d post-MI	↑ LVEF, ↑ wall motion score index, ↑ vascular density, ↑ areas of viable myocardium	[111]

BMC bone marrow cell; *d* day; *G-CSF* granulocyte colony-stimulating factor; *h* hour; *IV* intravenous; *LV* left ventricular; *LVEDD* LV end-diastolic diameter; *LVEDV* LV end-diastolic volume; *LVEF* LV ejection fraction; *MI* myocardial infarction; *SC* subcutaneous

4.2 G-CSF for Cardiomyopathy in Animals

Moving beyond the acute myocardial infarction model, the utility of G-CSF in the setting of both ischemic and nonischemic chronic cardiomyopathy has been evaluated in several studies (Table 2). The efficacy of G-CSF was tested in a model of chronic hibernating myocardium (dysfunction due to a prolonged reduction in blood supply) in pigs [47]. G-CSF was administered for 1 week and the animals were followed for 2 months. G-CSF therapy resulted in increased myocardial vascular density, reduced fibrosis, and decreased apoptosis within the ischemic zone, with attendant improvement in global left ventricular function [47]. The reparative efficacy of G-CSF in heart failure of ischemic origin was further tested in mice with established left ventricular remodeling and dysfunction at 12 weeks after myocardial infarction [48]. A smaller dose of G-CSF (10 μg/kg/day) was administered 5 days/week over a period of 4 weeks (in contrast to 5–10 consecutive days in other studies). This prolonged regimen increased G-CSFR expression in cardiomyocytes, reduced fibrosis by increasing the expression of MMP-2 and MMP-9, induced hypertrophy in viable cardiomyocytes, and improved left ventricular function [48]. Further, G-CSF therapy was associated with activation of STAT3 and expression of the transcription factor GATA-4 and other sarcomeric proteins (myosin heavy chain, troponin I, and desmin), suggesting direct and G-CSFR-mediated benefits of G-CSF independent of those derived from mobilized bone marrow cells [48]. However, in more recent studies, both short-term high-dose and long-term low-dose G-CSF regimens failed to produce any significant improvement in left ventricular function, infarct size, and hypertrophy in a rat model of postinfarct ischemic cardiomyopathy [49].

The benefits of G-CSF in nonischemic dilated cardiomyopathy were documented in a hamster model of autophagic dilated cardiomyopathy [50]. G-CSF therapy attenuated left ventricular remodeling and improved function and increased survival. Such benefits were further confirmed in a subsequent study in a model of doxorubicin-induced dilated cardiomyopathy [51]. G-CSF therapy started after 2 weeks of cessation of doxorubicin and continued for 8 days decreased cardiomyocyte apoptosis and the expression of apoptotic mediators such as Fas with significant improvement in left ventricular function and hemodynamic parameters [51].

4.3 Combination Cytokine Therapy with Other Mobilizing Agents in Animals

It is well known that different mobilizing agents release bone marrow cells with considerably different phenotypic characteristics and biologic behaviors. Although the molecular reasons for such selectivity and synergism remain unclear, this phenomenon has been used to augment mobilization.

Table 2 Animal studies of G-CSF therapy in cardiomyopathy

Species	Model	G-CSF dose	Duration	Results	References
C57BL/6 mouse	Doxorubicin-induced cardiomyopathy	50 μg/kg/d	*Early group*: immediately after doxorubicin injection for 8 d *Delayed group*: 3 wk after doxorubicin injection for 8 d	↓ Mortality, ↓ cardiac toxicity, myocardial regeneration +	[112]
C57BL/6 mouse	Permanent coronary occlusion	10 μg/kg/d	12 wk after MI, on the first 5 d of each wk, continued for 4 wks	↓ Infarct size, ↓ Fibrosis, ↑ LVEF, ↑ Hemodynamics, ↑ Cardiomyocyte size	[48]
Swine	Ameroid-induced chronic coronary occlusion	10 μg/kg/d	Immediately after MI for 7 d	↓ LV dilation, ↓ Cardiac fibrosis, ↓ Apoptosis, ↑ LVEF, ↑ Hemodynamics, Neovascularization +	[47]
UM-X7.1 hamster	Autosomal recessive cardiomyopathy	10 μg/kg/d	5 d/wk from 15 to 30 wk of age	↓ Mortality, ↓ LV remodeling, ↓ Fibrosis, ↑ LVEF, ↑ Cardiomyocyte size	[50]
Wistar rat	Doxorubicin-induced cardiomyopathy	50 μg/kg/d	2 wk after doxorubicin injection for 8 d	↓ Apoptosis, ↑ LVEF	[51]
C57BL/6 mouse	Doxorubicin-induced cardiomyopathy	100 μg/kg/d	Immediately after doxorubicin injection for 5 d	↓ LV dilation, ↓ Fibrosis, ↓ Inflammation, ↑ LVEF, ↑ Hemodynamics	[113]
Wistar rat	Permanent coronary occlusion	*Protocol I*: 50 μg/kg/d *Protocol II*: 10 μg/kg/d	4 wk after MI, rats were assigned to two protocols. *Protocol I*: SC. G-CSF daily for 7 d *Protocol II*: G-CSF for 4 wk on first 5 d of each wk	↔ LV fractional shortening, ↔ Hemodynamics, ↔ Infarct size, ↔ Hypertrophy	[49]

d day; *G-CSF* granulocyte colony-stimulating factor; *LV* left ventricular; *LVEF* LV ejection fraction; *MI* myocardial infarction; *SC* subcutaneous; *wk* week

4.3.1 G-CSF and SCF

In the first study of infarct repair [6], G-CSF was administered in combination with SCF, which is an excellent mobilizing agent, especially when used in combination with G-CSF [52]. SCF (c-kit ligand or Steel factor) binds to c-kit (CD117, a type III receptor tyrosine kinase) expressed on the surface of a range of hematopoietic progenitors and mast cells, and influences hematopoiesis at early stages [53, 54]. Consistent with its known synergism with colony-stimulating factors, administration of SCF along with G-CSF has shown to increase the number of circulating pluripotent hematopoietic stem cells by 250-fold [55]. Accordingly, in the study by Orlic et al. [6], G-CSF and SCF were injected once daily for 5 days before coronary ligation followed by an additional 3 days of treatment. Twenty-seven days later, the therapy showed reduced mortality, reduced infarct size, and improved left ventricular function, which were associated with increased homing of bone marrow cells to the infarcted myocardium and formation of new cardiomyocytes and vascular structures. Directionally concordant observations were made in a subsequent study [9], in which G-CSF and SCF were injected in a clinically relevant model of ischemia/reperfusion with cytokine therapy initiated 4 h after reperfusion.

The ability of the G-CSF and SCF combination therapy along with vascular endothelial growth factor-2 (VEGF-2) gene transfer to improve myocardial function in ischemic cardiomyopathy was examined in a swine model of chronic ischemia in the circumflex artery territory [56]. Administration of G-CSF and SCF in combination with intramyocardial *VEGF-2* gene transfer improved all indices of myocardial perfusion and function. However, cytokine therapy alone was ineffective in improving cardiac function. Also using a combination of G CSF and SCF started 4 h after an acute myocardial infarction, Norol et al. [57] reported an increase in myocardial blood flow and endothelial cell differentiation in a nonhuman primate model, albeit without any significant improvement in left ventricular function or infarct size.

4.3.2 G-CSF and Flt3 Ligand

Flt3 ligand (FL) is a transmembrane protein that binds to flt3/flk2 (CD135), a type III receptor tyrosine kinase, which is primarily expressed in the most primitive hematopoietic progenitor cells as well as pluripotent hematopoietic stem cells [58]. FL alone or in combination with other factors, including interleukins and colony-stimulating factors; induces the proliferation and mobilization of murine and human hematopoietic progenitor cells as well as lymphoid dendritic cells [58, 59]. Studies in mice have shown that although G-CSF and FL used alone exert modest effects to mobilize progenitors to accelerate hematopoietic recovery, they exhibit a striking synergy when used in combination [60–62]. Accordingly, we compared the efficacy of cytokine therapy with G-CSF alone versus G-CSF + SCF and G-CSF + FL in a mouse model of ischemia/reperfusion injury [9]. In comparison with other groups, the combination of G-CSF + FL was superior in attenuating left ventricular remodeling, improving left ventricular function, and mobilizing $Lin^{-}/Sca\text{-}1^{+}/c\text{-}kit^{+}$ bone marrow cells into the peripheral circulation. Furthermore, these beneficial effects of G-CSF + FL were sustained during long-term follow-up (up to 48 weeks) [63]. The likely explanation for the superiority of the G-CSF + FL combination in cardiac repair may lie in its enhanced efficacy in mobilizing primitive bone marrow cells in greater numbers [9].

4.3.3 G-CSF and Erythropoietin

Erythropoietin (EPO) is a hematopoietic growth factor that promotes the proliferation, differentiation, and survival of erythroid progenitors in the bone marrow [64]. Recent studies have shown that administration of recombinant EPO in the setting of acute myocardial injury can reduce infarct size and apoptosis and improve left ventricular function [65–68]. Aside from its direct cardioprotective effects, EPO can also increase the number of circulating endothelial progenitor cells and increase myocardial vascularity [69–71]. Given the cardioprotective and mobilizing properties of both agents, in a recent study Yeghiazarians et al. [72] employed the optimal dose of G-CSF and EPO combination in infarcted mice. While left ventricular function was stabilized in all treated groups with less scarring, increased homing of bone marrow cells, and reduced number of apoptotic cells, the results in the G-CSF + EPO combination were not significantly superior to those with either cytokine alone [72]. Similar results have been obtained in a large animal model of acute myocardial infarction, with the exception of a greater improvement in diastolic function with combination therapy [73, 74]. Further mechanistic studies are certainly warranted to improve the molecular basis of these observations, and to formulate superior cytokine combinations.

5 Mechanisms of Cardiac Repair with G-CSF

The cardiac reparative effects of G-CSF were initially thought to originate largely from bone marrow cells homed into the injured myocardium. Recent studies indicate participation of a number of additional mechanisms, including inhibition of apoptosis, induction of angiogenesis, modulation of extracellular matrix, direct cardioprotective effects of G-CSF, and favorable modulation of the myocardial matrix and constituent cells (Fig. 2). These insights gleaned from basic studies can be potentially utilized to improve outcomes of G-CSF therapy in the clinical setting.

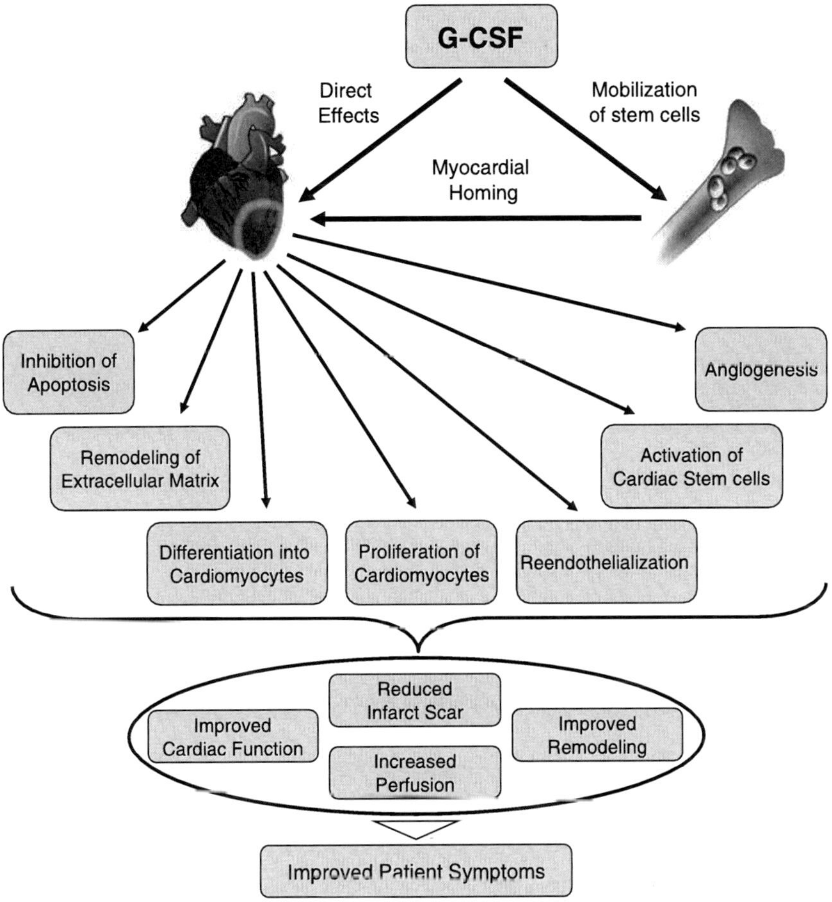

Fig. 2 Bone marrow cell-dependent and -independent cardioprotective and reparative actions of G-CSF. G-CSF-mobilized bone marrow cells can home to the heart to initiate myocardial repair via several potential mechanisms: differentiation into cells of cardiac lineages; promotion of angiogenesis and reendothelialization; effects on resident cardiac progenitors and myocytes leading to cellular proliferation; favorable modulation of the extracellular matrix; and activation of antiapoptotic signaling. In addition to or overlapping with these bone marrow cell-mediated effects, G-CSF can also exert direct angiogenic and cytoprotective effects in the infarcted myocardium. Together these events reduce infarct scar, increase perfusion, improve remodeling, and improve left ventricular function, eventually leading to improvement in patient symptoms. *BMC* bone marrow cell, *G-CSF* granulocyte colony-stimulating factor

5.1 *Mobilization of Bone Marrow Cells*

Although mobilization of bone marrow cells followed by myocardial homing and differentiation into cardiomyocytes and vascular cells have been reported by several groups [6, 9], these observations were made primarily with combination cytokine therapy (G-CSF + SCF and G-CSF + FL), while G-CSF as a single agent was relatively ineffective [9]. Additional studies with G-CSF have generated conflicting evidence regarding the ability of homed bone marrow cells to undergo cardiomyocytic differentiation and thereby contribute toward functional and structural improvement after myocardial infarction [75–77]. In one study by Askari et al. [78], G-CSF therapy with transplantation of SDF-1-expressing cardiac fibroblasts into the periinfarct area induced homing of $CD117^+$ bone marrow cells to the myocardium with resulting improvement in cardiac function, although G-CSF alone was unable to induce bone marrow cell homing. In another study by Adachi et al. [79], G-CSF therapy resulted in increased infiltration of the infarcted myocardium by bone marrow-derived side-population cells. Similarly, G-CSF administration after myocardial infarction has been shown to induce myocardial homing and cardiac differentiation of mobilized cardiomyogenic cells [76]. Importantly, in the study by Misao et al. [80], administration of AMD3100, a specific inhibitor of CXCR4, reduced myocardial homing of $CXCR4^+$ bone marrow cells and abolished the reparative effects of G-CSF therapy, underscoring the importance of bone marrow cell homing in G-CSF-induced infarct repair. Together, these results indicate that myocardial recruitment of G-CSF-mobilized bone marrow cells is necessary, at least in part, for the repair of infarct by G-CSF.

5.2 *Direct Cardioprotective and Antiapoptotic Effects*

Because G-CSFR is expressed in cardiomyocytes and endothelial cells [10, 11], systemically administered G-CSF can potentially modulate myocyte intracellular signaling independent of the bone marrow cell homing process. In support of this notion, a direct cardioprotective effect of G-CSF was reported in a hamster model of autophagic dilated cardiomyopathy [50]. Interestingly, while ineffective in reducing myocyte apoptosis, G-CSF protected myocytes against autophagic cellular degradation via activation of the JAK/STAT pathway, thus mitigating the anatomical and functional progression to heart failure [50].

In addition, G-CSF has been shown to induce an acute "postconditioning-like" effect in myocardial ischemia/reperfusion injury [81]. These investigators used an isolated-perfused rat heart model that excluded the contribution of bone marrow cells entirely. In this model, administration of G-CSF at the onset of reperfusion activated the Akt/endothelial nitric oxide synthase (eNOS) signaling cascade resulting in increased nitric oxide production and reduction in infarct size [81]. Furthermore, the involvement of endothelial cells in G-CSF-mediated cardioprotection was suggested by the blunting of the infarct-sparing effects by *N*-nitro-L-arginine methyl ester (L-NAME), an inhibitor of nitric oxide synthase.

The G-CSFR-mediated activation of JAK2, STAT1, and STAT3 in cardiomyocytes and endothelial cells [10, 11] also contributes to the direct cardioprotective effects of G-CSF in the infarcted heart through mitigation of cardiomyocyte apoptosis [10]. In one study by Harada et al. [10], H_2O_2-induced reduction in Bcl-2 was prevented by pretreatment with G-CSF, with resulting inhibition of cardiomyocyte apoptosis. In vivo, in transgenic mice expressing a dominant negative mutant STAT3 (dnSTAT3-Tg), G-CSF therapy failed to improve left ventricular function despite an increased number of circulating *c*-kit$^+$/Sca-1$^+$ bone marrow cells. The benefits of G-CSF were further corroborated by observations made in a Langendorff-perfused heart model of ischemia/reperfusion injury [10]. These results suggest that inhibition of apoptosis in the myocardium via activation of the JAK/STAT pathway significantly contributes to the overall beneficial effects of G-CSF in infarcted hearts.

5.3 Angiogenesis and Effects on the Myocardial Matrix

Recent studies have identified a number of additional beneficial actions of G-CSF relevant to the ischemic myocardium. Reports by Ohki et al. [82] and Cappocia et al. [83] have shown that G-CSF-stimulated monocytes and neutrophils induce new vessel formation in ischemic tissues. Furthermore, G-CSF can increase the expression of SDF-1 in the infarcted myocardium [80] and intercellular adhesion molecule-1 (ICAM-1) in arterioles in the infarct border zone [12], both of which can promote the recruitment of circulating leuckocytes and CXCR4$^+$ cells [80] with consequent increase in myocardial vascularity. Consistent with these observations, proangiogenic properties of G-CSF have been demonstrated in a number of studies in both small animal [10–12, 44, 80, 84] and large animal [8] models of myocardial infarction. It is conceivable that the G-CSF-induced neovascularization serves not only to secure blood supply to the ischemic myocardium but also to improve postinfarct remodeling.

A growing body of evidence suggests that G-CSF also modulates the various constituents of the myocardial extracellular matrix in a favorable fashion. In the postinfarct myocardium, G-CSF has been shown to increase the levels of MMP-1 and MMP-9 with accelerated resorption of necrotic tissue and reduction in granulation and scar tissue [7, 80]. The increased mRNA expression of procollagen type I and type III and TGF-β1 in the infarcted area may represent another mechanism for infarct scar reduction with G-CSF [85]. A study by Fujita et al. [86] provides additional evidence supporting G-CSF-induced modulation of the myocardial cellular composition. In this study in GFP chimeric mice, post-myocardial infarction G-CSF treatment for 10 days increased homing of GFP$^+$ bone marrow cells in the infarct area. Although these GFP$^+$ cells differentiated into vimentin$^+$ and α-SMA$^+$ myofibroblasts, cardiac function and remodeling improved [86].

Additional beneficial actions of G-CSF in this regard include an anti-inflammatory effect by inhibiting the production or activity of inflammatory mediators such as IL-1β, IL-12, tumor necrosis factor (TNF)-α, and (IFN)-γ [87, 88]. Since continued inflammation may influence myocardial remodeling adversely, the anti-inflammatory activities of G-CSF may favorably impact the recovery of cardiac structure and function after myocardial infarction. Finally, G-CSF treatment has been shown to reduce the incidence of ventricular arrhythmias by increasing the connexin-43 expression in the peri-infarct zone and modulating the function of gap junctions in cardiomyocytes [11]. Collectively, the above evidence indicates that G-CSF is able to exert multifarious beneficial actions on the infarcted myocardium in addition to the bone marrow cell-mediated effects.

6 Clinical Trials of G-CSF Therapy for Cardiac Repair

The need for novel therapies for ischemic heart disease, coupled with the promising results of animal studies, led quickly to the evaluation of the safety and efficacy of G-CSF in patients with acute myocardial infarction as well as cardiomyopathy. The early trials used variable regimens of G-CSF in unselected patients, producing divergent results and underscoring the acute need for an improved understanding of the mechanistic basis of cardiac repair with G-CSF.

6.1 G-CSF Therapy in Patients with Acute Myocardial Infarction

A summary of clinical trials of G-CSF therapy in patients with acute myocardial infarction is provided in Table 3. In the first MAGIC cell randomized trial [89], clinically stable patients with acute myocardial infarction were randomized into three groups: control, G-CSF alone (10 μg/kg for 4 days before percutaneous coronary invention [PCI]), and a combination of G-CSF and intracoronary infusion of G-CSF-mobilized peripheral blood stem cells. Despite improvements in ejection fraction, left ventricular end-systolic volume, exercise capacity, and myocardial perfusion, an increased rate of in-stent restenosis was noted in G-CSF-treated patients after 6 months of follow-up [89] leading to premature termination of the trial.

In the phase 1 FIRSTLINE-AMI trial, 50 patients with acute ST-elevation myocardial infarction (STEMI) were randomly assigned to receive a 6-day course of 10 μg/kg/day G-CSF starting at 90 min after reperfusion or standard therapy alone and were followed for 1 year. G-CSF therapy improved left ventricular function and enhanced infarct wall thickening in systole without any increase in the rate of restenosis [90, 91]. In a nonrandomized trial by Kuethe et al. [92], 14 patients with acute myocardial infarction were treated with G-CSF (10 μg/kg/day) for 7 days starting 2 days after PCI. Treated patients exhibited significant improvements in left ventricular regional wall motion, ejection fraction, and myocardial perfusion after 3 months with no major adverse effects. A favorable trend in left ventricular ejection fraction (LVEF) and left ventricular end-diastolic volume (LVEDV) was observed in treated patients at 6 months in the study by Valgimigli et al [93], who randomized patients with STEMI to receive a lower dose

Table 3 Clinical trials using G-CSF in acute myocardial infarction

References and clinical trial	Study design	No. of patients	G-CSF regimen	Follow-up	Study outcomes
[89], (MAGIC cell trial)	Randomized	Cell infusion = 10 G-CSF = 10 Control = 7	10 μg/kg/d for 4 d	6 mo	↑ Exercise capacity, ↑ myocardial perfusion, ↑ LVEF, angiogenesis +
[114]	Observational	G-CSF = 13	10 μg/kg/d for 10 d	3 mo	↑ LVEF, BMC mobilization +
[90, 91], (FIRSTLINE-AMI)	Randomized	G-CSF = 25 Control = 25	10 μg/kg/d for 6 d	12 mo	↑ Regional wall motion in infarct zone, ↑ metabolic activity in infarct zone at 4 months, ↑ resting LVEF, ↓ LV remodeling
[93]	Randomized, placebo-controlled	G-CSF = 10 Control = 10	5 μg/kg/d for 4 d	6 mo	↑ LVEF, ↓ LVEDV, EPC mobilization +
[92]	Nonrandomized, open-label	G-CSF = 14 Control = 9	10 μg/kg/d for 7 d	3 mo	↑ Regional wall motion and perfusion, ↑ LVEF
[115], (Rigenera)	Randomized	G-CSF = 14 Control = 27	10 μg/kg/d for 5 d	4–6 mo	↑ LVEF, ↓ LVEDV, ↓ LVESV
[116]	Randomized	G-CSF = 12 (AMI), Control = 12 (AMI)	10 d (titrated to WBC count of 30,000/μl)	1 mo	↑ LVEF, ↑ regional wall motion, $CD34^+$ BMC mobilization +
[96]	Pilot, dose-escalation, randomized	G-CSF = 12 Control = 6	5 and 10 μg/kg/d for 5 d	1 mo	↔ LV function, BMC mobilization +
[15], (STEMMI)	Randomized, double-blind, placebo-controlled	G-CSF = 39 Control = 39	10 μg/kg/d for 6 d	6 mo	↔ LVEF, BMC mobilization +
[16], (REVIVAL-2)	Randomized, double-blind, placebo-controlled	G-CSF = 56 Control = 58	10 μg/kg/d for 5 d	4–6 mo	↔ LVEF, ↔ infarct size, BMC mobilization +

(continued)

Table 3 (continued)

References and clinical trial	Study design	No. of patients	G-CSF regimen	Follow-up	Study outcomes
[97], (G-CSF-STEMI)	Randomized, double-blind, placebo-controlled	G-CSF = 23 Control = 21	10 μg/kg/d for 5 d	12 mo	↔ LVEF, ↔ LV size, BMC mobilization +, ↑ myocardial perfusion (early)
[94]	Randomized, placebo-controlled	G-CSF = 18 Control = 22	2.5 μg/kg/d for 5 d	6 mo	↑ LVEF, ↔ LVEDV, BMC mobilization +
[105], (STEM-AMI)	Randomized, placebo-controlled	G-CSF = 24 Control = 25	5 μg/kg bid for 5 d	6 mo	↓ Infarct size, improved remodeling, ↔ LVEF

bid twice a day; *BMC* bone marrow cell; *EDV* end-diastolic volume; *ESV* end-systolic volume; *EF* ejection fraction; *EPC* endothelial progenitor cell; *mo* month; *G-CSF* granulocyte colony-stimulating factor; *LV* left ventricle; *WBC* white blood cell of G-CSF or placebo. The feasibility and safety of an even lower dose of G-CSF (2.5 μg/kg for 5 days) for left ventricular dysfunction after acute myocardial infarction were evaluated in a subsequent single-blind randomized clinical trial [94], which enrolled patients with a total occlusion of the left anterior descending coronary artery and successful reperfusion. Compared with data at 4 days, the ejection fraction was greater and left ventricular end-systolic volume tended to be smaller in the G-CSF group after 6 months, with no significant difference in the incidence of restenosis between groups [94]. In another randomized trial (The Rigenera Study) that enrolled patients with large anterior myocardial infarction with ejection fraction <50%, improved left ventricular ejection fraction and LVEDV were noted in the G-CSF-treated group after a median follow-up duration of 5 months [95]. However, in the dose escalation study Ellis et al. [96], failed to demonstrate improvement in left ventricular function with G-CSF therapy after 30 days.

The favorable outcomes from relatively smaller trials were not confirmed in subsequent randomized controlled trials that enrolled greater number of patients. In the double-blind placebo-controlled STEMMI trial [15], patients with STEMI received G-CSF (10 μg/kg/day for 6 days) or placebo after PCI <12 h after symptom onset. Although treatment with G-CSF was safe and well tolerated, it did not lead to improvement in left ventricular function or infarct size [15]. The double-blind placebo-controlled REVIVAL-2 trial [16] was similar in design to STEMMI, except that G-CSF treatment was initiated 5 days after PCI compared with 1–2 days in STEMMI. A total of 114 patients were randomized to placebo or G-CSF. While G-CSF therapy resulted in successful mobilization of bone marrow cells, no significant improvement in left ventricular function or infarct size was noted after 6 months [16]. In the double-blind placebo-controlled phase 2 G-CSF-STEMI trial [97], 44 patients with subacute STEMI undergoing late revascularization (≥6 h and <7 days after symptom onset) were randomized to receive G-CSF (10 μg/kg/day for 5 days) or placebo. No significant differences in ejection fraction, infarct size, myocardial perfusion, left ventricular end-diastolic volume (LVEDV), and left ventricular end systolic volume (LVESV) were noted during follow-up up to 12 months [97, 98].

The above differences in outcomes in relatively small clinical trials have been evaluated in several meta-analyses of pooled data [13, 14]. In the meta-analysis by Abdel-Latif et al. [13], which included eight eligible randomized controlled trials ($n = 385$ patients) of G-CSF therapy in patients with acute myocardial infarction, no significant improvement in ejection fraction, infarct size, LVEDV, or LVESV was observed in G-CSF-treated patients compared with controls. The observations were similar in analyses that included randomized controlled trials with acute myocardial infarction as well as cardiomyopathy. However, data from stratified analyses suggested that G-CSF might benefit acute myocardial infarction patients with impaired left ventricular function (ejection fraction <50%) at baseline, and if G-CSF is administered early after myocardial infarction. Importantly, G-CSF therapy appeared to be safe and well tolerated with no significant increase in adverse events, including in-stent restenosis. These safety data are consistent with a meta-analysis of individual patient data ($n = 106$ patients) by Ince et al. [99], which did not show an increased risk of coronary re-stenosis with G-CSF therapy after PCI in acute myocardial infarction. Similarly, in a meta-analysis of ten randomized controlled trials [14], although G-CSF therapy appeared to be safe, no significant benefit in left ventricular function or infarct size after acute myocardial infarction and reperfusion was observed in treated patients compared with controls.

6.2 G-CSF Therapy in Patients with Chronic Myocardial Ischemia

The benefits of G-CSF in the setting of chronic ischemic heart disease have been evaluated in several clinical trials [100–103] (Table 4). In one study [100], G-CSF therapy (5 μg/kg/day for 6 days) increased circulating $CD34^+$ cells and reduced angina attacks in patients with severe occlusive coronary artery disease without improving myocardial perfusion or function. In another study [101], similar mobilization of bone marrow cells with endothelial progenitor potential was observed with G-CSF treatment in patients with CAD and angina; however, there was no objective improvement in cardiac function or perfusion. Similarly disappointing results were also obtained in a study that evaluated the combination of intramyocardial injection of VEGF-A_{165} plasmid followed by G-CSF injection in patients with severe chronic ischemic heart disease [102]. Despite an

Table 4 Clinical trials using G-CSF in chronic myocardial ischemia

References	Study design	No. of patients	G-CSF regimen	Follow-up	Study outcomes
[117]	Nonrandomized	G-CSF = 16 Control = 15	10 μg/kg/d for 5 d	3 mo	↔ LVEF, ↔ regional wall motion, BMC mobilization +
[100]	Nonrandomized	G-CSF = 13 Control = 16	5 μg/kg/d for 6 d	2 mo	Improved clinical symptoms, ↓ LVEF, ↔ myocardial perfusion
[102]	Nonrandomized	G-CSF + VEGF = 16 VEGF = 16 Placebo = 16	10 μg/kg/d for 6 d	3 mo	↔ LVEF, ↔ myocardial perfusion, ↔ LVEDV
[103]	Nonrandomized	G-CSF = 16 Control = 8	480 μg bid (titrated) for 4 × 10-d courses	6 mo	↓ NYHA class, ↑ 6-min walk distance, BMC mobilization +

bid twice a day; *BMC* bone marrow cell; *EF* ejection fraction; *G-CSF* granulocyte colony-stimulating factor; *LV* left ventricular; *NYHA* New York Heart Association; *VEGF* vascular endothelial growth factor increase in circulating $CD34^+$ cells, this combinatorial approach failed to improve clinical symptoms and myocardial function and perfusion. However, repeated G-CSF administration (10 days × 4 cycles) adjusted by bone marrow cell mobilization improved NYHA functional class and 6-min walking distance in patients with dilated or ischemic cardiomyopathy without significant changes in echocardiographic parameters. However, two patients with ischemic cardiomyopathy complained of increased angina, and one of them experienced ventricular fibrillation eventually followed by death due to multiorgan failure, raising concerns regarding safety of G-CSF in patients with ischemic cardiomyopathy [103].

Collectively, the results do not tend to support a beneficial impact of G-CSF on chronically ischemic and failing myocardium. However, the current evidence with G-CSF in ischemic heart disease and heart failure is largely based on nonrandomized trials with relatively few patients, and the possibility that modifications in G-CSF regimens, combination therapy, and patient selection may improve the outcomes in patients with heart failure cannot be excluded.

7 Future Perspectives

Our understanding of how adult cells repair unrelated organs has grown exponentially over the past decade. It is now evident that many more variables influence this process of cellular reconstitution than once thought. Indeed, a growing body of evidence supports the efficacy of bone marrow cell therapy for cardiac repair [42, 43], and in this light, the failure of G-CSF therapy to improve cardiac structure and function may be viewed as a failure to optimize the factors that influence bone marrow cell mobilization, homing, timing of therapy, duration of therapy, selection of patients, etc. In this regard, the meta-analysis of pooled data identified a beneficial impact of G-CSF therapy on left ventricular function in patients in whom G-CSF therapy was initiated earlier [13]. Our meta-analysis also identified a significant improvement in left ventricular function with G-CSF therapy in patients with worse function at baseline [13]. This observation regarding left ventricular ejection fraction is consistent with other studies that reported greater improvement in cardiac function with bone marrow cell therapy in patients with worse left ventricular function at baseline [104]. Consistent with this observation, in a recent randomized controlled trial [105], G-CSF therapy reduced infarct size and improved left ventricular remodeling (assessed by MRI) in patients with anterior STEMI. Importantly, the left ventricular ejection fraction at baseline in patients included in this study was <45%, and G-CSF therapy was initiated within 12 h after PCI [105]. Thus, although meta-analyses can serve as a guide, the true impact of these variables on outcomes of G-CSF treatment cannot be ascertained in a meta-analysis, and appropriately designed larger randomized trials with sufficiently long follow-up will be necessary instead.

The outcomes of cardiac repair with G-CSF may also improve with a combinatorial approach with other cytokines, which may induce more vigorous and/or more selective mobilization of bone marrow cells. Although data from animal studies suggest superior cardiac repair with the addition of other agents to G-CSF, such as SDF-1 [78], SCF [6, 9], and FL [9], clinical data from the use of these regimens are not available. The combination of sitagliptin (a dipeptidyl peptidase-IV inhibitor) with G-CSF in patients with acute myocardial infarction is currently being tested in a phase 3 trial (SITAGRAMI) [106], and the safety and efficacy of intracoronary infusion of mobilized stem cells with G-CSF and darbepoietin alfa are being evaluated in the MAGIC Cell-5-Combicytokine Trial (NCT00501917). This is logical, as erythropoietin has been shown to confer cardioprotective benefits in the setting of myocardial ischemic injury [65, 66]. However, the results of G-CSF and EPO combination therapy in pigs have failed to show additional benefits over single agents [73, 74], and the intracoronary infusion of mobilized cells may make the difference.

8 Conclusions

Accumulating evidence from animal studies supports the ability of G-CSF to induce myocardial repair by both bone marrow cell-dependent and cell-independent mechanisms. The early clinical trials of G-CSF therapy for cardiac repair have yielded mixed results, and meta-analyses of pooled data have shown no significant benefit with G-CSF monotherapy for cardiac repair in unselected patients with acute myocardial infarction [13, 14]. Although the reasons underlying this failure of translation remain poorly understood, further modifications in G-CSF regimens and patient selection may improve outcomes in future trials. Therapy with G-CSF in combination with other cytokines has also shown promise in animal studies, and is being currently evaluated in patients with acute myocardial infarction.

Acknowledgments This publication was supported in part by NIH grants R01 HL-89939, R21 HL-89737, R37 HL-55757, and P01 HL-78825, and the Maureen and Marvin Dunn endowment.

References

1. Arai KI, Lee F, Miyajima A, Miyatake S, Arai N, Yokota T (1990) Cytokines: coordinators of immune and inflammatory responses. Annu Rev Biochem 59:783–836
2. Pimentel E (1990) Colony-stimulating factors. Ann Clin Lab Sci 20:36–55
3. Lapidot T, Petit I (2002) Current understanding of stem cell mobilization: the roles of chemokines, proteolytic enzymes, adhesion molecules, cytokines, and stromal cells. Exp Hematol 30:973–981
4. Papayannopoulou T (2004) Current mechanistic scenarios in hematopoietic stem/progenitor cell mobilization. Blood 103:1580–1585
5. Anderlini P, Donato M, Chan KW et al (1999) Allogeneic blood progenitor cell collection in normal donors after mobilization with filgrastim: the MD Anderson Cancer Center experience. Transfusion 39:555–560

6. Orlic D, Kajstura J, Chimenti S et al (2001) Mobilized bone marrow cells repair the infarcted heart, improving function and survival. Proc Natl Acad Sci U S A 98:10344–10349
7. Minatoguchi S, Takemura G, Chen XH et al (2004) Acceleration of the healing process and myocardial regeneration may be important as a mechanism of improvement of cardiac function and remodeling by postinfarction granulocyte colony-stimulating factor treatment. Circulation 109:2572–2580
8. Iwanaga K, Takano H, Ohtsuka M et al (2004) Effects of G-CSF on cardiac remodeling after acute myocardial infarction in swine. Biochem Biophys Res Commun 325:1353–1359
9. Dawn B, Guo Y, Rezazadeh A et al (2006) Postinfarct cytokine therapy regenerates cardiac tissue and improves left ventricular function. Circ Res 98:1098–1105
10. Harada M, Qin Y, Takano H et al (2005) G-CSF prevents cardiac remodeling after myocardial infarction by activating the Jak-Stat pathway in cardiomyocytes. Nat Med 11:305–311
11. Kuhlmann MT, Kirchhof P, Klocke R et al (2006) G-CSF/SCF reduces inducible arrhythmias in the infarcted heart potentially via increased connexin43 expression and arteriogenesis. J Exp Med 203:87–97
12. Deindl E, Zaruba MM, Brunner S et al (2006) G-CSF administration after myocardial infarction in mice attenuates late ischemic cardiomyopathy by enhanced arteriogenesis. FASEB J 20:956–958
13. Abdel-Latif A, Bolli R, Zuba-Surma EK, Tleyjeh IM, Hornung CA, Dawn B (2008) Granulocyte colony-stimulating factor therapy for cardiac repair after acute myocardial infarction: a systematic review and meta-analysis of randomized controlled trials. Am Heart J 156:216–226
14. Zohlnhofer D, Dibra A, Koppara T et al (2008) Stem cell mobilization by granulocyte colony-stimulating factor for myocardial recovery after acute myocardial infarction: a meta-analysis. J Am Coll Cardiol 51:1429–1437
15. Ripa RS, Jorgensen E, Wang Y et al (2006) Stem cell mobilization induced by subcutaneous granulocyte-colony stimulating factor to improve cardiac regeneration after acute ST-elevation myocardial infarction: result of the double-blind, randomized, placebo-controlled stem cells in myocardial infarction (STEMMI) trial. Circulation 113:1983–1992
16. Zohlnhofer D, Ott I, Mehilli J et al (2006) Stem cell mobilization by granulocyte colony-stimulating factor in patients with acute myocardial infarction: a randomized controlled trial. JAMA 295:1003–1010
17. Avalos BR (1996) Molecular analysis of the granulocyte colony-stimulating factor receptor. Blood 88:761–777
18. Touw IP, van de Geijn GJ (2007) Granulocyte colony-stimulating factor and its receptor in normal myeloid cell development, leukemia and related blood cell disorders. Front Biosci 12:800–815
19. Duhrsen U, Villeval JL, Boyd J, Kannourakis G, Morstyn G, Metcalf D (1988) Effects of recombinant human granulocyte colony-stimulating factor on hematopoietic progenitor cells in cancer patients. Blood 72:2074–2081
20. Molineux G, Pojda Z, Hampson IN, Lord BI, Dexter TM (1990) Transplantation potential of peripheral blood stem cells induced by granulocyte colony-stimulating factor. Blood 76:2153–2158
21. Liu F, Poursine-Laurent J, Link DC (1997) The granulocyte colony-stimulating factor receptor is required for the mobilization of murine hematopoietic progenitors into peripheral blood by cyclophosphamide or interleukin-8 but not flt-3 ligand. Blood 90:2522–2528
22. Liu F, Poursine-Laurent J, Link DC (2000) Expression of the G-CSF receptor on hematopoietic progenitor cells is not required for their mobilization by G-CSF. Blood 95:3025–3031
23. Kronenwett R, Martin S, Haas R (2000) The role of cytokines and adhesion molecules for mobilization of peripheral blood stem cells. Stem Cells 18:320–330
24. Nervi B, Link DC, DiPersio JF (2006) Cytokines and hematopoietic stem cell mobilization. J Cell Biochem 99:690–705

25. Levesque JP, Takamatsu Y, Nilsson SK, Haylock DN, Simmons PJ (2001) Vascular cell adhesion molecule-1 (CD106) is cleaved by neutrophil proteases in the bone marrow following hematopoietic progenitor cell mobilization by granulocyte colony-stimulating factor. Blood 98:1289–1297
26. Levesque JP, Hendy J, Takamatsu Y, Williams B, Winkler IG, Simmons PJ (2002) Mobilization by either cyclophosphamide or granulocyte colony-stimulating factor transforms the bone marrow into a highly proteolytic environment. Exp Hematol 30:440–449
27. Petit I, Szyper-Kravitz M, Nagler A et al (2002) G-CSF induces stem cell mobilization by decreasing bone marrow SDF-1 and up-regulating CXCR4. Nat Immunol 3:687–694
28. Levesque JP, Hendy J, Winkler IG, Takamatsu Y, Simmons PJ (2003) Granulocyte colony-stimulating factor induces the release in the bone marrow of proteases that cleave c-KIT receptor (CD117) from the surface of hematopoietic progenitor cells. Exp Hematol 31: 109–117
29. Levesque JP, Hendy J, Takamatsu Y, Simmons PJ, Bendall LJ (2003) Disruption of the CXCR4/CXCL12 chemotactic interaction during hematopoietic stem cell mobilization induced by GCSF or cyclophosphamide. J Clin Invest 111:187–196
30. Levesque JP, Liu F, Simmons PJ et al (2004) Characterization of hematopoietic progenitor mobilization in protease-deficient mice. Blood 104:65–72
31. Masure S, Proost P, Van Damme J, Opdenakker G (1991) Purification and identification of 91-kDa neutrophil gelatinase. Release by the activating peptide interleukin-8. Eur J Biochem 198:391–398
32. Kjeldsen L, Sengelov H, Lollike K, Nielsen MH, Borregaard N (1994) Isolation and characterization of gelatinase granules from human neutrophils. Blood 83:1640–1649
33. Pruijt JF, Fibbe WE, Laterveer L et al (1999) Prevention of interleukin-8-induced mobilization of hematopoietic progenitor cells in rhesus monkeys by inhibitory antibodies against the metalloproteinase gelatinase B (MMP-9). Proc Natl Acad Sci USA 96:10863–10868
34. Pelus LM, Bian H, King AG, Fukuda S (2004) Neutrophil-derived MMP-9 mediates synergistic mobilization of hematopoietic stem and progenitor cells by the combination of G-CSF and the chemokines GRObeta/CXCL2 and GRObetaT/CXCL2delta4. Blood 103:110–119
35. Kollet O, Dar A, Shivtiel S et al (2006) Osteoclasts degrade endosteal components and promote mobilization of hematopoietic progenitor cells. Nat Med 12:657–664
36. Christopherson KW, Cooper S, Hangoc G, Broxmeyer HE (2003) CD26 is essential for normal G-CSF-induced progenitor cell mobilization as determined by CD26−/− mice. Exp Hematol 31:1126–1134
37. Vagima Y, Avigdor A, Goichberg P et al (2009) MT1-MMP and RECK are involved in human CD34+ progenitor cell retention, egress, and mobilization. J Clin Invest 119:492–503
38. Nagasawa T, Kikutani H, Kishimoto T (1994) Molecular cloning and structure of a pre-B-cell growth-stimulating factor. Proc Natl Acad Sci U S A 91:2305–2309
39. Semerad CL, Christopher MJ, Liu F et al (2005) G-CSF potently inhibits osteoblast activity and CXCL12 mRNA expression in the bone marrow. Blood 106:3020–3027
40. Dlubek D, Drabczak-Skrzypek D, Lange A (2006) Low CXCR4 membrane expression on CD34(+) cells characterizes cells mobilized to blood. Bone Marrow Transplant 37:19–23
41. Kim HK, De La Luz Sierra M, Williams CK, Gulino AV, Tosato G (2006) G-CSF down-regulation of CXCR4 expression identified as a mechanism for mobilization of myeloid cells. Blood 108:812–820
42. Abdel-Latif A, Bolli R, Tleyjeh IM et al (2007) Adult bone marrow-derived cells for cardiac repair: a systematic review and meta-analysis. Arch Intern Med 167:989–997
43. Dawn B, Abdel-Latif A, Sanganalmath SK, Flaherty MP, Zuba-Surma EK (2009) Cardiac repair with adult bone marrow-derived cells: the clinical evidence. Antioxid Redox Signal 11:1865–1882
44. Ohtsuka M, Takano H, Zou Y et al (2004) Cytokine therapy prevents left ventricular remodeling and dysfunction after myocardial infarction through neovascularization. FASEB J 18:851–853

45. Takahama H, Minamino T, Hirata A et al (2006) Granulocyte colony-stimulating factor mediates cardioprotection against ischemia/reperfusion injury via phosphatidylinositol-3-kinase/Akt pathway in canine hearts. Cardiovasc Drugs Ther 20:159–165
46. Werneck-de-Castro JP, Costa ESRH, de Oliveira PF et al (2006) G-CSF does not improve systolic function in a rat model of acute myocardial infarction. Basic Res Cardiol 101: 494–501
47. Hasegawa H, Takano H, Iwanaga K et al (2006) Cardioprotective effects of granulocyte colony-stimulating factor in swine with chronic myocardial ischemia. J Am Coll Cardiol 47:842–849
48. Li Y, Takemura G, Okada H et al (2006) Treatment with granulocyte colony-stimulating factor ameliorates chronic heart failure. Lab Invest 86:32–44
49. Louzada RA, Oliveira PF, Cavalcanti-de-Albuquerque JP et al (2010) Granulocyte-colony stimulating factor treatment of chronic myocardial infarction. Cardiovasc Drugs Ther 24:121–130
50. Miyata S, Takemura G, Kawase Y et al (2006) Autophagic cardiomyocyte death in cardiomyopathic hamsters and its prevention by granulocyte colony-stimulating factor. Am J Pathol 168:386–397
51. Hou XW, Son J, Wang Y et al (2006) Granulocyte colony-stimulating factor reduces cardiomyocyte apoptosis and improves cardiac function in adriamycin-induced cardiomyopathy in rats. Cardiovasc Drugs Ther 20:85–91
52. Menendez P, Caballero MD, Prosper F et al (2002) The composition of leukapheresis products impacts on the hematopoietic recovery after autologous transplantation independently of the mobilization regimen. Transfusion 42:1159–1172
53. Costa JJ, Demetri GD, Harrist TJ et al (1996) Recombinant human stem cell factor (kit ligand) promotes human mast cell and melanocyte hyperplasia and functional activation in vivo. J Exp Med 183:2681–2686
54. Broudy VC (1997) Stem cell factor and hematopoiesis. Blood 90:1345–1364
55. Bodine DM, Seidel NE, Gale MS, Nienhuis AW, Orlic D (1994) Efficient retrovirus transduction of mouse pluripotent hematopoietic stem cells mobilized into the peripheral blood by treatment with granulocyte colony-stimulating factor and stem cell factor. Blood 84: 1482–1491
56. Kawamoto A, Murayama T, Kusano K et al (2004) Synergistic effect of bone marrow mobilization and vascular endothelial growth factor-2 gene therapy in myocardial ischemia. Circulation 110:1398–1405
57. Norol F, Merlet P, Isnard R et al (2003) Influence of mobilized stem cells on myocardial infarct repair in a nonhuman primate model. Blood 102:4361–4368
58. Lyman SD (1998) Biologic effects and potential clinical applications of Flt3 ligand. Curr Opin Hematol 5:192–196
59. Ashihara E, Shimazaki C, Sudo Y et al (1998) FLT-3 ligand mobilizes hematopoietic primitive and committed progenitor cells into blood in mice. Eur J Haematol 60:86–92
60. Molineux G, McCrea C, Yan XQ, Kerzic P, McNiece I (1997) Flt-3 ligand synergizes with granulocyte colony-stimulating factor to increase neutrophil numbers and to mobilize peripheral blood stem cells with long-term repopulating potential. Blood 89:3998–4004
61. Sudo Y, Shimazaki C, Ashihara E et al (1997) Synergistic effect of FLT-3 ligand on the granulocyte colony-stimulating factor-induced mobilization of hematopoietic stem cells and progenitor cells into blood in mice. Blood 89:3186–3191
62. Neipp M, Zorina T, Domenick MA, Exner BG, Ildstad ST (1998) Effect of FLT3 ligand and granulocyte colony-stimulating factor on expansion and mobilization of facilitating cells and hematopoietic stem cells in mice: kinetics and repopulating potential. Blood 92:3177–3188
63. Sanganalmath SK, Stein AB, Guo Y et al (2009) The beneficial effects of postinfarct cytokine combination therapy are sustained during long-term follow-up. J Mol Cell Cardiol 47: 528–535
64. Fisher JW (2003) Erythropoietin: physiology and pharmacology update. Exp Biol Med (Maywood) 228:1–14

65. Calvillo L, Latini R, Kajstura J et al (2003) Recombinant human erythropoietin protects the myocardium from ischemia-reperfusion injury and promotes beneficial remodeling. Proc Natl Acad Sci USA 100:4802–4806
66. Parsa CJ, Matsumoto A, Kim J et al (2003) A novel protective effect of erythropoietin in the infarcted heart. J Clin Invest 112:999–1007
67. Cai Z, Semenza GL (2004) Phosphatidylinositol-3-kinase signaling is required for erythropoietin-mediated acute protection against myocardial ischemia/reperfusion injury. Circulation 109:2050–2053
68. Moon C, Krawczyk M, Ahn D et al (2003) Erythropoietin reduces myocardial infarction and left ventricular functional decline after coronary artery ligation in rats. Proc Natl Acad Sci USA 100:11612–11617
69. Heeschen C, Aicher A, Lehmann R et al (2003) Erythropoietin is a potent physiologic stimulus for endothelial progenitor cell mobilization. Blood 102:1340–1346
70. Prunier F, Pfister O, Hadri L et al (2007) Delayed erythropoietin therapy reduces post-MI cardiac remodeling only at a dose that mobilizes endothelial progenitor cells. Am J Physiol Heart Circ Physiol 292:H522–H529
71. Westenbrink BD, Lipsic E, van der Meer P et al (2007) Erythropoietin improves cardiac function through endothelial progenitor cell and vascular endothelial growth factor mediated neovascularization. Eur Heart J 28:2018–2027
72. Yeghiazarians Y, Khan M, Angeli FS et al (2010) Cytokine combination therapy with long-acting erythropoietin and granulocyte colony stimulating factor improves cardiac function but is not superior than monotherapy in a mouse model of acute myocardial infarction. J Card Fail 16:669–678
73. Angeli FS, Amabile N, Shapiro M et al (2010) Cytokine combination therapy with erythropoietin and granulocyte colony stimulating factor in a porcine model of acute myocardial infarction. Cardiovasc Drugs Ther 24:409–420
74. Quinn CM, Dawn B (2010) G-CSF and erythropoietin combination therapy for infarct repair: two plus two equals two? Cardiovasc Drugs Ther 24:369–371
75. Fukuhara S, Tomita S, Nakatani T et al (2004) G-CSF promotes bone marrow cells to migrate into infarcted mice heart, and differentiate into cardiomyocytes. Cell Transplant 13:741–748
76. Kawada H, Fujita J, Kinjo K et al (2004) Nonhematopoietic mesenchymal stem cells can be mobilized and differentiate into cardiomyocytes after myocardial infarction. Blood 104: 3581–3587
77. Deten A, Volz HC, Clamors S et al (2005) Hematopoietic stem cells do not repair the infarcted mouse heart. Cardiovasc Res 65:52–63
78. Askari AT, Unzek S, Popovic ZB et al (2003) Effect of stromal-cell-derived factor 1 on stem-cell homing and tissue regeneration in ischaemic cardiomyopathy. Lancet 362:697–703
79. Adachi Y, Imagawa J, Suzuki Y et al (2004) G-CSF treatment increases side population cell infiltration after myocardial infarction in mice. J Mol Cell Cardiol 36:707–710
80. Misao Y, Takemura G, Arai M et al (2006) Importance of recruitment of bone marrow-derived CXCR4+ cells in post-infarct cardiac repair mediated by G-CSF. Cardiovasc Res 71: 455–465
81. Ueda K, Takano H, Hasegawa H et al (2006) Granulocyte colony stimulating factor directly inhibits myocardial ischemia-reperfusion injury through Akt-endothelial NO synthase pathway. Arterioscler Thromb Vasc Biol 26:e108–e113
82. Ohki Y, Heissig B, Sato Y et al (2005) Granulocyte colony-stimulating factor promotes neovascularization by releasing vascular endothelial growth factor from neutrophils. FASEB J 19:2005–2007
83. Capoccia BJ, Shepherd RM, Link DC (2006) G-CSF and AMD3100 mobilize monocytes into the blood that stimulate angiogenesis in vivo through a paracrine mechanism. Blood 108:2438–2445
84. Kanellakis P, Slater NJ, Du XJ, Bobik A, Curtis DJ (2006) Granulocyte colony-stimulating factor and stem cell factor improve endogenous repair after myocardial infarction. Cardiovasc Res 70:117–125

85. Sugano Y, Anzai T, Yoshikawa T et al (2005) Granulocyte colony-stimulating factor attenuates early ventricular expansion after experimental myocardial infarction. Cardiovasc Res 65:446–456
86. Fujita J, Mori M, Kawada H et al (2007) Administration of granulocyte colony-stimulating factor after myocardial infarction enhances the recruitment of hematopoietic stem cell-derived myofibroblasts and contributes to cardiac repair. Stem Cells 25:2750–2759
87. Hartung T (1998) Anti-inflammatory effects of granulocyte colony-stimulating factor. Curr Opin Hematol 5:221–225
88. Boneberg EM, Hareng L, Gantner F, Wendel A, Hartung T (2000) Human monocytes express functional receptors for granulocyte colony-stimulating factor that mediate suppression of monokines and interferon-gamma. Blood 95:270–276
89. Kang HJ, Kim HS, Zhang SY et al (2004) Effects of intracoronary infusion of peripheral blood stem-cells mobilised with granulocyte-colony stimulating factor on left ventricular systolic function and restenosis after coronary stenting in myocardial infarction: the MAGIC cell randomised clinical trial. Lancet 363:751–756
90. Ince H, Petzsch M, Kleine HD et al (2005) Prevention of left ventricular remodeling with granulocyte colony-stimulating factor after acute myocardial infarction: final 1-year results of the Front-Integrated Revascularization and Stem Cell Liberation in Evolving Acute Myocardial Infarction by Granulocyte Colony-Stimulating Factor (FIRSTLINE-AMI) Trial. Circulation 112:I73–I80
91. Ince H, Petzsch M, Kleine HD et al (2005) Preservation from left ventricular remodeling by front-integrated revascularization and stem cell liberation in evolving acute myocardial infarction by use of granulocyte-colony-stimulating factor (FIRSTLINE-AMI). Circulation 112:3097–3106
92. Kuethe F, Figulla HR, Herzau M et al (2005) Treatment with granulocyte colony-stimulating factor for mobilization of bone marrow cells in patients with acute myocardial infarction. Am Heart J 150:115
93. Valgimigli M, Rigolin GM, Cittanti C et al (2005) Use of granulocyte-colony stimulating factor during acute myocardial infarction to enhance bone marrow stem cell mobilization in humans: clinical and angiographic safety profile. Eur Heart J 26:1838–1845
94. Takano H, Hasegawa H, Kuwabara Y et al (2007) Feasibility and safety of granulocyte colony-stimulating factor treatment in patients with acute myocardial infarction. Int J Cardiol 122:41–47
95. Leone AM, Galiuto L, Garramone B et al (2007) Usefulness of granulocyte colony-stimulating factor in patients with a large anterior wall acute myocardial infarction to prevent left ventricular remodeling (the rigenera study). Am J Cardiol 100:397–403
96. Ellis SG, Penn MS, Bolwell B et al (2006) Granulocyte colony stimulating factor in patients with large acute myocardial infarction: results of a pilot dose-escalation randomized trial. Am Heart J 152:1051.e9–1051.e14
97. Engelmann MG, Theiss HD, Hennig-Theiss C et al (2008) Autologous bone marrow stem cell mobilization induced by granulocyte colony-stimulating factor after subacute ST-segment elevation myocardial infarction undergoing late revascularization: final results from the G-CSF-STEMI (Granulocyte Colony-Stimulating Factor ST-Segment Elevation Myocardial Infarction) trial. J Am Coll Cardiol 48:1712–1721
98. Engelmann MG, Theiss HD, Theiss C et al (2010) G-CSF in patients suffering from late revascularised ST elevation myocardial infarction: final 1-year-results of the G-CSF-STEMI Trial. Int J Cardiol 144:399–404
99. Ince H, Valgimigli M, Petzsch M et al (2008) Cardiovascular events and re-stenosis following administration of G-CSF in acute myocardial infarction: systematic review and meta-analysis. Heart 94:610–616
100. Wang Y, Tagil K, Ripa RS et al (2005) Effect of mobilization of bone marrow stem cells by granulocyte colony stimulating factor on clinical symptoms, left ventricular perfusion and function in patients with severe chronic ischemic heart disease. Int J Cardiol 100:477–483

101. Hill JM, Bartunek J (2006) The end of granulocyte colony-stimulating factor in acute myocardial infarction? Reaping the benefits beyond cytokine mobilization. Circulation 113:1926–1928
102. Ripa RS, Wang Y, Jorgensen E, Johnsen HE, Hesse B, Kastrup J (2006) Intramyocardial injection of vascular endothelial growth factor-A165 plasmid followed by granulocyte-colony stimulating factor to induce angiogenesis in patients with severe chronic ischaemic heart disease. Eur Heart J 27:1785–1792
103. Huttmann A, Duhrsen U, Stypmann J et al (2006) Granulocyte colony-stimulating factor-induced blood stem cell mobilisation in patients with chronic heart failure–feasibility, safety and effects on exercise tolerance and cardiac function. Basic Res Cardiol 101:78–86
104. Schachinger V, Erbs S, Elsasser A et al (2006) Intracoronary bone marrow-derived progenitor cells in acute myocardial infarction. N Engl J Med 355:1210–1221
105. Achilli F, Malafronte C, Lenatti L et al (2010) Granulocyte colony-stimulating factor attenuates left ventricular remodelling after acute anterior STEMI: results of the single-blind, randomized, placebo-controlled multicentre STem cEll Mobilization in Acute Myocardial Infarction (STEM-AMI) Trial. Eur J Heart Fail 12:1111–1121
106. Theiss HD, Brenner C, Engelmann MG et al (2010) Safety and efficacy of SITAgliptin plus GRanulocyte-colony-stimulating factor in patients suffering from Acute Myocardial Infarction (SITAGRAMI-Trial)–rationale, design and first interim analysis. Int J Cardiol 145:282–284
107. Li Y, Fukuda N, Yokoyama S et al (2006) Effects of G-CSF on cardiac remodeling and arterial hyperplasia in rats. Eur J Pharmacol 549:98–106
108. Beohar N, Flaherty JD, Davidson CJ et al (2007) Granulocyte-colony stimulating factor administration after myocardial infarction in a porcine ischemia-reperfusion model: functional and pathological effects of dose timing. Catheter Cardiovasc Interv 69:257–266
109. Ieishi K, Nomura M, Kawano T et al (2007) The effect of G-CSF in a myocardial ischemia reperfusion model rat. J Med Invest 54:177–183
110. Cheng Z, Ou L, Liu Y et al (2008) Granulocyte colony-stimulating factor exacerbates cardiac fibrosis after myocardial infarction in a rat model of permanent occlusion. Cardiovasc Res 80:425–434
111. Angeli FS, Smith C, Amabile N et al (2010) Granulocyte colony stimulating factor in myocardial infarction with low ejection fraction. Cytokine 51:278–285
112. Tomita S, Ishida M, Nakatani T et al (2004) Bone marrow is a source of regenerated cardiomyocytes in doxorubicin-induced cardiomyopathy and granulocyte colony-stimulating factor enhances migration of bone marrow cells and attenuates cardiotoxicity of doxorubicin under electron microscopy. J Heart Lung Transplant 23:577–584
113. Li L, Takemura G, Li Y et al (2007) Granulocyte colony-stimulating factor improves left ventricular function of doxorubicin-induced cardiomyopathy. Lab Invest 87:440–455
114. de Suarez lezo J, Torres A, Herrera I et al (2005) Effects of stem-cell mobilization with recombinant human granulocyte colony stimulating factor in patients with percutaneously revascularized acute anterior myocardial infarction. Rev Esp Cardiol 58:253–261
115. Leone AM, Rutella S, Bonanno G et al (2006) Endogenous G-CSF and CD34+ cell mobilization after acute myocardial infarction. Int J Cardiol 111:202–208
116. Suzuki K, Nagashima K, Arai M et al (2006) Effect of granulocyte colony-stimulating factor treatment at a low dose but for a long duration in patients with coronary heart disease. Circ J 70:430–437
117. Hill JM, Syed MA, Arai AE et al (2005) Outcomes and risks of granulocyte colony-stimulating factor in patients with coronary artery disease. J Am Coll Cardiol 46:1643–1648

Index

G. Molineux et al. (eds.), *Twenty Years of G-CSF*, Milestones in Drug Therapy,
DOI 10.1007/978-3-0348-0218-5, © Springer Basel AG 2012

E

F

G